McFarlin Library
WITHDRAWN

Advanced Petroleum Reservoir Simulation

Scrivener Publishing
3 Winter Street, Suite 3
Salem, MA 01970

Scrivener Publishing Collections Editors

James E. R. Couper	Richard Erdlac
Rafiq Islam	Pradip Khaladkar
Norman Lieberman	Peter Martin
W. Kent Muhlbauer	Andrew Y. C. Nee
S. A. Sherif	James G. Speight

Publishers at Scrivener
Martin Scrivener (martin@scrivenerpublishing.com)
Phillip Carmical (pcarmical@scrivenerpublishing.com)

Advanced Petroleum Reservoir Simulation

M. Rafiqul Islam
S.H. Moussavizadegan
S. Mustafiz
J.H. Abou-Kassem

Scrivener

Copyright © 2010 by Scrivener Publishing LLC. All rights reserved.

Co-published by John Wiley & Sons, Inc. Hoboken, New Jersey, and Scrivener Publishing LLC, Salem, Massachusetts
Published simultaneously in Canada

No part of this publication may be reproduced, stored in a retrieval system, or transmitted in any form or by any means, electronic, mechanical, photocopying, recording, scanning, or otherwise, except as permitted under Section 107 or 108 of the 1976 United States Copyright Act, without either the prior written permission of the Publisher, or authorization through payment of the appropriate per-copy fee to the Copyright Clearance Center, Inc., 222 Rosewood Drive, Danvers, MA 01923, (978) 750-8400, fax (978) 750-4470, or on the web at www.copyright.com. Requests to the Publisher for permission should be addressed to the Permissions Department, John Wiley & Sons, Inc., 111 River Street, Hoboken, NJ 07030, (201) 748-6011, fax (201) 748-6008, or online at http://www.wiley.com/go/permission.

Limit of Liability/Disclaimer of Warranty: While the publisher and author have used their best efforts in preparing this book, they make no representations or warranties with respect to the accuracy or completeness of the contents of this book and specifically disclaim any implied warranties of merchantability or fitness for a particular purpose. No warranty may be created or extended by sales representatives or written sales materials, The advice and strategies contained herein may not be suitable for your situation. You should consult with a professional where appropriate. Neither the publisher nor author shall be liable for any loss of profit or any other commercial damages, including but not limited to special, incidental, consequential, or other damages.

For general information on our other products and services or for technical support, please contact our Customer Care Department within the United States at (800) 762-2974, outside the United States at (317) 572-3993 or fax (317) 572-4002.

Wiley also publishes its books in a variety of electronic formats. Some content that appears in print may not be available in electronic formats. For more information about Wiley products, visit our web site at www.wiley.com.

For more information about Scrivener products please visit www.scrivenerpublishing.com.

Cover design by Russell Richardson.

Library of Congress Cataloging-in-Publication Data:

ISBN 978-0-470-625811

Printed in the United States of America

10 9 8 7 6 5 4 3 2 1

Contents

Foreword		xiii
Introduction		xv

1. **Reservoir Simulation Background** — 1
   1.1 Essence of Reservoir Simulation — 1
   1.2 Assumptions Behind Various Modeling Approaches — 5
   1.3 Material Balance Equation — 5
      1.3.1 Decline Curve — 6
      1.3.2 Statistical Method — 6
      1.3.3 Analytical Methods — 7
      1.3.4 Finite Difference Methods — 8
      1.3.5 Darcy's Law — 11
   1.4 Recent Advances in Reservoir Simulation — 12
      1.4.1 Speed and Accuracy — 12
      1.4.2 New Fluid Flow Equations — 13
      1.4.3 Coupled Fluid Flow and Geo-mechanical Stress Model — 16
      1.4.4 Fluid Flow Modeling Under Thermal Stress — 17
   1.5 Future Challenges in Reservoir Simulation — 18
      1.5.1 Experimental Challenges — 18
      1.5.2 Numerical Challenges — 20
         1.5.2.1 Theory of Onset and Propagation of Fractures Due to Thermal Stress — 20
         1.5.2.2 2-D and 3-D Solutions of the Governing Equations — 20
         1.5.2.3 Viscous Fingering During Miscible Displacement — 20
         1.5.2.4 Improvement in Remote Sensing and Monitoring Ability — 21

			1.5.2.5	Improvement in Data Processing Techniques	21
		1.5.3	\multicolumn{2}{l	}{Remote Sensing and Real-time Monitoring}	22
			1.5.3.1	Monitoring Offshore Structures	23
			1.5.3.2	Development of a Dynamic Characterization Tool (Based on Seismic-while-drilling Data)	24
			1.5.3.3	Use of 3-D Sonogram	24
			1.5.3.4	Virtual Reality (VR) Applications	25
			1.5.3.5	Intelligent Reservoir Management	26

1.6 Economic Models Based on Futuristic Energy Pricing Policies 27
1.7 Integrated System of Monitoring, Environmental Impact and Economics 29

2. Reservoir Simulator-input/output 31
2.1 Input and Output Data 32
2.2 Geological and Geophysical Modeling 34
2.3 Reservoir Characterization 37
    2.3.1 Representative Elementary Volume, REV 38
    2.3.2 Fluid and Rock Properties 41
        2.3.2.1 Fluid Properties 42
            2.3.2.1.1 Crude Oil Properties 43
            2.3.2.1.2 Natural Gas Properties 45
            2.3.2.1.3 Water Content Properties 46
    2.3.3 Rock Properties 47
2.4 Upscaling 52
    2.4.1 Power Law Averaging Method 53
    2.4.2 Pressure-solver Method 54
    2.4.3 Renormalization Technique 56
    2.4.4 Multiphase Flow Upscaling 57
2.5 Pressure/Production data 60
    2.5.1 Phase Saturations Distribution 61
2.6 Reservoir Simulator Output 62
2.7 History-matching 65
    2.7.1 History-matching Formulation 68
    2.7.2 Uncertainty Analysis 71

Contents　　vii

　　　　　2.7.2.1　Measurement Uncertainty　71
　　　　　2.7.2.2　Upscaling Uncertainty　74
　　　　　2.7.2.3　Model Error　75
　　　　　2.7.2.4　The Prediction Uncertainty　76
　　2.8　Real-time Monitoring　77

3. **Reservoir Simulators: Problems, Shortcomings, and Some Solution Techniques**　83
　　3.1　Multiple Solutions in Natural Phenomena　85
　　　　3.1.1　Knowledge Dimension　88
　　3.2　Adomian Decomposition　103
　　　　3.2.1　Governing Equations　105
　　　　3.2.2　Adomian Decomposition of Buckley–Leverett Equation　108
　　　　3.2.3　Results and Discussions　110
　　3.3　Some Remarks on Multiple Solutions　113

4. **Mathematical Formulation of Reservoir Simulation Problems**　115
　　4.1　Black Oil Model and Compositional Model　116
　　4.2　General Purpose Compositional Model　118
　　　　4.2.1　Basic Definitions　118
　　　　4.2.2　Primary and Secondary Parameters and Model Variables　120
　　　　4.2.3　Mass Conservation Equation　123
　　　　4.2.4　Energy Balance Equation　126
　　　　4.2.5　Volume Balance Equation　132
　　　　4.2.6　The Motion Equation in Porous Medium　133
　　　　4.2.7　The Compositional System of Equations and Model Variables　138
　　4.3　Simplification of the General Compositional Model　141
　　　　4.3.1　The Black Oil Model　141
　　　　4.3.2　The Water Oil Model　143
　　4.4　Some Examples in Application of the General Compositional Model　146
　　　　4.4.1　Isothermal Volatile Oil Reservoir　146
　　　　4.4.2　Steam Injection Inside a Dead Oil Reservoir　149

		4.4.3	Steam Injection in Presence of Distillation and Solution Gas	150

5. The Compositional Simulator Using the Engineering Approach — 155

- 5.1 Finite Control Volume Method — 156
  - 5.1.1 Reservoir Discretization in Rectangular Coordinates — 157
  - 5.1.2 Discretization of Governing Equations — 158
    - 5.1.2.1 Components Mass Conservation Equation — 159
    - 5.1.2.2 Energy Balance Equation — 167
  - 5.1.3 Discretization of Motion Equation — 170
- 5.2 Uniform Temperature Reservoir Compositional Flow Equations in a 1-D Domain — 172
- 5.3 Compositional Mass Balance Equation in a Multidimensional Domain — 178
  - 5.3.1 Implicit Formulation of Compositional Model in Multi-Dimensional Domain — 180
  - 5.3.2 Reduced Equations of Implicit Compositional Model in Multidimensional Domain — 183
  - 5.3.3 Well Production and Injection Rate Terms — 186
    - 5.3.3.1 Production Wells — 186
    - 5.3.3.2 Injection Wells — 188
  - 5.3.4 Fictitious Well Rate Terms (Treatment of Boundary Conditions) — 189
- 5.4 Variable Temperature Reservoir Compositional Flow Equations — 193
  - 5.4.1 Energy Balance Equation — 193
  - 5.4.2 Implicit Formulation of Variable Temperature Reservoir Compositional Flow Equations — 197
- 5.5 Solution Method — 201
  - 5.5.1 Solution of Model Equations Using Newton's Iteration — 202
- 5.6 The Effects of Linearization — 207
  - 5.6.1 Case I: Single Phase Flow of a Natural Gas — 208

		5.6.2	Effect of Interpolation Functions and Formulation	214
		5.6.3	Effect of Time Interval	215
		5.6.4	Effect of Permeability	217
		5.6.5	Effect of Number of Gridblocks	217
		5.6.6	Spatial and Transient Pressure Distribution Using Different Iinterpolation Functions	219
		5.6.7	CPU Time	222
		5.6.8	Case II: An Oil/Water Reservoir	224
6.	**A Comprehensive Material Balance Equation for Oil Recovery**			**245**
	6.1	Background		245
	6.2	Permeability Alteration		248
	6.3	Porosity Alteration		249
	6.4	Pore Volume Change		251
	6.5	A Comprehensive MBE with Memory for Cumulative Oil Recovery		252
	6.6	Numerical Simulation		255
		6.6.1	Effects of Compressibilities on Dimensionless Parameters	257
		6.6.2	Comparison of Dimensionless Parameters Based on Compressibility Factor	258
		6.6.3	Effects of M on Dimensionless Parameter	259
		6.6.4	Effects of Compressibility Factor with M Values	259
		6.6.5	Comparison of Models Based on RF	260
		6.6.6	Effects of M on MBE	262
	6.7	Appendix 6A: Development of an MBE for a Compressible Undersaturated Oil Reservoir		264
		6.7.1	Development of a New MBE	265
		6.7.2	Conventional MBE	272
		6.7.3	Significance of C_{epm}	274
		6.7.4	Water Drive Mechanism with Water Production	275
		6.7.5	Depletion Drive Mechanism with No Water Production	276

7. **Modeling Viscous Fingering During Miscible Displacement in a Reservoir** — 277
   - 7.1 Improvement of the Numerical Scheme — 277
     - 7.1.1 The Governing Equation — 279
     - 7.1.2 Finite Difference Approximations — 281
       - 7.1.2.1 Barakat–Clark FTD Scheme — 281
       - 7.1.2.2 DuFort–Frankel Scheme — 283
     - 7.1.3 Proposed Barakat–Clark CTD Scheme — 284
       - 7.1.3.1 Boundary Conditions — 285
     - 7.1.4 Accuracy and Truncation Errors — 285
     - 7.1.5 Some Results and Discussion — 286
     - 7.1.6 Influence of Boundary Conditions — 293
   - 7.2 Application of the New Numerical Scheme to Viscous Fingering — 295
     - 7.2.1 Stability Criterion and Onset of Fingering — 295
     - 7.2.2 Base Stable Case — 296
     - 7.2.3 Base Unstable Case — 302
     - 7.2.4 Parametric Study — 309
       - 7.2.4.1 Effect of Injection Pressure — 309
       - 7.2.4.2 Effect of Overall Porosity — 314
       - 7.2.4.3 Effect of Mobility Ratio — 317
       - 7.2.4.4 Effect of Longitudinal Dispersion — 320
       - 7.2.4.5 Effect of Transverse Dispersion — 324
       - 7.2.4.6 Effect of Aspect Ratio — 327
     - 7.2.5 Comparison of Numerical Modeling Results with Experimental Results — 330
       - 7.2.5.1 Selected Experimental Model — 330
       - 7.2.5.2 Physical Model Parameters — 331
       - 7.2.5.3 Comparative Study — 332
       - 7.2.5.4 Concluding Remarks — 336

8. **Towards Modeling Knowledge and Sustainable Petroleum Production** — 339
   - 8.1 Essence of Knowledge, Science, and Emulation — 339
     - 8.1.1 Simulation vs. Emulation — 340
     - 8.1.2 Importance of the First Premise and Scientific Pathway — 342
     - 8.1.3 Mathematical Requirements of Nature Science — 344
     - 8.1.4 The Meaningful Addition — 348

		8.1.5	"Natural" Numbers and the Mathematical Content of Nature	350
	8.2	The Knowledge Dimension		354
		8.2.1	The Importance of Time as the Fourth Dimension	354
		8.2.2	Towards Modeling Truth and Knowledge	362
	8.3	Examples of Linearization and Linear Thinking		363
	8.4	The Single-Parameter Criterion		365
		8.4.1	Science Behind Sustainable Technology	366
		8.4.2	A New Computational Method	366
			8.4.2.1 The Currently Used Model	366
			8.4.2.2 Towards Achieving Multiple Solutions	372
	8.5	The Conservation of Mass and Energy		374
		8.5.1	The Avalanche Theory	375
		8.5.2	Aims of Modeling Natural Phenomena	380
		8.5.2	Challenges of Modeling Sustainable Petroleum Operations	382
	8.6	The Criterion: The Switch that Determines the Direction at a Bifurcation Point		386
		8.6.1	Some Applications of the Criterion	388
	8.7	The Need for Multidimensional Study		396
	8.8	Assessing the Overall Performance of a Process		399
	8.9	Implications of Knowledge-Based Analysis		406
		8.9.1	A General Case	407
		8.9.2	Impact of Global Warming Analysis	410
	8.10	Examples of Knowledge-Based Simulation		413

9. Final Conclusions   **421**

Appendix A   User's Manual for Multi-Purpose Simulator for Field Applications (MPSFFA, Version 1–15)   **423**

    A.1 Introduction   423
    A.2 The Simulator   423
    A.3 Data File Preparation   425
       A.3.1 Format Procedure A   426
       A.3.2 Format Procedure B   427
       A.3.3 Format Procedure C   427

	A.3.4 Format Procedure D	427
	A.3.5 Format Procedure E	428
A.4	Description of Variables Used in Preparing a Data File	428
A.5	Instructions to Run Simulator and Graphic Post Processor on PC	439
A.6	Limitations Imposed on the Compiled Versions	441
A.7	Example of a Prepared Data File	442

References 447
Index 463

Foreword

Petroleum is still the world's most important source of energy and reservoir performance and petroleum recovery are often based on assumptions that bear little relationship to reality. Not all equations written on paper are correct and are only dependent on the assumptions used. In reservoir simulation, the principle of *garbage in and garbage out* is well known leading to systematic and large errors in the assessment of reservoir performance. This book presents the shortcomings and assumptions of previous methods. It then outlines the need for a new mathematical approach that eliminates most of the shortcomings and spurious assumptions of the conventional approach. The volume will provide the working engineer or graduate student with a new, more accurate, and more efficient model for a very important aspect of petroleum engineering: reservoir simulation leading to prediction of reservoir behavior.

Reservoir simulation studies are very subjective and vary from simulator to simulator. Currently available simulators only address a very limited range of solutions for a particular reservoir engineering problem. While benchmarking has helped accept differences in predicting petroleum reservoir performance, there has been no scientific explanation behind the variability that has puzzled scientists and engineers. For a modeling process to be accurate, the input data have to be accurate for the simulation results to be acceptable. The requirements are that various sources of errors must be recognized and data integrity must be preserved.

Reservoir simulation equations have an embedded variability and multiple solutions that are in line with physics rather than spurious mathematical solutions. To this end, the authors introduce mathematical developments of new governing equations based on in-depth understanding of the factors that influence fluid flow in porous media under different flow conditions leading to a series of workable mathematical and numerical techniques are also presented that allow one to achieve this objective.

This book provides a readable and workable text to counteract the errors of the past and provides the reader with an extremely useful predictive tool for reservoir development.

Dr. James G. Speight

Introduction

Petroleum is still the world's most important source of energy, and, with all of the global concerns over climate change, environmental standards, cheap gasoline, and other factors, petroleum itself has become a hotly debated topic. This book does not seek to cast aspersions, debate politics, or take any political stance. Rather, the purpose of this volume is to provide the working engineer or graduate student with a new, more accurate, and more efficient model for a very important aspect of petroleum engineering: reservoir simulation.

The term, "knowledge-based," is used throughout as a term for our unique approach, which is very different from past approaches and which we hope will be a very useful and eye-opening tool for engineers in the field. We do not intend to denigrate other methods, nor do we suggest by our term that other methods do not involve "knowledge." Rather, this is simply the term we use for our approach, and we hope that we have proven that it is more accurate and more efficient than approaches used in the past.

It is well known that reservoir simulation studies are very subjective and vary from simulator to simulator. While SPE benchmarking has helped accept differences in predicting petroleum reservoir performance, there has been no scientific explanation behind the variability that has frustrated many policy makers and operations managers and puzzled scientists and engineers. In this book, for the first time, reservoir simulation equations are shown to have embedded variability and multiple solutions that are in line with physics rather than spurious mathematical solutions. With this clear description, a fresh perspective in reservoir simulation is presented.

Unlike the majority of reservoir simulation approaches available today, the "knowledge-based" approach in this book does not stop at questioning the fundamentals of reservoir simulation but offers solutions and demonstrates that proper reservoir simulation should be transparent and empower decision makers rather than creating a black box. Mathematical developments of new governing equations based on in-depth understanding of the factors that influence fluid flow in porous media under different flow conditions are introduced. The behavior of flow through matrix and fractured systems in the same reservoir, heterogeneity and fluid/rock properties interactions, Darcy and non-Darcy flow are among the issues that are thoroughly addressed. For the first time, the fluid memory factor is introduced with a functional form. The resulting governing equations are solved without linearization at any stage. A series of clearly superior mathematical and numerical techniques are also presented that allow one to achieve this objective.

In our approach, we present mathematical solutions that provide a basis for systematic tracking of multiple solutions that are inherent to non-linear governing equations. This is possible because the new technique is capable of solving non-linear equations without linearization. To promote the new models, a presentation of the common criterion and procedure of reservoir simulators currently in use is provided. The models are applied to difficult scenarios, such as in the presence of viscous fingering, and results are compared with experimental data. It is demonstrated that the currently available simulators only address a very limited range of solutions for a particular reservoir engineering problem. Examples are provided to show how our approach extends the currently known solutions and provides one with an extremely useful predictive tool for risk assessment.

The Need for a Knowledge-Based Approach

In reservoir simulation, the principle of GIGO (Garbage in and garbage out) is well known (latest citation by Rose, 2000). This principle implies that the input data have to be accurate for the simulation results to be acceptable. The petroleum industry has established itself as the pioneer of subsurface data collection (Abou-Kassem et al, 2006). Historically, no other discipline has taken so much care in making sure input data are as accurate as the latest technology would allow. The recent plethora of technologies dealing with subsurface mapping, real time monitoring, and high speed data transfer is evidence of the fact that input data in reservoir simulation are not the weak link of reservoir modeling.

However, for a modeling process to be knowledge-based, it must fulfill two criteria, namely, the source has to be true (or real) and the subsequent processing has to be true (Zatzman and Islam, 2007). The source is not a problem in the petroleum industry, as a great deal of advances have been made on data collection techniques. The potential problem lies within the processing of data. For the process to be knowledge-based, the following logical steps have to be taken:

- Collection of data with constant improvement of the data acquisition technique. The data set to be collected is dictated by the objective function, which is an integral part of the decision-making process. Decision making, however, should not take place without the abstraction process. The connection between objective function and data needs constant refinement. This area of research is one of the biggest strengths of the petroleum industry, particularly in the information age.
- The gathered data should be transformed into information so that it becomes useful. With today's technology, the amount of raw data is so huge, the need for a filter is more important than ever before. It is important, however, to select a filter that doesn't skew the data set toward a certain decision. Mathematically, these filters have to be non-linearized (Mustafiz et al, 2008). While the concept of non-linear filtering is not new, the existence of non-linearized models is only beginning to be recognized (Abou-Kassem et al, 2006).
- Information should be further processed into "knowledge" that is free from preconceived ideas or a "preferred decision." Scientifically, this process must be free from information lobbying, environmental activism, and other biases. Most current models include these factors as an integral part of the decision-making process (Eisenack et al, 2007), whereas a scientific knowledge model must be free from those interferences as they distort the abstraction process and inherently prejudice the decision-making. Knowledge gathering essentially puts information into the big picture.

For this picture to be free from distortion, it must be free from non-scientific maneuvering.
- Final decision-making is knowledge-based, only if the abstraction from the above three steps has been followed without interference. Final decision is a matter of Yes or No (or True or False or 1 or 0) and this decision will be either knowledge-based or prejudice-based. Figure i.1 shows the essence of knowledge-based decision-making.

The process of aphenomenal or prejudice-based decision-making is illustrated by the inverted triangle, proceeding from the top down (Fig. i.2). The inverted representation stresses the inherent instability and unsustainability of the model. The source data from which a decision eventually emerges already incorporates their own justifications, which are then massaged by layers of opacity and disinformation.

The disinformation referred to here is what results when information is presented or recapitulated in the service of unstated or unacknowledged ulterior intentions (Zatzman and Islam, 2007). The *methods* of this disinformation achieve their effect by presenting evidence or raw data selectively, without disclosing either the fact of such selection or the criteria guiding the selection. This process of selection obscures any distinctions between the data coming from nature or from any all-natural pathway, on the one hand, and data from unverified or untested observations on the other. In social science, such maneuvering has been well known, but the recognition of this aphenomenal (unreal) model is new in science and engineering (Shapiro et al, 2007).

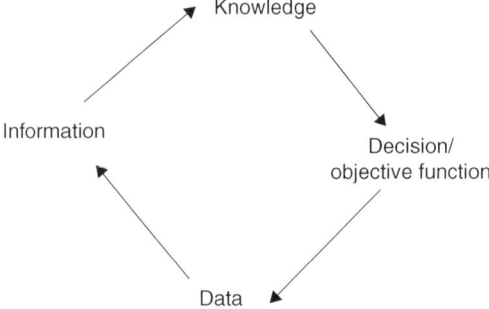

Figure i.1 The knowledge model and the direction of abstraction.

INTRODUCTION   xix

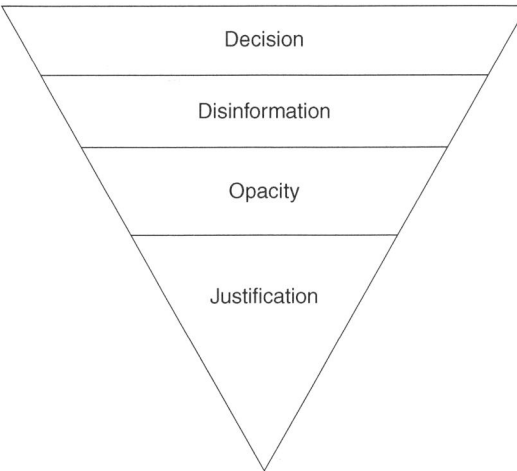

Figure i.2 Aphenomenal decision-making.

Summary of Chapters

Chapter 1 presents the background of reservoir simulation, as it has been developed in the last five decades. This chapter also presents the shortcomings and assumptions of previous methods. It then outlines the need for a new mathematical approach that eliminates most of the short-comings and spurious assumptions of the conventional approach.

Chapter 2 presents the requirements for data input in reservoir simulation. It highlights various sources of errors in handling such data. It also presents guidelines for preserving data integrity with recommendations for data processing.

Chapter 3 presents the solutions to some of the most difficult problems in reservoir simulation. It gives examples of solutions without linearization and elucidates how the knowledge-based approach eliminates the possibility of coming across spurious solutions that are common in the conventional approach. It highlights the advantage of solving governing equations without linearization and demarks the degree of errors committed through linearization, as done in the conventional approach.

Chapter 4 presents a complete formulation of black oil simulation for both isothermal and non-isothermal cases, using the engineering approach. It demonstrates the simplicity and clarity of the engineering approach.

Chapter 5 presents a complete formulation of compositional simulation, using the engineering approach. It shows how very complex and long governing equations are amenable to solutions without linearization using the knowledge-based approach.

Chapter 6 presents a comprehensive formulation of the material balance equation (MBE) using the memory concept. Solutions of the selected problems are also offered in order to demonstrate the need of recasting the governing equations using fluid memory. This chapter shows how a significant error can be committed in terms of reserve calculation and reservoir behavior prediction if the comprehensive formulation is not used.

Chapter 7 uses the example of miscible displacement as an effort to model enhanced oil recovery (EOR). A new solution technique is presented and its superiority in handling the problem of viscous fingering is discussed.

Chapter 8 highlights the future needs of the knowledge-based approach. A new combined mass and energy balance formulation is presented. With the new formulation, various natural phenomena related to petroleum operations are modeled. It is shown that with this formulation one would be able to determine the true cause of global warming, which in turn would help develop sustainable petroleum technologies. Finally, this chapter shows how the criterion (trigger) is affected by the knowledge-based approach. This caps the argument that the knowledge-based approach is crucial for decision-making.

Chapter 9 concludes major findings and recommendations of this book.

The Appendix is the manual for the 3D, 3-phase reservoir simulation program. This program is available for download from **www.scrivenerpublishing.com**.

1
Reservoir Simulation Background

The Information Age is synonymous with knowledge. If, however proper science is not used, information alone cannot guarantee transparency, which is the pre-condition to Knowledge. Proper science requires thinking or imagination with conscience, the very essence of humanity. Imagination is necessary for anyone wishing to make decisions based on science and always begins with visualization – actually, another term for simulation. Even though there is a commonly held misconception that physical experimentation precedes scientific analysis, the truth is simulation is the first one that has to be worked out even before designing an experiment. This is why the petroleum industry puts so much emphasis on simulation studies.

The petroleum industry is known to be the biggest user of computer models. More importantly, unlike other big-scale simulations, such as space research and weather models, petroleum models do not have an option of verifying with real data. Because petroleum engineers do not have the luxury of launching a "reservoir shuttle" or a "petroleum balloon" to roam around the reservoir, the task of modeling is the most daunting. Indeed, from the advent of computer technology, the petroleum industry pioneered the use of computer simulations in virtually all aspects of decision-making. From the golden era of the petroleum industry, very significant amounts of research dollars have been spent to develop some of the most sophisticated mathematical models ever used. Even as the petroleum industry transits through its "middle age" in a business sense and the industry no longer carries the reputation of being the "most aggressive investor in research," oil companies continue to spend liberally for reservoir simulation studies and even for developing new simulators.

1.1 Essence of Reservoir Simulation

Today, practically all aspects of reservoir engineering problems are solved with reservoir simulators, ranging from well testing to prediction of enhanced oil recovery. For every application, however, there is a custom-designed simulator. Even though, quite often, 'comprehensive', 'all-purpose', and other denominations are used to describe a company simulator, every simulation study is a unique process, starting from the reservoir description to the final analysis of results. Simulation is the art of combining physics, mathematics, reservoir engineering, and computer programming to develop a tool for predicting hydrocarbon reservoir performance under various operating strategies.

Figure 1.1 depicts the major steps involved in the development of a reservoir simulator (Odeh, 1982). In this figure, the *formulation* step outlines the basic assumptions inherent to the simulator, states these assumptions in precise mathematical terms, and applies them to a control volume in the reservoir. Newton's approximation is used to render these control volume equations into a set of coupled, nonlinear partial differential equations (PDEs) that describe fluid flow through porous media (Ertekin et al, 2001). These PDEs are then discretized, giving rise to a set of non-linear algebraic equations. Taylor series expansion is used to discretize the governing PDEs. Even though this procedure has been the standard in the petroleum industry for decades, only recently Abou-Kassem et al (2006) pointed out that there is no need to go through this process of expressing in PDE, followed by discretization. In fact, by setting up the algebraic equations directly, one can make the process simple and yet maintain accuracy (Mustafiz et al, 2008). The PDEs derived during the *formulation* step, if solved analytically, would give reservoir pressure, fluid saturations, and well flow rates as continuous functions of space and time. Because of the highly nonlinear nature of the PDEs, analytical techniques cannot be used and solutions must be obtained with numerical methods. In contrast to analytical solutions, numerical solutions give the values of pressure and fluid saturations only at discrete points in the reservoir and at discrete times. *Discretization* is the process of converting PDEs into algebraic equations. Several numerical methods can be used to discretize the PDEs; however, the most common approach in the oil industry today is the finite-difference method. To carry out discretization, a PDE is written for a given point in space at a given time level. The choice of time level (old time level, current time level, or the intermediate time level) leads to the explicit, implicit, or Crank-Nicolson formulation method. The discretization process results in a system of nonlinear algebraic equations. These equations generally cannot be solved with linear equation solvers and linearization of such equations becomes a necessary step before solutions can be

Reservoir Simulation Background

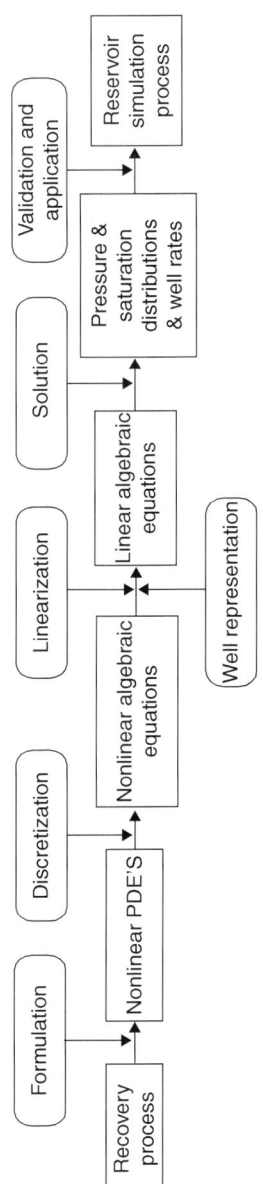

Figure 1.1 Major steps used to develop reservoir simulators (redrawn from Odeh, 1982).

obtained. *Well representation* is used to incorporate fluid production and injection into the nonlinear algebraic equations. *Linearization* involves approximating nonlinear terms in both space and time. Linearization results in a set of linear algebraic equations. Any one of several linear equation solvers can then be used to obtain the *solution*. The solution comprises of pressure and fluid saturation distributions in the reservoir and well flow rates. *Validation* of a reservoir simulator is the last step in developing a simulator, after which the simulator can be used for practical field applications. The validation step is necessary to make sure that no error was introduced in the various steps of development and in computer programming.

It is possible to bypass the step of formulation in the form of PDEs and directly express the fluid flow equation in the form of nonlinear algebraic equation as pointed out in Abou-Kassem et al (2006). In fact, by setting up the algebraic equations directly, one can make the process simple and yet maintain accuracy. This approach is termed the "Engineering Approach" because it is closer to the engineer's thinking and to the physical meaning of the terms in the flow equations. Both the engineering and mathematical approaches treat boundary conditions with the same accuracy if the mathematical approach uses second order approximations. The engineering approach is simple and yet general and rigorous.

There are three methods available for the discretization of any PDE: the Taylor series method, the integral method, and the variational method (Aziz and Settari, 1979). The first two methods result in the finite-difference method, whereas the third results in the variational method. The "Mathematical Approach" refers to the methods that obtain the nonlinear algebraic equations through deriving and discretizing the PDEs. Developers of simulators relied heavily on mathematics in the mathematical approach to obtain the nonlinear algebraic equations or the finite-difference equations. A new approach that derives the finite-difference equations without going through the rigor of PDEs and discretization and that uses fictitious wells to represent boundary conditions has been recently presented by Abou-Kassem et al (2006). In addition, it results in the same finite-difference equations for any hydrocarbon recovery process. Because the engineering approach is independent of the mathematical approach, it reconfirms the use of central differencing in space discretization and highlights the assumptions involved in choosing a time level in the mathematical approach.

1.2 Assumptions Behind Various Modeling Approaches

Reservoir performance is traditionally predicted using three methods, namely, 1) Analogical; 2) Experimental, and 3) Mathematical. The analogical method consists of using mature reservoir properties that are similar to the target reservoir to predict the behavior of the reservoir. This method is especially useful when there is a limited available data. The data from the reservoir in the same geologic basin or province may be applied to predict the performance of the target reservoir. Experimental methods measure the reservoir characteristics in the laboratory models and scale these results to the entire hydrocarbon accumulations. The mathematical method applied basic conservation laws and constitutive equations to formulate the behavior of the flow inside the reservoir and the other characteristics in mathematical notations and formulations. The two basic equations are the material balance equation or continuity equation and the equation of motion or momentum equation. These two equations are expressed for different phases of the flow in the reservoir and combine to obtain single equations for each phase of the flow. However, it is necessary to apply other equations or laws for modeling enhanced oil recovery. As an example, the energy balance equation is necessary to analyze the reservoir behavior for the steam injection or *in situ* combustion reservoirs.

The mathematical model traditionally includes material balance equation, decline curve, statistical approaches and also analytical methods. The Darcy's law is almost always used in all of available reservoir simulators to model the fluid motion. The numerical computations of the derived mathematical model are mostly based on the finite difference method. All these models and approaches are based on several assumptions and approximations that may produce erroneous results and predictions.

1.3 Material Balance Equation

Material balance equation is known to be the classical mathematical representation of the reservoir. According to this principle, the amount of material remaining in the reservoir after a production time interval is equal to the amount of material originally present in the reservoir minus the amount of material removed from the reservoir due to production plus the amount of material added to the reservoir due to injection. This equation describes the fundamental physics of the production scheme of the reservoir. There are several assumptions in the material balance equation:

- Rock and fluid properties do not change in space;
- Hydrodynamics of the fluid flow in the porous media is adequately described by Darcy's law;
- Fluid segregation is spontaneous and complete;
- Geometrical configuration of the reservoir is known and exact;
- PVT data obtained in the laboratory with the same gas-liberation process (flash vs. differential) are valid in the field;
- Sensitive to inaccuracies in measured reservoir pressure. The model breaks down when no appreciable decline occurs in reservoir pressure, as in pressure maintenance operations.

1.3.1 Decline Curve

The rate of oil production decline generally follows one of the following mathematical forms: exponential, hyperbolic and harmonic. The following assumptions apply to the decline curve analysis

- The past processes continue to occur in the future;
- Operation practices are assumed to remain same.

1.3.2 Statistical Method

In this method, the past performance of numerous reservoirs is statistically accounted for to derive the empirical correlations, which are used for future predictions. It may be described as a 'formal extension of the analogical method'. The statistical methods have the following assumptions:

- Reservoir properties are within the limit of the database;
- Reservoir symmetry exists;
- Ultimate recovery is independent of the rate of production.

In addition, Zatzman and Islam (2007) recently pointed out a more subtle, yet far more important shortcoming of the statistical methods. Practically, all statistical methods assume that two or more objects based on a limited number of tangible expressions makes it legitimate to comment on the underlying science. It is equivalent to stating if effects show a reasonable correlation, the causes can also be correlated. As Zatzman and Islam (2007) pointed out, this poses a serious problem as, in absence of time space correlation (pathway rather than end result), anything can be correlated with anything, making the whole process of scientific investigation spurious. They make their point by showing the correlation between global warming (increases) with a decrease in the number of pirates. The absurdity of the statistical process becomes evident by drawing this analogy. Shapiro et al (2007) pointed out another severe limitation of the statistical method. Even though they commented on the polling techniques used in various surveys, their comments are equally applicable in any statistical modeling. They wrote:

"Frequently, opinion polls generalize their results to a U.S. population of 300 million or a Canadian population of 32 million on the basis of what 1,000 or 1,500 "randomly selected" people are recorded to have said or answered. In the absence of any further information to the contrary, the underlying theory of mathematical statistics and random variability assumes that the individual selected "perfectly" randomly is no more nor less likely to have any one opinion over any other. How perfect the randomness may be determined from the "confidence" level attached to a survey, expressed in the phrase that describes the margin of error of the poll sample lying plus or minus some low single-digit percentage "nineteen times out of twenty," i.e. a confidence level of 0.95. Clearly, however, assuming *in the absence of any knowledge otherwise* a certain state of affairs to be the case, *viz.*, that the sample is random and no one opinion is more likely than any other, seems more useful for projecting horoscopes than scientifically assessing public opinion."

1.3.3 Analytical Methods

In most of the cases, the fluid flow inside the porous rock is too complicated to solve analytically. These methods can apply to some simplified model. However, this solution can be applied as the bench mark solution to validate the numerical approaches.

1.3.4 Finite Difference Methods

Finite difference calculus is a mathematical technique, which is used to approximate values of functions and their derivatives at discrete points, where they are not known. The history of differential calculus dates back to the time of Leibnitz and Newton. In this concept, the derivative of a continuous function is related to the function itself. Newton's formula is the core of differential calculus and suffers from the approximation that the magnitude and direction change independently of one another. There is no problem in having separate derivatives for each component of the vector or in superimposing their effects separately and regardless of order.

That is what mathematicians mean when they describe or discuss Newton's derivative being used as a "linear operator". Following this, comes Newton's difference-quotient formula. When the value of a function is inadequate to solve a problem, the rate at which the function changes, sometimes becomes useful. Therefore, the derivatives are also important in reservoir simulation. In Newton's difference-quotient formula, the derivative of a continuous function is obtained. This method relies implicitly on the notion of approximating instantaneous moments of curvature, or infinitely small segments, by means of straight lines. This alone should have tipped everyone off that his derivative is a linear operator precisely because, and to the extent that, it examines change over time (or distance) within an already established function (Islam, 2006). This function is applicable to an infinitely small domain, making it non-existent. When, integration is performed, however, this non-existent domain is assumed to be extended to finite and realistic domains, making the entire process questionable.

The publication of the book, *Principia Mathematica* by Sir Isaac Newton at the end of the 17th century has been the most significant development in European-centered civilization. It is also evident that some of the most important assumptions of Newton were just as aphenomenal (Zatzman and Islam, 2007a). By examining the first assumptions involved, Zatzman and Islam (2007) were able to characterize Newton's laws as aphenomenal for three reasons: that they 1) remove time-consciousness; 2) recognize the role of 'external force'; and 3) do not include the role of first premise. In brief, Newton's law ignores, albeit implicitly, all intangibles from nature science. Zatzman and Islam (2007) identified the most significant contribution of Newton in mathematics as the famous definition of

Reservoir Simulation Background

the derivative as the limit of a difference quotient involving changes in space or in time as small as anyone might like, but not zero, *viz.*

$$\frac{d}{dt}f(t) = \lim_{\Delta t \to 0} \frac{f(t + \Delta t) - f(t)}{\Delta t} \qquad (1.1)$$

Without regards to further conditions being defined as to when and where differentiation would produce a meaningful result, it was entirely possible to arrive at "derivatives" that would generate values in the range of a function at points of the domain where the function was not defined or did not exist. Indeed, it took another century following Newton's death before mathematicians would work out the conditions – especially the requirements for continuity of the function to be differentiated within the domain of values – in which its derivative (the name given to the ratio-quotient generated by the limit formula) could be applied and yield reliable results. Kline (1972) detailed the problems involving this breakthrough formulation of Newton. However, no one in the past did propose an alternative to this differential formulation, at least not explicitly. The following figure (Fig. 1.2) illustrates this difficulty.

In this figure, economic index (it may be one of many indicators) is plotted as a function of time. In nature, all functions are very similar.

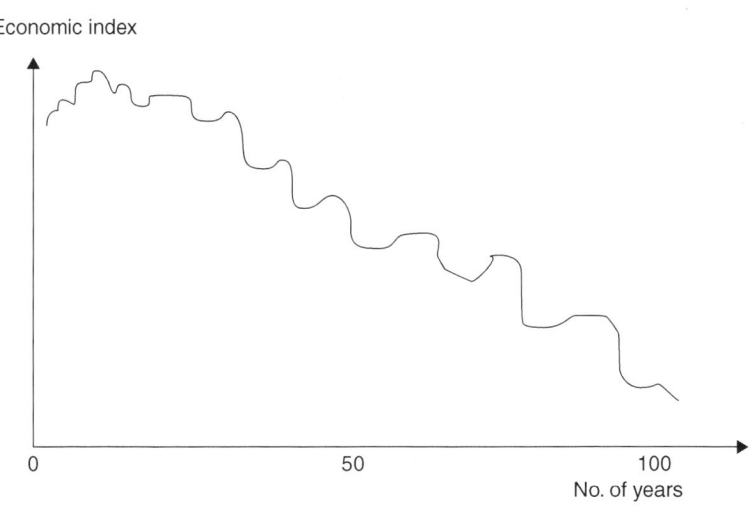

Figure 1.2 Economic wellbeing is known to fluctuate with time (adapted from Zatzman et al, 2009).

They do have local trends as well as global trends (in time). One can imagine how the slope of this graph on a very small time frame would be quite arbitrary and how devastating it would be to take that slope to a long-term. One can easily show how the trend emerging from Newton's differential quotient would be diametrically opposite to the real trend.

The finite difference methods are extensively applied in the petroleum industry to simulate the fluid flow inside the porous medium. The following assumptions are inherent to the finite difference method.

1. The relationship between derivative and the finite difference operators, e.g., forward difference operator, backward difference operator and the central difference operator is established through the Taylor series expansion. The Taylor series expansion is the basic element in providing the differential form of a function. It converts a function into a polynomial of infinite order. This provides an approximate description of a function by considering a finite number of terms and ignoring the higher order parts. In other words, it assumes that a relationship between the operators for discrete points and the operators of the continuous functions is acceptable.
2. The relationship involves truncation of the Taylor series of the unknown variables after few terms. Such truncation leads to accumulation of error. Mathematically, it can be shown that most of the error occurs in the lowest order terms.
    a. The forward difference and the backward difference approximations are the first order approximations to the first derivative.
    b. Although the approximation to the second derivative by central difference operator increases accuracy because of a second order approximation, it still suffers from the truncation problem.
    c. As the spacing size reduces, the truncation error approaches to zero more rapidly. Therefore, a higher order approximation will eliminate the need of same number of measurements or discrete points.

It might maintain the same level of accuracy; however, less information at discrete points might be risky as well.
3. The solutions of the finite difference equations are obtained only at the discrete points. These discrete points are defined either according to block-centered or point distributed grid system. However, the boundary condition, to be specific, the constant pressure boundary, may appear important in selecting the grid system with inherent restrictions and higher order approximations.
4. The solutions obtained for grid-points are in contrast to the solutions of the continuous equations.
5. In the finite difference scheme, the local truncation error or the local discretization error is not readily quantifiable because the calculation involves both continuous and discrete forms. Such difficulty can be overcome when the mesh-size or the time step or both are decreased leading to minimization in local truncation error. However, at the same time the computational operation increases, which eventually increases the round-off error.

1.3.5 Darcy's Law

Because practically all reservoir simulation studies involve the use of Darcy's law, it is important to understand the assumptions behind this momentum balance equation. The following assumptions are inherent to Darcy's law and its extension:

- The fluid is homogenous, single-phase and Newtonian;
- No chemical reaction takes place between the fluid and the porous medium;
- Laminar flow condition prevails;
- Permeability is a property of the porous medium, which is independent of pressure, temperature and the flowing fluid;
- There is no slippage effect; e.g., Klinkenberg phenomenon;
- There is no electro-kinetic effect.

1.4 Recent Advances in Reservoir Simulation

The recent advances in reservoir simulation may be viewed as:

- Speed and accuracy;
- New fluid flow equations;
- Coupled fluid flow and geo-mechanical stress model; and
- Fluid flow modeling under thermal stress.

1.4.1 Speed and Accuracy

The need for new equations in oil reservoirs arises mainly for fractured reservoirs as they constitute the largest departure from Darcy's flow behavior. Advances have been made in many fronts. As the speed of computers increased following Moore's law (doubling every 12 to 18 months), the memory also increased. For reservoir simulation studies, this translated into the use of higher accuracy through inclusion of higher order terms in Taylor series approximation as well as great number of grid blocks, reaching as many as a billion blocks. The greatest difficulty in this advancement is that the quality of input data did not improve at par with the speed and memory of the computers. As Fig. 1.3 shows, the data gap remains possibly the biggest challenge in describing a reservoir. Note that the inclusion of large number of grid blocks makes the prediction more arbitrary than that predicted by fewer blocks, if the number of input data points is not increased proportionately. The problem is particularly acute when fractured formation is being modeled. The problem of reservoir cores being smaller than the representative elemental volume (REV) is a difficult one, which is more accentuated for fractured formations that have a higher REV. For fractured formations, one is left with a narrow band of grid blocks, beyond which solutions are either meaningless (large grid blocks) or unstable (too small grid blocks). This point is elucidated in Fig. 1.4. Figure 1.4 also shows the difficulty associated with modeling with both too small or too large grid blocks. The problem is particularly acute when fractured formation is being modeled. The problem of reservoir cores being smaller than the representative elemental volume (REV) is a difficult one, which is more accentuated for fractured formations that have a higher REV. For fractured formations, one is left with a narrow band of grid blocks, beyond which

Reservoir Simulation Background

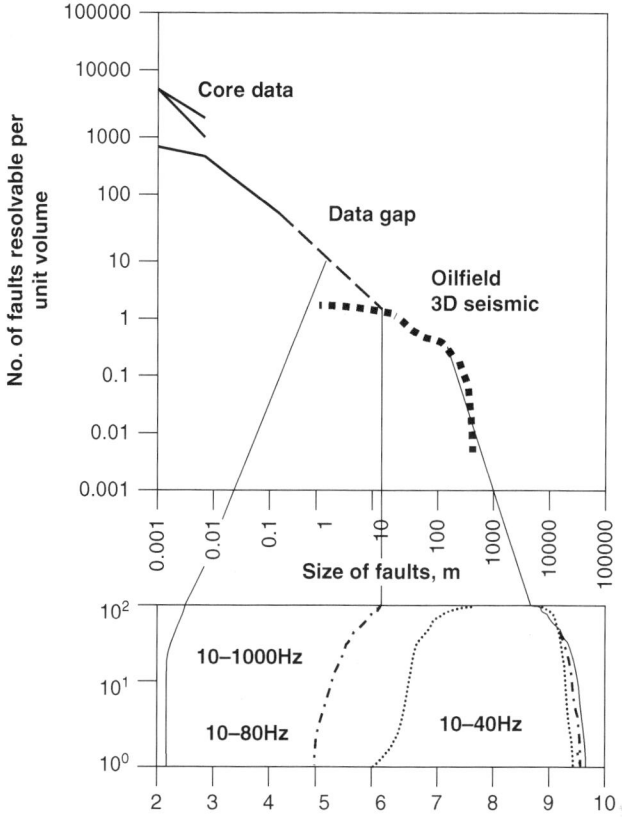

Figure 1.3 Data gap in geophysical modeling (after Islam, 2001).

solutions are either meaningless (large grid blocks) or unstable (too small grid blocks).

1.4.2 New Fluid Flow Equations

A porous medium can be defined as a multiphase material body (solid phase represented by solid grains of rock and void space represented by the pores between solid grains) characterized by two main features: that a Representative Elementary Volume (REV) can be determined for it, such that no matter where it is placed within a domain occupied by the porous medium, it will always contain both a persistent solid phase and a void space. The size of the REV is such that parameters that represent the distributions of the void space and the solid matrix within it are statistically meaningful.

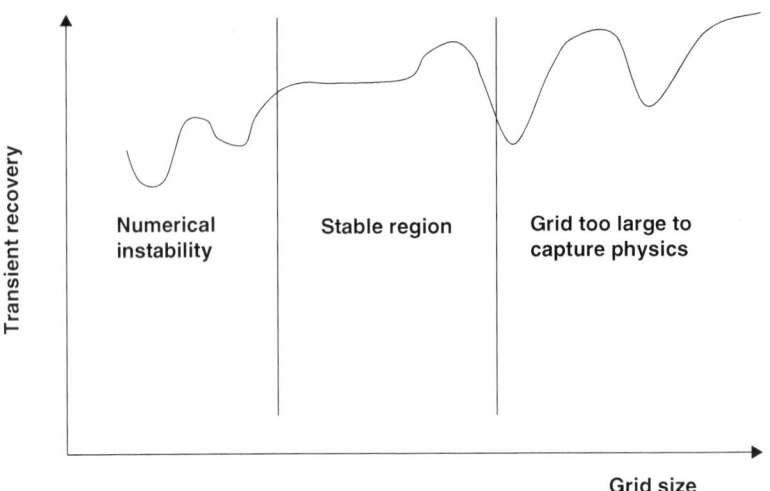

Figure 1.4 The problem with the finite difference approach has been the dependence on grid size and the loss of information due to scaling up (from Islam, 2002).

Theoretically, fluid flow in porous medium is understood as the flow of liquid or gas or both in a medium filled with small solid grains packed in homogeneous manner. The concept of heterogeneous porous medium then introduced to indicate properties change (mainly porosity and permeability) within that same solid grains packed system. An average estimation of properties in that system is an obvious solution, and the case is still simple.

Incorporating fluid flow model with a dynamic rock model during the depletion process with a satisfactory degree of accuracy is still difficult to attain from currently used reservoir simulators. Most conventional reservoir simulators do not couple stress changes and rock deformations with reservoir pressure during the course of production and do not include the effect of change of reservoir temperature during thermal or steam injection recoveries. The physical impact of these geo-mechanical aspects of reservoir behavior is neither trivial nor negligible. Pore reduction and/or pore collapse leads to abrupt compaction of reservoir rock, which in turn cause miscalculations of ultimate recoveries, damage to permeability and reduction to flow rates and subsidence at the ground and well casings damage. In addition, there are many reported environmental impacts due to the withdrawal of fluids from underground reservoirs.

Using only Darcy's law to describe hydrocarbon fluid behavior in petroleum reservoirs when high gas flow rate is expected or when encountered in an highly fractured reservoir is totally misleading. Nguyen (1986) has showed that using standard Darcy flow analysis in some circumstances can over-predict the productivity by as much as 100 percent.

Fracture can be defined as any discontinuity in a solid material. In geological terms, a fracture is any planar or curvy-planar discontinuity that has formed as a result of a process of brittle deformation in the earth's crust. Planes of weakness in rock respond to changing stresses in the earth's crust by fracturing in one or more different ways depending on the direction of the maximum stress and the rock type. A fracture can be said to consist of two rock surfaces, with irregular shapes, which are more or less in contact with each other. The volume between the surfaces is the fracture void. The fracture void geometry is related in various ways to several fracture properties. Fluid movement in a fractured rock depends on discontinuities, at a variety of scales ranging from micro-cracks to faults (in length and width). Fundamentally, describing flow through fractured rock involves describing physical attributes of the fractures: fracture spacing, fracture area, fracture aperture and fracture orientation and whether these parameters allow percolation of fluid through the rock mass. Fracture parameters also influence the anisotropy and heterogeneity of flow through fractured rock. Thus the conductivity of a rock mass depends on the entire network within the particular rock mass and is thus governed by the connectivity of the network and the conductivity of the single fracture. The total conductivity of a rock mass depends also on the contribution of matrix conductivity at the same time.

A fractured porous medium is defined as a portion of space in which the void space is composed of two parts: an interconnected network of fractures and blocks of porous medium, the entire space within the medium is occupied by one or more fluids. Such a domain can be treated as a single continuum, provided an appropriate REV can be found for it.

The fundamental question to be answered in modeling fracture flow is the validity of the governing equations used. The conventional approach involves the use of dual-porosity, dual permeability models for simulating flow through fractures. Choi et al (1997) demonstrated that the conventional use of Darcy's law in both fracture and matrix of the fractured system is not adequate.

Instead, they proposed the use of the Forchheimer model in the fracture while maintaining Darcy's law in the matrix. Their work, however, was limited to single-phase flow. In future, the present status of this work can be extended to a multiphase system. It is anticipated that gas reservoirs will be suitable candidates for using Forchheimer extension of the momentum balance equation, rather than the conventional Darcy's law. Similar to what was done for the liquid system (Cheema and Islam, 1995); opportunities exist in conducting experiments with gas as well as multiphase fluids in order to validate the numerical models. It may be noted that in recent years several dual-porosity, dual-permeability models have been proposed based on experimental observations (Tidwell and Robert, 1995; Saghir et al, 2001).

1.4.3 Coupled Fluid Flow and Geo-mechanical Stress Model

Coupling different flow equations has always been a challenge in reservoir simulators. In this context, Pedrosa et al (1986) introduced the framework of hybrid grid modeling. Even though this work was related to coupling cylindrical and Cartesian grid blocks, it was used as a basis for coupling various fluid flow models (Islam and Chakma, 1990; Islam, 1990). Coupling flow equations in order to describe fluid flow in a setting, for which both pipe flow and porous media flow prevail continues to be a challenge (Mustafiz et al, 2005).

Geomechanical stresses are very important in production schemes. However, due to strong seepage flow, disintegration of formation occurs and sand is carried towards the well opening. The most common practice to prevent accumulation as followed by the industry is to take filter measures, such as liners and gravel packs. Generally, such measures are very expensive to use and often, due to plugging of the liners, the cost increases to maintain the same level of production. In recent years, there have been studies in various categories of well completion including modeling of coupled fluid flow and mechanical deformation of medium (Vaziri et al, 2002). Vaziri et al (2002) used a finite element analysis developing a modified form of the Mohr–Coulomb failure envelope to simulate both tensile and shear-induced failure around deep wellbores in oil and gas reservoirs. The coupled model was useful in predicting the onset and quantity of sanding. Nouri et al (2006) highlighted the experimental part of it in addition to a numerical analysis and measured the severity of sanding in terms of rate and duration. It should be noted that these studies (Nouri et al, 2002; Vaziri et al, 2002 and Nouri et al, 2006) took into account the elasto-plastic stress-strain relationship with strain softening to capture sand production in a more realistic manner. Although, at present these studies lack validation with field data, they offer significant insight into the mechanism of sanding and have potential in smart-designing of well-completions and operational conditions.

Recently, Settari et al (2006) applied numerical techniques to calculate subsidence induced by gas production in the North Adriatic. Due to the complexity of the reservoir and compaction mechanisms, Settari (2006) took a combined approach of reservoir and geo-mechanical simulators in modeling subsidence. As well, an extensive validation of the modeling techniques was undertaken, including the level of coupling between the fluid flow and geo-mechanical solution. The researchers found that a fully coupled solution had an impact only on the aquifer area, and an explicitly coupled technique was good enough to give accurate results. On grid issues, the preferred approach was to use compatible grids in the reservoir domain and to extend that mesh to geo-mechanical modeling. However, it was also noted that the grids generated for reservoir simulation are often not suitable for coupled models and require modification.

In fields, on several instances, subsidence delay has been noticed and related to over consolidation, which is also termed as the threshold effect (Merle et al, 1976; Hettema et al, 2002). Settari et al (2006) used the numerical modeling techniques to explore the effects of small levels of over-consolidation in one of their studied fields on the onset of subsidence and the areal extent of the resulting subsidence bowl. The same framework that Settari et al (2006) used can be introduced in coupling the multiphase, compositional simulator and the geo-mechanical simulator in future.

1.4.4 Fluid Flow Modeling Under Thermal Stress

The temperature changes in the rock can induce thermo-elastic stresses (Hojka et al, 1993), which can either create new fractures or can alter the shapes of existing fractures, changing the nature of the primary mode of production. It can be noted that the thermal stress occurs as a result of the difference in temperature between injected fluids and reservoir fluids or due to the Joule Thompson effect. However, in the study with unconsolidated sand, the thermal stresses are reported to be negligible in comparison to the mechanical stresses (Chalaturnyk and Scott, 1995). A similar trend is noticeable in the work by Chen et al (1995), which also ignored the effect of thermal stresses, even though a simultaneous modeling of fluid flow and geomechanics is proposed.

Most of the past research has been focused only on thermal recovery of heavy oil. Modeling subsidence under thermal recovery technique (Tortike and Farouq Ali, 1987) was one of the early attempts that considered both thermal and mechanical stresses in their formulation. There are only few investigations that attempted to capture the onset and propagation of fractures under thermal stress. Recently, Zekri et al (2006) investigated the effects of thermal shock on fractured core permeability of carbonate formations of UAE reservoirs by conducting a series of experiments. Also, the stress-strain relationship due to thermal shocks was noted. Apart from experimental observations, there is also the scope to perform numerical simulations to determine the impact of thermal stress in various categories, such as water injection, gas injection/production etc. More recently, Hossain et al (2009) showed that new mathematical models must be introduced in order to include thermal effects combined with fluid memory.

1.5 Future Challenges in Reservoir Simulation

The future development in reservoir modeling may be looked at different aspects. These are may be classified as:

- Experimental challenges;
- Numerical Challenges; and
- Remote sensing and real-time monitoring.

1.5.1 Experimental Challenges

The need of well designed experimental work in order to improve the quality of reservoir simulators cannot be over-emphasized. Most significant challenges in experimental design arise from determining rock and fluid properties. Eventhough progress has been made in terms of specialized core analysis and PVT measurements, numerous problems persist due to difficulties associated with sampling techniques and core integrity. Recently, Belhaj et al (2006) used a 3-D spot gas pearmeameter to measure permeability at any spot on the surface of the sample, regardless of the shape and size. Moreover, a mathematical model was derived to describe the flow pattern associated with measuring permeability using the novel device.

In a reservoir simulation study, all relevant thermal properties including coefficient of thermal expansion, porosity variation with temperature, and thermal conductivity need to be measured in case such information are not available. Experimental facilities e.g., double diffusive measurements, transient rock properties; point permeability measurements can be very important in fulfilling the task. In this regard, the work of Belhaj et al (2006) is noteworthy.

In order to measure the extent of 3-D thermal stress, a model experiment is useful to obtain temperature distribution in carbonate rock formation in the presence of a heat source. Examples include microwave heating water-saturated carbonate slabs in order to model only conduction and radiation. An extension to the tests can be carried out to model thermal stress induced by cold fluid injection for which convection is activated. The extent of fracture initiation and propagation can be measured in terms of so-called damage parameter. Time-dependent crack growth still is an elusive topic in petroleum applications (Kim and van Stone, 1995). The methodology outlined by Yin and Liu (1994) can be considered to measure fracture growth. The mathematical model can be developed following the numerical method developed by Wang and Maguid (1995). Young's modulus, compressive strength, and fracture toughness are important for modeling the onset and propagation of induced fracture for the selected reservoir. Incidentally, the same set of data is also useful for designing hydraulic fracturing jobs (Rahim and Holditch, 1995).

The most relevant application of double diffusive phenomena, involving thermal and solutal transfer is in the area of vapor-extraction (VAPEX) of heavy oil and tar sands. From the early work of Roger Butler, numerous experimental studies have been reported. Some of the latest ones are reported by the research group of Gu at Petroleum Technology Research Centre (PTRC) in Canada. Despite making great advances (e.g. Yang and Gu, 2005; Tharanivasan et al, 2004), proper characterization of such complex phenomena continues to be a formidable challenge.

1.5.2 Numerical Challenges

1.5.2.1 Theory of Onset and Propagation of Fractures Due to Thermal Stress

Fundamental work needs to be performed in order to develop relevant equations for thermal stresses. Similar work has been initiated by Wilkinson et al (1997), who used finite element modeling to solve the problem. There has been some progress in the design of material manufacturing for which in situ fractures and cracks are considered to be fatal flaws. Therefore, formulation of complete equations is required in order to model thermal stress and its effect in petroleum reservoirs. It is to be noted that this theory deals with only the transient state of the reservoir rock.

1.5.2.2 2-D and 3-D Solutions of the Governing Equations

In order to determine fracture width, orientation, and length under thermal stresses as a function of time, it is imperative to solve the governing equations first in 2-D. The finite difference is the most accepted technique to develop the simulator. An extension of the developed simulator to the cylindrical system is useful in designing hydraulic fractures in thermally active reservoirs. The 3-D solutions are required to determine 3-D stresses and the effects of permeability tensor. Such simulation will provide one with the flexibility of determining fracture orientation in the 3-D model and guide as a design tool for hydraulic fracturing. Although the 3-D version of the hydraulic fracturing model can be in the framework put forward earlier (Wilkinson et al, 1997), differences of opinion exist as to how thermal stress can be added to the in situ stress equations.

1.5.2.3 Viscous Fingering During Miscible Displacement

Viscous fingering is believed to be dominant in both miscible and immiscible flooding and of much importance in a number of practical areas including secondary and tertiary oil recovery. However, modeling viscous fingering remains a formidable task. Only recently, researchers from Shell have attempted to model viscous fingering with the chaos theory. Islam (1993) has reported in a series of publications that periodic and even chaotic flow can be captured properly by solving the governing partial differential equations with improved accuracy (Δx^4, Δt^2). This needs to be demonstrated for viscous fingering. The tracking of chaos (and hence viscous fingering) in a miscible displacement system can be further enhanced by studying phenomena that onset fingering in a reservoir. It eventually will lead to developing operating conditions that would avoid or minimize viscous fingering. Naami et al (1999) conducted both experimental and numerical modeling of viscous fingering in a 2-D system. They modeled both the onset and propagation of fingers by solving governing partial differential equations. Recent advances in numerical schemes (Aboudheir et al, 1999; Bokhari and Islam, 2005) can be suitably applied in modeling of viscous fingering. The scheme proposed by Bokhari and Islam (2005) is accurate in the order of Δx^4 in space and Δt^2 in time. Similar approaches can be extended for tests in a 3-D system in future. Modeling viscous fingering using finite element approach has been attempted as well (Saghir et al, 2000).

1.5.2.4 *Improvement in Remote Sensing and Monitoring Ability*

It is true that there is skepticism about the growing pace of applying 4-D seismic for enhanced monitoring (Feature-*First break*, 1997), yet the advancement in the last decade assures that the on-line monitoring of reservoirs is not an unrealistic dream (Islam, 2001). Strenedes (1995) reported that the average recovery factor from all the fields in the Norwegian sector increased from 34–39% over 2–3 years, due to enhanced monitoring. The needs for an improved technique was also emphasized to face the challenges of declining production in North Sea.

One of the most coveted features in present reservoir studies is to develop advanced technologies for real-time data transmission for both down-hole and wellhead purposes (chemical analysis of oil, gas, water, and solid) from any desired location. This research can lead to conducting real-time control of various operations in all locations, such as in the wellbore, production string and pipelines remotely. However, a number of problems need to be addressed to make advances in remote sensing and monitoring.

1.5.2.5 *Improvement in Data Processing Techniques*

The first stage of data collection follows immediate processing. Even though great deal of care is taken for collection of rock and fluid samples, the importance of improving data processing technique is seldom felt. Of course, errors in core data may enter due to measurement errors in the laboratory and/or during sample collection, but the most important source of error lies within processing the data.

Data can be from fluid analysis (e.g. PVT), core analysis, geophone data, real-time monitoring data, wellhead data, or others. Great difficulties arise immediately, as practically all processors are linear. Recently, Panawalage et al (2004a, 2004b) showed how non-linear modeling can be used to reverse absolute permeability information from raw data. This work has been advanced further to permeability tensor by Mousavizadegan et al (2006a, 2006b). Even though great advances have been made in laboratory measurements of permeability data and the possibility of in situ permeability measurement is not considered to be unrealistic (Khan and Islam, 2007), processing of such data with a non-linear solver is at its infancy.

In processing sonic data, the wave equations are solved in order to reverse calculate reservoir properties. Most commonly used wave equation is Maxwell's equation. While the original form of this equation is non-linear, due to the lack of a truly non-linear solver, this equation is linearized, leading to the determination of coefficients that have limited application to say the least.

1.5.3 Remote Sensing and Real-time Monitoring

The conventional seismic technology has a resolution of 20 m for the reservoir region. While this resolution is sufficient for exploration purposes, it falls short of providing meaningful results for petroleum field development, for which 1 m resolution is necessary to monitor changes (with 4-D seismic) in a reservoir. For the wellbore, a resolution of 1 mm is necessary. This can also help detect fractures near the wellbore. The current technology does not allow one to depict the reservoir, the wellbore, or the tubular with acceptable resolution (Islam, 2001). In order to improve resolution within a wellbore, acoustic response need to be analyzed. In addition, fiber-optic detection of multiphase flow can be investigated. Finally, it will be possible to develop a data acquisition system that can be used as a real-time monitoring tool, once coupled with a signal processor. Recently, Zaman et al (2006) used a laser spectroscope to detect paraffin in paraffin-contaminated oil samples. After passing through the oil sample, the laser light was detected by a semi-conductor photodiode, which, in turn, converted the light signal into electric voltage. In their study, the paraffin concentrations ranged between 20% and 60% wt and a thickness of 1 and 10 mm. They developed a 1-D mathematical model to describe the process of laser radiation attenuation within the oil sample based on energy balance. Furthermore, the problem was numerically solved with reasonable agreement with experimental results. Their model can be used to predict the net laser light and the amount of light absorbed per unit volume at any point within the oil sample. The mathematical model was extended to different oil production schemes to determine the local rate of absorption in an oil layer under different working environments.

1.5.3.1 Monitoring Offshore Structures

In order to remain competitive in today's global economic environment, owners of civil structures need to minimize the number of days their facilities are out of service due to maintenance, rehabilitation or replacement. Indicators of structural system performance are needed for the owner to allocate resources toward repair, replacement or rehabilitation of their structures. To quantify these system performance measures requires structural monitoring of large civil structures while in service (Mufti et al, 1997). It is, therefore, important to develop a structural monitoring system that will integrate:

   a. Fiber optic sensor systems;
   b. Remote monitoring communication systems;
   c. Intelligent data processing system;
   d. Damage detection and modal analysis system; and
   e. Non-destructive evaluation system.

It will be more useful if the monitoring device is capable of detecting signs of stress corrosion cracking. A system of fiber optic-based sensor and remote monitoring communication will allow not only monitoring of the internal operating pressures but also the residual stress levels, which are suspected for the initiation and growth of near-neutral pH stress corrosion cracking. Finally, the technology can be applied in real-time in monitoring offshore structures. Along this line of research, the early detection of precipitation of heavy organics such as paraffin, wax, resin, asphaltene, diamondoid, mercaptdans, and organometallic compounds, which can precipitate out of the crude oil solution due to various forces causing blockage in the oil reservoir, well, pipeline and in the oil production and processing facilities is worth mentioning (Zaman et al, 2004). Zaman et al (2004) utilized a solid detection system by light transmittance measurement for asphaltene detection, photodiode for light transmittance measurement for liquid wax, detection, and ultrasound and strain gauge solid wax detection. Such an attempt, if effectively used, has the potential to reduce pigging (the common commercial term for cleaning the pipeline) and in turn, the maintenance cost considerably.

1.5.3.2 Development of a Dynamic Characterization Tool (Based on Seismic-while-drilling Data)

A dynamic reservoir characterization tool is needed in order to introduce real-time monitoring. This tool can use the inversion technique to determine permeability data. At present, cuttings need to be collected before preparing petrophysical logs. The numerical inversion requires the solution of a set of non-linear partial differential equations. Conventional numerical methods require these equations to be linearized prior to solution (discussed early in this chapter). In this process, many of the routes to final solutions may be suppressed (Mustafiz et al, 2008a) while it is to be noted that a set of non-linear equations should lead to the emergence of multiple solutions. Therefore, it is important that a nonlinear problem is investigated for multiple-value solutions of physical significance.

1.5.3.3 Use of 3-D Sonogram

This feature illustrates the possibility of using 3-D sonogram for volume visualization of the rock ahead of the drill bit. In order to improve resolution and accuracy of prediction ahead of the drill bit, the 3-D sonogram technique will be extremely beneficial. The latest in ultrasound technology offers the ability to generate images in 4-D (time being the 4th dimension). In preparation to this task, a 3-D sonogram can be adopted to detect composition of fluid through non-invasive methods. Note that such a method is not yet in place in the market. Also, there is the potential of coupling 3-D sonogram with sonic while drilling in near future.

This coupling will allow one to use drilling data to develop input data for the simulator with high resolution. Availability and use of a sophisticated compositional 3-D reservoir simulator will pave the way to developing real-time reservoir modeling – a sought after goal in the petroleum industry for some time.

1.5.3.4 Virtual Reality (VR) Applications

In the first phase, the coupling of an existing compositional, geomechanical simulator with the VR machine is required. Time travel can be limited to selected processes with limited number of wells primarily. Later time travel can expand as the state-of-the-art in simulation becomes more sophisticated.

Describing petroleum reservoirs is considered to be more difficult than landing man on the moon. Indeed, reservoir engineers have the difficult task of conducting reservoir design without ever going for a site inspection. This application is aimed at creating a virtual reservoir that can undergo various modes of petroleum production schemes (including thermal, chemical, and microbial enhanced oil recovery or "EOR"). The authors comprehend that in future the virtual reservoir, in its finished form, will be coupled with virtual production and separation systems. A virtual reservoir will enable one to travel through pore spaces at the speed of light while controlling production/injection schemes at the push of a button. Because time travel is possible in a virtual system, one does not have to wait to see the impact of a reservoir decision (e.g. gas injection, steam huff-and-puff) or production problems (e.g. wellbore plugging due to asphaltene precipitation).

The use of virtual reality in petroleum reservoir is currently being discussed only in the context of 3-D visualization (Editorial, 1996). A more useful utilization of the technique, of course, will be in reservoir management, offshore monitoring, and production control. While a full-fledged virtual reservoir is still considered to be a tool for the future, one must concentrate on physics and mathematics of the development in order to ensure that a virtual reservoir does not become a video game. Recently, several reports have appeared on the use of virtual reality in platform systems, and even production networking (Editorial, 1996). An appealing application of virtual reality lies in the areas of replacing expensive laboratory experiments with computational fluid dynamics models. However, petroleum-engineering phenomena are still so poorly described (mathematically) that replacing laboratory experiments will lead to gross misunderstanding of dominant phenomena. Recently, Statoil has developed a virtual reality machine that would simulate selected phenomena in the oilfield. Similarly, Norsk Hydro has reported the virtual modeling of a cave. The reservoir simulator behind the machine, however, is only packed with rudimentary calculations. More advanced models have been used in drilling and pipelines.

Even though the concept is novel, the execution of the described plan can be realistic in near future. The reservoir data (results as well as the reservoir description) will be fed into an ultra-fast data acquisition system. The key here is to solve the reservoir equations so fast that the delay between data generation and the data storage/distribution unit is not "felt" by a human. The data acquisition system could be coupled with digital/analog converters that will transform signals into tangible sensations. These output signals should be transferred to create visual, thermal, acoustic, and piezometric effects. Therefore, this task should lead to coupling the virtual reality capability with a state-of-the-art reservoir model. When it becomes successful, it will not be a mere dream to extend the model to a vertical section of the well, as well as surface facilities.

1.5.3.5 Intelligent Reservoir Management

Intelligent systems can be utilized effectively to help both operators and design engineers to make decisions. The major goal of this management program is to develop a novel Knowledge Based Expert System that helps design engineers to choose a suitable EOR method for an oil reservoir. It should be a comprehensive expert system ES that integrates the environmental impacts of each EOR process into the technical and economical feasibility of different EORs.

Past intelligent reservoir management referred to computer or artificial intelligence. Recently, Islam (2006) demonstrated that computer operates quite differently from how humans think. He outlined the need for new line of expert systems that are based on human intelligence, rather than artificial intelligence. Novel expert systems embodying pro-nature features are proposed based on natural human intelligences (Ketata et al, 2005a, 2005b). These expert systems use human intelligence which is opposite to artificial intelligence. In these publications, authors attempted to include the knowledge of non-European races who had a very different approach to modeling. Also, based on Chinese abacus and quipu (Latin American ancient tribe), Ketata et al (2006a; 2006b) developed an expert system that can be characterized as the first expert system without using the conventional computer counting system. These expert systems provide the basis of an intelligent, robust, and efficient computing tool.

Because all natural phenomena are non-linear, we argued that any acceptable computational technique must produce multiple solutions. With this objective, Islam (2006) developed a new computational method that finds dynamic derivatives of any function and also solves set of non-linear equations. More recently, Mousavizadegan et al (2007), proposed a new technique for finding invariably multiple solutions to every natural equation. These techniques essentially create a cloud of data points and the user can decide which ones are most relevant to a certain application.

Another significant aspect of "intelligence" was addressed by Ketata et al (2006a, 2006b). This aspect involves the redefining of zero and infinity. It is important to note that any discussion of human intelligence cannot begin without the mathematics of intangibles, which include proper understanding of these concepts.

Finally, a truly intelligent reservoir model should be able to model chaos. It is recognized that "chaos" is the interrelated evolutionary order of nature elements. It is the science of objects and systems in nature. A new chaos theory has been developed by Ketata et al (2006c).

1.6 Economic Models Based on Futuristic Energy Pricing Policies

There is a distinct need to integrate energy pricing and economic models with a reservoir simulator. The energy pricing policy is one of the most complex and sensitive global issues. With growing worldwide concern about environment and conservation of nature, the economic models must reflect them through futuristic, green-energy policies. The economics models should have the following features, which are often ignored in economic models. They are (Khan and Islam, 2006; 2007):

   a. Short-term and long-term impact of oil production on agriculture, livestock, fisheries and others affecting the food chain;
   b. Intangible cost of groundwater and air pollution resulting from petroleum activities;
   c. Clean-up cost of accidental oil spills;
   d. Costs related to inherently deficient engineering design;
   e. Costs related to political constraints on energy pricing.

The comprehensive economic model will allow one to evaluate a project based on its merit from economic as well as environmental and social values. Even though political constraints constitute a part of social models, it is well known that petroleum policies are more acutely dependent on world politics (Baade et al, 1990; Stickey, 1993). Therefore, political constraints should be identified as the primary variable in energy pricing (Zatzman and Islam, 2005).

The new model must also focus on the integration of pricing and qualitative controls in the abatement of cost-effective reductions in energy related CO_2 emissions. Worldwide demand for energy is increasing faster than ever (Dung and Piracha, 2000). The demand for oil and gas is increasing faster than the growth rates of GDP in developing and developed countries. Petroleum is used as inputs for manufacturing, construction, agriculture, power generating, and service sectors. This is also a final consumer product. As demand for goods and services increases, the demand for oil will also increase. As GDP increases, the demand for oil as a final consumer product will also increase. Therefore, inefficient supply and demand management policies on petroleum can impose severe constraints on output, employment, standard of living, the backward and forward linkage effects and the growth of other dependent sectors of the domestic and global economy. The new model must develop a pricing policy that will ensure sustainable development while allowing certain qualitative controls which will minimize the average emissions of energy related CO_2. Such policies are feasible and can be achieved at minimum social cost (Howarth and Winslow, 1994). The qualitative control can be achieved through the employment of green technologies (Islam, 2000).

Zatzman and Islam (2007a) investigated the possibilities about the true potential of a given resource in the process of collecting during actual production information about dynamic changes in reservoir conditions in situ. The idea would be to eliminate much of the guesswork built into current demand-based modeling of energy prices. In general, these models project costs well into the future to "take care" of the margin of error incorporated in the guesswork. It can be considered a shift from the demand-based to the supply-based modeling, from control exercised downstream over national production companies upstream to a profound challenge against such control. According to the proposal by Zatzman and Islam (2007a), it is less meaningful to speak of "future energy price", and more meaningful to think in terms of another energy pricing model. The correction of guesswork with actual knowledge gathered in situ during production has a number of implications. It renders moot, or irrelevant, any previous assumption of resource scarcity or globally finite supply as a boundary condition, as the issue shifts increasingly away from the price of energy to how this price and its trend are modeled.

The state-of-the-art model needs to be employed to predict the future oil price so that producers, consumers, firms, governments and other stakeholders in the industry are able to plan a sustainable growth of output, employment and income. Also, the model should be able to decompose the effects of each of the variables in determining the oil price and computing the individual contribution. The behaviors of the following variables will determine equilibrium world price of oil:

 a. Excess demand;
 b. Market power;
 c. World commodity price index;
 d. Price expectations;
 e. Political power;
 f. Environmental cost;
 g. Taxes and tariffs, and
 h. Cost of political constraints.

1.7 Integrated System of Monitoring, Environmental Impact and Economics

Recently, there have been renewed efforts by the petroleum companies to integrate economics with reservoir simulation. As for example, Schlumberger has been conducting reservoir management and economic analysis through ECLIPSE and MERAK. MERAK can perform economic evaluation, decline analysis, and fiscal modeling that calculate the value of oil and gas properties in Canada, the United States and around the world. The merge of such a package with a state-of-the-art reservoir simulator can be used as research tool for validation of various concepts, such as cost of environmental impact, financial worth of bottom-line driven economy, and others. All engineering, economical, environmental, and socio-political constraints should be integrated through a global optimization package (Finley et al, 1998).

The authors also envision that a global optimization package, in its matured form, will initiate the development of solutions to environmental problems, including risk analysis and decision support (Abdeh-Kolahchi et al, 2009; Hossain, 2008; Khan et al, 2009). An exposure assessment of the airborne and waterborne pollutants during CO_2 and water injection processes respectively can be conducted to study their impact on marine environment and offshore workers (Khan and Islam, 2007). The study on the evaluation of pollutants in conjunction with fluid flow injection will be helpful in creating a multiple-response experimental design. The rationale for conducting experiments will be to introduce a non-bias approach from both the process operation and environmental aspects. Most importantly, the goal should be to introduce the principles of *pollution prevention* (P2) at the research level in order to reduce the need for *"end-of-pipe"* mitigation strategies during implementation. Also the exposure assessment and treatment mitigation phase can be incorporated into the virtual reservoir model, as described previously. Recently, Lakhal et al (2007) outlined how petroleum operations can be rendered 'green' by following what they termed the 'Olympic' model.

Finally, whenever the community in question could be viewed as having been marginalized, and left behind by economic events, there is a perception that little is being done to mitigate these contamination problems, in contrast to large cities where such problems receive close attention, leading to the perception of an *"environmental injustice"* regarding such pollution. However, with the increasing awareness of human rights and justice in all aspects of life, the general public seems to be committed to have such environmental injustice dealt with in a fair and equitable manner (Zatzman and Islam, 2007a).

2
Reservoir Simulator-input/output

Hydrocarbons accumulate in underground trap forms by structural and/or stratigraphic features. The oil and gas accumulations usually occur in porous and permeable portions of beds which are normally sandstone, limestone or dolomite formations. The part of the trap that is oil and gas productive is termed as "reservoir". A reservoir may hydraulically connect to various volumes of water-bearing rock called aquifers. There are many reservoirs located in large sedimentary basins and share a common aquifer. In numerous cases, the entire trap is filled with hydrocarbon and the trap and the reservoir are the same. Oil and gas are produced from wells drilled into underground porous rocks. The ensemble of wells draining a common oil accumulation or source or the surface area defined by the well distribution is denoted as *"oil field"*. Most oil fields do not produce more than 45% of the oil-in-place, even after enhanced oil recovery (EOR) schemes have been applied (Islam, 2001). It may be due to bypassing a significant portion of the original reserves by the applied displacement techniques. Finding this missing oil can lead to significant economic windfalls because the infrastructure for additional oil recovery is already in place and the cost of production is likely to be minimal. This complies with the ultimate goal underlying the development of the science of reservoir engineering to attain a maximum efficiency in the exploitation of oil- and gas-bearing reservoirs or in other words, to obtain maximum recovery of oil and gas at a minimum cost.

The simulation of petroleum reservoirs is an essential practice in the development of more efficient techniques to increase hydrocarbon recovery and considered the main tool for modern reservoir management. For optimal reservoir management, it's critical to determine the reserves, recovery factors and economic limits as quickly as possible, but that's a difficult job. Using reservoir simulation, engineers are able to forecast a range of production and depletion scenarios based on different variables. This greatly improves decision-making up-front, before money is spent to drill new wells, establish infrastructure and surface facilities and above all damaging the reservoir and losing great deal of production.

Reservoir simulators allow engineers and geoscientists to build dynamic models that predict the movement of oil and gas flowing in reservoirs under *in situ* conditions.

However, all beneficial aspects of a reservoir simulation are affected by the data that are fed into the simulators, input data, and also a correct recording of the performance of the simulated and real systems, output data. The system consists of reservoir, wells and the other facilities. The reservoir simulator is normally constructed on very highly uncertain input data regarding the rock and fluid properties. The input data should be delineated based on the output information from the real and simulator systems during the production life of the reservoir. It will help to adapt an optimum strategy to recover the hydrocarbons.

The intent of this chapter is to provide insight of the related issues beyond the body of the simulator and its construction. Once an in-house simulator is built or a commercial simulator is to be used, what are the associated concerns that the user in specific should be wary about? Key features of input and output apprehensions will be discussed.

2.1 Input and Output Data

Reservoir simulators use mathematical expressions, usually in partial differential equations forms, to model the flow behavior of oil, water and gas inside reservoirs. The reservoir is divided into discrete gridblocks, and powerful computers are used to compute the changes in conditions over many discrete time intervals. Simulated performance of the field is compared to actual data recorded in the real system. In the oil industry, the predictive capacity of a simulator is adapted such that the predicted results approach measured production data. This process is generally referred to as *"history-matching"*. The history-matching process is carried out due to the uncertainty involved in input data and also the uncertain adopted fluid flow model. A schematic layout of history-match of the real and simulated data is given in Fig. 2.1. The production data are used to update the model parameters and therefore, the history-matching process could be considered as a closed-loop control process.

A schematic illustration of the geological and geophysical activities is also depicted in Fig. 2.1. The initial geological data are usually obtained by the seismic surveys, well logging, core sampling.

Reservoir Simulator-input/output

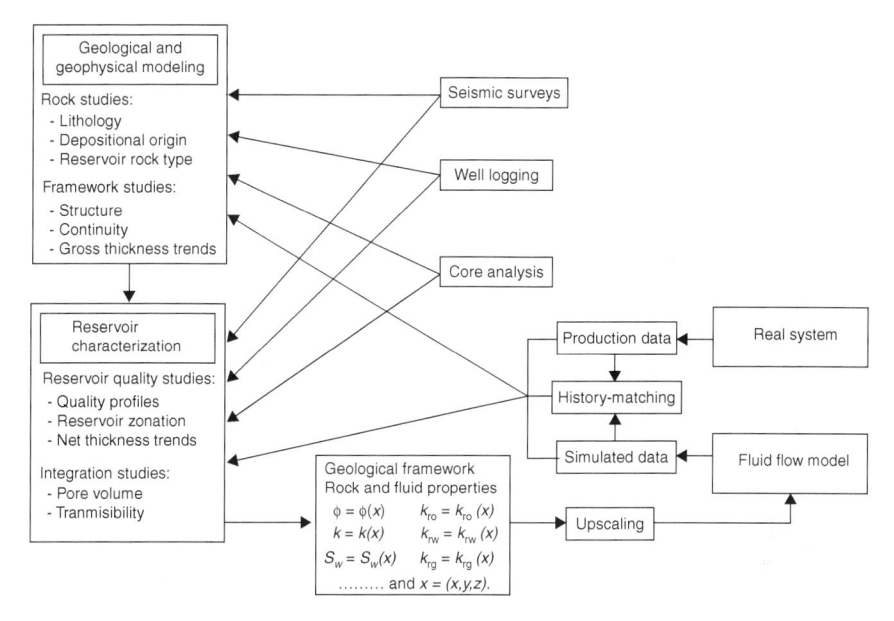

Figure 2.1 A schematic description of history-matching and geological activities regarding to a reservoir.

These data are delineated with the information obtained during the oil and gas production through the history-matching process. The effectiveness of the procedure depends on the duration of updating input data and the history-matching process. The traditional history-matching process is usually performed on a campaign basis, typically after period of years. In many cases, the matching techniques are usually ad-hoc and involve manual adjustment of model parameters, instead of systematic parameter updating.

The Fig. 2.1 may also be considered as schematic diagrams to illustrate the input and output data at different levels of reservoir modeling. The first sets of data are generated through the geological and geophysical studies. The geological and geophysical model based on the core analysis and well logging with application of seismic survey model is a preliminary identification of a hydrocarbon reservoir. The fluid flow model can be used to test physical model against production performance of the reservoir. The adjustments are made to the reservoir characterization model until a match is achieved.

Reservoir fluid flow modeling is a mathematical description of the fluid flow through the porous rock. The objective of developing the reservoir flow modeling is to optimize the hydrocarbon recovery. The mathematical modeling is transferred into a numerical formulation to find quantitative description for the fluid flow. The numerical description is usually constructed on a discretized reservoir consisting of many units. The size of a unit should allow using average properties throughout. However, many factors such as the capabilities of computing facilities and the applied numerical schemes play an important role in selection of the size of a representative computing unit or simply a grid. A detailed analysis indicates that for an accurate reservoir engineering analysis, geostatistical models should have information of 1 m scale. This is the only scale that would satisfy the representative elemental volume (REV) requirement of an enhanced oil recovery (EOR) system. This scale length is orders of magnitude higher than that of core samples and at least an order of magnitude lower than the conventional seismic data. This data gap constitutes the weakest link between geophysical information and reservoir engineering (Islam, 2001). This constitutes the problem of up- and downscaling.

Upscaling is an averaging process from one scale to a larger scale. It is one of the most challenging problems in modeling of a petroleum reservoir. A medium property is observed at one scale on a particular support (volume) of measurement, but the value of that property is needed on a different volume size at a different location. In reservoir modeling, the upscaling process may be divided into two levels. The first level is scaling up the laboratory measurement on core samples, perhaps a few centimeters in length and diameter, to geophysical cells of several meters. The second level is referring to scale up from geological fine grid to the fluid flow simulation grid. The upscaling in reservoir simulation is mainly referred to as the second level of upscaling process.

2.2 Geological and Geophysical Modeling

The first sets of data are generated through the geological and geophysical studies. The reservoirs are normally thousands of feet under the ground level, the size, the shape and the constituent of that reservoir are uncertain. The process of finding an oil-bearing rock and estimating the quality and quantity of the rock and fluid is referred to as the geological and geophysical activities. Geologists and geophysicists provide information about the reservoir and its contents that will lead to *"geological and geophysical modeling"* and then *"reservoir characterization"*.

The geologist and geophysicists usually concentrate on the rock attributes in four stages (Harris, 1975):

a. rock studies; to establish lithology and to determine depositional environment, and to distinguish the reservoir rock from non-reservoir rock;
b. constructing a geological framework; to establish the structural style and to determine the three dimensional continuity character and gross-thickness trends of the reservoir rock;
c. reservoir-quality studies; to determine the framework variability of the reservoir rock in term of porosity, permeability, and capillary properties (the aquifers surrounding the field are similarly studied); and
d. integration studies; to develop the hydrocarbon pore volume and fluid transmissibility pattern in three dimensions.

The first two tasks may be categorized under the geological and geophysical modeling of the reservoir. The other two may be classified as reservoir characterization, Fig. 2.1.

The first critical area in the process of constructing a comprehensive reservoir model is building a geological framework of the structure and reservoir architecture from seismic data and well logs (if the oil field has been penetrated by wells). This geological model represents all major geological features that may affect the connectivity of the reservoir. This stage of reservoir studies comes out with a geological framework of the structure and reservoir architecture from the different sources of measurement mainly from the seismic data. However, all major tools and techniques of formation evaluation (mud logging, wireline logging, etc.) should be viewed only as a means to construct a reliable model of the reservoir rock in the subsurface. The geological modeling represents all major geological features (faults, flow barriers, compartments, pinch outs, etc.) that are likely to affect the connectivity of the reservoir.

The distribution of the reservoir- and non-reservoir-rock types and of the reservoir fluids determine the geometry of the model and influence the type of the model to be used. For example, the number and scale of the shale (or dense carbonate) breaks in the physical framework determine the continuity of the reservoir facies and influence the vertical and horizontal dimensions of each cell. Seismic processing is often targeting structure interpretation. For attribute inversion, more high frequency energy should be captured and relative amplitude should be preserved. In practice, reprocessing is costly; therefore filter and reverse-filter may need to be applied to prepare the seismic data for inversion. Since well logs and geological interpretations are usually available, this preparation should take well logs and geological information into account.

The geological framework is the basic inputs in the reservoir characterization and then simulation of fluid flow inside the porous media. This provides the structural style and reservoir architecture, the connectivity of the reservoir rock and the gross thickness trends of the pay. Reservoir architecture includes the gross geometrical structure of the reservoir and its physical dimensions. The architecture also belongs to the characteristics of contiguous water-bearing reservoir and the uniformity or variability of the producing section within the reservoir. The application of 3D seismic interpretation play a greater role in the development of the early stage and to find out the architecture of the reservoir where limited data are available. Gross geometrical structure of the reservoir represents the shape of the reservoir whether it is in a regular (rectangular, circular, cylindrical etc.) or any irregular shape. The underground petroleum trap structure is totally dependent on the sedimentation procedure, age of the rock, type of rock, and the geographical location of the reservoir. The use of modern technology is capable to give a general idea of the structure. However, it is still an issue and uncertainty remains to figure out the geometrical structure of the reservoir. Reservoir physical dimension is also an uncertain issue because of the uncertainty behind the prediction of gross geometrical structure. The physical dimensions of a reservoir mean length, width, height or radius and depth of the reservoir. For any reservoir engineer/geologist, the initial task is to calculate the area or volume of the reservoir.

The geological model that is constructed in such a way is a static model based on initial condition measurements and surveys and bears a substantial uncertainty. In other words, this is preliminary reservoir rock type identification that is based on the data obtained in seismic surveys and well loggings. The model should have the ability to convert into a dynamic model to be modified and to capture all possible variation of geological properties using the production data.

2.3 Reservoir Characterization

The geologists and geophysicists are collaborating to produce a complete map of the reservoir and to characterize it. To portray a reservoir, integration of all data with various qualities and quantities in a consistent manner is required to describe rock and fluid properties through the reservoir. Effective formation evaluation requires the integrated use of every piece of available data. This part of the study denotes as *"reservoir characterization"*.

The basic inputs in the process of reservoir characterization are the geological framework, well log data, core analysis data, seismic amplitude data, acoustic impedance data, well test data and any other data that can be correlated to rock and fluid properties. The interpretation results from all wells drilled into the reservoir combined with the seismic data previously acquired to construct a three-dimensional (3-D) model of the reservoir, as shown in Fig. 2.2. This model not only depicts the shape of the reservoir, but details properties of the rocks and fluids as well. However, the rock and fluid properties should be anticipated in a consistent manner that depends on the quality and quantity of the available data. For

Figure 2.2 A detailed description of a reservoir.

the reservoirs with many wells penetrations and consequently considerable well loggings data, the seismic survey data with high degrees of uncertainties may not relied on considerably but in reservoirs with limited number of penetrating wells such as offshore reservoirs, the seismic data plays an important role to characterize the reservoir.

The reservoir characterization should provide the main parameters of the reservoir fluid flow modeling. These parameters are the rock and reservoir fluid properties, i.e. porosity, permeability, density, viscosity, pressure, temperature, etc., and the variation of them through the reservoir. Almost all the reservoir rocks are highly variable in their properties. The heterogeneity of reservoir matrix may be classified into two main categories. The first is the lithological heterogeneity and the second source of heterogeneity is attributed to inherent spatial rock variability, which is the variation of rock properties from one point to another in space due to the different deposition condition and different loading histories. The first question arising is how to treat the fluid and the matrix to obtain a complete description of them. It is very complicated, if not impossible, to describe in an exact manner the geometry of the internal solid surfaces that bound the flow domain inside a porous medium. The geometry of the solid can not be defined by stating the equations that describe the surface bounding the fluid. Therefore, certain macroscopic (or average) geometric properties, such as porosity, are employed as parameters describing or actually reflecting geometry of a porous matrix. The second question is regarding to the size of a representative elementary volume (REV) that reflects the heterogeneity of the solid matrix.

2.3.1 Representative Elementary Volume, REV

If we direct our attention to the fluid contained in porous space, we also encounter a lot of difficulties to describe the phenomena associated with the fluid itself, such as motion, mass transport. If we consider the fluid itself, it consists of many molecules. To describe the motion of the fluid, a large number of equations should be provided to solve the problem at the molecular level. Since the final goal is to describe the fluid motion in porous media, it is necessary to have a higher level and treat the fluid as continua. The concept of a particle is essential to the treatment of fluids as continua. A particle is an ensemble of many molecules contained in a small volume.

The size of a particle should be larger than the mean free path of a single molecule and, however, be sufficiently small as compared to the considered fluid domain that by averaging fluid and flow properties over the molecules included in it, meaningful values, i.e. values relevant to the description of bulk fluid properties, will be obtained. The mean values are then related to some centriod of the particle and then, at every point in the domain occupied by a fluid, we have a particle possessing definite dynamic and kinematic properties.

The concept of the fluid continuum and the definition of density as an example of the fluid properties are illustrated in Fig. 2.3. It shows that the particle size (elementary volume) should be of order of magnitude of the average distance λ between the molecules (mean free path of molecules). However, to capture the fluid in homogeneity, the size of particle should be less than an upper limit. This may be characterized by the length, L (or L_x, L_y and L_z in the direction of three coordinates). The length L may be considered as characteristic length for the macroscopic changes in fluid properties such as density. The volume L^3 (or $L_x L_y L_z$) may be used as the upper limit for the particle volume, ΔU_i.

If we turn our attention to the solid medium confining the fluid, the size of a representative elementary porous medium volume

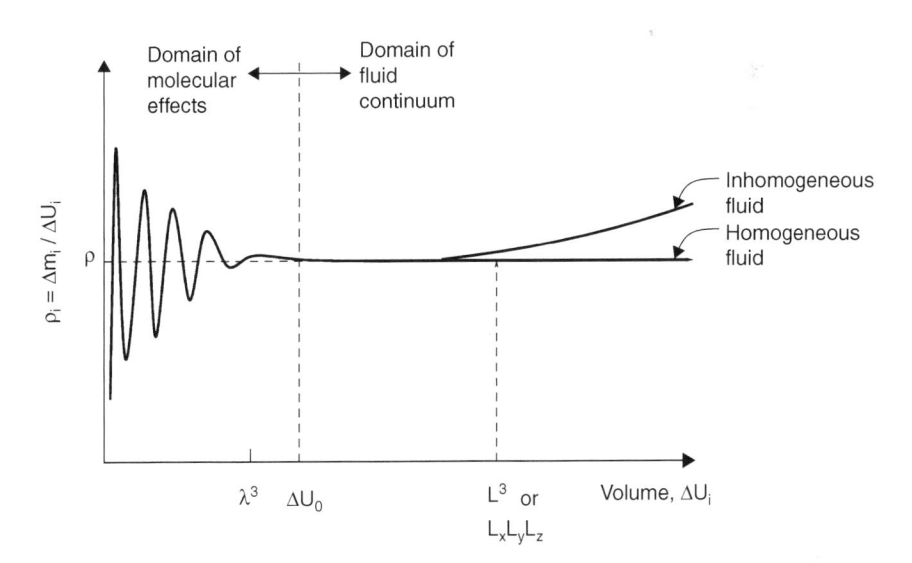

Figure 2.3 The fluid continuum domain and the variation of fluid density (modified from Bear, 1972).

around a point P should be determined. The size of representative elementary volume (REV) should be much smaller than the size of entire fluid flow domain to associate the resulting averages with a point P. On the other hand, it must be larger than the size of a single pore that it includes a sufficient number of pores to permit the meaningful statistical average required in the continuum concept (Bear, 1972). The definition of REV and representative elementary property (REP) is illustrated in Fig. 2.4. The property of the medium and more importantly permeability that is a property of medium and fluid both, should be averaging around a volume ΔU_0 as REV. The representative elementary volume may be defined as the volume ΔU_0 at which the fluctuation in the REP is not significant beyond it.

The representative elementary volume (REV) may be defined as a critical averaging volume beyond which there is no significant fluctuation in the representative property as the addition of extra voids or solids has a minor effect on the averaged property. The Fig. 2.5 shows the result of a thought experiment. It is impossible to carry out actual measurements over enough scales to confirm the behavior shown in this plot. However, it indicates that the variability of a specified property of the rock matrix is varied erratically at a small scale, less than the REV, has a zero (or small) variability at an intermediate scale, in the range of REV. The variation of the

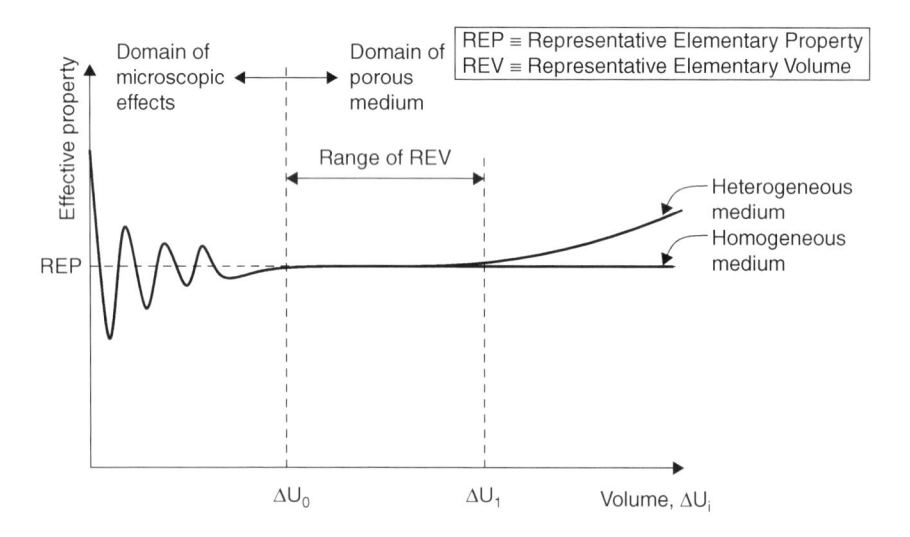

Figure 2.4 Definition of representative elementary volume (REV) and representative elementary property (REP), from Bear (1975).

Reservoir Simulator-input/output 41

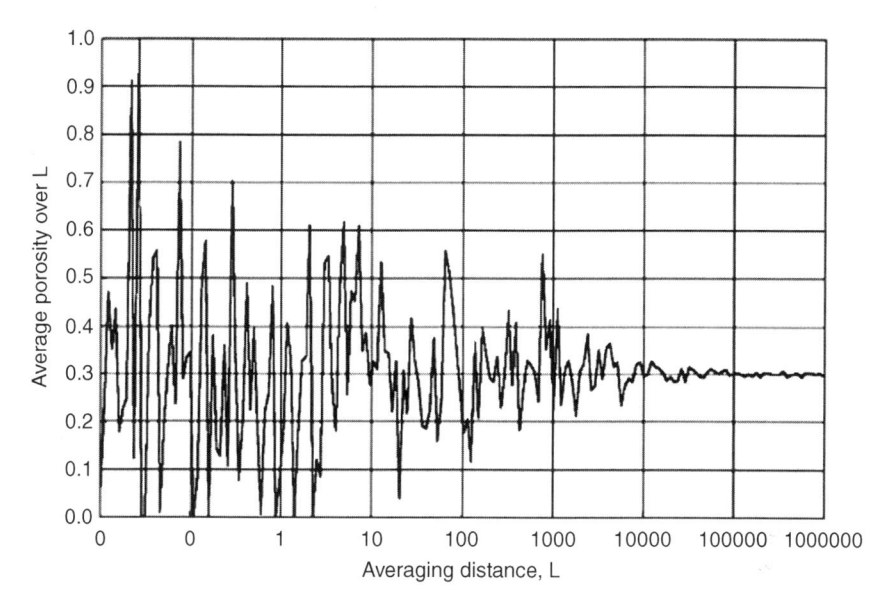

[Figure: plot of Average porosity over L vs Averaging distance, L]

Figure 2.5 The variation of average porosity as a function one-dimensional proxy, L, of the support (measurement) volume (From Lake and Srinivasan, 2004).

specified property such as porosity will be increased with increasing the scale larger than REV. These results are consistent with the general observation. The variability of the average porosity in a set of 1" diameter core plugs is not zero and this variability is certainly smaller than the variability of porosity on the scale of several μm (say 100 μm). The statistical investigation using the variance of the mean of the porosity by Lake and Srinivasan (2004) indicates the same trend for small scale. However, they do not find any region of stable porosity at intermediate even though the parameters were chosen to emphasize such stability, Fig. 2.5. This shows that the only way to obtain a stable average is for the averaging volume to exceed the largest scale of heterogeneity. This is not a satisfactory outcome. It says that the REV is the largest scale of heterogeneity in the field. Since, any cell-by cell computation or any measurement below the REV scale, there is some uncertainty in the modeled rock properties.

2.3.2 Fluid and Rock Properties

We discussed the approach in defining the fluid and rock properties. The fluid is considered to be a continuum media and the rock properties are averaging on a representative elementary volume (REV) in spite of the fact the existence of the REV is tenuous, because it has never been identified in real media. The size of the REV is related to how to locally correlate the property on the pore (microscopic) scale. It should be large enough for statistically meaningful averaging. If we follow the traditional definition of REV as given in Fig. 2.4, it should be small enough to avoid heterogeneity. As a very rough number, the typical REV size is somewhere around 100–1000 grain diameters. It should be emphasized that despite all ambiguities, the notion of REV is essential to allow us to use continuous mathematics. Our main focus is on all properties needed in the flow computations. These properties formed a central part of the inputs to a numerical simulator.

2.3.2.1 Fluid Properties

Petroleum deposits vary widely in properties. The bulk of the chemical compound presents are hydrocarbons that are comprised of hydrogen and carbon. A typical crude oil contains hundreds of different chemical compounds and normally is separated into crude fractions according to the range of boiling points of the compound included in each fraction. Hydrocarbons may be gaseous, liquid or solid at normal pressure and temperature depending on the number and arrangement of the carbon atoms in the molecules. Hydrocarbons are moved from gaseous state to solid state with increasing the number of carbon atoms. The compound with up to four carbons are gaseous, with four to twenty in a liquid state and those with more than twenty carbon atoms are solid. Liquid mixtures, such as crude oils, may contain gaseous or solid compounds or both in solution. A number of non-hydrocarbons may occur in crude oils and gases such as sulfur (S), Nitrogen (N), and oxygen (O).

The petroleum reservoirs may be classified according to the state of the fluid compound and divided into two broad categories of oil and gas reservoirs. These may be subdivided into different groups according to the hydrocarbon compounds, the initial temperature and pressure of the reservoir and so on. The phase diagrams can be applied to express the behavior of the reservoir fluid in a graphical form. The pressure-temperature diagram is one that can be applied to illustrate the reservoir fluid behavior and to classify reservoirs. A typical phase diagram of a multi-components compound is given in Fig. 2.6. In this diagram, the critical point is the point at which

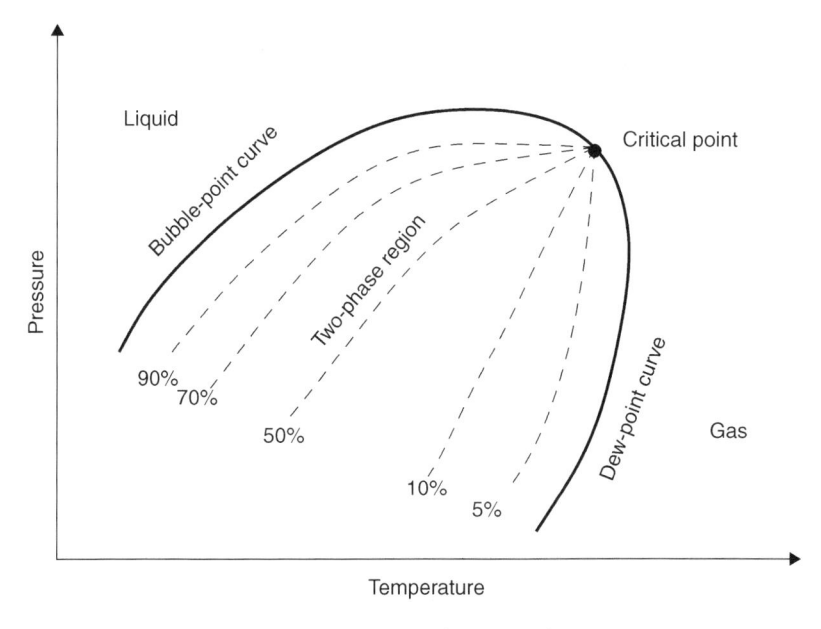

Figure 2.6 A typical pressure-temperature diagram of a ordinary crude oil.

all properties of the liquid and gaseous states are identical. If the reservoir temperature T_R is less than the critical point temperature T_C, the reservoir is called an oil reservoir. The reservoir is called a gas reservoir if $T_R > T_C$. A classification of the oil and gas reservoir is given in Fig. 2.7. The parameter denoting P_R is the initial reservoir pressure and P_{BP} is the bubble-point pressure. The gas-oil ratio is denoted by GOR in the diagram and the notation T_{cr} is the maximum temperature that the liquid phase may exist.

A complete description of the reservoir fluid properties is necessary to predict the pressure variation inside the reservoir and the other flow parameters. The reservoir rock usually contains some amount of water along with the hydrocarbon compounds. The fluid physical properties of primary interest in reservoir mathematical modeling are:

2.3.2.1.1 Crude Oil Properties
- *Specific gravity, γ_o*

$$\gamma_o = \frac{\rho_o}{\rho_w} \tag{2.1}$$

Advanced Petroleum Reservoir Simulation

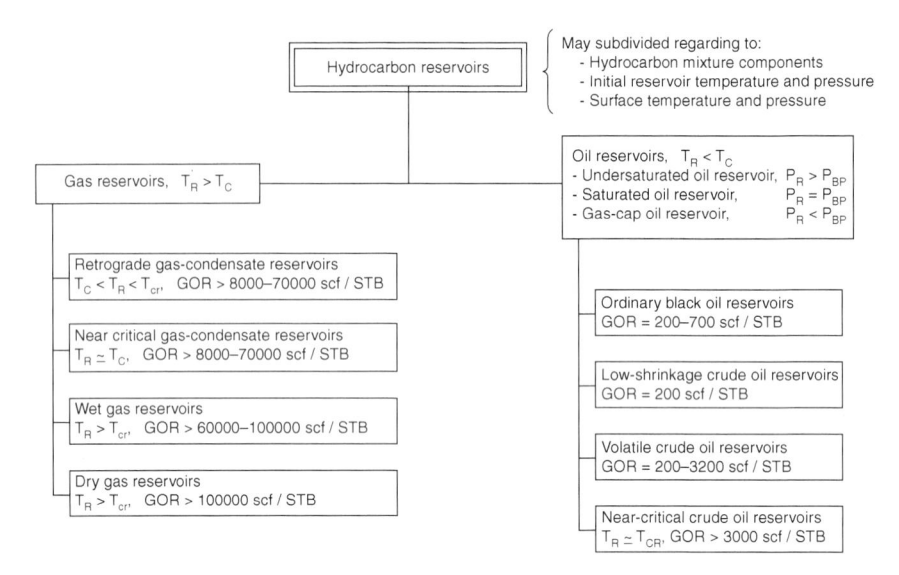

Figure 2.7 A classification of hydrocarbon reservoirs.

ρ_o, ρ_w are both at standard condition (i.e. $P = 14.696$ psi, $T = 520^{\circ R}$ according to SPE).
- *Specific gravity of the solution gas*, γ_g
- *Gas solubility*, R_S $R_S = f(P, T, \gamma_o, \gamma_g)$
- *Bubble-point pressure*, P_b $P_b = f(R_s, T, \gamma_o, \gamma_g)$
- *Oil formation volume factor*, $B_o = f(R_s, T, \gamma_o, \gamma_g)$

$$B_o = \frac{(V_o)_{P,T}}{(V_o)_{Std.}} \tag{2.2}$$

- *Isothermal compressibility coefficient*,

$$c_o = \frac{1}{\rho_o}\left(\frac{\partial \rho_o}{\partial P}\right)_T = -\frac{1}{B_o}\left(\frac{\partial B_o}{\partial P}\right)_T \tag{2.3}$$

for under-saturated crude oil system,

$$c_o = -\frac{1}{B_o}\left(\frac{\partial B_o}{\partial P}\right)_T + \frac{B_g}{B_o}\frac{\partial R_S}{\partial P} \tag{2.4}$$

for saturated crude oil system.
- *Oil density, ρ_o*

$$\rho_o = \rho_{o0} \exp\left[c_o (P - P_0)\right] \quad (2.5)$$

Oil may be considered as slightly compressible fluid. If the series expansion of exponential function is used and consider only the linear part of it, the oil density may express as:

$$\rho_o = \rho_{o0}\left[1 + c_o (P - P_0)\right] \quad (2.6)$$

where P_0 is a reference pressure.
- *Crude oil viscosity, μ_o*
- *Surface tension, σ*

2.3.2.1.2 Natural Gas Properties
- *Apparent molecular weight, M_a*

$$M_a = \sum_{i=1} y_i M_i \quad (2.7)$$

where y_i is the mole fraction and M_i is the molecular weight of ith component.
- *Specific gravity, γ_g*

$$\gamma_g = \frac{\rho_g}{\rho_{air}}, \quad \gamma_g = \frac{M_a}{28.96} \quad (2.8)$$

- *Compressibility factor (gas deviation factor), z*

$$z = \frac{V_{actual}}{V_{Ideal}}, \quad z = f(y_i, P, T) \quad (2.9)$$

where V_{Ideal} is the gas volume if it is treated as an ideal gas.

- *Isothermal gas compressibility coefficient,* c_g

$$c_g = -\frac{1}{v}\left(\frac{\partial v}{\partial P}\right)_T, \quad c_g = \frac{1}{P} - \frac{1}{z}\left(\frac{\partial z}{\partial P}\right)_T \qquad (2.10)$$

where v is the specific volume of the gas.
- *Gas formation volume factor,* B_g

$$B_g = \frac{V_{P,T}}{V_{Std}} \qquad (2.11)$$

where V_{Std} is the volume of the gas at standard condition.
- *Gas expansion factor,* E_g

$$E_g = \frac{1}{B_g} \qquad (2.12)$$

- *Viscosity,* μ_g $\quad \mu_g = f(yi, P, T)$

2.3.2.1.3 Water Content Properties
- *Formation volume factor,* B_w
- *Gas solubility,* R_{Sw}
- *Compressibility coefficient,* C_w
- *Viscosity,* μ_w

The fluid properties and pressure-volume-temperature (PVT) and phase-equilibrium behavior are measured and studied in laboratories to characterize the reservoir fluids and to evaluate volumetric performance at various pressure levels. Different laboratory tests are used to measure the fluid sample properties. The primary tests are involved in measurement of the specific-gravity, gas-oil ratio, viscosity and so on. These are routine field (on-site) tests. There are several other laboratory tests such as compositional analysis tests, constant-composition expansion tests and differential liberation tests to obtain:

- the compositional description of the fluid;
- saturation pressure (bubble-point or dew-point pressure);

- isothermal compressibility coefficients;
- compressibility factors of the gas phase;
- total hydrocarbon volume as a function of pressure;
- amount of gas in solution as a function of pressure;
- the shrinkage in the oil volume as a function of pressure;
- properties of the evolved gas including the composition of the liberated gas;
- the gas compressibility factor;
- the gas specific gravity;
- density of the remaining oil as a function of pressure; and so on.

The sampling pressure may be different from the actual reservoir pressure. In these cases, the PVT measured data should be adjusted to reflect the actual reservoir situations.

2.3.3 Rock Properties

The basic inputs in the process of reservoir characterization are the geological framework, well log data, core analysis data, seismic amplitude data, acoustic impedance data, well test data and any other data that can be correlated to rock and fluid properties. The interpretation results from all wells drilled into the reservoir are combined with the seismic data previously acquired to construct a three-dimensional (3-D) model of the reservoir, as shown in Fig. 2.2. This model not only depicts the shape of the reservoir, but details properties of the rocks and fluids as well. However, the rock and fluid properties should be anticipated in a consistent manner that depends on the quality and quantity of the available data. For the reservoirs with many wells penetrations and consequently considerable well loggings data, the seismic survey data with high degrees of uncertainties may not be relied on considerably but in reservoirs with limited number of penetrating wells such as offshore reservoirs, the seismic data play an important role to characterize the reservoir.

- *Porosity*, ϕ – Petroleum reservoirs usually have heterogeneous porosity distribution; i.e. porosity changes with location. A reservoir is homogeneous if porosity is constant independent of location. Porosity depends on reservoir pressure because of solid and pore compressibility. It may be defined as two types of porosities: absolute porosity ϕ_a; and effective porosity ϕ. The effective porosity is used in reservoir engineering calculations.

$$\phi = \frac{\text{interconnected pore volume}}{\text{bulk volume}}$$

- *Saturation, S_o, S_g, S_w*

$$S_o = \frac{V_o}{\phi V_b}, \quad S_g = \frac{V_g}{\phi V_b}, \quad S_w = \frac{V_w}{\phi V_b} \tag{2.13}$$

where V_o is the volume of oil, V_b is the bulk volume of the rock, V_g is the volume of gas and V_w is the volume of water.

$$S_o + S_g + S_w = 1 \tag{2.14}$$

- *Permeability, k* – Permeability is the capacity of the rock to transmit fluid through its connected pores when the same fluid fills all the interconnected pores. The flux vector, the rate of fluid for unit area, may be given by using Darcy's law

$$\mathbf{q} = -\left(\frac{1}{\mu}\right) k(\mathbf{x}) \cdot \nabla \Phi, \tag{2.15}$$

where: $\mathbf{q}$ is the flux vector; $\mathbf{x} = (x,y,z)$; and $\nabla \Phi = \nabla(P + \rho g z)$. The notation $k(\mathbf{x})$ is permeability tensor that is a directional rock property.

$$k(\mathbf{x}) = \begin{bmatrix} k_{xx} & k_{xy} & k_{xz} \\ k_{yx} & k_{yy} & k_{yz} \\ k_{zx} & k_{zy} & k_{zz} \end{bmatrix} \tag{2.16}$$

The diagonal terms, such as k_{xx}, represent the directional flow, the flow in the same direction as the pressure gradient. The off-diagonal terms, such as k_{xy}, represent a cross-flow, the flow perpendicular to the pressure gradient. If the reservoir coordinates coincide with the principal directions of permeability, then permeability can be represented by k_{xx}, k_{yy}, k_{zz}. The diagonal terms may be simply denoted by k_x, k_y, k_z. The reservoir is described as having isotropic permeability distribution if $k_x = k_y = k_z$.

- *Relative permeability*, k_{ro}, k_{rg}, k_{rw} – The relative permeability is defined for simultaneous flow of two or more immiscible fluids in a porous medium. The relative permeability is the ability of a porous medium to transmit a fluid at a point, if part of the porous formation at that point is occupied by other fluids.

$$k_{ro} = \frac{k_o}{k}, \quad k_{rg} = \frac{k_g}{k}, \quad k_{rw} = \frac{k_w}{k} \qquad (2.17)$$

Where k_{ro}, k_{rg} & k_{rw} are the relative permeability of oil, gas and water, respectively, k_o, k_g & k_w are the effective permeability of oil, gas and water, respectively, and k is the absolute permeability of the rock matrix. The relative permeability is a function of the fluids saturation. For the general case of three phase flow it may be written that:

$$k_{rw} = f(S_w), \quad k_{rg} = f(S_g), \quad k_{ro} = f(S_w, S_g) \qquad (2.18)$$

A typical diagram for the variation of relative permeability of two phase flow of oil and water is given in Fig. 2.8. The water as the wetting fluid ceases to flow at relatively large saturation that is called the irreducible water saturation S_{iw} or connate water saturation S_{cw}. The oil as the non-wetting fluid also ceases to flow at saturation more than zero, connate oil saturation S_{co}.

- *Capillary pressure*, P_c – The discontinuity of pressure at the interface of two immiscible fluids is the capillary pressure.

$$P_c = P_{nwp} - P_{wp} \qquad (2.19)$$

Where P_{nwp} is the non-wetting phase, i.e. oil phase in water/oil flow, and P_{wp} is the wetting phase pressure, i.e. water phase in water/oil flow. The capillary pressure is also a function of the fluids saturation, i.e. $P_c = P_c(S_w)$ in water/oil flow.

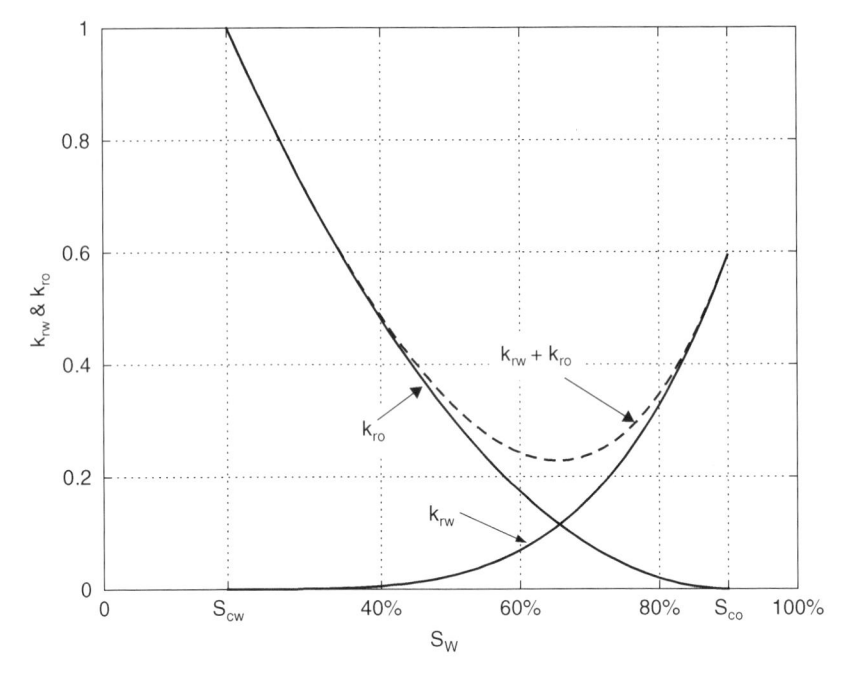

Figure 2.8 Typical relative permeability curves to water and oil.

The notations and the dimensional units of the quantities that are characterized as reservoir fluid and rock and appeared in flow equations are given in Table 2.1 for different unit systems.

The above properties and the others that are not mentioned are normally obtained through the core analysis test. This information should be combined with the well log data and the other available data to characterize the reservoir rock in a consistent manner. Geostatistical methods may be applied to integrate the sparse data and to propagate reservoir properties in a manner that is statistically coherent and consistent. These algorithms are applied to model spatial autocorrelation, to create continuous maps based on the data for the areal and vertical sections and to simulate random realizations (data set) based on a given spatial autocorrelation model. The reservoir is depicted using a very small grid to anticipate the heterogeneity of the reservoir properties. It is the description of the reservoir on the *"geo-cellular model"*. However, the description honors the known and inferred statistics of the reservoir and suffers a considerable degree of uncertainty. The geological description of the -cellular grids in many cases are not feasible.

Reservoir Simulator-input/output

Table 2.1 The symbols and units of rock, fluid and flow properties

Quantity	Symbol	System of units		
		Customary Unit	SI Unit	Lab Unit
Length	$x, y, z, \ldots$	ft	m	cm
Area	$A, A_x, A_y, \ldots$	ft^2	m^2	cm^2
density	ρ_o, ρ_w, ρ_g	lb/ft^3	kg/m^3	g/cm^3
Specific gravity	r_o, r_g	………	……..	……..
Gas solubility	R_s, R_{sw}	scf/STB	??	??
Bubble point pressure	P_b	psi	Pa	atm
Dew-point pressure	P_{dp}	psi	Pa	atm
Formation volume factor	B_o, B_w, B_g	RB/scf	m^3/std m^3	cm^3/std cm^3
Compressibility coef.	C_o, C_w, C_g, C_ϕ	………	………	………
Viscosity	μ_o, μ_w, μ_g	Cp	kg/m·sec	cp
Surface tension	$\sigma_o, \sigma_w, \sigma_g$			
Molecular weight	$M_a, M_i, \ldots$			
Mole fraction	y_i			
Porosity	ϕ	………	………	………
Permeability	k	D, mD	m^2	D, mD
Relative permeability	k_{ro}, k_{rw}, k_{rg}	………	………	………
Saturation	S_o, S_w, S_g	………	………	………
Capillary pressure	CP	psi	Pa	atm
Oil flow rate	q_o	STB/day	std m^3/day	std cm^3/day
gas flow rate	q_g	scf/day	std m^3/day	std cm^3/day
Volumetric velocity	u_o, u_w, u_g	RB/dat.ft^2	m/day	cm/sec
Time	t	day	day	sec

2.4 Upscaling

One of the challenging problems in the description of a heterogeneous medium is the problem of averaging and upscaling from one scale to another. A medium property is observed at one scale on a particular support (volume) of measurement, but the value of that property is needed on a different volume size at a different location. There are two levels of upscaling. This is the first level of upscaling that a small set of laboratory measurements need to be interpreted at the scale of the reservoir. The laboratory measurement on core samples, perhaps a few centimeters in length and diameter, should be scaled up to geophysical cells of several meters.

Another upscaling problem refers to scale up the geological fine grid to the simulation grid. Geostatistical methods are capable of providing many more values than can be easily accommodated in reservoir fluid flow simulators; hence the geo-statistically derived values are often upscaled (Paterson et al, 1996). This may called the second level of upscaling. It is a process that scales properties of fine grid to a coarse grid, such that the fluid flow behavior in the two systems is similar (Qi and Hesketh, 2007). Geological models may include $O(10^7 - 10^8)$ cells for a typical reservoir, whereas practical industrial models typically contain $O(10^5 - 10^6)$ grid blocks (depending on the type of model), with the model size often determined such that the simulation can be run in a reasonable time frame (i.e. overnight) on the available hardware (Gerritsen and durlofsky, 2005).

The upscaling in reservoir simulation is mainly referred to scale up from a geological cell grid to a simulation grid, the second level of upscaling. It is one of the first challenges in the reservoir engineering simulations and is subject to intensive research activity (Zhang et al, 2006; Farmer, 2002; Christie et al, 2001 and Christie, 1996). It should be indicated that upscaling increases the uncertainty of the simulation output. The key point is to minimize the amount of upscaling that must be done. In many cases, the parameters may not be fixed through history-matching when the coarse grids are applied. It is necessary to refine the simulation scale to find a better match between the simulation model and real system.

An upscaling algorithm assigns suitable values for porosity, permeability, water saturation, and the other fluid flow properties and functions to a simulation grid based on the reservoir characterization.

The major methods in upscaling are: power-law averaging methods, arithmetic-, geometric- and harmonic-mean technique; renormalization techniques; pressure-solver method; tensor method; and pseudo-function technique. For the Additive parameters such as porosity, fluid saturations and more generally volume/weight percentages of various phases, the arithmetic averaging method gives a reasonable and fast equivalent number. The main problem in upscaling is related to the non-additive properties like the effective permeability. These properties are not intrinsic characters of the heterogeneous medium. They depend on the boundary condition and the distribution of the heterogeneities which depend on the volume being considered (Begg et al, 1989).

2.4.1 Power Law Averaging Method

The most obvious approaches of upscaling are the application of mean-values. These techniques are very fast, but suffer from some limitations in applicability. There are several different kinds of calculations of the mean value.

- Arithmetic mean-value:

$$k_e = \frac{1}{n}\left(\sum_{i=1}^{n} k_i\right) = \frac{1}{n}(k_1 + \cdots + k_n) \qquad (2.20)$$

- Geometric mean-value:

$$k_e = \left(\prod_{i=1}^{n} k_i\right)^{1/n} = \sqrt[n]{k_1 \cdot k_2 \ldots \cdot k_n} \qquad (2.21)$$

- Harmonic mean-value:

$$k_e = n\left(\sum_{i=1}^{n} \frac{1}{k_i}\right)^{-1} = \frac{n}{\frac{1}{k_1} + \cdots + \frac{1}{k_n}} \qquad (2.22)$$

Here, k_i is the permeability of a fine grid-block and k_e is the effective permeability of a coarse-grid block (i.e. a simulation grid).

The arithmetic averaging method may be used in single phase flow for horizontal flow and the geometric averaging method may be suitable for vertical flow. Journal et al, (1986) derived a model for the effective permeability for shaley reservoirs. They demonstrate by experiments and simulations that flow conditions do not depend on the detail of the permeability spatial distribution, but rather on the spatial connectivity of the extreme permeability values, either low such as impervious shale barriers, or high such as open fractures. They assumed the flow is single phase and steady state and approximated the permeability filed by of two extreme modes of k_{sh} and k_{ss}. They established the following model:

$$k_e = \left[r_{sh} k_{sh}^w + (1 - r_{sh}) k_{ss}^w \right]^{1/w} \qquad (2.23)$$

Where:
$r_{sh} \equiv$ volumetric portion of shale
$k_{sh} \equiv$ permeability of shale
$k_{ss} \equiv$ permeability of sandstone
$w \equiv$ power-average

The power-average w mainly depends on the shale distribution. A power-average with a low power $w = 0.12$ gives a good approximation for the effective permeability in shaley reservoir with $r_{sh} \in [0, 0.5]$.

2.4.2 Pressure-solver Method

This method is applied for single phase flow. The effective permeability, k_e, of a heterogeneous medium is the permeability of an equivalent homogeneous medium that give the same flux for the same boundary conditions. Thus, it may be derived by equating expressions for the flux through a volume of the real heterogeneous medium with the flux through an equal volume of the similar homogeneous medium with the same boundary conditions. A steady-state, single-phase incompressible flow can be modeled by using the continuity equation and the Darcy's law,

$$\nabla \cdot \mathbf{q} = 0, \quad \mathbf{q} = -k(\mathbf{x}) \cdot \nabla p \qquad (2.24)$$

where a unit viscosity is assumed, $\mathbf{x} = (x,y,z)$ and $k(\mathbf{x})$ is the permeability tensor. A matrix equation may be set up by combining the continuity equation and the Darcy's law.

$$\nabla \cdot k(\mathbf{x})\nabla p = 0 \qquad (2.25)$$

The directional effective permeability in each direction is calculated by assuming no flow condition along the sides. This combined equation is solved for P_I at the inlet and P_o at the outlet and sum the flux in a given direction such as x-direction. The effective permeability is given by using the Darcy's law for the equivalent homogeneous medium with equivalent volume, $k_e^x = -\Delta x^{q/A}$.

Begg et al (1989) obtained the effective permeability in the vertical direction. They modeled the heterogeneous medium by describing the permeability distribution, k_{ijk}, on a fine-scale grid (see Fig. 2.9), and solved the governing equation for the pressure, p_{ijk}, in each grid block. Darcy's law is then used to give the flux out of each block on the outlet face. They found the total flux through the model by summing the flux of each block on the outlet face. The flux through an equivalent homogeneous medium is also given by Darcy's law. Equating the two expressions for the flux gives the effective permeability in vertical direction.

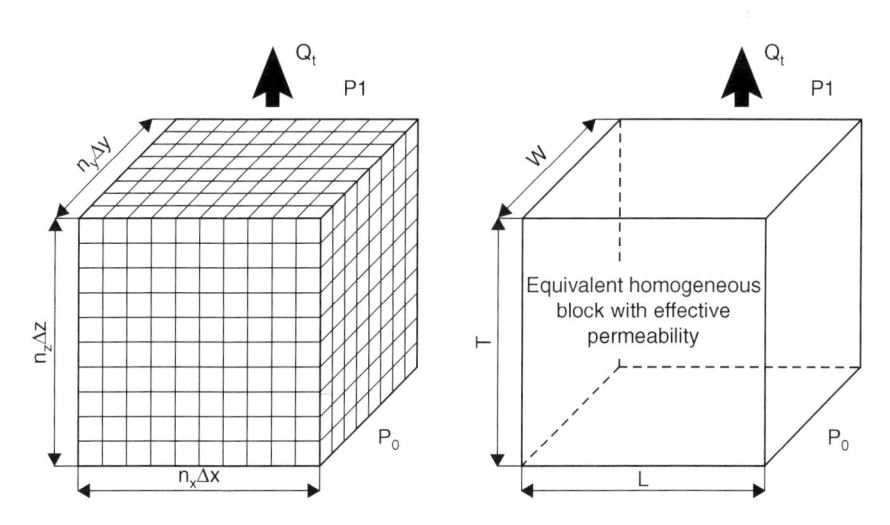

Figure 2.9 Pressure-solver method illustration (redrawn from Begg et al, 1989).

$$k_e^v = \frac{n_z+1}{n_x n_y (p_l - p_o)} \sum_{i=1}^{n_x} \sum_{j=1}^{n_y} k_{ijn_z}(p_{ijn_z} - p_o) \qquad (2.26)$$

This gives only the directional effective permeability. The full-tensor effective permeability can be obtained by assuming periodic boundary conditions, for a complete description see Pickup et al (1994).

2.4.3 Renormalization Technique

The renormalization technique is a stepwise averaging procedure and offers a faster, but less accurate, technique to calculate the effective permeability. The essence of the renormalization procedure is very simple and is to replace a single upscaling step from the fine grid to the coarse grid with a series of steps which pass from the fine grid to the coarse grid through a series of increasingly coarse intermediate grids. In other words, the properties are first distributed on a fine grid, then the effective property is calculated for a group of cells on this fine grid to give a cell value on a coarser grid. This process is continued until the original grid is reduced to a single grid block. This gives an approximation to the effective property value for the original grid. The unit cell can be selected differently to do a coarsening procedure. The very simple case that can be computed analytically is to adopt a unit cell comprised 2×2 (or $2 \times 2 \times 2$ in three dimensions) grid blocks. A schematic illustration of the method is depicted in Fig. 2.10 for 2D. The effective permeability, k_e, of a block of four grid is:

$$k_e = \frac{4(k_1+k_3)(k_2+k_4)[k_2k_4(k_1+k_3)+k_1k_3(k_2+k_4)]}{[k_2k_4(k_1+k_3)+k_1k_3(k_2+k_4)](k_1+k_2+k_3+k_4)} \\ +3(k_1+k_2)(k_3+k_4)(k_1+k_3)(k_2+k_4) \qquad (2.27)$$

This gives an approximation to the effective property value for the original grid. This has enabled to carry out calculation of effective permeability for extremely large grids. The effective permeability A complete description of the method and the application of it for single- and double phase flow can be found in King et al (1989, 1993).

Reservoir Simulator-input/output

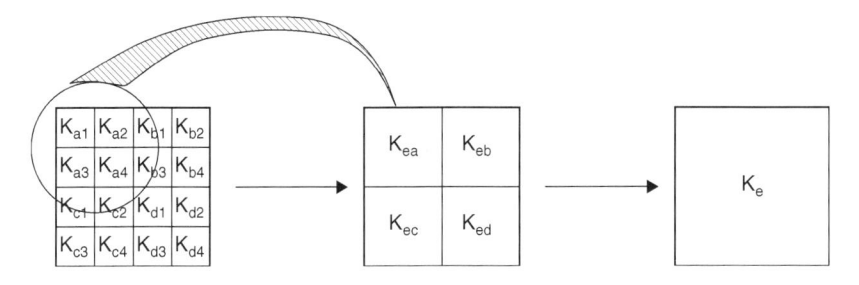

Figure 2.10 Schematic diagram of upscaling using renormalization technique.

2.4.4 Multiphase Flow Upscaling

The simulation of multiphase flow is more complicated by the fact that it is necessary to consider three forces: viscous, capillary and gravity. The effective permeability tensor will be different for each phase and will depend on phase saturation as well as on the balance between these forces (i.e. the viscous/capillary and viscous/gravity scaling groups) and on the boundary conditions. The pseudo-function techniques are used for multiphase fluid flow upscaling. These methods are referred to as pseudoization in conventional reservoir simulation practice. The main objective is to represent the behavior of small-scale multi-phase fluid mechanics and heterogeneity in coarse grid simulation models. The pseudo-functions (e.g. pseudo relative permeability) incorporate the interaction between the fluid mechanics and the heterogeneity as well as correcting for numerical dispersion. One common use of pseudo functions is to reduce the number of gridblocks, and sometimes even reduce the dimension of the problem, such as reducing a 3-D field case model to a 2-D cross-sectional model. By doing so, we hope to retain fine grid information while carrying out coarse grid simulations. The pseudo-functions may be categorized by the scaling groups under which the fine grid displacement was carried out (i.e. viscous/capillary and viscous/gravity ratio and certain geometrical scaling groups) and are valid for the boundary conditions relevant to the particular flows. If there in non-uniform coarsening to simulate the flow of different region with different rate, it is necessary to use different pseudo curves for different parts of the reservoir.

There are many pseudoization techniques described in the literature. Figure 2.11 shows two adjacent course gridblocks

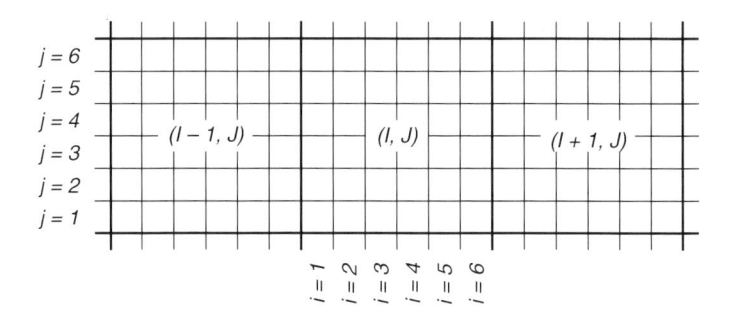

Figure 2.11 The coarse gridblocks with the fine sub-grids.

together with the fine sub-grids. The sub-grids may be labeled by i, j (and k for 3D) and the coarse grid may be denoted by I, J (and K for 3D). The power law averaging methods are adopted for the additive properties such as, saturations, density, viscosity. The pseudo-function can be computed using the upscaling Darcy's Law.

Most pseudo-function techniques use a pore-volume weighted average to determine the saturation.

$$S_p^{I,J} = \frac{\sum_i \sum_j (V\phi S_p)_{i,j}}{\sum_i \sum_j (V\phi)_{i,j}} \qquad (2.28)$$

The viscosity and the density may be upscaled with the pore-volume weighted average.

$$\mu_p^{I,J} = \frac{\sum_i \sum_j (V\phi \mu_p)_{i,j}}{\sum_i \sum_j (V\phi)_{i,j}}, \quad \rho_p^{I,J} = \frac{\sum_i \sum_j (V\phi \rho_p)_{i,j}}{\sum_i \sum_j (V\phi)_{i,j}} \qquad (2.29)$$

or the rate weighted average.

$$\mu_p^{I,J} = \frac{\sum_j (q_p \mu_p)_j}{\sum_j (q_p)_j}, \quad \rho_p^{I,J} = \frac{\sum_j (q_p \rho_p)_j}{\sum_j (q_p)_j} \qquad (2.30)$$

It is not clear which averaging should be used to find better results. The upscaled pressure can be computed using: (a) the pore-volume weighted average; or (b) the permeability weighted average (Kyte and Berry, 1975).

a. $$p_p^{I,J} = \frac{\sum_j \left[V\phi\left(p_p + \rho_p g H\right)\right]_j}{\sum_j \left(V\phi\right)_j}$$ (2.31)

b. $$p_p^{I,J} = \frac{\sum_j \left[k_{rp} k h \left(p_p + \rho_p g H\right)\right]_j}{\sum_j \left(k_{rp} k h\right)_j}$$ (2.32)

Hence, the pseudo relative permeability for phase p can be obtained from Darcy's law.

$$k_{rp}^{I,J} = \frac{-\mu_p \sum_j \left(q_p\right)_j}{T^{I,J} \left(\Delta p_p - \rho_p g \Delta H\right)^{I,J}}$$ (2.33)

Where T is transmissibility and computed by,

$$T^{I,J} = \left(\frac{kH}{\Delta X}\right)^{I,J} = \sum_j \left(\frac{1}{\sum_i \left(\frac{\Delta x_i}{k_i h_i}\right)}\right)$$ (2.34)

H & h are the height of the coarse and fine grid, respectively, and X & x are the coarse and fine grid length, respectively.

There are many other methods to do upscaling. For more detailed descriptions see Christie (1996), Cao and Aziz (1999), Qi and Hesketh (2005) and Lohne and Virnovsky (2006). However, it is not clear which approaches should be used and there is no measure of quality for any upscaling routine.

2.5 Pressure/Production data

Production data are mainly fluids production flow rates of individual wells and eventually the total flow rate of the reservoir. Cumulative production of each fluid produced from the reservoir with production time is also another form of production data. Gas-oil ratio and water cut are part of production data as well. Production data as a function of elapsed production time should be the easiest task and most accurate relatively to other reservoir data. This type of data however, is the most important tool that describes the reservoir behavior and can very well predict the reservoir production behavior. The economic limits, the need for improving recovery technique and abandoned time can be determined by utilizing decline curve analysis technique.

Surface tanks or separators are where production rates of produced fluids can be measured. The surface pressure can be used to calculate the bottom-hole flowing pressure; this pressure can be measured directly as well using down-hole installed gauges. It is necessary to measure and record well/reservoir productions; we need to calculate the overall hydrocarbon reservoir production capacity to assist the management in planning economic and marketing strategies.

It is necessary to analyze production data on a daily basis; such data would help determine if there is any impulsive change in the ratios of produced fluids that may indicate certain production problems. Therefore, immediate remediation has to be undertaken. Increase in water production may indicate that the water table has reached higher level perforations or leaking water from a top water formation broke to the well. Increase in gas-oil ratio could mean that the gas cap expanded and reached the top perforations. The gradual drop of production sends a clear message to the reservoir management staff that something has to be done to keep production steady before losing reservoir energy. Maybe pressure maintenance strategy, installing production pumps, gas-lift program are to be implemented depending on the reservoir condition and economic feasibility. It may be wise to consider an EOR technique to save the reservoir and maximize its production.

Loading production history data to the simulator, draws bubble maps of production on top of geological prediction maps, and offers cross plot between production and geological model predictions for verification. One has to be fully alerted of the implications of faulty production data and the catastrophic consequences caused to the history-match process and to the overall simulator's predictions.

2.5.1 Phase Saturations Distribution

Initially the reservoir is in a state of equilibrium and reservoir fluids distribution is kept constant. Once the reservoir is disturbed by drilling and consequently production, that state of equilibrium is no longer valid. The common belief is that the reservoir initially saturated by water, after the oil migration, the water is expelled from the reservoir and replaced by oil that fills the pore spaces within the reservoir rock. Practically, water saturation will never get to zero percent, some of the water "residual water saturation" will be trapped in small pores and pore throats. Because hydrocarbons are lighter than water, over long geological time the fluids in the reservoir will segregate according to their density. Gas, if any, will be occupying the top portion of the reservoir then the oil at lower level and finally water will occupy the bottom portion.

By calculating water saturation with resistivity log and capillary pressure, water contact and free water level can be determined or verified. Water contact is for the determination of water-free production, and is actually an average number. Between water contact and free water level, which may be as thick as a thousand feet or more, a well may produce water, oil, or gas in different proportions depending on the elevation, availability of oil, and porosity. Understanding and modeling of water contact and free water level is important for reserve estimation and production planning when a large area is not covered by wells.

Once reservoir fluids are produced or fluids are injected into the reservoir, it is important to understand and track the movement of fluids. Reservoir simulation as a conventional practice is widely used to predict fluid flow in a reservoir. The performance of the simulation model can be improved with production history-matching. Fluid saturation within the reservoir is very much affected by the physical properties of these fluids, the reservoir rock properties and the reservoir pressure/stress *in situ* conditions. Obviously, fractures contained by the reservoir plays a major role in fluid saturations and movements. Rock wettability and capillarity are the main actors that control residual saturations and multiphase flow pattern in the reservoir.

In order to manage a reservoir we need to be able to predict how fluids will flow in response to the production and injection of fluids. Reservoir simulation allows us to predict fluid flow. Before production, a reservoir is characterized using data from geology, seismic exploration, logging, well tests, core and laboratory work. During production, pressure, saturation and temperature change, and the change is predicted by reservoir simulation. Dynamic data collected during production and 4-D seismic surveys can be used to improve reservoir simulation. As a first step, reservoir simulation parameters are adjusted so that the model predictions approximate the bottom-hole pressure (BHP), water cut, gas oil ratio (GOR) and other field observations. Reservoir parameters can be further adjusted so that the predicted changes in seismic response approximate the results from 4-D seismic surveys. We are currently searching for candidate 4-D research sites.

Unfortunately, production data do not tell much about the fluid saturations distribution inside the reservoir. Well logging data (resistivity log, sonic log, SP log, etc.) can be used to determine the fluid saturations surrounding the wellbore. Recently, 4-D seismic surveys have been used to observe fluid changes and movement in the reservoir (Sonneland et al, 1997). Four-D seismic survey results provide additional constraints that can be used to improve reservoir simulation models.

Both static (initial) and dynamic (post drilling) fluid saturations behaviors has to be understood and well studied before loaded to the reservoir simulator. Such data is very critical and sensitive in determining hydrocarbon in-place, hydrocarbon reserves and flow rates. Two-phase and three-phase models are based on such data, their predictions can be totally misleading if this data is not accurate enough. With the reservoir characterization model ready, the reservoir simulation model can be built to predict fluid changes during the reservoir futuristic production span and useful scenarios can be produced to decide the optimum recovery scheme.

2.6 Reservoir Simulator Output

Reservoir simulators, either the in-house built or the commercially used ones, agree on at least two issues. The first is that a numerical solution approach (finite differences or finite elements in most cases) has to be employed to produce a solution for the model, and second, the output is nothing but huge data in a digital form – in many cases hundreds of pages. It is impossible to grasp a meaning of the output without a graphical representation of this data. That is why the output of modern simulators always displayed in plotting or graphical, very often in 3-D representation.

Input variables for the model include the latest production data, as well as subsurface geologic data gathered from all wells drilled into the reservoir. Surface geology studies and seismic data create the three-dimensional shape of the reservoir – consisting of a big number of grid blocks in the simulation in most cases. Usually, the simulator is including a code written specifically to convert input and output data from simulator raw data to graphical mode.

For every grid block of the reservoir we include data describing each geologic property (rock properties, etc.), fluid properties and a dynamic parameter (commonly called transmissibility). This process is done at an arbitrary initial time (may be one minute, one hour or one day, etc.), and at the second time segment the same description given to each grid block in the previous time-step is to be updated. The interaction of flow parameters of the grid blocks is also included by imposing suitable boundary conditions at the interface between any adjacent blocks. With this much detail it is difficult to form an image of the reservoir without 3-D plots. The images provide an instant picture of the whole reservoir and its features, which is extremely informative. Some simulators equipped with a simulator animation processor capable of transforming data with hundreds of time segments and run them like a video.

Before the reservoir simulation era, our vision to the reservoir in the physical sense is only drawn by those reservoir rock and fluid samples that we retrieve from drilled wells, in addition to the tales of geologists about what happened through millions of years. The introduction of reservoir simulation technology supported by improved robustness and speed solver computers enabled us to see our reservoir in three dimensions. We can imagine how the reservoir has been formed and how the reservoir is behaving as we continue producing. The output of modern reservoir simulators can be a very fancy 3-D graphical configuration of the reservoir that enables us to visualize how the reservoir is shaped and how it performs.

With the increased accessibility to reservoir simulators, more oil and gas property evaluations are being substantiated by reservoir simulation studies. Consequently, the need to incorporate reservoir simulation and review industrial requirements for simulation studies have increased in the last two decades.

A quality production forecast is essential to a realistic reserve evaluation. Production forecasts are usually obtained through conventional reservoir engineering calculations, particularly decline curve analysis and material balance. Reservoirs undergo a series of different depletion mechanisms in their production life that cause the decline trend to shift, often to the extent that the decline trend is no longer predictable. Using a reservoir model to simulate reservoir behavior under different operation strategies becomes necessary for forecasting future performance or devising optimum development scenarios to improve recovery.

Output from seismic simulation is analyzed to investigate the changes in geological characteristics of reservoirs. The output is also processed to guide future oil reservoir simulations. Seismic simulations produce output that represents the traces of sound waves generated by sound sources and recorded by receivers on a three-dimensional grid over many time steps. One analysis of seismic datasets involves mapping and aggregating traces onto a three-dimensional volume through a process called seismic imaging. The resulting three-dimensional volume can be used for visualization or to generate input for reservoir simulators. The seismic imaging process is an example of a larger class of operations referred to as generalized reductions. Processing for generalized reductions consist of three main steps: Retrieving data items of interest; Applying application-specific transformation operations on the retrieved input items; and Mapping the input items to output items and aggregating, in some application specific way, all the input items that map to the same output data item. Most importantly, aggregation operations involve commutative and associative operations, i.e. the correctness of the output data values does not depend on the order input data items are aggregated. There are methods for performing generalized reductions in distributed and heterogeneous environments. These methods include: 1) a replicated accumulator strategy, in which the data structures to maintain intermediate results during data aggregation are replicated on all the machines in the environment, 2) a partitioned accumulator strategy, in which the intermediate data structures are partitioned among the processing nodes, and 3) a hybrid strategy, which combines the partitioned and replicated strategies to improve performance in a heterogeneous environment.

The degree of changes in seismic response to production within the reservoir is highly dependant on the physical properties of the reservoir rocks. Accurate modeling of these properties from existing data will allow consideration of their effect on the 4-dimensional response. Modeled acoustic/elastic properties include pore volume (Vp), Solid volume (Vs), bulk density, bulk Poisson's ratio, and reflectivity, both at normal incidence and non-vertical incidence. The rock physics model will allow variation in fluid saturation, fluid properties, and reservoir pressure, allowing fluid substitution modeling to represent realistic reservoir changes.

The use of dynamic reservoir simulators helps in generating different production scenarios, quantifies oil/gas reserves in the reservoir and predicts future oil, gas and water productions. These important simulation outputs are very crucial in determining production strategies and implementations of improved recovery techniques. It can be used to assess the effect of different reservoir scenarios on the 4D response. Scenarios modeled will determine expectations of reservoir production changes. Examples include, but are not limited to:

   a. Saturation change representing water influx from injection well or a natural aquifer.
   b. Pressure change in areas where reservoir pressure is not maintained by injection and/or an aquifer.
   c. Fluid property changes representing the release of solution gas as reservoir pressure falls below the bubble point.
   d. A combination of saturation, pressure, and fluid property changes as predicted by simulation of reservoir production.
   e. Changes in the overburden or reservoir physical rock properties brought about by production induced *in situ* stress changes.

2.7 History-matching

The initial (static) geological modeling and reservoir characterization carry a large degree of uncertainty. The field information is usually sparse and noisy. Some parts of data are obtained from cores of a few centimeters in length and diameter. This is about

$10^{-17} - 10^{-16}$ of a normal reservoir volume. The uncertainty may be quite large for the rock properties in inter-wells. Consequently, the production profile associated with the development schemes cannot be predicted exactly. The standard procedure is transferring the static model into a dynamic model by constraining the model to data representative of the chosen recovery scheme in the form of oil, water and gas production rates and pressure. In contrast to the static data (e.g. geometry and geology) obtained prior to the inception of production, the dynamic data are a direct measure of the reservoir response to the recovery process in application. *History-matching* incorporates dynamic data in the generation of reservoir models and leads to quantification of errors and *uncertainty analysis* in forecasting.

The primary objective of a history-matching is to improve and to validate the reservoir simulation model parameters. History-matching involves adjusting model parameters with the aim of obtaining a model output as close as possible to the production (history) data. In general, the use of initial static input data leads to a set of output that does not match with the historical production data to an acceptable accuracy level. To improve the quality of the match, an iterative procedure should be applied to adjust the initial simulation data systematically. The procedure may be enumerated as:

a. set the objectives of the history-matching;
b. selection of the history-matching method;
c. specify the data and the criteria for matching;
d. specify the reservoir data that should be adjusted;
e. run the simulation model with the available data and compute the desired output data for history-matching;
f. compare the simulated and historical data;
g. change the initial input data;
h. repeat the steps e to g to find the desired level of match based the specified criteria; and
i. finding the uncertainty envelope and the confident intervals (uncertainty analysis).

There may be several beneficial aspects of and secondary objectives of adjusting the initial description of the reservoir. The history-matching process leads to better understanding of the process occurring in the reservoir; identify the level of supporting aquifers; identify the areas of bypass; to find out paths of fluid migration;

and so on. The history-matching process may be carried out manually or automatically using computer logic to adjust the reservoir data. The manual history-matching involves running the reservoir simulation model for the historical period and compare with the real system by a reservoir engineer. The reservoir engineer will adjust the simulation parameters in an effort and running the simulator several times to improve the match. It may be several trials and errors to find the most appropriate data to reach some degrees of history-matching. The result is a model such that the difference between the performance of the model and the history of a reservoir is minimized according to the judgment of the field reservoir engineers.

Automatic history-matching techniques use some computer logic to adjust the reservoir data based on the historical production data and fall into two categories: deterministic and stochastic methods. Deterministic methods are based in the inverse problem theory, while the stochastic methods, in few words, mimic the trial and error approach of the manual history-matching procedure. The most efficient deterministic methods are the gradient-based techniques such as Steepest Descent, Gauss–Newton and Levenberg–Marquardt algorithms. They are called gradient-based methods because they need to compute the gradients of the mathematical model with respect to the parameters (permeability, porosity, or any other property that can be parameterized) in order to minimize the objective function. The methods involve calculating sensitivity parameters, the variation of the predicted quantity (i.e. the production rates) with respect to the parameters, and require the differentiation of the mathematical model with respect to the model parameters. These sensitivity parameters are then used to construct a Hessian matrix whose inverse gives a good approximation to the covariance matrix that enters into the model misfit calculation (Lépine et al, 1999). These methods have a very fast convergence rate to a optimal set of parameters but in some situations, these algorithms may not converge or converge to a local minimum of the mathematical model.

The most common stochastic methods are the ones based on simulated annealing and genetic algorithms. These methods do not need to compute gradients, but their convergence rate is much slower than that of the gradient methods. On other hand, their computational implementation is much easier than the implementation of a deterministic algorithm. Simulated annealing and genetic algorithms are classified as global optimization algorithms because theoretically they always reach the global minimum of the objective function. In practice, the number of iterations is limited, and so, the global minimum may not be reached (Portella and Prais, 1999).

2.7.1 History-matching Formulation

The goal is to determine a spatial distribution a set of model parameters m such that $y_s(m) - y_o \approx 0$ where, y_s is the model output and y_o is historical observed data. In other words, the history-matching process is that of finding a zero to a multivariate time-varying function, namely, the reservoir simulator. The history-matching is a nonlinear inverse problem. It is ill-posed because it does not fulfill the Hadamard classical requirements of existence, uniqueness and stability, under data perturbations, of a solution (Hadamard, 1923). Therefore, there exist families of models that will match the observed data to some degree of accuracy. Although model updating has been a topic of active research for the past few decades, existing algorithms have only had a limited success in applications to real reservoirs. Since the history-matching is an inherently ill-posed problem, additional sources of information such as prior knowledge in terms of geological constraints have to be utilized in order to generate a reliable set of model parameters and reduce the uncertainty envelope.

Solving the forward model $y_o = y_s(m)$ is usually very time-consuming, and it may take several hours to find a single evaluation in practical cases. Therefore, the maximum likelihood and a reasonable estimate of the dispersion of the distribution around it must often be determined. It is generally accomplished through least-squares techniques when both the probability of the model parameters and the observed data are assumed to be Gaussian. Under these assumptions, the history-matching process reduces to minimize a misfit function

$$S(m) = \sum_{k=1}^{n_p} \frac{1}{2\sigma_k^2} \left[y_{ks}(m) - y_{ko} \right]^2 + \left(m - m_{prior} \right)^T \left(C_m^{prior} \right)^{-1} \left(m - m_{prior} \right) \quad (2.35)$$

where σ_k is the standard deviation of kth observation parameter the same as the oil rate or BHP, $y_{ks}(m)$ is a predicted quantity such as the oil rate of bottom hole pressure, y_{ko} is an observed quantity (historical production data), n_p is the number of production parameters and x is the reservoir parameter vector. The matrix $[C_m^{prior}]$ is the prior covariance of the reservoir parameters. The goal is finding a set of reservoir parameters by minimizing the misfit function $S(m)$. The minimum of the misfit function may be found by gradient techniques such as Gauss–Newton, steepest descent, etc. It should be emphasized that several sets of reservoir and model parameters m may fulfill the criteria for the minimum value of the misfit function.

The predicted value y_{ks} and the misfit is not really only a function of reservoir parameters. It is also function of dynamic state of the reservoir x (i.e. pressure, saturation, etc.) and also the control variables u such as the injection rate, well rates, etc. The set of reservoir parameters m should be updated in view of the all fluid flow and geological modeling constraints. The general formulation may be written in the form

$$\min_{m^n}\left[S(m, u^n, x^{n+1}) = \sum_{n=0}^{N-1} \sum_{k=1}^{n_p} \frac{1}{\sigma_k^2} \left[y_{ks}(m, u^n, x^{n+1}) - y_{ko} \right]^2 \right.$$
$$\left. + (m - m_{prior})^T \left[C_m^{prior} \right]^{-1} (m - m_{prior}) \right]$$
$$\text{and} \quad \forall n = (0, 1, \ldots, N-1)$$

subject to:

$g^n(m, u^n, x^n, x^{n+1}) = 0 \quad \text{and} \quad \forall n = (0, 1, \ldots, N-1)$

$x^0 = x_0 \quad$ initial condition

$m \in$ Geological consistent realizations (2.36)

Here, m are the model parameters to be estimated, and m_{prior} and $[C_m^{prior}]^{-1}$ are the prior estimate and prior covariance matrix of m, respectively.

The actual adjustment of reservoir data during history-matching process is carried out in two stage procedure, namely pressure match and saturation match. In pressure match stage, the remaining reserves are of primary concern. The well-voidage rates are the most appropriate data for the history-matching process. It ensures that the volume of reservoir voidage computed by the reservoir model is identical to the reservoir fluid removed during the historical period. During this stage, the average reservoir pressure throughout time, regional pressure gradients, and well pressures are also matched. The most commonly adjusted reservoir parameters are aquifer size, pore volume (PV), and total system compressibility in matching of the average reservoir pressure. The material balance studies and aquifer-influx studies can aid in performing the data adjustment. The aquifer connectivity, reservoir $k_h h$, transmissibility across faults, and regional PVs are the most common data that are adjusted to match the regional pressures and pressure gradients.

The second stage of history-matching is the saturation match and is the most time consuming part of the process. The primary objective in the saturation-matching stage is to adjust data to match gas oil ratio (GOR) and water oil ratio (WOR). If the gas oil ratio (GOR) and WOR can be matched, it will ensure matching of oil rates in an oil reservoir and gas rates in a gas reservoir and voidage rates. The well histories that are matched are water cuts, GORs and breakthrough times. It is necessary to incorporate all knowledge of reservoir geology and recovery process in this stage of history-matching. The selection of parameters to be adjusted during the history-matching process should be made judiciously and with the aid of filed reservoir engineers and geologists. The most commonly used parameters are aquifer size and strength, $k_h h$, k_v/k_h, k_{ro}, k_{rg} & k_{rw} and the presence of vertical permeable barriers. There are no specific rules for choosing which reservoir parameters to adjust during the history-matching process. It depends on the situation and the setting objectives for the history-matching process.

In most of available literature, independent of the method used for the history-matching, there is usually an assumption that there exists a simple unique solution at the "correct" model. This therefore neglects the inherent non-uniqueness of the solution of the underlying inverse problem and, consequently, leads to the assumption that a good history-matched model is a good representation of the reservoir and therefore gives a good forecast. In other words, the history-matching is non-unique and the forecast production profiles are uncertain. We should investigate a parameter space instead of one single parameter for a certain condition. However, it is necessary to quantify the error in forecasting and find the uncertainty envelope.

2.7.2 Uncertainty Analysis

Uncertainty is a measure of the "goodness" of a result. The result can be a measurement quantity, a mathematical or a numerical value. Without such a measure, it is impossible to judge the fitness of the value as a basis for making decisions relating to scientific and industrial excellence. Evaluation of uncertainty is an ongoing process that can consume time and resources and needs the knowledge of data analysis techniques and particularly statistical analysis. Therefore, it is important for personnel who are approaching uncertainty analysis for the first time to be aware of the resources required and to carefully lay out a plan for data collection and analysis.

The sources of the uncertainty may be classified as:

1. random measurement, systematic (bias) uncertainty and lack of representativeness;
2. upscaling process; and
3. model uncertainty.

A structured approach is needed to estimate uncertainties of different sources. The requirements include:

- A method of determining uncertainties in individual terms in the reservoir simulation;
- A method of aggregating the uncertainties of individual terms.

2.7.2.1 Measurement Uncertainty

There are a number of basic statistical concepts and terms that are central for estimating the uncertainty attributed to random measurement. The process of estimating uncertainties is based on certain characteristics of the variable of interest (input quantity) as estimated from its corresponding data set. The ideal information includes:

- arithmetic mean (mean) of the data set;
- standard deviation of the data set (the square root of the variance);
- standard deviation of the mean (the standard error of the mean);
- probability distribution of the data; and

- covariance of the input quantity with other input quantities.

Measurement systems consist of the instrumentation, the procedures for data acquisition and reduction, and the operational environment, e.g. laboratory, large scale specialized facility, and *in situ*. Measurements are made of individual variables, x_i to obtain a result, R which is calculated by combining the data for various individual variables through data reduction equations

$$R = R(x_1, x_2, \ldots, x_n) \tag{2.37}$$

Each of the measurement systems used to measure the value of an individual variable x_i is influenced by different elemental error sources. The effects of these elemental errors are manifested as bias errors (estimated by B_i) and precision errors (estimated by P_i) in the measured values of the variable x_i. These errors in the measured values then propagate through the data reduction equation, thereby generating the bias, B_R, and precision, P_R, errors in the experimental result, r.

The total uncertainty in the result R is the root sum-square (RSS) of the bias B_R and precision limits P_R.

$$U_R^2 = B_R^2 + P_R^2 \tag{2.38}$$

Where:

$$B_R^2 = \sum_{i=1}^{I} \theta_i^2 B_i^2 + 2 \sum_{i=1}^{I} \sum_{k=i+1}^{I-1} \theta_i \theta_k B_{ik} \tag{2.39}$$

$$P_R = \sum_{i=1}^{I} (\theta_i P_i)^2 \tag{2.40}$$

for a single test; and

$$P_R \approx P_{\bar{R}} = \sum_{i=1}^{I} (\theta_i P_{\bar{i}})^2 \quad P_{\bar{i}} = w_i S_{\bar{i}} \quad S_{\bar{i}} = \left[\sum_{k=1}^{M} \frac{(x_k - \bar{x}_i)^2}{M-1} \right]^{1/2} \tag{2.41}$$

for multiple tests measurement. The notations:

B_i is the variance of the bias error distribution of variable x_i;

B_{ik} is the covariance of the bias distribution of variables x_i and x_k;

$\theta_i = \frac{\partial R}{\partial x_i}$ is the sensitivity coefficient;

$S_{\bar{i}}$ is the standard deviation of M sample results of variable x_i;

M is the number of measurement set; and

w_i is a weight function that is dependent on the number of measurement test.

One complication with reservoir related measurements is that independent repeated measurements that reduce the experimental uncertainty are usually difficult to obtain. It is the case especially for the relative permeability because flooding of the core sample with water and oil may change wetting properties. This implies that the relative uncertainties in the relative permeability curves are determined by the random errors in the measurements of differential pressure and of oil and water rates. However, most of the parameters in a reservoir model are not obtained directly from the measurement and inferred through model interpretation and a reduction equation.

The other source of uncertainty is associated with lack of complete correspondence between conditions associated with the available data and the conditions associated with real reservoir condition. Random sampling is usually not feasible since the wells are drilled far from random and core plugs are often taken from the most homogeneous parts of a core. For this reason, classical estimation theory falls short. By geological skill it is still possible to choose core plugs that are representative for a certain geological environment (litho-faces), but it is well known from statistical experience that nonrandom sampling may have serious pitfalls. In addition, the samples represent only a minor fraction of the total reservoir volume, so the differences between the sample and the population means and variances may be quite substantial.

Another complication is that laboratory measurements are performed on core material that has been contaminated by drilling fluids. Also, the transport to the surface will expose the core to varying temperature and pressure. When the core reaches the laboratory, its state may have changed substantially from its origin. So, even if core floods are performed under reservoir pressure and temperature and with (recombined) reservoir fluids, one should expect deviations from the original reservoir conditions. The representation of fluid samples is often questionable because it is difficult to obtain a representative *in situ* sample and bring it to the surface without any leakage of gas. Another favorable situation is when the reservoir has a gas oil contact, and separator samples are available for both reservoir oil and gas. Again, very accurate samples for the oil and the gas zone can be obtained.

Direct reservoir measurements like logs, well tests, and tracer tests are representative for the reservoir properties in the volumes they contact. However, they measure responses on very different scales. While most logs explore volumes comparable to those of core plugs, well and tracer test responses frequently are averages over hundreds of meters. The reservoir simulation model should comply with the rate and pressure data from well tests and tracer production profiles. Traditionally such tests have been applied in the modeling of large-scale "architecture" like reservoir boundaries, major faults, layering, and channeling. Recently, methods have been developed to utilize well test 17 and tracer test 18 information directly in stochastic reservoir description. In such applications, the consequences of the uncertainties in the measured test responses may be less significant than the high degree of nonuniqueness in the interpretations (Bu and Damsleth, 1996).

2.7.2.2 *Upscaling Uncertainty*

The main issue in upscaling is related to the upscaling of non-additive properties such as permeability and relative permeability. There are different methods for upscaling of the rock and fluid flow properties. Some of them are mentioned in section 2.5. The upscaling method may be classified as single-phase and multi-phase upscaling. In the single-phase upscaling, the absolute permeability is only upscaled from a geo-cellular grid to the simulation grid. The relative permeability is considered identical for both scales. The relative permeability and capillary pressure are also upscaled along with the absolute permeability from a fine grid into a coarser grid. In many cases, the coarse block properties are obtained by considering only the fine grid scale region corresponding to the target coarse block.

It is called as the local upscaling. The global upscaling is referred to the case that the entire fine scale model is solved and the solution is applied to obtain the coarse scale behavior. However, there are other categories as extended local upscaling, a border region around the coarse grid is also taken into account, and quasi-global upscaling in which an approximate solution of entire flow region is adopted to derive the coarse blocks behavior. The local upscaling methods do not consider the permeability connectivity.

The most important detrimental effects of the upscaling are: homogenization of the medium; and coarsening of the computational grid. The permeability field is made smoother due to the homogenization and the truncation error is increased by using larger computational girds. Error introduced due to the homogenization may be referred to as the "loss of heterogeneity error" and due to the coarsening of the computational grid may be referred to as the "discretization error" (Sablok and Aziz, 2005). The combination of these two gives the total error due to the upscaling process. The total error may be small due to the opposing effect of these two types of errors contributed to the total upscaling error. As mentioned in REV that the representative volume should be as large as the dimension of the field, in the low level upscaling the discretization error may be dominated while the loss of heterogeneity error may dominate for high level upscaling. Li et al (1995) show that the local upscaling techniques that treat the coarse grid cells as independent, provide more uncertain result than the global upscaling methods.

2.7.2.3 Model Error

The model error includes the approximation in mathematical representation, the solution method that is mostly numeric and also the model of error estimation. The mathematical formulations in most of the available simulators are based on the material balance equation and Darcy's law. These two fundamental equations do not mimic the reality of the flow inside the porous rock. The assumptions behind them are expressed in an earlier section. However, the final solution is nonlinear that is complicated to solve analytically and the solution of it should be obtained with numerical methods. There are some analytical solutions for a few special problems in reservoir engineering such as Buckley–Leveret flow that suffer a quite large number of assumptions and does not show the real situation in the fluid flow inside the reservoir rock. The most applied methods are the finite difference and finite element methods. The finite difference method is based on the Taylor series expansion. The nonlinear parts of the Taylor series expansion are normally neglected to approximate the derivation. If we consider that the normal size of a grid block is in order of $10\,m$, the neglecting of the nonlinear part of the Taylor series expansion produce a substantial uncertainty. Therefore, if we assume that reservoir model, the geological and characterization model were exactly correct in every detail, we cannot get exact prediction of the reservoir performance due to the fact the reservoir fluid flow model does not mimic reality.

We may have fixed certain aspects of our reservoir model (for example, the reservoir geometry) and only attempted to predict uncertainty with regard to other parameters such as porosity and permeability. Even if we get those parameters exactly correct, we may not get an exact prediction because of errors in the fixed aspects of the model. This second source of model error could in principle be removed by including all possible parameters in the uncertainty analysis but this is never feasible in practice. The true model error is virtually impossible to quantify, so in practice it has to be neglected. The size of the model error depends strongly on the parameterization of the model, and one may hope that if this is well chosen, the model error will be small (Lépine et al, 1999).

2.7.2.4 The Prediction Uncertainty

When the parameters are set according to the production data, the next step is obtaining the uncertainty envelope and the confidence region. This is important in forecasting the reservoir behavior and decision making process. If the set of parameters m^0 are the most likely values after history-matching, the uncertainty envelope may be quantified by

$$[C_y] = [A][C_x][A]^T \qquad (2.42)$$

where $[C_y]$ is the covariance matrix of the predicted quantity, $[A]$ is the sensitivity matrix whose elements are $A_{ij} = \partial y_i / \partial x_j$ and $[C_x]$ is the covariance matrix of the reservoir and model parameters. The first task is to specify the parameters that used for uncertainty analysis.

They may be the same as those of the history-matching problem or may be different from them. When the parameters are specified, the most likely history-matching values of them using the objective function is computed and then the sensitivity matrix is obtained. The model and reservoir parameters covariance can be obtained using the Hessian matrix

$$H_{ij} = \frac{\partial^2 E(\mathbf{x})}{\partial x_i \partial x_j}\bigg|_{\mathbf{x}=\mathbf{x}^0} \tag{2.43}$$

where $H_{ij} = \frac{\partial^2 E(\mathbf{x})}{\partial x_i \partial x_j}\big|_{\mathbf{x}=\mathbf{x}^0}$. This provides the uncertainty in the reservoir parameters. The variance of the predicted values is given by the corresponding diagonal element of the covariance matrix. The confident intervals can be obtained from this if its probability distribution is assumed to be Gaussian.

2.8 Real-time Monitoring

The history-matching process can be considered as a closed-loop process. The production data from the real system are adopted to modify the reservoir and model parameters and to use for future prediction. The history-matching process is usually carried out after a period of years. The traditional history-matching involves manual adjustment of model parameters and is usually ad-hoc. According to Brouwer et al (2004), the drawbacks of a traditional history-matching are:

   a. It is usually only performed on a campaign basis, typically after periods of years.
   b. The matching techniques are usually ad-hoc and involve manual adjustment of model parameters, instead of systematic parameter updating.
   c. Uncertainties in the state variables, model parameters and measured data are usually not explicitly taken into account.
   d. The resulting history-matched models often violate essential geological constraints.

e. The updated model may reproduce the production data perfectly, but have no predictive capacity, because it has been over-fitted by adjusting a large number of unknown parameters using a too small number of measurements.

To adopt an optimum production strategy and to produce oil and gas in challenging physical environments such as deepwater reservoirs and oil-bearing formation in the Arctic, it is required to update the model more frequently and systematically. If the model is also smart, it will produce real-time data. Smart field technologies are currently generating significant interest in the petroleum industry, primarily because it is estimated that their implementation could increase oil and gas reserves by 10–15% (Severns, 2006). This help to optimize the reservoir performance under geological uncertainty and also incorporate dynamic information in real-time and reduce uncertainty. A schematic layout of a smart reservoir modeling is depicted in Fig. 2.12 (redrawn from Brouwer et al, 2004). It is a real-time closed-loop to control the reservoir behavior and to attain an optimum production. This figure illustrates a true closed-loop optimal control approach that shifts

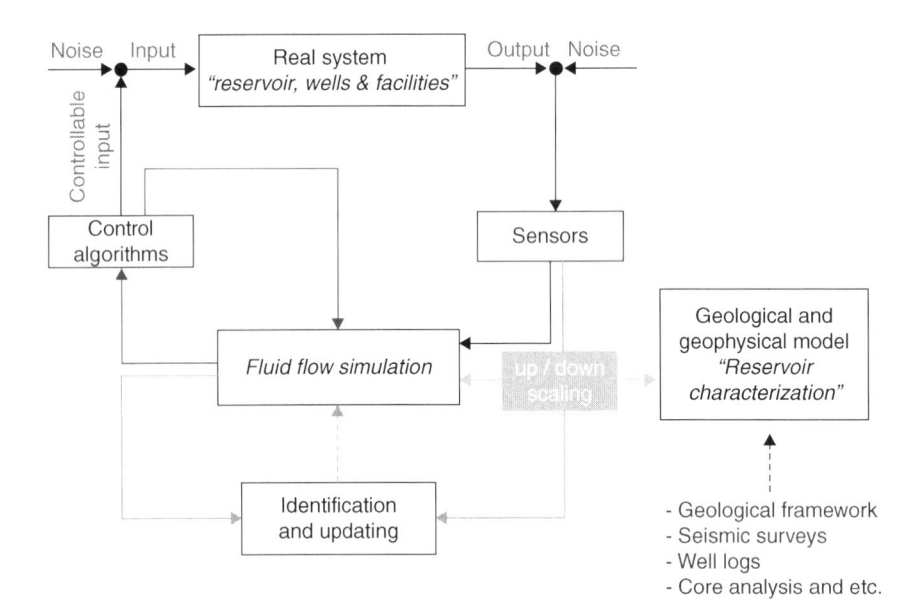

Figure 2.12 Schematic layout of a comprehensive smart reservoir modeling (redrawn from Brouwer et al, 2004).

from a campaign-based ad-hoc history-matching to a more frequent systematic updating of system models, based on data from different sources, while honoring geological constraints and the various sources of uncertainty.

This simulation model relates the simulated output data with a cost function. The cost function, designated by $J(u)$, might be net present value (NPV) or cumulative oil produced. The notation u is the vector of control variables, such as well rates, bottom hole pressures (BHP). The closed-loop process initiates with an optimization loop (marked in blue in Fig. 2.12). The control variables are set by minimizing or maximizing the cost function. The optimizing process must be performed on an uncertain simulation model. The control variables, i.e. well rates, BHP, are then applied to the real system, i.e. reservoir, wells and facilities, as input over the control step. These inputs impact the outputs from the real system. The new measured output data (production information) are applied to updated model parameters to reduce the uncertainty. This is called the history-matching process that is called the model-updating loop, marked in red in Fig. 2.12. The process repeated over the life of the reservoir to obtain an optimized performance of the system. The closed-loop approach for efficient real-time optimization consists of three key components: efficient optimization algorithms, efficient model updating algorithms, and techniques for uncertainty propagation.

The optimization process can be described in general form according to Sarma et al (2005) as:

$$\max_{u^n} \left[J\left(u^0, u^1, \cdots, u^{N-1}\right) = \phi\left(x^N\right) + \sum_{n=0}^{N-1} L^n \left(x^{n+1}, u^n, m\right) \right]$$

and $\forall n \in (0, 1, \cdots, N-1)$

Subject to:

$g^n\left(x^{n+1}, x^n, u^n, m\right) = 0$ and $\forall n \in (0, 1, \ldots, N-1)$

$x^0 = x_0$ Initial condition of the system (reservoir, wells and the facilities)

$c^n\left(x^{n+1}, x^n, u^n, m\right) \leq 0$ and $\forall n \in (0, 1, \ldots, N-1)$

$Au^n \leq b$ and $\forall n \in (0, 1, \ldots, N-1)$

$LB \leq u^n \leq UB$ and $\forall n \in (0, 1, \ldots, N-1)$ \hfill (2.44)

Here, n is the control step index and N is the total number of control steps, x^n refers to the dynamic states of the system, such as pressures, saturations, compositions, etc., and m is the model parameters (permeability, porosity, etc.). The first equation is the cost function that should be maximized by controlling the variable vector u^n. The cost function subjected to a set of constraints consists of: the set of reservoir simulation equations g^n for each grid block at each time step, initial system condition and additional constraints for the controls including – nonlinear inequality path constraints, linear path constraints, and bounds on the controls. These path constraints must be satisfied at every time step. These last constraints can be maximum injection rate constraint, maximum water cut constraint, maximum liquid production rate constraint, etc. The function $\phi(x^N)$ is the dynamic states of the last control step (e.g. abandonment cost). The function L^n involves the quantity related to the well parameters.

This problem that is the same as history-matching problem may be solved using stochastic algorithms, such as genetic algorithm and simulated annealing algorithm, or deterministic algorithm, i.e. gradient-based methods. The deterministic methods are generally very efficient, requires few forward model evaluations and also guarantees reduction of the objective function at each iteration, but only assures local optima for non-convex problems (Gill et al, 1982). However, the computation of the gradients of the objective function with respect to the control variables are complicated to obtain analytically and should be used with numerical methods to compute them. The number of required gradient depends on the well number and well variables, i.e. BHP, that is taken into account for the optimization process. The total number is the product of the well number and the number of well variables.

The most efficient algorithms for calculating gradients are the adjoint techniques especially for a large number of controls, as the algorithm is independent of the number of controls. The adjoint model equations are obtained from the necessary conditions of optimality of the optimization problem defined by Eqs. (2.44). The essence of the theory is that the cost function along with all the constraints can be written equivalently in the form of an augmented cost function using the a set of Lagrange multipliers. The result is a modified objective function. The solution procedure and more detailed descriptions may be found in Brouwer (2004) and Sarma et al, (2006).

The model-updating, automatic history-matching, and the uncertainty analysis are the other two key points in closed-loop real-time optimization systems. These two have been already explained in subsections 2.7.1 and 2.7.2.

3
Reservoir Simulators: Problems, Shortcomings, and Some Solution Techniques

All real objects are nonlinear and all natural phenomena are multidimensional. The parameters involved in a natural phenomenon are not independent of each other and the variation of each of them causes others to be affected. The natural phenomena are chaotic, not in the conventional sense of being arbitrary and/or unpredictable, but in the sense that they are always non-linear, making them prone to multiple solutions with classical mathematics. They are also irreversible in time, which means reproducibility is absurd unless one knows the exact time function. Exact time function is beyond the scope of human knowledge today and is not likely to be known in foreseeable future. We are unaware of the equations that truly govern natural phenomena and also the procedures to solve those equations exactly. Often several key simplifications are posed to get rid of nonlinearities and find a numerical description of a natural phenomenon.

Not surprisingly, petroleum reservoir engineering problems are known to be inherently nonlinear. Consequently, solutions to the complete multiphase flow equations have been principally attempted with numerical methods. The assumption here is analytical methods are not capable of handling non-linear problems. However, simplified forms of the problem have been solved some 60 years ago, when the well known Buckley–Leverett formulation was introduced. Ever since that pioneer work that neglected the capillary term, this formulation has been widely accepted in the petroleum industry. By using the method of characteristic, the multiphase one-dimensional fluid flow was solved. However, the resulting solution was a triple-valued one for a significant region. For decades, the existence of multiple solutions was considered to be the result of nonlinearity. Buckley and Leverett introduced shock utilizing the concept of material balance and two decades later, when numerical solutions were possible, it was discovered that the triple-value problem disappeared if the complete flow equation, including the capillary pressure form, is solved. Even though this demonstration was possible because of numerical solution schemes, it should not mean, numerical methods are free from linearization. In fact, every numerical solution imposes linearization at some point of the solution scheme. Therefore, a numerical technique cannot be used to definitively state the origin of multiple solutions or to claim that the solution is exact.

In the current chapter, we applied several polynomials and simultaneous equations of two variables as a model for a natural phenomenon in which the other parameters are kept constant. It is shown that they produce multiple solutions regardless if that the solutions are realistic or not and the number of solutions depends on the degree of nonlinearity of the equation. From the study it can be inferred that a phenomenon with only two variables produces more than one solution and, therefore, a multi-variable phenomenon surely has multiple solutions. This is true as long as we do not have a mathematical or numerical tool that can predict exact solution to a multi-variable problem.

In this work, we also applied Adomian Decomposition Method (ADM) in solution of nonlinear equation of Buckley–Leverett problem to introduce new approaches in reservoir simulations. The Adomian decomposition method is capable of solving nonlinear partial differential equations without any linearization assumptions. Even though the technique itself is not free from assumptions of known functionality (equivalent to knowing the exact function that describes a natural phenomenon), The ADM offers the best hope of finding solutions without *a priori* linearization of governing equations. This technique is used to unravel the true nature of the one-dimensional, two-phase flow. With this technique, we demonstrate that the Buckley–Leverett 'shock' is neither necessary nor an accurate portrayal of the displacement process. By using the ADM, the solution profile observed through numerous experimental studies was rediscovered.

Here, we are going to account for the possibilities of the existence of multiple solutions in reservoir engineering problems. It is well recognized that the flow inside porous media is nonlinear and may produce different situations, owing to the presence of many parameters that influence the process. The other aim is to introduce new approaches such as the ADM in simulation of petroleum problems. We applied the ADM in solution of two phase flow of water and oil.

3.1 Multiple Solutions in Natural Phenomena

All mathematical models of the real physical problems are nonlinear. The nonlinearity is related to the interaction and inclusion of different parameters involved in a physical problem. Several key assumptions are posed to get rid of nonlinearities and find out a numerical description of the problem. Some simplifications may also be imposed during the numerical evaluation of a problem. In this way, a problem is forced to have a single solution ignoring the possibility of having multiple solutions. In addition, not all of the available methods are capable of predicting multiple solutions. In fact, until now, a systematic method for determining multiple solutions is limited to three variables.

The general development of the set of governing equations always proceeds the same way for any material. A set of conservation laws is usually applied in integral form to a finite mass of material. Typical 'laws' express the conservation of mass, momentum, and energy. It is asserted that the 'laws' are true and the problems become that of solving the constitutive relationship of the 'law'. These equations are then converted to a local form and are cast in the form of partial differential equations. These differential equations cannot be solved in a general way for the details of the material motion. In order to close the system, the next step, always, is to specify the material response. The mathematical conditions are usually referred to as the constitutive relations. The last step is to combine these constitutive relations with the local form of the balance equations. The combination of these two sets of relations is called the field equations which are the differential equations governing the material of interest.

We are unaware of the mathematical model that truly simulates a natural phenomenon. The available models are based on several assumptions. For examples, there are many models that describe different fluid flows. The most general equations in fluid mechanics are the Navier–Stokes equations. The assumptions in derivation the Navier–Stokes equation are:

- The fluid is a continuum media; it indicates that we deal with a continuous matter.
- The fields of interest such as pressure, velocity, density, temperature, etc., are piecewise continuous functions of space and time.

- The fluid is Newtonian; a further, and very strong, restriction used is a linear stress-rate of strain relationship.

'Continuous' means, there should be no boundary. Even quarks are not continuous. In fact, unless the size of the constitutive particles is zero, there cannot be any continuity. For any variable to be continuous in space, the above requirement of zero size must apply. For a variable to be continuous in time, the notion of 'piecewise' is absurd. Both space and time domains are continuous and must extend to infinity for 'conservation of mass' to hold true. There is not a single linear object in nature, let alone a linear relationship. In reality, there is not a single Newtonian fluid. The assumption of linear stress-rate of strain relationship is *aphenomenal*, Zatzman and Islam (2007); Khan and Islam (2006) as the steady state assumption, in which the time dimension is eliminated.

A general model that explains completely the fluid motion and describes the nonlinearity due to the turbulence and chaotic motion of a fluid flow has not been developed so far. The solution for a turbulent flow is usually obtained based on the Navier–Stokes equations that are not developed for such a flow.

The numerical description is also found based on some simplification and linearization during the solution process. After the first linearization of the process itself by imposing 'laws' to forcibly describe natural phenomena, further linearization is involved during solution schemes Mustafiz et al (2008). All analytical methods impose linearization by dropping nonlinear terms, which is most often accomplished by neglecting terms or by imposing a fictitious boundary condition. Numerical techniques, on the other hand, impose linearization through discretization (Taylor series expansion), followed by linear matrix solving.

The existence of multiple solutions can be found in numerous problems. The occurrence of multiple solutions in solving the TSD-Euler equation was examined by Nixon (1989) and it was found that such solutions exist for a small range of Mach numbers and airfoil thickness. Nixon (1989) also found that a vorticity flux on the airfoil surface can enhance the appearance of multiple solutions.

We also observe the presence of multiple solutions, which depend on the pathway, in material processing operations. The existence of multiple roots in isothermal ternary alloys was discovered by Coates and Kirkaldy (1971) and was further explored by Maugis et al (1996). Coriell et al (1998) continued investigation on one-dimensional similarity solutions during solidification/melting of a binary alloy. Their study, to some extent, was analogous to the isothermal ternary system, except that the phases were then solid and liquid and temperature played the role of one of the components of the ternary. The diffusivity equation was used to express the variation of temperature and concentration of fluid and solid in time and space. The equation was transferred to an ordinary differential equation using the similarity technique and the existence of multiple similarity solutions for the solidification/melting problem was noticed. These results corresponded to significantly different temperature and composition profiles. Recently, a computational procedure to find the multiple solutions of convective heat transfer was proposed by Mishra and DebRoy (2005). In this approach, the conventional method of numerical solution was combined with a real number genetic algorithm (GA). These led the researchers to find a population of solutions and search for and obtain multiple set of input variable, all of which gave the desired specific output.

The existence of multiple solutions was investigated in separation technology using membrane separators by Tiscareno–Lechuga (1999). The author discussed conditions about occurrence of multiple solutions when the mole fraction of a component with intermediate permeability was specified as a design variable. When the pressure in the permeate chamber was significantly lower than that of the rententent, the conditions turned to be simpler and were expressed through equations, which involved only the composition of the feed and the permeability of the membrane.

We take into account some bivariate polynomials of different degree as a token-model for a natural phenomenon. It is assumed that the other contributing parameters of the model of the bivariate polynomial are constant. The number of solutions depends on the degree of the nonlinearity of the polynomial. The solutions are obtained using the Newton method and presented in graphical form for a limited region.

Some nonlinear simultaneous equations are also taken into account and the solutions of them are obtained with the Newton and Adomian decomposition methods. Our objective is to show that conventional techniques do not generate multiple solution, for instance, ADM even though that is a very powerful method in solution of nonlinear equations cannot produce multiple solutions. We also proposed a new scheme to show the feasibility of generating multiple solutions.

3.1.1 Knowledge Dimension

Dimensions provide the existence and imagination of the universe. It may be defined as the elements or factors making up a complete personality or entity. The dimensions are unique (each dimension has unique properties that makes it different from others), codependent (the dimensions are equally dependent to each other for their existence) and transcendence (dimensions have the ability of extending or lying beyond what it would otherwise be able to do).

Knowledge is synonymous to truth and reflects information about the properties, which exist in objects, events or facts. Knowledge explains about the physical properties, (which are observable and measurable), date, history, theories, opinions and etc. It does not have time, space, mass and energy. Knowledge is a dimension for phenomena and may be possible to measure it by bytes of information.

Knowledge can be obtained through the physical and/or mathematical simulation of a phenomenon. The physical simulation is carried on by geometrical, kinematical and dynamical scaling up or down of a problem. In many cases, it is not possible to obtain a complete physical simulation and therefore, the experimental results are based on several assumptions. The mathematical simulation is obtained by finding the governing equation and the internal relations between the involving parameters. Since any phenomenon is affected by a number of factors, the attempt of finding the truth greatly relies on how closely these factors are addressed. It is observed that all physical phenomena when described by governing equations involve several assumptions. It is understandable that as we reduce the number of assumptions, we reach closer to the truth.

The multi-dimensionality is another aspect of the knowledge dimension. As time is passed, knowledge is increased and grown up. It may be considered as one dimensional knowledge. The next step is the consciousness, which is knowledge of the knowledge. It may be considered as the second dimension of the knowledge dimension. Each dimension is naturally independent and therefore it may let that independent knowledge to enter. These are indications that there is no limitation or beginning and end for the knowledge dimension. This multi-dimensionality of the knowledge dimension indicate that there may be also a range of solutions for a specific phenomenon due to the fact that there are different factors that contribute to that phenomenon.

If we consider a pure material and plot the curve for melting or freezing of it in a certain pressure, there is a constant temperature during the freezing or melting process. If an isomorphous alloy, which consists of an arbitrary composition of components A and B, is taken into account, the freezing or melting process is to take place in a range of temperature that is dependent on the composition of the alloy and the pressure, Fig. 3.1. Therefore, we are dealing with a range of temperature instead of a constant temperature. Another, interesting point is that during the freezing and melting process the concentrations of the equilibrium liquid or solid phases are changing and varying in a certain range dependent on the final concentration of liquid or solid state. This is more pronounced for an alloy of more components.

There should be a population of solutions for a problem related to a natural phenomenon, dependent on the number and the behavior of involved parameters. There are many situations that the variation of parameters involving a natural phenomenon is approximated with a polynomial function Bjorndalen (2002), Bjorndalen et al (2005) and Mustafiz et al (2008). This is because we do not know the governing equation for that phenomenon. The number of solutions for such a function depends on the nonlinearity of that function. There are also roots that are not real. All of the roots of such a polynomial indicate that we should expect different solutions regardless of the

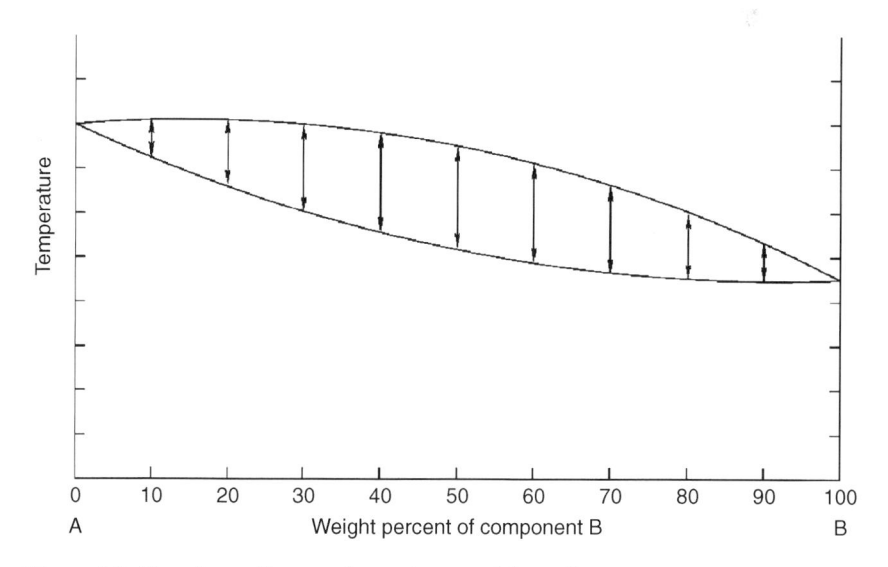

Figure 3.1 The phase diagram for an isomorphism alloy.

physical significance of them. It also indicates that if we can represent the natural data as a polynomial function at any given time, we can determine the roots, all of which should be considered for reporting some natural phenomena. Thus, roots including the imaginary ones are considered as the multiple solutions.

We take into account three bivariate polynomials and solve them for all possible roots. These polynomials are a third, two fourths and a fifth degree polynomials with two variables x and y.

Example 1. The first polynomial is a third degree bivariate polynomial.

$$4y^3 - 3y^2 - 2y + 2 = -5x^3 + 4x^2 + 3x + 1 \qquad (3.1)$$

The solution of this third degree polynomial is shown in Figs. 3.2 to 3.4. The real roots of the bivariate polynomial are depicted in Fig. 3.2. This polynomial gives three real roots in a limited range of x and y. In general, the polynomial has three complex roots at each constant real value of x and y. These roots can not be shown in a single graph. These are depicted in Fig. 3.3 when the variable y is a fixed real number and in Fig. 3.4 for a fixed real value of x.

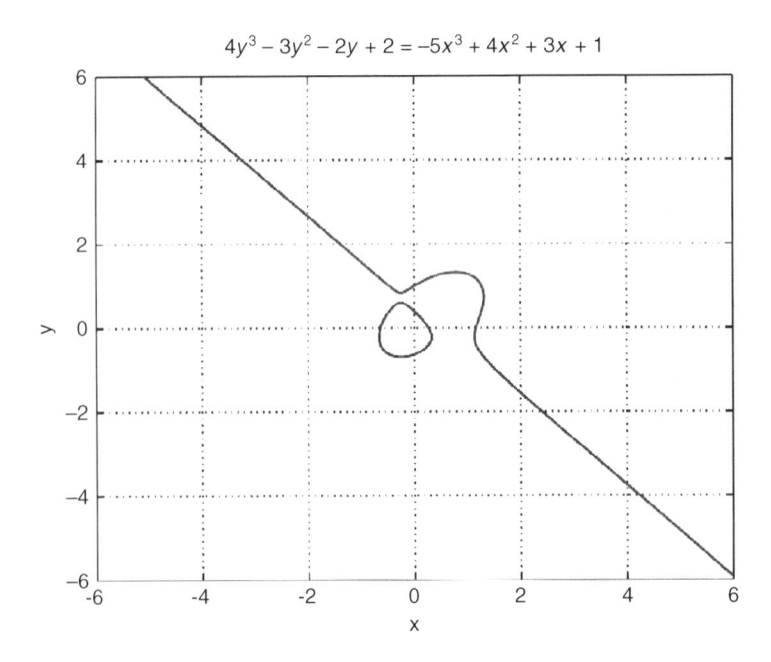

Figure 3.2 The graph of a third degree polynomial.

Reservoir Simulator

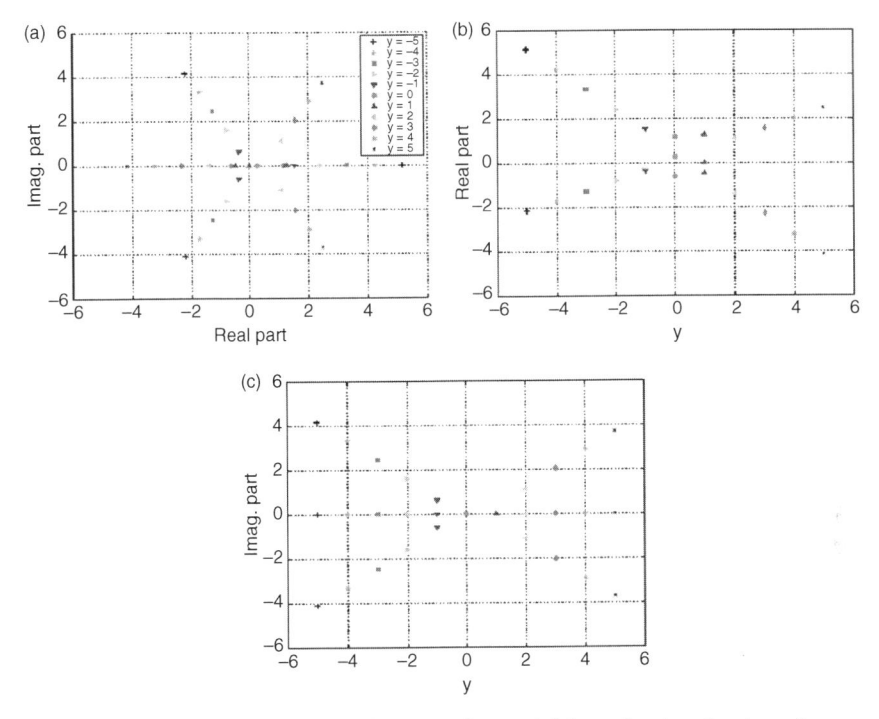

Figure 3.3 The roots of the third degree polynomial for a fixed real value of y.

It is indicated that with such a simple nonlinear problem we are dealing with a population of solutions that some of them may not be tangible. However, it does not mean that the intangible solutions are not natural. A complex number consists of a real part and an imaginary part. In many cases, it was understood that the only real part described the real world. The later applications of the complex number in different branches of sciences such as quantum mechanics, control theory, fluid dynamics and signal analysis reveal that nature has no preference for the real number and the imaginary part being just as physical as the real part.

Example 2. The second example is a fourth degree polynomial.

$$5y^4 + 4y^3 - 3y^2 - 2y + 2 = 6x^4 - 5x^3 - 4x^2 + 3x - 2 \quad (3.2)$$

The roots of the fourth degree polynomial are given in Figs. 3.5 to 3.7. In general, this polynomial should have four roots for each constant value of the variables. Four real roots for x can be found in

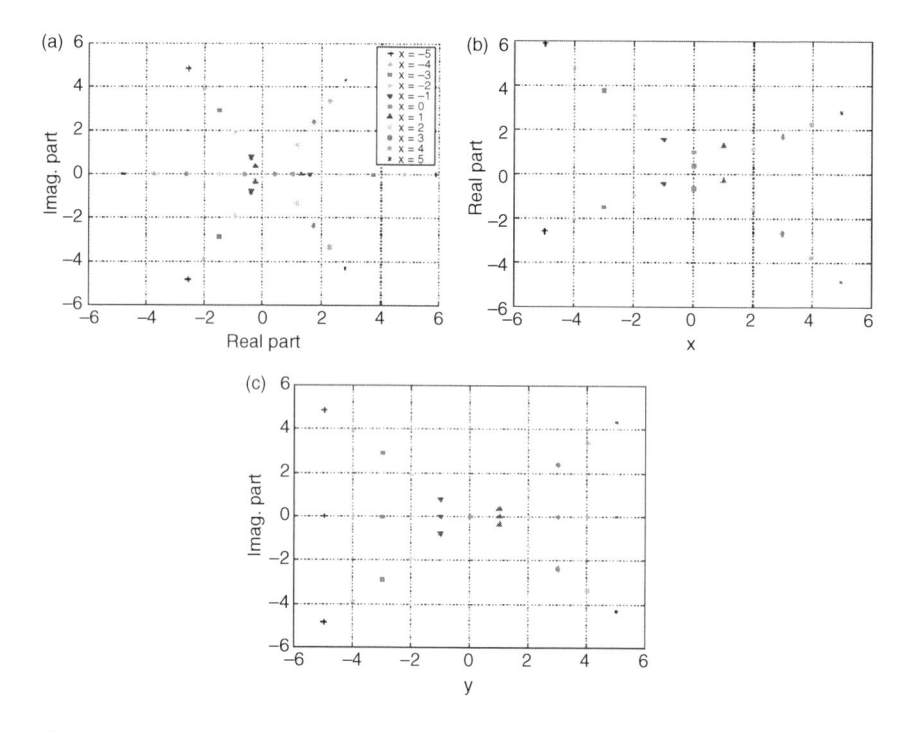

Figure 3.4 The roots of the third degree polynomial for a fixed real value of x.

a limited range of the variable y. It does not have four real roots for the variable y for a fixed real value of x. At most two real roots for the variable y can be obtained if x has a real value. All of the imaginary roots for y = 0, ±1, ±2, ±3, ±4, ±5 is given in Fig. 3.6. The same graphs are shown for the roots of the fourth degree polynomial in Fig. 3.7 when x = 0, ±1, ±2, ±3, ±4, ±5.

Example 3. The next example is a fifth degree polynomial.

$$6y^6 + 5y^4 + 4y^3 + 3y^2 + 2y + 1 = 7x^5 + 6x^4 + 5x^3 + 4x^2 + 3x + 2 \quad (3.3)$$

The real roots of this polynomial are depicted in Figs. 3.8 to 3.10. This fifth degree polynomial has a real root for every real value of x and y. All roots of the polynomial are shown in the complex planes in Figs. 3.9-a and 3.10-a for y, x = 0, ±1, ±2, ±3, ±4, ±5, respectively. The figures show that each two of the four complex roots are complex conjugate.

Reservoir Simulator 93

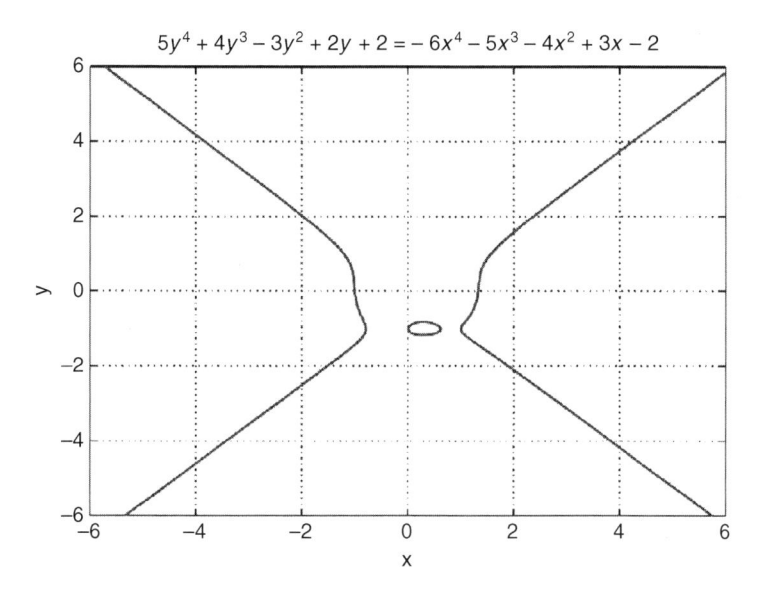

Figure 3.5 The graph of the real value of the first fourth degree polynomial.

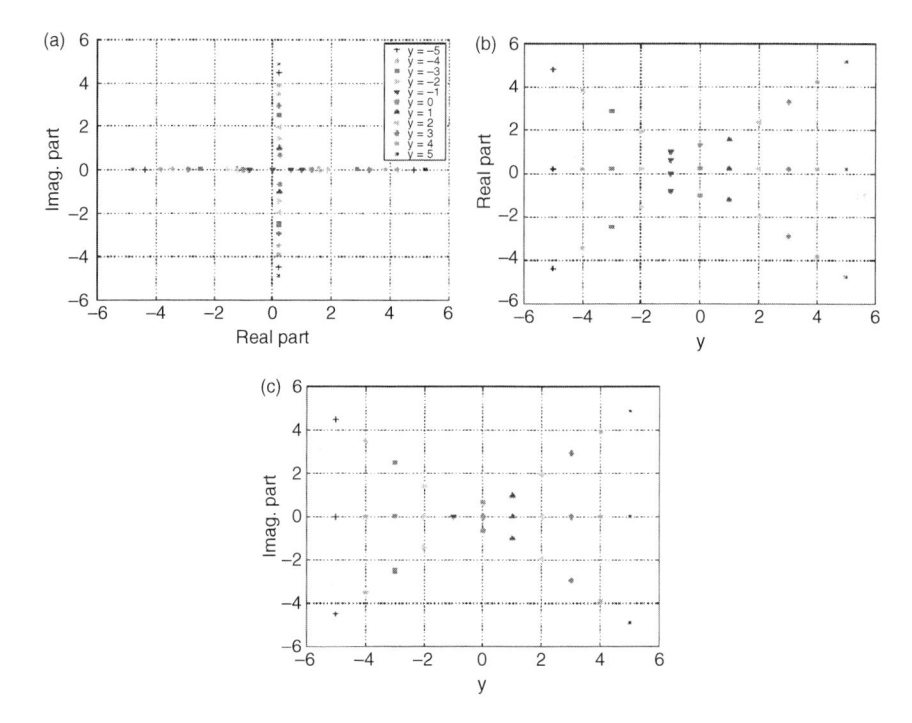

Figure 3.6 The roots of the first fourth degree polynomial for a fixed real value of y.

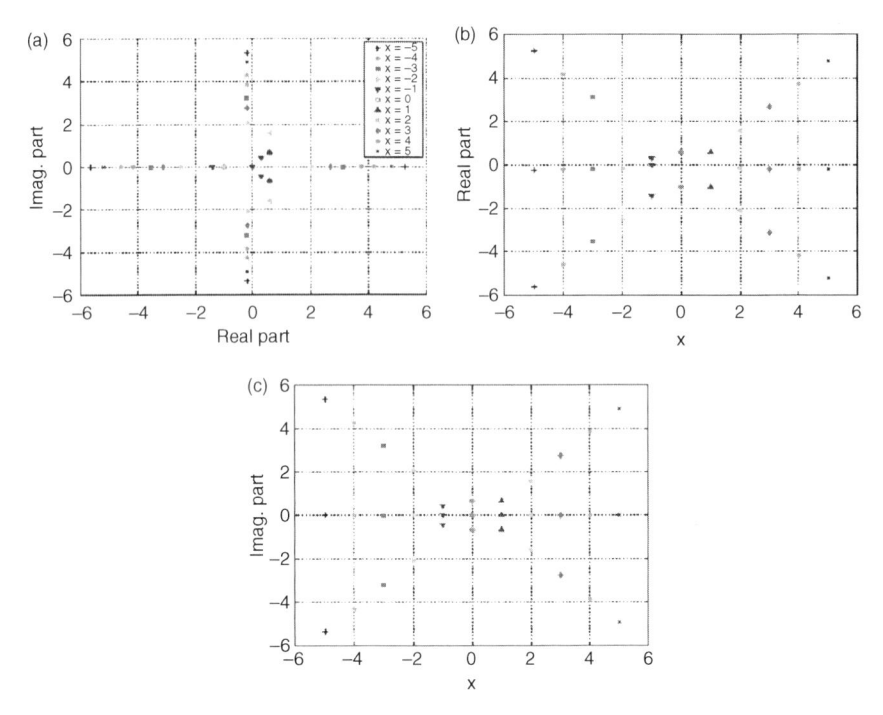

Figure 3.7 The roots of the first fourth degree polynomial for a fixed real value of x.

The variation of a parameter causes other parameters to be affected and prescribes some changes during the process. This indicates that all of the contributing parameters in a natural phenomenon are dependent on each other. These may be explained with some nonlinear simultaneous equations. The solution for these systems of nonlinear function is obtained mostly by the numerical methods. However, in many cases, the restriction of the applied method may limit obtaining all of the solutions possible.

The well known method in solution of nonlinear algebraic and simultaneous equations is the Newton method. The main restriction of this method is that the initial value for starting the iteration should be near the exact solution. If the initial guess is far from the exact solution it may result in a divergent iterations. The Adomian Decomposition Method (ADM) is a powerful method that can be used to obtain the solutions of systems of nonlinear simultaneous equations. ADM was first proposed by a North

Reservoir Simulator

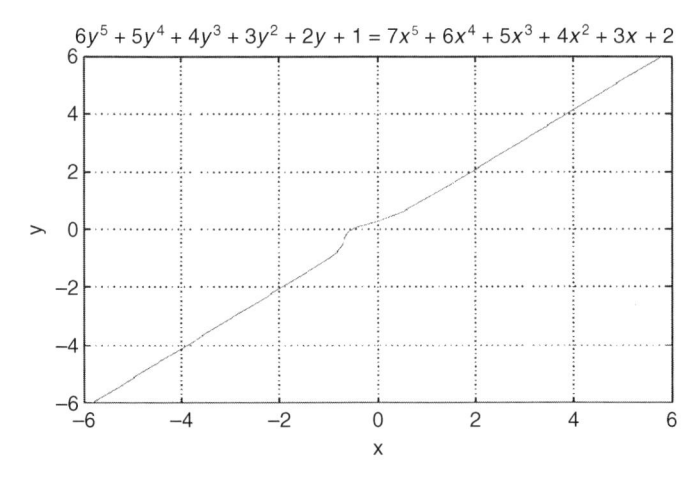

Figure 3.8 The graph of the real roots of the fifth degree polynomial.

American physicist, G. Adomian (1923–1996). The method is well addressed in Mousavizadegan et al (2006) to discuss the limitations of ADM for partial differential equations and Mustafiz et al (2008) to solve the Buckley-Leverett equation with the effect of the capillary pressure.

The ADM solution is obtained in a series form while the nonlinear term is decomposed into a series in which the terms are calculated recursively using Adomian polynomials. A simultaneous algebraic equation with independent variables is taken into account. It may written that

$$f_i(x_1, x_2, ..., x_n) = 0 \quad \text{and} \quad i = 1, 2, ..., n \qquad (3.4)$$

Each equation can be solved for an independent variable as

$$x_i = a_i + g_i(x_1, x_2, ..., x_n) \quad \text{and} \quad i = 1, 2, ..., n \qquad (3.5)$$

The solution may be expressed as a series solution as

$$x_i = \sum_{i=0}^{\infty} x_{i,j} \quad \text{and} \quad i = 1, 2, ..., n \qquad (3.6)$$

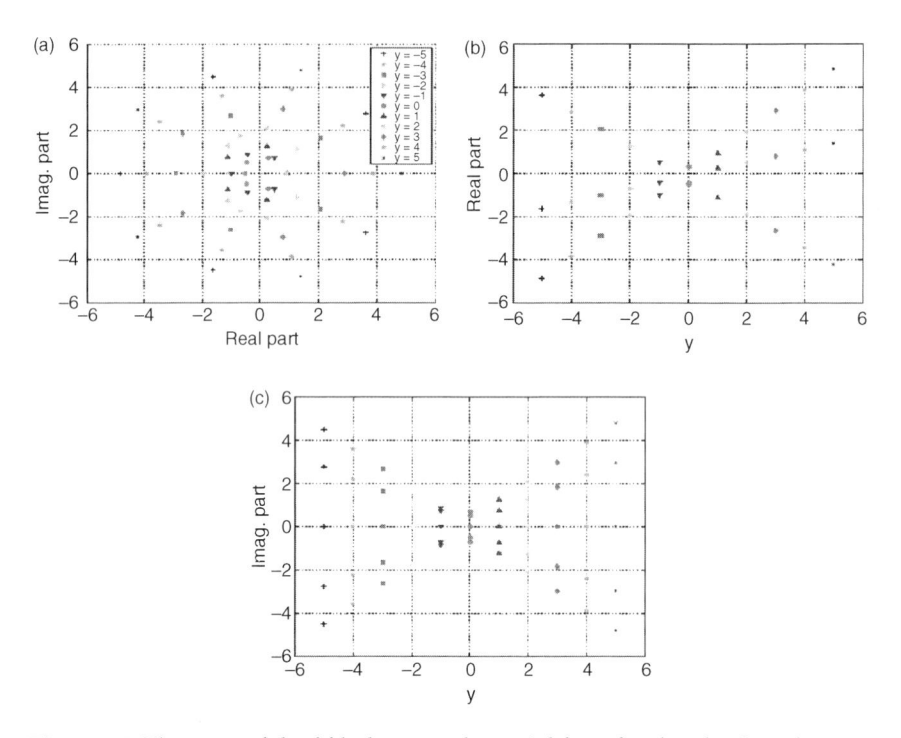

Figure 3.9 The roots of the fifth degree polynomial for a fixed real value of y.

The components of the series solution are obtained using the Adomian decomposition method in the form

$$\sum_{i=0}^{\infty} x_{i,j} = a_i + \sum_{k=0}^{\infty} A_{i,k} \qquad (3.7)$$

where

$$x_{i,0} = a_i, \; x_{i,1} = A_{i,0}, \; x_{i,2} = A_{i,1}, ..., \; x_{i,m} = A_{i,m-1}, \; x_{i,m+1} = A_{i,m}, ... \qquad (3.8)$$

The term $A_{i,m}$ is obtained by using the Adomian polynomial that is given in the following form.

$$A_{i,m} = \frac{1}{m!} \left[\frac{d^m}{d\lambda^m} g_i \left(\sum_{k=0}^{\infty} \lambda^k x_{1,k}, \sum_{k=0}^{\infty} \lambda^k x_{2,k}, ..., \sum_{k=0}^{\infty} \lambda^k x_{n,k} \right) \right]_{\lambda=0}$$

$$\text{and} \quad m = 1, 2, ..., \infty \quad \text{for} \quad i = 1, 2, ..., n \qquad (3.9)$$

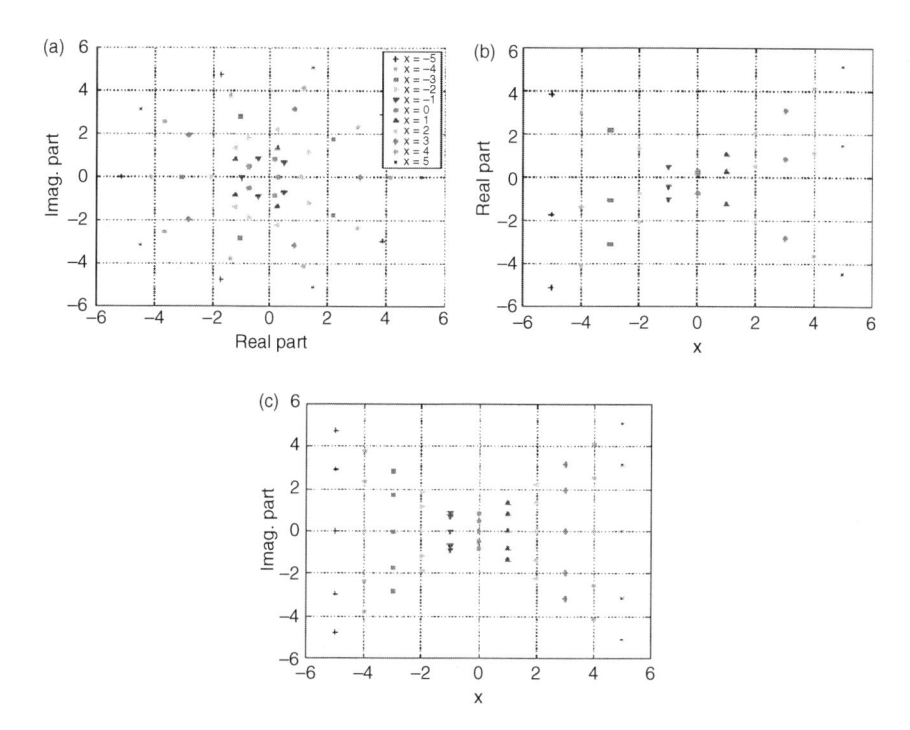

Figure 3.10 The roots of the fifth order polynomial for a fixed real value of x.

These are the first three elements of the Adomian polynomials.

$$A_{i,0} = g_i(x_{1,0}, x_{2,0}, ..., x_{n,0})$$

$$A_{i,1} = x_{1,1} \frac{\partial}{\partial x_1} g_i(x_{1,0}, x_{2,0}, ..., x_{n,0}) + \cdots + x_{n,1} \frac{\partial}{\partial x_n} g_i(x_{1,0}, x_{2,0}, ..., x_{n,0})$$

$$A_{i,2} = x_{1,2} \frac{\partial}{\partial x_1} g_i(x_{1,0}, x_{2,0}, ..., x_{n,0}) + \cdots + x_{n,2} \frac{\partial}{\partial x_n} g_i(x_{1,0}, x_{2,0}, ..., x_{n,0})$$

$$+ \frac{1}{2} x_{1,1}^2 \frac{\partial^2}{\partial x_1^2} g_i(x_{1,0}, x_{2,0}, ..., x_{n,0}) + \cdots$$

$$+ \frac{1}{2} x_{n,1}^2 \frac{\partial^2}{\partial x_n^2} g_i(x_{1,0}, x_{2,0}, ..., x_{n,0})$$

$$+ x_{1,1} x_{2,1} \frac{\partial^2}{\partial x_1 x_2} g_i(x_{1,0}, x_{2,0}, ..., x_{n,0}) + \cdots$$

$$+ x_{1,1} x_{n,1} \frac{\partial^2}{\partial x_1 x_n} g_i(x_{1,0}, x_{2,0}, ..., x_{n,0})$$

$$+ x_{2,1}x_{3,1}\frac{\partial^2}{\partial x_2 x_3}g_i(x_{1,0}, x_{2,0}, \ldots, x_{n,0}) + \cdots$$

$$+ x_{2,1}x_{n,1}\frac{\partial^2}{\partial x_2 x_n}g_i(x_{1,0}, x_{2,0}, \ldots, x_{n,0}) + \cdots$$

$$+ x_{n-1,1}x_{n,1}\frac{\partial^2}{\partial x_2 x_3}g_i(x_{1,0}, x_{2,0}, \ldots, x_{n,0}) \qquad (3.10)$$

The rest is lengthy and more complicated. MATLAB is used to compute the elements of $A_{i,j}$. The elements of the series solution x_i are obtained according to (3.6).

Example 4. The first nonlinear system of equation is

$$\begin{cases} x^2 - 10x + 4y^2 + 9 = 0 \\ xy^2 + x - 10y + 5 = 0 \end{cases} \qquad (3.11)$$

The real solutions can be obtained by plotting the equations as given in Fig. 3.11. The plot indicates that there are two common real roots for this nonlinear SAE.

We use the ADM to find the solution for (3.11). The series solutions are obtained using the equations (3.4) to (3.10) with different numbers of the elements. The computations are carried on using MATLAB. The solutions and the errors with different number of elements are

$x = 1.21244$ & $y = 0.669277$ if $i = 4$ $E_1 = 0.1373$ & $E_2 = 0.0627$
$x = 1.23491$ & $y = 0.679418$ if $i = 8$ $E_1 = 0.0223$ & $E_2 = 0.0108$
$x = 1.23802$ & $y = 0.680857$ if $i = 12$ $E_1 = 0.0046$ & $E_2 = 0.0023$
$x = 1.23916$ & $y = 0.681396$ if $i = 16$ $E_1 = 0.0011$ & $E_2 = 0.0006$
$x = 1.23933$ & $y = 0.681474$ if $i = 20$ $E_1 = 0.0003$ & $E_2 = 0.0001$

where E1 and E2 are the deviation of the first and second equation from zero. However, it gives a good approximation for the solution of this system of simultaneous equations.

The deficiency of the ADM is that it is not able to give the second solution as seen from Fig. 3.12. The other restriction is that the Adomian polynomial does not always give a convergent series

Reservoir Simulator

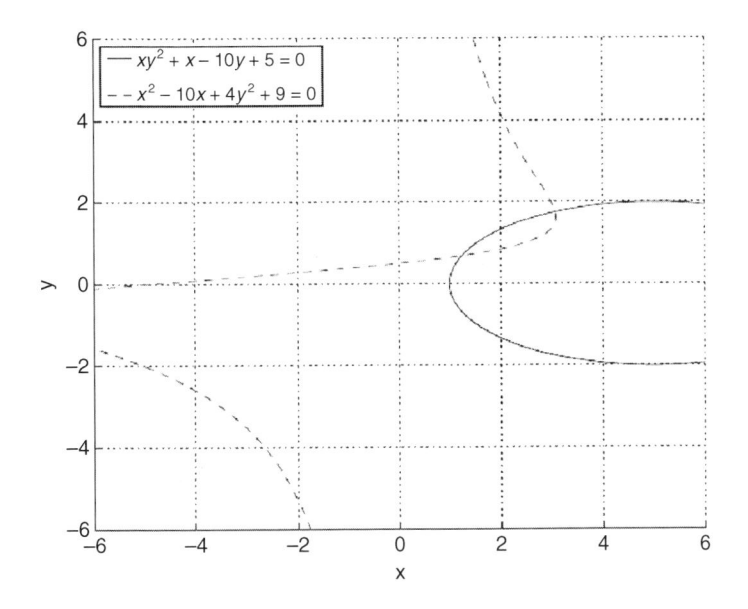

Figure 3.11 The graphs of the simultaneous equations.

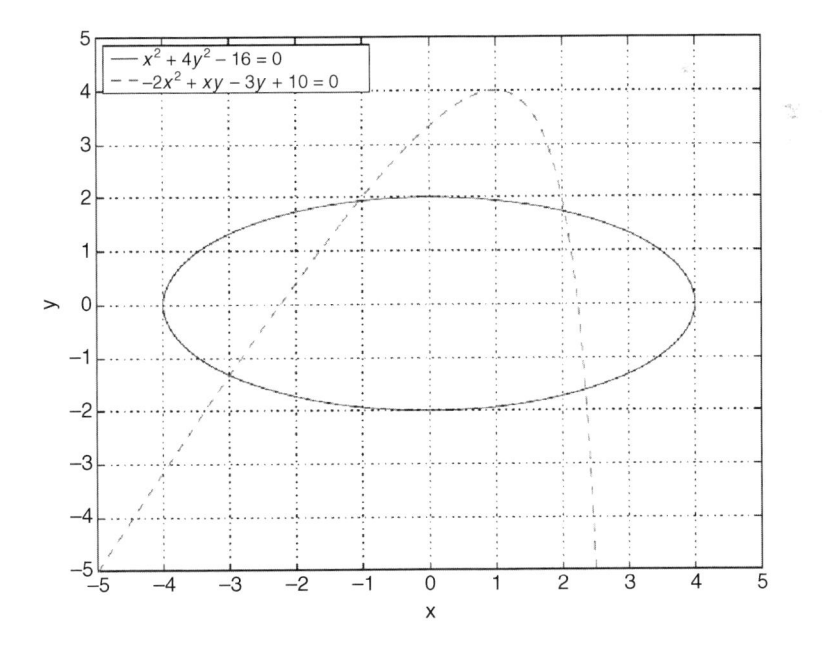

Plot legend: $xy^2 + x - 10y + 5 = 0$; $x^2 - 10x + 4y^2 + 9 = 0$

Figure 3.12 The graphs of the simultaneous equations.

Plot legend: $x^2 + 4y^2 - 16 = 0$; $-2x^2 + xy - 3y + 10 = 0$

solution. It depends on the type of the equations and the first term of the series solution. It is necessary sometimes to change the form of the equations to a get a convergent series solution for the problem.

Example 5. The second nonlinear SAE is

$$\begin{cases} x^2 + 4y^2 - 16 = 0 \\ -2x^2 + xy - 3y + 10 = 0 \end{cases} \quad (3.12)$$

This system of simultaneous equations has four real common roots as shown in Fig. 3.12.

These roots are computed with ADM. The system of equations is written in the form

$$\begin{cases} x = \pm\sqrt{0.5(xy - 3y + 10)} \\ y = \pm 0.5\sqrt{16 - x^2} \end{cases} \quad (3.13)$$

for each of the variables. It is assumed that $x = \sum_{i=0}^{\infty} x_i$ and $y = \sum_{i=0}^{\infty} y_i$.

Using ADM, it is set that $x_0 = 0$ and $y_0 = 0$. The elements of the series solutions are obtained using the Adomian polynomial as given in (3.9).

$$x_m = \frac{1}{m!} \left\{ \frac{\partial^{m-1}}{\partial \lambda^{m-1}} \left[\pm \sqrt{0.5 \left(\sum_{k=0}^{\infty} \lambda^k x_m \sum_{k=0}^{\infty} \lambda^k y_k - 3 \sum_{k=0}^{\infty} \lambda^k y_k + 10 \right)} \right] \right\}_{\lambda=0}$$

for $m = 1, 2, \ldots$.

$$y_m = \frac{1}{m!} \left\{ \frac{\partial^{m-1}}{\partial \lambda^{m-1}} \left[\pm 0.5 \sqrt{16 - \sum_{k=0}^{\infty} \left(\lambda^k x_k \right)^2} \right] \right\}_{\lambda=0} \quad \text{for } m = 1, 2, \ldots. \quad (3.14)$$

The computations are carried out and the solution are obtained when the series solution are truncated to $i = 10$. The solutions are

$$x = 2.0426 \ \& \ y = 1.7144; \quad x = -1.0435 \ \& \ y = 1.9316;$$
$$x = -2.9794 \ \& \ y = -1.3193; \text{ and } x = 2.3554 \ \& \ y = -1.6235.$$

which are very accurate. The more accurate result can be found with increasing the number of the series solution elements. The multiple solutions can be found if the solution is arranged in proper form. It can be considered as one of the most challenging tasks in application of ADM. A proper arrangement should be selected to obtain a convergent result as well as the multiple solutions.

Example 6. The third example is a third degree system of two simultaneous equations of the variables x and y.

$$\begin{cases} x^3 + y^3 - 10x - 5 = 0 \\ x^3 - y^3 - 15y^2 + 2 = 0 \end{cases} \quad (3.15)$$

The equations are plotted in Fig. 3.13. It is realized that there are four real solutions for this system of nonlinear equations. The solution for the variables may be arranged in the form

$$\begin{cases} x = -0.2 + 0.1(x^3 + y^3) \\ y = \pm\sqrt{\dfrac{1}{15}(2 + x^3 - y^3)} \end{cases} \quad (3.16)$$

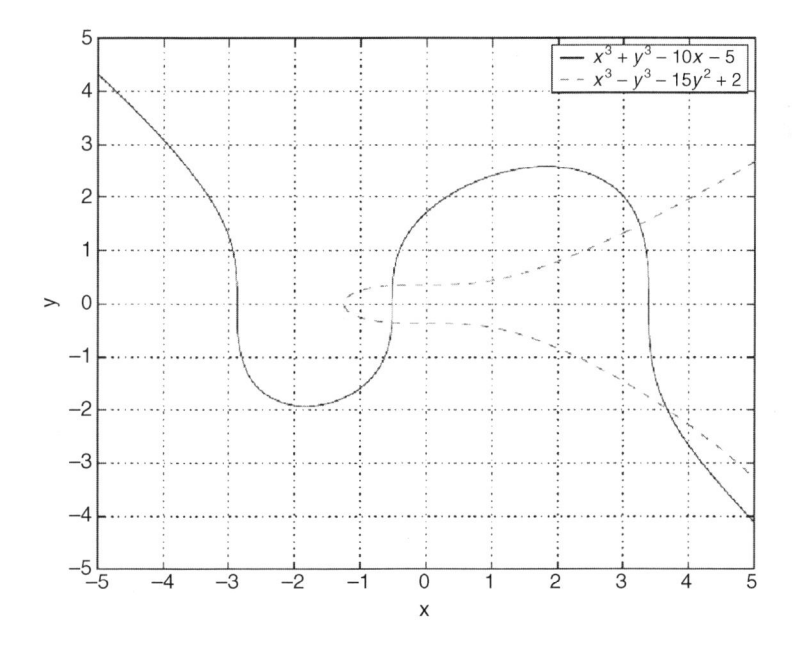

Figure 3.13 The graphs of the simultaneous equations.

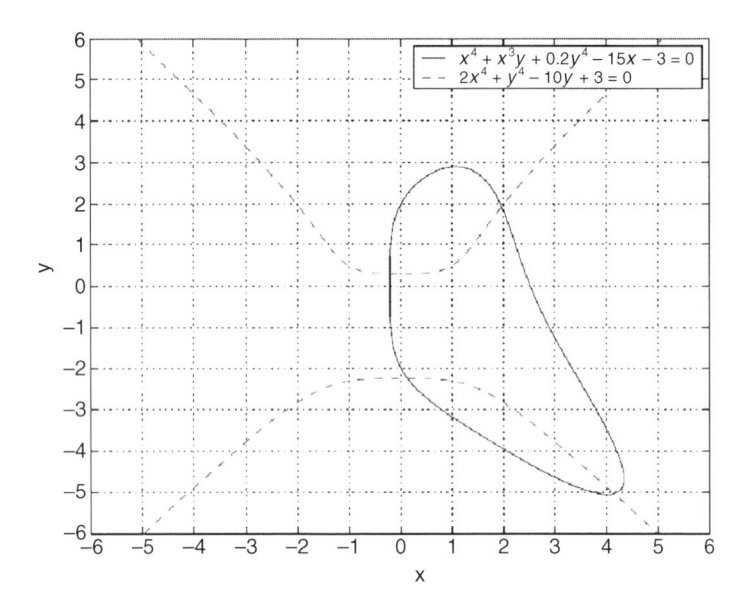

Figure 3.14 The graphs of the simultaneous equations.

Using ADM, the solutions are expressed in the series form as (3.6). The elements of the series solution for x and y are computed by taking into account that $x_0 = -0.5$ and $y_0 = 0$; the rest are computed using Adomian polynomial (3.9). The arrangement of the SAE in the form of (3.15) gives two of the common real roots that are ($x = -0.5089$ & $y = 0.3489$) and ($x = -0.5185$ & $y = 0.3565$). This arrangement in the form of (3.15) does not give all the common real roots of the SAE (3.14). The other solutions may be obtained with another arrangement for x and y. We try different arrangements for the variables but almost all of them result in a divergent series.

Example 7. The next example is a fourth degree system of simultaneous equations.

$$\begin{cases} x^4 + x^3 y + \frac{1}{5} y^4 - 15x - 3 = 0 \\ 2x^4 - y^4 - 10y + 3 = 0 \end{cases} \tag{3.17}$$

There are four common real roots for this system of equations as shown in Fig. 3.14. This system of equations are rearranged in the form

$$\begin{cases} x = -\dfrac{1}{5} + \dfrac{1}{15}(x^4 + x^3 y + \dfrac{1}{5} y^4) \\ y = \dfrac{3}{10} + \dfrac{1}{10}(2x^4 - y^4) \end{cases} \quad (3.18)$$

for x and y variables. The only solution that can be obtained with the ADM and the arrangement (3.17) is $x = -0.19995$ & $y = 0.29952$ where the series solution is truncated to $i = 5$.

3.2 Adomian Decomposition

Since Adomian (1984) proposed his decomposition technique, the Adomian decomposition method (ADM) has gained significant interest among researchers, particularly in the fields of physics and mathematics. They applied this method to many deterministic and stochastic problems (Adomian, 1986; Eugene, 1993; and Adomian 1994). In this method, the governing equation is transformed into a recursive relationship and the solution appears in the form of a power series. The ADM has emerged as an alternative method to solve various mathematical models including algebraic, differential, integral, integro-differential, partial differential equations (PDEs) and systems, higher-order ordinary differential equations, and others.

Guellal and Cherruault (1995) utilized the method to solve an elliptical boundary value problem with an auxiliary condition. Laffez and Abbaoui (1996) used it in modeling thermal exchanges associated to drilling wells. The application of this method is also noticeable in medical research; in which the decomposition technique is used to solve differential system of equations in modeling of HIV immune dynamics (Adjedj, 1999). (Wazwaz A., 2001) and Wazwaz and El-Sayed (2001) reported that the ADM could be useful in solving problems without considering linearization, perturbation, or unjustified assumptions that may alter the nature of the problem under investigation. Also there have been suggestions that this method can be more advantageous over numerical methods by providing analytic, verifiable, rapidly convergent approximations, which add insight into the character and the behavior of the solution as obtainable in the closed form solution El-Sayed and Abdel-Aziz (2003). In particular, they noticed the strength of the ADM in handling nonlinear problems in terms of rapid convergence. Biazar and Ebrahimi (2005) also expressed similar notions about rapid convergence and further added the advantage of the technique in terms of considerable savings in computation time when they attempted to solve hyperbolic equations.

Even though, petroleum problems are some of the most intriguing candidates for the ADM, there have been sparse attempts to utilize the method in the petroleum problems. Only recently, Mustafiz et al (2005) reported the decomposition of the non-Darcy flow equations in porous media. In a different paper, Mustafiz and Islam (2005) transformed the diffusivity equations in well-test applications into canonical forms.

The Buckley–Leverett (1942) analysis is considered to be the first pioneer work in the study of linear displacement of a fluid by another fluid. The solution of their displacement study on two-phase fluid excluded the effect of capillary and gave multiple results for saturation at a given position. Realizing the fact that such co-existence of multiple saturation values is physically unrealistic, Buckley and Leverett (1942) used the fundamental concept of material-balance to explain the shock. However, the lack of theoretical justification of shock was a major constraint in understanding the displacement phenomenon more vividly than before, and would not be re-evaluated until the next significant work published by Holmgren and Morse (1951). They utilized the Buckley–Leverett theory to calculate the average water saturation at breakthrough and explained dispersion as a consequence of capillary effects.

To solve the displacement equation including capillary as well as gravity, Fayers and Sheldon (1959) attempted a Lagrangian approach. They, however, did not succeed to determine the time required to obtain a particular saturation, which later was explained by Bentsen (1978) revealing the fact that the distance traveled by zero saturation is governed by a separate equation. Bentsen also noted that at slower injection rates, the input boundary condition of constant normalized saturation that Fayers and Sheldon used was incorrect in formulation. Also, there have been numerical investigations in the past to solve the displacement equation. Hovanessian and Fayers (1961) were able to avoid profiles of multiple valued saturations by considering the capillary pressure.

The Adomian decomposition method is applied to solve the nonlinear Buckley–Leverett equation. The solution for the water saturation is expressed in a series form. The base element of the series solution is obtained using the solution of the linear part of the Buckley–Leverett equation without the effect of the capillary pressure using the characteristic method. The other elements of the series solution are obtained recursively using the Adomian polynomial. The modification of the ADM in selection of the base element of the series solution makes the ADM feasible in solution of the nonlinear Buckley–Leverett equation. The computation is carried out for a reservoir of certain properties and initial conditions.

3.2.1 Governing Equations

The Buckley–Leverett equation is given by

$$\frac{\partial S_w}{\partial t} + \frac{q}{A\phi}\frac{\partial f_w}{\partial S_w}\frac{\partial S_w}{\partial x} = 0 \tag{3.19}$$

where f_w is expressed in the form

$$f_w = \left(\frac{1}{1+\frac{k_{ro}\mu_w}{k_{rw}\mu_o}}\right)\left\{1 + \frac{Akk_{ro}}{q\mu_o}\left[\frac{\partial P_c}{\partial x} - (\rho_w - \rho_o)g\sin\alpha\right]\right\}. \tag{3.20}$$

This equation indicates that the fractional flow rate of water depends on reservoir characteristics, water injection rate, viscosity and direction of flow. The effect of capillary pressure, P_c, which appears in the fractional flow equation, on saturation profiles is important, since these profiles affect the ultimate economic oil recovery (Bentsen, 1978). The ratio of effective permeability to viscosity is defined as the mobility which is shown for water and oil, respectively.

$$\lambda_w = \frac{kk_{rw}}{\mu_w} \quad \text{and} \quad \lambda_o = \frac{kk_{ro}}{\mu_o}. \tag{3.21}$$

The mobility ratio of oil to water is defined by

$$M = \frac{\lambda_o}{\lambda_w} = \frac{k_{ro}\mu_w}{k_{rw}\mu_o}. \tag{3.22}$$

The assumptions associated with the Buckley–Leverett equation, according to Ertekin et al (2001) are:

- oil and water phases are assumed incompressible;
- the porous medium is assumed incompressible, which implies that porosity is constant;
- injection and production are taken care of by means of boundary conditions, which indicate that there is no external sink or source in the porous medium;
- the cross-sectional area that is open to flow is constant;
- the saturation-constraint equation for two-phase flow is valid; and
- the fractional flow of water is dependent on water saturation only.

The expression of the fractional flow rate of water in (3.20) suggests that (3.19) is a nonlinear differential equation and it is due to the effect of capillary pressure. In the simplest case of horizontal flow and neglecting the effects of capillary pressure variation along the reservoir, the expression for f_w in (3.20) is simplified to a linear differential equation, which is given by

$$f_w = \frac{1}{1+M}. \qquad (3.23)$$

The water saturation distribution along the reservoir can be found at different time steps by knowing the injection flow rate, the initial water saturation distribution, and the variation of fractional water flow rate. The water is normally the wetting fluid in the water-oil two phase flow systems. The relative permeability of the water and oil for a specific reservoir are obtained with the drainage and imbibitions process on the core in the laboratory. However, in this chapter, the variation of relative permeability to water and oil as a function of water saturation are obtained using the following empirical relationships respectively.

$$k_{rw} = a_1 S_{wn}^{n1} \qquad (3.24)$$

Reservoir Simulator

$$k_{ro} = a_2(1-S_{wn})^{n2} \tag{3.25}$$

where the normalized water saturation, S_{wn}, is defined as

$$S_{wn} = \frac{S_w - S_{wi}}{1 - S_{wi} - S_{or}}. \tag{3.26}$$

If the effects of capillary pressure are included in a horizontal reservoir, (3.27) takes the form

$$f_w = \frac{1}{1+\frac{k_{ro}\mu_w}{k_{rw}\mu_o}}\left[1 + \frac{Akk_{ro}}{q\mu_o}\frac{\partial P_c}{\partial x}\right]. \tag{3.27}$$

Capillary pressure at any point is directly related to the mean curvature of the interface, which, in turn, is a function of saturation. Therefore, it can be safely assumed that capillary pressure is only a function of water saturation. By applying the chain rule,

$$P_c = f(S_w) \quad \rightarrow \quad \frac{\partial P_c}{\partial x} = \frac{\partial P_c}{\partial S_w}\frac{\partial S_w}{\partial x} \tag{3.28}$$

and by incorporating (3.26) and (3.27) in (3.19), the following partial differential equation is obtained

$$\frac{\partial S_w}{\partial t} + \frac{q}{A\phi}\frac{\partial}{\partial S_w}\left(\frac{1}{1+\frac{k_{ro}\mu_w}{k_{rw}\mu_o}}\right)\frac{\partial S_w}{\partial x} + \frac{kk_{ro}}{\mu_o\phi}\frac{\partial}{\partial S_w}\left(\frac{1}{1+\frac{k_{ro}\mu_w}{k_{rw}\mu_o}}\right)\frac{\partial P_c}{\partial S_w}\left(\frac{\partial S_w}{\partial x}\right)^2$$
$$+ \frac{k}{\mu_o\phi}\left(\frac{1}{1+\frac{k_{ro}\mu_w}{k_{rw}\mu_o}}\right)\frac{\partial k_{ro}}{\partial S_w}\frac{\partial P_c}{\partial S_w}\left(\frac{\partial S_w}{\partial x}\right)^2 + \frac{kk_{ro}}{\mu_o\phi}\left(\frac{1}{1+\frac{k_{ro}\mu_w}{k_{rw}\mu_o}}\right)\frac{\partial^2 P_c}{\partial S_w^2}\left(\frac{\partial S_w}{\partial x}\right)^2 = 0$$
$$\tag{3.29}$$

Equation (3.28) is a nonlinear partial differential equation and the nonlinearity arises because of the inclusion of capillary pressure in it.

3.2.2 Adomian Decomposition of Buckley–Leverett Equation

In the Adomian decomposition method, the solution of a given problem is considered as a series solution. Therefore, the water saturation is expressed as

$$S_w(x,t) = \sum_{n=0}^{\infty} S_{wn}. \tag{3.30}$$

If a functional equation is taken into account for water saturation, it can be written in the form

$$\sum_{n=0}^{\infty} S_{wn} = f(x,t) + \sum_{n=0}^{\infty} A_n. \tag{3.31}$$

where $f(x,t)$ is a given function and A_n $(S_{w0}, S_{w1}, S_{w2}, ..., S_{wn})$ or, simply, A_n's are called Adomian polynomials. The Adomian polynomials are expressed as

$$A_n = \frac{1}{n!} \frac{d^n}{d\lambda^n} \left[N\left(\sum_{i=0}^{\infty} \lambda^i S_{wn}(x,t)_i \right) \right]_{\lambda=0} \tag{3.32}$$

where N is an operator and λ is a parameter. The elements of the series solution for $S_{wn}(x,t)$ are obtained recursively by

$$S_{w0}(x,t) = f(x,t),\ S_{w1}(x,t) = A_0,\ ...\ S_{wk}(x,t) = A_{k-1}\ .... \tag{3.33}$$

By this arrangement, the linear and nonlinear part of the functional (3.29) is replaced by a known function using the recursive (3.31). By integrating (3.28) with respect to t

$$S_w = S_w(x,0) - \int_0^t \left[\frac{q}{A\phi} \frac{\partial}{\partial S_w} \left(\frac{1}{1 + \frac{k_{ro}\mu_w}{k_{rw}\mu_o}} \right) \frac{\partial S_w}{\partial x} \right] dt$$

$$- \int_0^t \left[\frac{kk_{ro}}{\mu_o \phi} \frac{\partial}{\partial S_w} \left(\frac{1}{1 + \frac{k_{ro}\mu_w}{k_{rw}\mu_o}} \right) \frac{\partial P_c}{\partial S_w} \left(\frac{\partial S_w}{\partial x} \right)^2 \right]

$$+\frac{k}{\mu_o\phi}\left(\frac{1}{1+\frac{k_{ro}\mu_w}{k_{rw}\mu_o}}\right)\frac{\partial k_{ro}}{\partial S_w}\frac{\partial P_c}{\partial S_w}\left(\frac{\partial S_w}{\partial x}\right)^2$$

$$+\frac{kk_{ro}}{\mu_o\phi}\left(\frac{1}{1+\frac{k_{ro}\mu_w}{k_{rw}\mu_o}}\right)\frac{\partial^2 P_c}{\partial S_w^2}\left(\frac{\partial S_w}{\partial x}\right)^2\Bigg]dt. \tag{3.34}$$

Comparing (3.34) with (3.31) and taking into account of (3.33), the elements of the water saturation series are obtained

$$S_{w0}(x,t) = S_w(0,x) - \int_0^t \left[\frac{q}{A\phi}\frac{\partial}{\partial S_{w0}}\left(\frac{1}{1+\frac{k_{ro}\mu_w}{k_{rw}\mu_o}}\right)\frac{\partial S_{w0}}{\partial x}\right]dt$$

$$S_{w1}(x,t) = -\int_0^t\left[\frac{kk_{ro}}{\mu_o\phi}\frac{\partial}{\partial S_{w0}}\left(\frac{1}{1+\frac{k_{ro}\mu_w}{k_{rw}\mu_o}}\right)\frac{\partial P_c}{\partial S_{w0}} + \frac{k}{\mu_o\phi}\left(\frac{1}{1+\frac{k_{ro}\mu_w}{k_{rw}\mu_o}}\right)\frac{\partial k_{ro}}{\partial S_{w0}}\frac{\partial P_c}{\partial S_{w0}}\right.$$

$$\left.+\frac{kk_{ro}}{\mu_o\phi}\left(\frac{1}{1+\frac{k_{ro}\mu_w}{k_{rw}\mu_o}}\right)\frac{\partial^2 P_c}{\partial S_{w0}^2}\right]\left(\frac{\partial S_{w0}}{\partial x}\right)^2 dt$$

$$S_{w2}(x,t) = -\frac{d}{d\lambda}\int_0^t\left[\frac{kk_{ro}}{\mu_o\phi}\frac{\partial}{\partial(S_{w0}+S_{w1})}\left(\frac{1}{1+\frac{k_{ro}\mu_w}{k_{rw}\mu_o}}\right)\frac{\partial P_c}{\partial(S_{w0}+S_{w1})}\right.$$

$$+\frac{k}{\mu_o\phi}\left(\frac{1}{1+\frac{k_{ro}\mu_w}{k_{rw}\mu_o}}\right)\frac{\partial k_{ro}}{\partial(S_{w0}+S_{w1})}\frac{\partial P_c}{\partial(S_{w0}+S_{w1})}$$

$$\left.+\frac{kk_{ro}}{\mu_o\phi}\left(\frac{1}{1+\frac{k_{ro}\mu_w}{k_{rw}\mu_o}}\right)\frac{\partial^2 P_c}{\partial(S_{w0}+S_{w1})^2}\right]\left(\frac{\partial(S_{w0}+\lambda S_{w1})}{\partial x}\right)^2 dt$$

$$S_{w3}(x,t) = -\frac{1}{2!}\frac{d^2}{d\lambda^2}\int_0^t\left[\frac{kk_{ro}}{\mu_o\phi}\frac{\partial}{\partial(S_{w0}+S_{w1}+S_{w2})}\left(\frac{1}{1+\frac{k_{ro}\mu_w}{k_{rw}\mu_o}}\right)\right.$$

$$\times\frac{\partial P_c}{\partial(S_{w0}+S_{w1}+S_{w2})} + \frac{k}{\mu_o\phi}\left(\frac{1}{1+\frac{k_{ro}\mu_w}{k_{rw}\mu_o}}\right)$$

$$\times\frac{\partial k_{ro}}{\partial(S_{w0}+S_{w1}+S_{w2})}\frac{\partial P_c}{\partial(S_{w0}+S_{w1}+S_{w2})} + \frac{kk_{ro}}{\mu_o\phi}\left(\frac{1}{1+\frac{k_{ro}\mu_w}{k_{rw}\mu_o}}\right)$$

$$\left.\times\frac{\partial^2 P_c}{\partial(S_{w0}+S_{w1}+S_{w2})^2}\right]\left(\frac{\partial(S_{w0}+\lambda S_{w1}+\lambda^2 S_{w2})}{\partial x}\right)^2 dt \tag{3.35}$$

3.2.3 Results and Discussions

The method is applied to a given reservoir with the initial water saturation of 0.18. This value is corresponding to the irreducible water saturation of the reservoir. Water is injected into the reservoir with a linear flow rate of 1 ft/day. The oil and water viscosities are 1.73 cp and 0.52 cp, respectively. The flow of the displaced phase (oil) ceases at $S_{oc} = 0.1$. The porosity of the medium is 0.25 with an absolute permeability of $k = 10$ md.

The normalized water saturation and the water and oil relative permeability are obtained using

$$S_{wn} = \frac{S_w - 0.18}{1 - 0.1 - 0.18}, \quad k_{rw} = 0.59439\, S_{wn}^{4}, \quad k_{ro} = \left(1 - S_{wn}\right)^2 \quad (3.36)$$

A typical plot of variation of relative permeability to water, k_{rw}, relative permeability to oil, k_{ro}, fractional flow curve, f_w and its derivative, df_w/dS_w are shown in Fig. 3.15. The capillary pressure data are also known as shown in Table 3.1.

The solution for the first base element of the series solution of the water saturation, S_{w0}, is obtained through the solution of the first equation in (3.35). It can be written by integrating from the both sides of the first equation in (3.35) and using (3.33) that

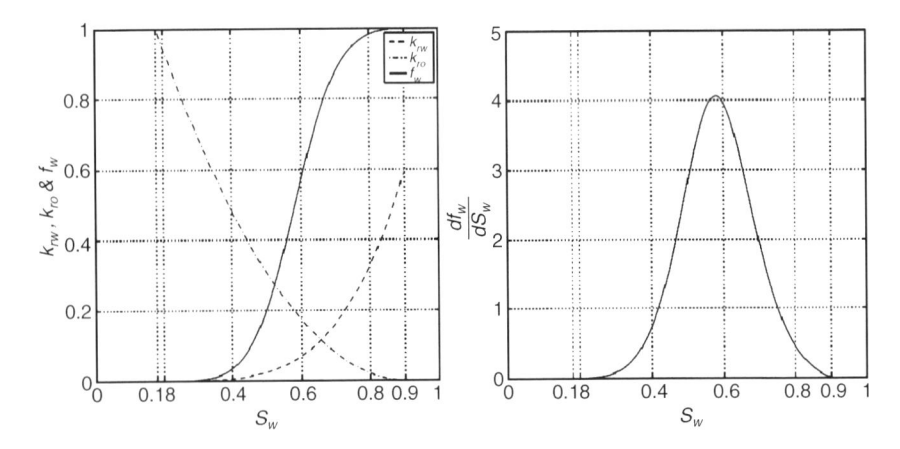

Figure 3.15 The variation of the water and oil relative permeability, the fractional water flow rate and the differentiation of fractional water flow rate with the water saturation.

Reservoir Simulator

Table 3.1 Capillary pressure data

S_{wn}	P_c [atm]	S_{wn}	P_c [atm]	S_{wn}	P_c [atm]	S_{wn}	P_c [atm]
0.00	3.9921	0.11	1.2036	0.30	0.3600	0.65	0.0550
0.01	3.5853	0.15	0.8745	0.36	0.2699	0.72	0.0350
0.02	3.1987	0.18	0.7010	0.42	0.1980	0.87	0.0100
0.05	2.2577	0.21	0.5709	0.48	0.1450	0.95	0.0027
0.08	1.6209	0.25	0.4592	0.56	0.0920	1.00	0.0000

$$\frac{\partial S_{w0}}{\partial t} + \frac{q}{A\phi} \frac{\partial}{\partial S_{w0}} \left(\frac{1}{1+M} \right) \frac{\partial S_{w0}}{\partial x} = 0. \tag{3.37}$$

This equation suggests that S_{w0} is constant along a direction that is called characteristic direction. The characteristic direction can be obtained by

$$\left(\frac{dx}{dt} \right)_{S_{w0}} = \frac{q}{A\phi} \frac{\partial}{\partial S_{w0}} \left(\frac{1}{1+M} \right)_t \tag{3.38}$$

that is the Buckley–Leverett frontal advance equation. Integrating respect to time, the distribution of S_{w0} is found in the form of

$$x(t, S_{w0}) = \frac{qt}{A\phi} \frac{\partial}{\partial S_{w0}} \left(\frac{1}{1+M} \right)_t. \tag{3.39}$$

The solution of Eq. 3.20 gives the variation of the S_{w0} along the reservoir at certain time. The distribution of S_{w0} along the reservoir is obtained based on the definition of the mobility ratio Eq. 3.4 and the relations for the relative permeability of water and oil in (3.35). It corresponds to the variation of the water saturation when the effect of the capillary pressure is ignored. The computations are carried out at different times of $t = 0.5, 1, 2, 5$ and 10 [days]. The solution for the water saturation without the capillary pressure effect shows the unrealistic physical situation that Buckley–Leverett mentioned in their pioneer paper with multiple-saturations at each distance (x-position) as given partly in Fig. 3.16. In order to avoid multiple saturation values at a particular distance, a saturation discontinuity at a distance, x_f is generally created in such a way that the areas ahead of the front and below the curve are equal to each other.

The other elements of the series solution for the water are obtained recursively using (3.35) and the solution for S_{w0}. The solution for S_{w1} is obtained using the solution of S_{w0} at a certain point for a given time. To induce the nonlinear dependence of capillary on saturation, during decomposition, the capillary pressure and its derivatives are approximated by using the cubic spline, as shown in Fig. 3.16. The interpolating splines are preferred over interpolating polynomials as they do not suffer from oscillations between the knots and are smoother and more realistic than linear splines (Ertekin et al, 2001). Such a technique is also used to observe the effects of linearization in pressure in reservoir flow equations Mustafiz et al (2006). The solutions for S_{w2} and S_{w3} are obtained recursively using (3.35). The computations shows that the series solution converges very fast and it is not necessary to go further than S_{w3}.

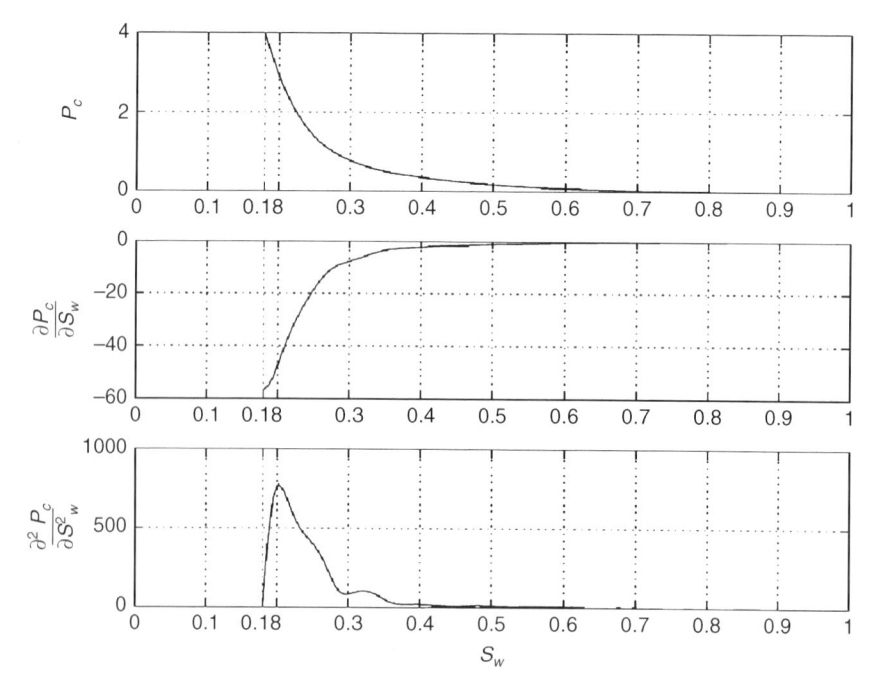

Figure 3.16 The capillary pressure variation and its first and second derivatives as a function of the water saturation.

Reservoir Simulator 113

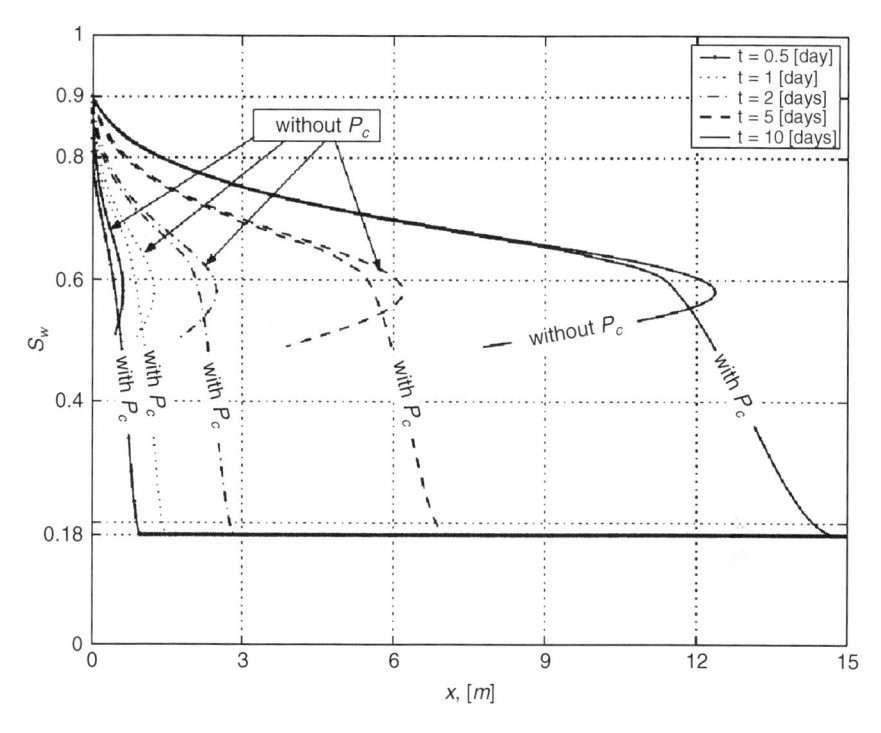

Figure 3.17 The water saturation distribution with and without the effect of capillary pressure using ADM.

The distribution of the water saturation along the reservoir is given in Fig. 3.17. The results of the water saturation without the effect of the capillary pressure are also depicted in Fig. 3.17 for the sake of comparison. This figure shows that by considering the effects of capillary pressure, it is possible to avoid unrealistic multiple saturation values. Moreover, the decomposition approach shows notable prediction of saturation profiles along the length. The gradual and mild changes observed in the saturation profile here, are perhaps less severe than the conventional prediction of shock-fronts.

3.3 Some Remarks on Multiple Solutions

Our objective is to discuss the possibility of the multiple solution for natural phenomena. In many cases, it is possible to express or approximate a natural phenomenon by a nonlinear polynomial.

This nonlinear polynomial has multiple solutions which depends on nonlinearity and number of variables. We explained about several bivariate polynomials of different degrees and show a population of solutions in the complex plane.

The other objective is to show the limitation of the methods in finding the complete solutions of a given nonlinear problem. The mathematical model for many real problems are given in the form of partial differential equations. In most of the cases, the solution of these PDEs are obtained by the numerical methods. The normal procedure is to recast the PDE in the form of nonlinear simultaneous algebraic equations (SAE). The solutions of this nonlinear SAE are obtained by some linearization during the process of computations. This may affect the quality as well as the quantity of the solutions.

We introduce several bivariate nonlinear SAEs with different degrees of nonlinearity. The real roots of these bivariate SAEs can be obtained graphically as shown in the previous sections. All of the SAEs are solved with Adomain decomposition method (ADM) and show the limitation of the ADM method, which is used as a sample method. However, other methods such as Newton's method may give all probable solutions but it is necessary to know the region of each solution to start a convergent iteration and find the roots. If the number of variables are increased, it will be a challenging task to find the region of all roots. Such investigation will enable us getting the complete picture of knowledge dimension, which may be useful in decision making.

To unravel the lure of finding a justifiable answer to the shock-front which is often less severe in the experiments than in the theoretical postulations, the Adomian decomposition method is investigated to solving the Buckley–Leverett equation. The solution by the ADM also assures that the nonlinear dependence of capillary pressure on saturation is maintained. The term, $S_{w0}(x, t)$, is prepared through the method of characteristics, which subsequently becomes useful for the purpose of evaluation of the other elements in water saturation series. The saturation at a point and at a given time is evaluated through recursion. The term $S_{w1}(x,t)$ is calculated using the solution of $S_{w0}(x,t)$ at any distance and time. The solutions for $S_{w2}(x,t)$ and $S_{w3}(x,t)$ are obtained using the results of $S_{w1}(x,t)$ and $S_{w2}(x,t)$, respectively. The computation is done up to the term $S_{w3}(x,t)$, as it is found that the series solution converges very quickly.

4
Mathematical Formulation of Reservoir Simulation Problems

From a mathematical point of view, a reservoir simulator consists of a set of partial differential equations and a set of algebraic equations, both with some appropriate initial and boundary conditions. The partial differential equations are obtained by considering the conversation of mass, momentum and energy and the algebraic equations describe the thermodynamic properties of the reservoir fluid and rock system. In isothermal cases, we deal with two basic equations: (1) the material balance or continuity equation; and (2) the equation of motion or momentum equation. These two equations provide the set of partial differential equations for the components of the reservoir fluids. It is a generally accepted assumption that the reservoir fluid is in a local and instant thermodynamic equilibrium and therefore, the number of mass conservation equations is identical to the number of fluid components.

When the hydrocarbon recovery takes place using heat injection in the form of steam injection, or *in situ* combustion, the temperature is not constant and an energy balance equation is required. In addition, chemical reaction such as vaporization, condensation, and cracking need to be accounted for. The simulator for chemical injections needs an additional equation which describes the concentration distribution of the chemical.

The reservoir is a three dimensional body and contains hydrocarbons that consists of different components and may be found in gaseous and/or liquid phases in conjunction with water. Hence, it is preferable to present the formulation in three dimensional (3D) with three phases (3P). However, in many cases a one or two dimensional (1D or 2D) and one or two phases (1P or 2P) modeling of the fluid flow is adequate to predict the dynamic behavior of a reservoir. The reservoir simulator may be categorized based on the number of components that are chosen to describe the displacement process. In many cases, a simplified approach of the hydrocarbon system by a pseudo-gas and a pseudo-oil component generally is accepted, and the thermodynamic properties of the given system depend only on the pressure. This method is called the *black-oil* approach and the number of components N_c (or the number of mass conservation equations) is identical to the number of phases N_p. Many enhanced oil recovery (EOR) processes are designed to take advantage of the phase behavior of multi-components fluid systems. In these cases, *compositional* models must be used to predict the dynamic behavior of the reservoir. This is the case for the reservoirs with highly volatile oil and in modeling depletion and/or cycling of retrograde gas-condensate reservoirs.

Darcy's law is used extensively in almost all existing reservoir simulators. The Darcy law is an empirical equation based on experimental observations of one-dimensional water flow through unconsolidated sands at low velocity. All the theoretical and experimental investigations indicate that Darcy's law is an approximation in describing the phenomenon of fluid flow in porous media, and is valid under a limited range of low velocities. With the growing demand to address specific reservoir flow behaviors related to particular production strategies, the need for modifying the currently used Darcy's model or developing new models to describe such circumstances became vital. In fractured reservoirs with a flow potential through matrix and fracture system, a satisfactory and meaningful solution cannot be achieved when Darcy's law is used. The investigation into the methodologies employed for solving this class of problems was motivated by numerous research papers and studies have been directed to this issue in an effort to generate suitable models that express fluid behavior in porous media within a wide spectrum of flow velocities considering both matrix and fracture contributions to fluid flow.

The general purpose compositional model in reservoir simulation is presented in the current chapter. The general case of multi-components fluid flowing in a three dimensional heterogeneous medium is taken into account and the basic mathematical formulations based on the conservation laws are presented. Some examples are presented and finally the black-oil model will be discussed.

4.1   Black Oil Model and Compositional Model

There are two different approaches to analyze the general case of three-dimensional three phase flow of fluid reservoir

Mathematical Formulation

through pored rock, the compositional and the black oil models. Generally, Crude oil contains some amount of dissolved gas and invariably occurs in conjunction with water. In many cases, it is acceptable to assume that the oil and gas compositions are fixed and the solubility of the gas in the oil depends on pressure only. And consequently, it is possible to consider a single oil "pseudo-component" and a single gas "pseudo-component." However, if oil and gas equilibrium compositions vary strongly as a function of space and time, a compositional formulation is needed that includes a larger number of components and appropriate equations of state.

It may be taken into account that there are n-hydrocarbon components present in the reservoir fluid in the form of oil or gas phases and a water component that can be found in the form of liquid (water) or vapor (gas) phase. Therefore, there may be n-component in oil phases and $(n + 1)$-component in gas phases and 1-component in water phase. When there is components transfer between phases, a fully compositional model should be used to analyze the reservoir fluid through the pored rock. Many EOR processes, including miscible gas injection, are specifically designed to take advantage of the phase behavior of multi-component fluid systems. Compositional modeling is also required in modeling depletion and/or cycling of retrograde reservoirs and reservoirs with highly volatile oils. In these cases, the phase compositions are away from the critical point, which simplifies the behavior of the fluid system.

The black-oil model is a simplified compositional model describing multiphase flow with mass interchange between phases in a porous medium. It consist three phases (gas, oil and water), can predict the compressibility and mass transfer effects, and can be used for a low-volatility system, consisting mainly of methane and heavy components, using data from a conventional differential vaporization test on reservoir oil samples. In this model it is assumed that no mass transfer between the water phase and the other two phases (gas and oil). In the hydrocarbon (gas-oil) system, only two components are considered: the oil component; and the gas component. The oil components (also called stock-tank oil) is the residual liquid at atmospheric pressure left after a differential vaporization, while the gas component is the remaining fluid in a porous medium.

4.2 General Purpose Compositional Model

A general case of a non-isothermal reservoir is taken into account. A mathematical description for the fluid flow and heat transferring in a permeable medium is obtained from:

- Conservation of mass;
- The conservation of energy;
- Conservation of momentum (motion equation) or the laws that describe the motion of fluid through the porous rock such as the Darcy law and Fick's laws;
- Equation of state and constitutive equations.

For formulating the conservation equations, it is unpractical to treat transport phenomena in porous media by referring only to the fluid continuum filling the void space, owing to our ignorance of the detailed geometry of the fluid-solid interfaces. The representative elementary volume (REV) is a fictitious continuum that the average values of variables and parameters over the REV are assigned to them as a mathematical point. The conservation equations are valid for any point in the fluid continuum filling the void space of a porous medium. However, we should formulate the conservation equations for REV or in other words the average value of the parameter over REV should be insert in those equations.

It is necessary to define some basic model parameters that are related to the fluid system and porous medium.

4.2.1 Basic Definitions

It is necessary to define some of the basic fluid parameters. It is assumed that the fluid is a continuum media. A fluid particle is defined as an ensemble of molecules included in a certain volume that is associated with REV of a fluid continuum. A petroleum reservoir fluid is generally a multi-components and multispecies (heterogeneous) fluid and it may be present in various phases. It is assumed that N_p phases are present in the reservoir. We label them with subscript j or refer to specific phases using the subscript g, o and w for gas, oil and water, respectively.

Components are designated by the subscript i. It is assumed that the reservoir fluid consists of N hydrocarbon components and a water component. The hydrocarbon component may be present in

Mathematical Formulation

oil and gas phase and the water component may be in the form of gas or water phase. Therefore, the fluid system consists of $N_C = N + 1$ components. The gas phase may have N_C components, the oil phase may consist of N components and the water phase has only one component. In other words, the water and oil are assumed to be immiscible and there is no mass transfer between these two phases: i.e., the hydrocarbon oil components do not dissolve in the water phase and the water component does not dissolve in the oil phase. Furthermore, the water phase does not transport any component other than the water component. Water component may evaporate into the gas phase.

An *extensive fluid property* depends on the mass of the fluid system such as volume, mass, momentum, kinetic energy and so on. With each extensive fluid property, we may associate an *intensive fluid property* that is defined as the amount of an extensive fluid property per unit mass of the fluid system. We may also introduce for each extensive property a density that is the amount of an extensive property per volume of the fluid system. If the amount of an extensive fluid property of the a-species in a multi-components fluid system is denoted by G_α, an *intensive fluid property* γ_α defined as the amount of entity G_a per unit mass of the fluid system and the density of G_a may be denoted by g_a that is the amount of entity G_a divided by the fluid system volume.

If we consider a species a of a fluid system composed of mixture of N chemical components, the mass of this componenets per unit volume of fluid (solution) may be denoted by $\rho_a = \frac{dm_a}{dU}$, where dm_a is the instantaneous mass of the a-species and dU is the volume of space occupied by the fluid system. The density of the fluid system is

$$\rho = \frac{dm}{dU} = \frac{\sum_{a=1}^{N} dm_a}{dU} = \sum_{a=1}^{N} \rho_a \qquad (4.1)$$

where dm is the total mass of the fluid system occupied by the volume dU. The mass fraction may be defined as the density of a-species to the total density of the fluid system.

$$y_a = \frac{\rho_a}{\rho}, \quad \text{and} \quad \sum_{a=1}^{N} y_a = 1 \qquad (4.2)$$

The velocity of a-species at a point P may be denoted by V_a and is the statistical average velocity within dU of the individual molecules of the a-species. Thus, it is a microscopic level parameter. The mass average velocity $\bar{V}_m$ is defined

$$\bar{V}_m = \frac{\sum_{\alpha=1}^{N}\rho_\alpha V_\alpha}{\rho} = \sum_{\alpha=1}^{N} y_\alpha V_\alpha \quad (4.3)$$

This mass average velocity may be interpreted as representation of momentum per unit mass of the flowing fluid. If the volume of the solution changes with concentration, the volume-average velocity $\bar{V}_u$ is used. It is defined as

$$\bar{V}_u = \sum_{\alpha=1}^{N} u_\alpha V_\alpha \quad (4.4)$$

where $u_\alpha = dU_\alpha/dU$ is the volumetric fraction of a-species. If the fluid is homogeneous it can be written $\bar{V}_\alpha = \bar{V}_m = \bar{V}_u$. For a non-homogeneous fluid, these velocities differ both in direction and in magnitude.

The reservoir fluid is multi-component and may be found in various phases. The density of the fluid phases may vary as a result of variations in the concentration of the various species comprising it. The phase density is defined as the mass of the phase j for a unit volume of the phase j.

$$\rho_j(p, m_i, T) = \frac{m_j}{U_j} \quad (4.5)$$

Where m_j and U_j are the mass and the volume of phase j, respectively. The density of each phase depends on the pressure (p), the phase components (m_i, $i = 1, \cdots, N$) and the temperature (T). The mass and volume of the phase j depend on the local thermodynamic equilibrium and is function of pressure, component masses and the temperature at a certain point, $U_j = U_j(p, m_i, T)$, $m_j = m_j(p, m_i, T)$.

4.2.2 Primary and Secondary Parameters and Model Variables

We assume that there is a local thermodynamic equilibrium. It is a generally accepted assumption in reservoir simulation. As a consequence, the continuum model can be visualized as a large number of differentially small subsystems. All the thermodynamic properties are defined for each subsystem at each instant of time, and each

Mathematical Formulation

subsystem is treated as a system in thermodynamic equilibrium, (Acs and Doleschall, 1985).

The degree of the freedom of a system in thermodynamic equilibrium according to Gibbs phase rule is $F = N_C + 2 - N_p$. The phase saturations must also be found and consequently, the unknown number of parameters will be $N_C + 1$. The independent parameters that determine all the others are *primary parameters*. It should be considered that $N_C + 1$ are independent intensive parameters that give no information about the system as a whole and an additional parameter is needed.

For the case of isothermal reservoir, the pressure p of the phase equilibrium and the mass of components m_i, $i = 1, ..., N_C$ are the most proper parameters to choose as primary parameters and these make the model more general. (The mole number instead of mass can be regarded as primary parameters but some adjustment is needed to compare with the black oil model.) For the non-isothermal cases, temperature T can take as primary parameters and find out the effect of them. By finding these primary variables and using algebraic equations such as empirical fluid property correlations or an equation of state, the equilibrium and *PVT* properties of the system can be determine uniquely.

The *secondary parameters* that depend on the primary variables may be designated as:

- $$y_{ij}(p, m_i, T) = \frac{m_{i,j}}{m_j}, \quad i = 1, \cdots, N_C \tag{4.6}$$

  Mass fraction of component i in phase j or mass concentration; ($m_{i,j}$ is the mass of components i in phase j and m_j is the mass of phase j.)

- $$\rho_j(p, m_i, T) = \frac{m_j}{U_j}, \quad i = 1, \cdots, N_C \tag{4.7}$$

  Phase density;

- $$U_j = U_j(p, m_i, T), \quad i = 1, \cdots, N_c \tag{4.8}$$

  The volume of phase j;

- $$U_f(p, m_i, T) = \sum_j U_j(p, m_i, T), \quad i = 1, \cdots, N_C \tag{4.9}$$

  Total reservoir fluid volume;

- $S_j(p,m_i,T) = U_j(p,m_i,T)/U_f(p,m_i,T)$, $i = 1,\cdots,N_c$ (4.10)
  Saturations; $(\sum_j S_j = 1$ and $j = 1,\cdots,N_p)$

- $k_j = k_j(p,m_i,T)$, $i = 1,\cdots,N_C$
  Relative permeability of phase j;

- $\mu_j = \mu_j(p,m_i,T)$, $i = 1,\cdots,N_C$
  Viscosity of phase j;

- $D_{ij} = D_{ij}(p,m_i,T)$, $i = 1,\cdots,N_C$
  Dispersion coefficient of component i in phase j; and

- $P_c(p,m_i,T) = p_j(p,m_i,T) - p(p,m_i,T)$ $i = 1,\cdots,N_c$ (4.11)
  Capillary pressure.

The variables y_{ij}, ρ_j, S_j are independent to the amount of the fluid and are intensive properties. However, the volume of phases and the total volume of the reservoir fluid are depend on the amount of the fluid mass and are extensive properties. The mass of the component i is also an extensive property and the intensive overall densities according to Acs, G. (1985) is defined as

$$r_i = \frac{m_i}{U_f(p,m_i,T)} = \sum_j \rho_j S_j y_{ij} \text{ for } j = 1,\cdots,N_p \quad (4.12)$$

The independent parameters p,T and the component masses $m_{ij} = 1, \ldots, N_C$ must be found to describe all the fluid properties. These are $N_C + 2$ independent variables that should be determined by applying the mass balance and energy balance equation plus an equation that describe the fluid motion inside the porous media.

The fluid mass balance equation is a system of differential equations that consists of N differential equations for N_C components of the reservoir fluid. There is also an energy balance equation. These N_C+1 equations are insufficient to determine $N + 2$ variables. The other equation is the volume balance equation. This equation indicate that the fluid reservoir is occupied the pore space of the rock volume at every point of time.

$$U_f(p,m_i,T) = U_p = \phi U_r \quad (4.13)$$

Where, U_p is the pore volume of the rock and U_r is the rock volume.

4.2.3 Mass Conservation Equation

The laws of conservation of mass for an arbitrary volume for any components of the reservoir fluid indicate that:

Flux of a component i through the boundary of an arbitrary volume + production of the component i from a source = Accumulation of thecomponent i in the specified volume.

This equation should be true for each component in the reservoir fluids. Most reservoirs contain water, oil and gas phases, each displaying limited solubility or miscibility in the others. These phases may contain some hydrocarbon and non hydrocarbon components. For example, the gas phase may contain methane, carbon dioxide, water vapor, hydrogen sulfide, propane, and some other hydrocarbon and non-hydrocarbon components. All components exist in other phases to some degree. In other words, the total amount of any components in a small reservoir volume are in the form of gas, oil and water phases and may transfer through the volume boundaries to the adjacent reservoir volumes or to the wells.

A small rectangular volume element of lengths dx, dy and dz along a Cartesian coordinate system is taken into account. The flux of mass through the boundaries is considered to occur by both convection and dispersion. The net mass flux of component i in the x-direction, as illustrated in Fig. 4.1, is:

$$\frac{\partial J_i}{\partial x} = \frac{\partial}{\partial x}(\rho_g u_g y_{i,g} + \rho_o u_o y_{i,o} + \rho_w u_w y_{i,w}) \\ - \frac{\partial}{\partial x}\left(\rho_g S_g D_{k,x}^{i,g} \frac{\partial y_{i,g}}{\partial x} + \rho_o S_o D_{k,x}^{i,o} \frac{\partial y_{i,o}}{\partial x} + \rho_w S_w D_{k,x}^{i,w} \frac{\partial y_{i,w}}{\partial x}\right)$$

(4.14)

Where:

ρ_g, ρ_o and ρ_w are the density of gas, oil and water phase, respectively;

$y_{i,g}$, $y_{i,o}$ and $y_{i,w}$ are the mass concentrations or the mass fraction of component i in gas, oil and water phase, respectively;

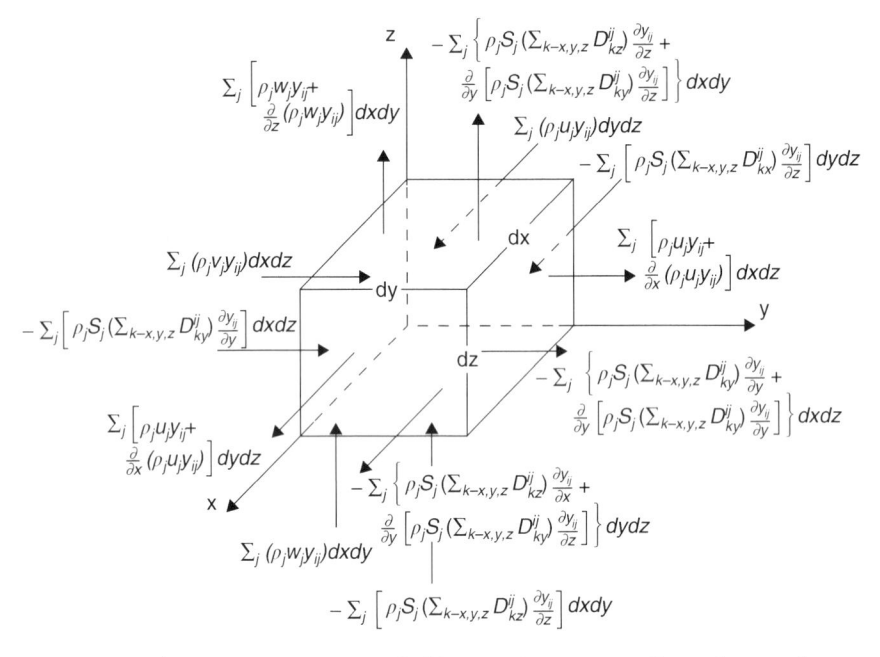

Figure 4.1 The convective and dispersive mass flux through an elementary volume.

- u_g, u_o and u_w are the gas, oil and water phase velocity in the x-direction, respectively;
- S_g, S_o and S_w are the gas, oil and water phase saturation, respectively; and $D_{k,x}^{i,g}$, $D_{k,x}^{i,o}$ and $D_{k,x}^{i,w}$ are the total coefficient of dispersivity for component i in phase gas, oil and water, respectively, in the x-direction.

The dispersion flux results from the velocity fluctuations through the representative elementary volume. It is different from the molecular diffusive flux that is neglected in formulation of the total mass flux. The dispersion coefficient according to Bear (1972) for an isotropic medium in Cartesian coordinate system is given by:

$$D_{k,l}^{i,j} = a_T |V_j| \delta_{k,l} + (a_L - a_T) V_{jk} V_{jl} / |V_j| \qquad (4.15)$$

Where:
- a_L is the dispersivity of the medium in the direction of the j-phase mean flow;
- a_T is the dispersivity of the medium in any axis perpendicular to the direction of the j-phase mean flow;

MATHEMATICAL FORMULATION

$|V_j|$ is the mean velocity of the phase j, in Cartesian coordinate system $|V_j| = \sqrt{V_{jx}^2 + V_{jy}^2 + V_{jz}^2}$ in which $u_j \equiv V_{jx}, v_j \equiv V_{jy}$ and $w_j \equiv V_{jz}$; and V_{jk} and V_{jl} are the velocity components of the mean velocity of the phase j in k- and l-direction, respectively.

As an example, the dispersive coefficient in the gas phase for the component i in the x-direction are:

$$D_{x,x}^{i,g} = a_T |V_g| + (a_L - a_T) V_{gx}^2 / |V_g| = a_T |V_g| + (a_L - a_T) u_g^2 / |V_g|$$
$$D_{y,x}^{i,g} = (a_L - a_T) V_{gx} V_{gy} / |V_g| = (a_L - a_T) u_g v_g / |V_g| \quad (4.16)$$
$$D_{z,x}^{i,g} = (a_L - a_T) V_{gx} V_{gz} / |V_g| = (a_L - a_T) u_g w_g / |V_g|.$$

And the total dispersion coefficient for component i in gas phase in the x-direction is:

$$D_{k,x}^{i,g} = D_{x,x}^{i,g} + D_{y,x}^{i,g} + D_{z,x}^{i,g}. \quad (4.17)$$

The net mass flux of the component i in the y- and z-direction, as illustrated in Fig. 4.1, can be given as:

$$\frac{\partial J}{\partial y} = \frac{\partial}{\partial y}(\rho_g v_g y_{i,g} + \rho_o v_o y_{i,o} + \rho_w v_w y_{i,w})$$
$$- \frac{\partial}{\partial y}\left(\rho_g S_g D_{k,y}^{i,g} \frac{\partial y_{i,g}}{\partial y} + \rho_o S_o D_{k,y}^{i,o} \frac{\partial y_{i,o}}{\partial y} + \rho_w S_w D_{k,y}^{i,w} \frac{\partial y_{i,w}}{\partial y}\right)$$

(4.18)

$$\frac{\partial J}{\partial z} = \frac{\partial}{\partial z}(\rho_g w_g y_{i,g} + \rho_o w_o y_{i,o} + \rho_w w_w y_{i,w})$$
$$- \frac{\partial}{\partial z}\left(\rho_g S_g D_{k,z}^{i,g} \frac{\partial y_{i,g}}{\partial z} + \rho_o S_o D_{k,z}^{i,o} \frac{\partial y_{i,o}}{\partial z} + \rho_w S_w D_{k,z}^{i,w} \frac{\partial y_{i,w}}{\partial z}\right)$$

(4.19)

where the coefficients and notations are the same as defined for the mass flux of component i in the x-direction.

The mass accumulation of the component *i* per pored space of the parallelepiped element is:

$$\frac{\partial M_i}{\partial t} = \frac{\partial}{\partial t}\left[\phi(\rho_g y_{i,g} S_g + \rho_o y_{i,o} S_o + \rho_w y_{i,w} S_w)\right] \quad (4.20)$$

The production rate of the component *i* may be given Q_i and therefore, the mass balance equation for the component *i* is:

$$\frac{\partial M_i}{\partial t} + \nabla \cdot J_i + q_i = 0 \quad (4.21)$$

or in expanded form

$$\frac{\partial}{\partial t}\left(\phi \sum_j \rho_j y_{i,j} S_j\right) + \nabla \cdot \sum_j \left(\rho_j y_{i,j} V_j - \rho_j S_j \underline{D}_{i,j} \nabla y_{i,j}\right) + q_i = 0, \quad i = 1, \cdots, n_c$$

(4.22)

where the subscript *j* is the fluid component phase and $\underline{D}_{i,j}$ is a second rank tensor relating to flux vector to $\nabla_{y_{i,j}}$. The notation q_i is the production rate per unit volume. The dispersive flux results from the velocity fluctuation through the REV.

4.2.4 Energy Balance Equation

The first law of thermodynamics states that energy is conserved and neither created nor destroyed and it is only transferred and converted to other types of energy. The laws of conservation of energy for an arbitrary volume of the reservoir fluid indicate that:

Flux of energy through the boundary of an arbitrary volume + Energy input from a source = Gain in internal energy.

The arbitrary volume may be considered as an infinitesimal rectangular parallelepiped of lengths *dx*, *dy* and *dz* along *x*-, *y*- and *z*- axis in a Cartesian coordinate system. The total energy, transferring through the representative elementary volume (REV) boundaries, consists of:

Mathematical Formulation

1. a part that is transferred through the mass transfer, convective term;
2. a part due to the heat transferred by conduction and radiation; and
3. the shear work, due to viscous stresses, occurs at the boundaries of the REV.

A) Shear work: The *shear work* on the surfaces of representative elementary volume which are solid confining walls is zero. The flow is approximately normal to the element dA at an inlet or outlet and hence, the only viscous-work term comes from the normal viscous stress $\tau_{nn} V_n \, dA$. At inlets and outlets, viscous normal stresses are extremely small and the viscous work volume is negligible.

B) Heat transfer: We neglect radiation and consider only heat conduction through the sides of the element. The total heat transfer may divide into two parts.

1. The heat conduction through the reservoir fluid is:

$$Q_f = -k_f \nabla T_f \qquad (4.23)$$

where k_f is the coefficient of thermal conductivity of the fluid and T_f is the reservoir fluid temperature.

2. The heat conduction through the rock matrix that may be given in the form:

$$Q_r = -k_r \nabla T_r \qquad (4.24)$$

where k_r is the coefficient of thermal conductivity of the rock and T_r is the reservoir rock matrix temperature.

The heat transfer in reservoir fluid due to the thermal conductivity in the x-, y- and z-directions are, see Fig. 4.2:

Net heat transfer in the x-direction
$$= -k_f \frac{\partial T_f}{\partial x} - \left[-k_f \frac{\partial T_f}{\partial x} + \frac{\partial}{\partial x}\left(-k_f \frac{\partial T_f}{\partial x} \right) dx \right] dy dz$$

$$= \frac{\partial}{\partial x}\left(k_f \frac{\partial T_f}{\partial x} \right) dx \, dy \, dz$$

$$(4.25)$$

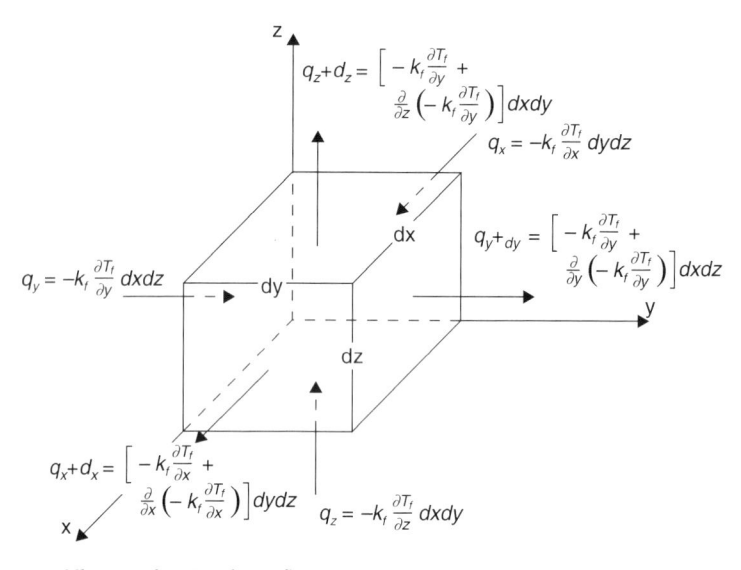

Figure 4.2 The conductive heat flux through an elementary volume.

$$\begin{aligned}\text{Net heat transfer} \\ \text{in the y-direction}\end{aligned} \quad \begin{aligned}&= -k_f \frac{\partial T_f}{\partial y} - \left[-k_f \frac{\partial T_f}{\partial y} + \frac{\partial}{\partial y}\left(-k_f \frac{\partial T_f}{\partial y}\right) dy\right] dx\, dz \\ &= \frac{\partial}{\partial y}\left(k_f \frac{\partial T_f}{\partial y}\right) dx\, dy\, dz \end{aligned} \quad (4.26)$$

$$\begin{aligned}\text{Net heat transfer} \\ \text{in the z-direction}\end{aligned} \quad \begin{aligned}&= -k_f \frac{\partial T_f}{\partial z} - \left[-k_f \frac{\partial T_f}{\partial z} + \frac{\partial}{\partial z}\left(-k_f \frac{\partial T_f}{\partial z}\right) dz\right] dx\, dy \\ &= \frac{\partial}{\partial z}\left(k_f \frac{\partial T_f}{\partial z}\right) dx\, dy\, dz \end{aligned} \quad (4.27)$$

The total heat flux due to thermal conductivity in the fluid reservoir may be given using the vector notation in the form;

$$\begin{aligned}\text{The net heat flux} &= \left[\frac{\partial}{\partial x}\left(k_f \frac{\partial T_f}{\partial x}\right) + \frac{\partial}{\partial y}\left(k_f \frac{\partial T_f}{\partial y}\right) + \frac{\partial}{\partial y}\left(k_f \frac{\partial T_f}{\partial z}\right) + \right] dx\, dy\, dz \\ &= \nabla \cdot (k_f \nabla T_f)\, dx\, dy\, dz. \end{aligned}$$

$$(4.28)$$

Mathematical Formulation

The net heat flux is proportional to the elementary volume. If the heat transfer for a unit volume of the element for the reservoir fluid is denoted by $\dot{q}_f$, it may be written that:

$$\dot{q}_f = \nabla \cdot (k_f \nabla T_f) \qquad (4.29)$$

The heat thermal conductivity through the reservoir rock matrix can be obtained in the same way as given for the reservoir fluid. The heat transfer for a unit volume for the rock matrix may be given in the form:

$$\dot{q}_r = \nabla \cdot (k_r \nabla T_r) \qquad (4.30)$$

where $\dot{q}_r$ is the heat thermal conductivity for a unit volume of the rock matrix. The total thermal conductivity for the REV can be given in the same form as:

$$\dot{q} = \nabla \cdot (k \nabla T) \qquad (4.31)$$

where k is the thermal conductivity of the representative elementary element (REV) and T is an average temperature of the reservoir fluid and the rock matrix. In most cases the rock and fluid are in the same temperature and there is no heat transfer between them. However in the case that there is some temperature difference between the rock and fluid, there some heat convection between the reservoir fluid and the rock matrix that can be stated by $h(T_s - T_f)$ where h is the heat convection coefficient.

C) Convective term in energy transferring: The flux of energy through the movement of the reservoir fluid is a vector quantity. It is shown the net energy transfer through the boundaries of the elementary volume with the reservoir fluid that cross the boundaries. The component of convective term of the energy flux in x-direction shows the difference of the energy of the reservoir fluid enter into the elementary volume minus the energy of the reservoir fluid exit the elementary volume boundaries along the x-axis. The energy of a unit volume of the total fluid system is:

$$\text{Total energy of a unit volume of the fluid system}, e_f = \sum_j \rho_j S_j (h_j + g \cdot R_j) \qquad (4.32)$$

where $j = 1, ..., n_p$ and n_p is the phase number and $R_j = x_j\hat{i} + y_j\hat{j} + z_j\hat{k}$. The notation h_j is the enthalpy of the phase j of the fluid system

$$h_j = \hat{u}_j + \frac{p_j}{\rho_j} \qquad (4.33)$$

where $\hat{u}_j$ is the internal energy of the fluid system. The enthalpy and internal energy of the fluid phase j may be given using the heat capacity at constant pressure C_p and the heat capacity at constant volume C_v.

$$h_j = C_p(T_f - T_0), \quad \hat{u}_j = C_v(T_f - T_0) \qquad (4.34)$$

where T_0 is a reference temperature. The notation z_j is the elevation respect to a reference point. It should be indicated that the kinetic energy of the fluid system is neglected here since it has a very small contribution to the energy balance of a reservoir.

The net energy flux by fluid system is the difference of the energy carrying by the fluid system into to elementary volume and the energy of the reservoir fluid exit from the boundaries. The convective term of the flux of energy that is denoted by E_{MF} here. It may be given as:

$$E_{MF} = \phi \left\{ \frac{\partial}{\partial x} \left| \sum_j \rho_j S_j (h_j + g \cdot R_j) u_j \right| + \frac{\partial}{\partial y} \left| \sum_j \rho_j S_j (h_j + g \cdot R_j) v_j \right| \right.$$
$$\left. + \frac{\partial}{\partial z} \left| \sum_j \rho_j S_j (h_j + g \cdot R_j) w_j \right| \right\} dx\, dy\, dz. \qquad (4.35)$$

It can be simplified by using the vector operators.

$$E_{MF} = \phi \left\{ \nabla \cdot \left[v \sum_j \rho_j S_j (h_j + g \cdot R_j) \right] \right\} dx\, dy\, dz$$
$$= \phi [\nabla \cdot (e_f V)] dx\, dy\, dz \qquad (4.36)$$

The gain of internal energy of an elementary volume is the local change of energy of the elementary volume and it is the variation of the energy of the whole system consisting of fluid and rock in

Mathematical Formulation

the elementary time period of ∂t. The internal energy per unit bulk volume is denoted by I_E. It is equal to:

$$I_E = \phi \sum_j \rho_j S_j (\hat{u}_j + g \cdot R_j) + (1-\phi)\rho_r C_r (T_r - T_0) \qquad (4.37)$$

The local change of internal energy of the whole system is the differential of I_E with respect to time.

$$\frac{\partial I_E}{\partial t} = \frac{\partial}{\partial t}\left[\phi \sum_j \rho_j S_j (\hat{u}_j + g \cdot R_j)\right] + \frac{\partial}{\partial t}\left[(1-\phi)\rho_r C_r (T_r - T_0)\right] \qquad (4.38)$$

The notations ρ_r and C_r are the density and the heat capacity of the rock matrix, respectively.

In many cases, the rock and the reservoir fluid temperature can be assumed to be identical, i.e. $T_f = T_r = T$. For such a case, the energy balance equation can be given in a single equation by using the average value for the rock and fluid properties. Using equations (4.32), (4.36) and (4.38), the energy balance equation may be given in the form:

$$\frac{\partial}{\partial t}\left[\phi \sum_j \rho_j S_j (\hat{u}_j + g \cdot R_j)\right] + \frac{\partial}{\partial t}\left[(1-\phi)\rho_r C_r (T - T_0)\right] - \nabla \cdot (k \nabla T)$$

$$- \phi \left\{ \nabla \cdot \left[\sum_j \rho_j S_j (h_j + g \cdot R_j) V_j \right] \right\} + e_s = 0 \qquad (4.39)$$

where e_s is the energy input from sources per unit volume.

In some cases, it is not possible to assume that the difference between the rock and reservoir fluid temperature is negligible. It is necessary to explicitly describe the heat transfer between the rock and fluid system. It is describe by $h(T_r - T_f)$ where h is the heat convection coefficient between the rock and the fluid. The energy balance equation defined separately for the rock and fluid reservoir and two sets of equations are employed to describe the phenomenon.

$$\frac{\partial}{\partial t}\left[\phi \sum_j \rho_j S_j (\hat{u}_j + g \cdot R_j)\right] - \nabla \cdot (k_f \nabla T_f) - \phi [\nabla \cdot (e_f V)] + h(T_r - T_f) + e_s = 0$$

For the reservoir fluid (4.40)

$$\frac{\partial}{\partial t}\left[(1-\phi)\rho_r C_r (T_r - T_0)\right] - \nabla \cdot (k_r \nabla T_r) - h(T_r - T_f) = 0$$

For the rock matrix (4.41)

4.2.5 Volume Balance Equation

So far, we have N_C mass balance equations and one or two sets of equations from the energy balance equation in the fluid reservoir and rock matrix. The unknowns are the pressure distribution p, the fluid system velocities (V_j for $j = 1, \ldots, N_C$), N_C component masses m_i and one or two temperature distributions for the reservoir fluid and rock. However, the velocity of the fluid system can be related to the pressure distribution using some kind of motion equation that describe the momentum conservation for a REV in a porous media. The most common formulation is the Darcy law. It provides a linear relation between the pressure distribution and the pressure variation along the reservoir. These relations will be discussed in detail in section 5.3.

If we do not consider the fluid reservoir velocities, the number of unknowns are ($N_C + 2$) in the case where $T_r = T_f$ or ($N_C + 3$) in the case where $T_r \neq T_f$, while the number of equations are ($N_C + 1$) or ($N_C + 2$) in the cases where $T_r = T_r$ and $T_r \neq T_f$, respectively. In other words, the number of equations is insufficient by one. The volume conservation equation as given in equation (4.13) is the extra equation that can be used to resolve the problem. This equation relates the reservoir fluid volume and its variation to the pore volume. This equation may be given in the

$$U_f(p, m_i, T) = U_p = \phi U_r. \tag{4.42}$$

The variation of them with respect to time should also be identical.

$$\frac{dU_f(p, m_i, T)}{dt} = \frac{d(\phi U_r)}{dt} \tag{4.43}$$

Taking into account that $m_i = \phi U_r r_i$ from the Eq. (4.12), it may be written that

$$\frac{dU_f(p, m_i, T)}{dt} = \frac{\partial U_f}{\partial p}\frac{\partial p}{\partial t} + U_r \frac{\partial U_f}{\partial m_1}\frac{\partial (\phi r_1)}{\partial t} + \cdots + U_r \frac{\partial U_f}{\partial m_{N_C}}\frac{\partial (\phi r_{N_C})}{\partial t} + \frac{\partial U_f}{\partial T}\frac{\partial T}{\partial t}.$$

(4.44)

This equation may be written in a more condensed form as:

$$\frac{dU_f(p, m_i, T)}{dt} = \frac{\partial U_f}{\partial p}\frac{\partial p}{\partial t} + U_r \sum_{i=1}^{N_c} \frac{\partial U_f}{\partial m_i}\frac{\partial(\phi r_i)}{\partial t} + \frac{\partial U_f}{\partial T}\frac{\partial T}{\partial t}. \quad (4.45)$$

It is assumed that the porosity is a function of pressure only and the variation of temperature has a negligible effect on the porosity.

$$\frac{d(\phi U_r)}{dt} = U_r \frac{\partial \phi}{\partial p}\frac{\partial p}{\partial t} \quad (4.46)$$

The ratio of $\frac{\partial U_f}{\partial m_i}$ is call the partial specific volume and designated by γ_i.

$$\gamma_i = \frac{\partial U_f}{\partial m_i} \quad (4.47)$$

The average fluid and rock compressibility and the average thermal expansion coefficient of the fluid system may be defined as:

$$\hat{c}_f = -\frac{1}{U_f}\frac{\partial U_f}{\partial p}, \quad \hat{c}_r = -\frac{1}{U_r}\frac{\partial U_r}{\partial p} = \frac{\partial \phi}{\partial p}, \quad \hat{c}_T = -\frac{1}{U_f}\frac{\partial U_f}{\partial T}. \quad (4.48)$$

Using (4.47) and (4.48) and substituting (4.45) and (4.46) in (4.43), the volume conservation equation will be:

$$(\phi \hat{c}_f + \hat{c}_r)\frac{\partial p}{\partial t} - \hat{c}_T \frac{\partial T}{\partial t} = \sum_{i=1}^{N_c} \gamma_i \frac{\partial(\phi r_i)}{\partial t} \quad (4.49)$$

This equation may be rewritten in the following form by using (4.21).

$$(\phi \hat{c}_f + \hat{c}_r)\frac{\partial p}{\partial t} - \hat{c}_T \frac{\partial T}{\partial t} + \sum_{i=1}^{N_c} \gamma_i (\nabla \cdot J_i + q_i) = 0 \quad (4.50)$$

4.2.6 The Motion Equation in Porous Medium

The equation of motion of a fluid is the momentum conservation equation of the fluid and actually is equivalent to Newton's second law of motion of continuous media. The equation of motion states the relation among the forces act on an arbitrary volume of the fluid. It may also interpret as an expression for the variation of pressure due to the motion of fluid system. The pressure variation for a unit volume of fluid may be given as:

$$\nabla p = \rho \Omega - \rho \frac{dV}{dt} + \nabla \cdot \tau_{ij} \tag{4.51}$$

On the right hand side, the term $\rho\Omega$ is the body forces per unit volume. We neglect all the body forces except the gravitational body force and therefore, it may be written equal to ρg. The notation g is the gravitation acceleration vector.

The second term on the right hand side is the product of density by the acceleration. The term $\frac{dV}{dt}$ is the acceleration of a fluid element. It is equal to:

$$\frac{dV}{dt} = \frac{\partial V}{\partial t} + (V \cdot \nabla)V \tag{4.52}$$

which consists of the local acceleration ($\frac{\partial V}{\partial t}$) and the convective acceleration $(V \cdot \nabla)V$. It is assumed that the convective term is small compare with the other forces and it is neglected.

$$(V \cdot \nabla)V \approx 0 \tag{4.53}$$

The last term on the right hand side, $\nabla \cdot \tau_{ij}$ is the viscosity force per unit volume of the fluid where τ_{ij} is the stress tensor. The viscous force that resists the motion of fluid system through porous medium is assumed to be proportional to the fluid velocity over:

$$F_v = -\frac{\mu}{B}V \tag{4.54}$$

Where, B is the conductance. Using (4.52) and (4.53) and neglect all the body forces except the gravitational force, the average equation of motion over a REV for an inhomogeneous fluid in laminar flow through a porous medium is:

$$V_j + \frac{B\rho}{\mu}\frac{\partial V_j}{\partial t} = -\frac{K_j}{\mu_j}(\nabla p - \rho_j g) \tag{4.55}$$

The notation V_j is the mass average velocity of phase j over a REV of porous medium. It may be replaced by the volumetric average velocity if the molecular diffusion mass flux is assumed to be much smaller than the convective mass flux.

The second term on the left hand side of (4.54) is related to the local acceleration of fluid particle. For flow that the local acceleration is small and the local inertia force can be neglected with respect to the pressure, gravitational and viscous forces, the motion equation can be written in the form:

$$V_j = -\frac{\underline{K_j}}{\mu_j}(\nabla p_j - \rho_j g) \qquad (4.56)$$

It is the general form of Darcy's law that can be applied for multiphase and three dimensional flow inside an anisotropic and heterogeneous porous medium. The notation V_j is the average volumetric velocity of phase j and $\underline{K_j}$ is the permeability tensor for the phase j. It may be expressed as multiplication of the absolute permeability tensor $\underline{K}$ by the the relative permeability of the phase j.

$$\underline{K_j} = \underline{K}\, k_{rj} \qquad (4.57)$$

The fluid potential is defined as $\Phi = p + \rho g \cdot R$. Using this definition, the Darcy law for the volumetric velocity of phase j may be given in the form:

$$V_j = -\frac{\underline{K_j}}{\mu_j}\left[\nabla p_j + \rho_j \nabla(g \cdot R_j)\right] = -\frac{\underline{K_j}}{\mu_j}\nabla \Phi_j \qquad (4.58)$$

The Darcy law is an experimental relation that is originally observed in a homogeneous medium for one dimensional homogeneous fluid. It is good for a laminar flow through a porous medium and therefore, the question is its range of validity. The Reynolds number is a dimensionless number expressing the ratio of inertia to the viscous forces and is applied to distinguish the laminar flow from the turbulent flow. It is normally defined as:

$$Re = \frac{Vd}{v} \qquad (4.59)$$

where V is the average volumetric velocity, d is a length dimension and v is the fluid kinematic viscosity. It is customary to use some representative dimension of the grains for the length dimesion d.

Collins (1961) suggests $d = (k/\phi)^{0.5}$ where k is the permeability and ϕ is the porosity. Wards applied $d = k^{0.5}$ for the representative grains dimension in definition of Reynolds number. *Darcy's law can be applied as long as the Reynolds number based on average grain diameter does not exceed some value between 1 and 10, (Re < 1 ÷ 10).* We call this Reynolds number as the *critical Reynolds number* and is denoted by Re_{cr}.

As Reynolds number increases beyond the critical Reynolds number, the relation between the volumetric velocity and the gradient of fluid potential is not linear any more as given by Darcy's law. This is represented in the schematic diagram in Fig. 4.3. It is assumed that the media is homogeneous and isotropic with a constant permeability k in this figure. The deviation from Darcy's law is clearly illustrated in the figure.

There are some nonlinear models that suggest a nonlinear relation between the fluid velocity and the fluid potential. The general form of this nonlinear model for one dimensional flow is:

$$-\frac{\partial p}{\partial x} + \rho g_x = Au + Bu^2 \qquad (4.60)$$

The additional quadratic term in the bulk velocity with the coefficient depending on the geometry of the pores was first suggested by Forchheimer (1931) to modify Darcy's equation. The usual form

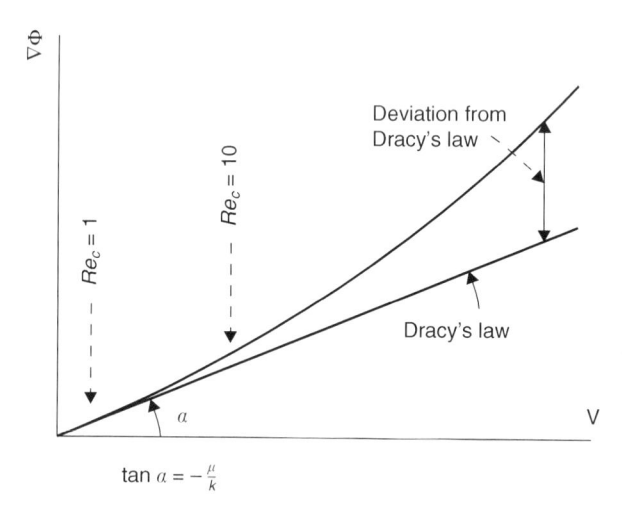

Figure 4.3 A schematic representation of Darcy and non-Darcy flow.

of *Forchheimer model* for one dimensional flow inside a porous medium in petroleum discipline is:

$$-\frac{\partial p}{\partial x} + \rho g_x = \frac{\mu}{k}u + \rho \beta u^2 \qquad (4.61)$$

where the coefficient β is a parameter in the Forchheimer equation for quantifying the non-Darcy flow effect. It may called an inertia resistance factor or non-Darcy flow coefficient.

Many researchers have attempted to clarify the physical meaning of the nonlinearity in the Forchheimer equation by recovering the Forchheimer equation from the Navier-Stokes equation using different techniques such as volume averaging, variation principles method, hybrid mixture theory, and two-scale homogenization. The deviation from Darcy's law in the Forchheimer equation is attributed to different phenomena such as the onset of flow separation (Irmay, 1958) the microscopic viscous force (Hassanizadeh and Gray, 1987), the microscopic inertia force that manifests itself in the interfacial drag force (Ruth and Ma, 1992). It was called a paradox (Nield, 2000) that the Forchheimer term, apparently independent of the viscosity, also represents viscous dissipation. The paradox can be resolved after recognizing that the Forchheimer drag term models essentially a form drag (or pressure drag) and involves the separation of boundary layers and wake formation behind solid obstacles (Joseph, 1982) It is inevitable on dimensional grounds that Darcy's term is proportional to the fluid viscosity, whereas the Forchheimer term does not explicitly involve the viscosity. Nevertheless, both terms are intrinsically linked and the division is not arbitrary.

Despite the diverse opinions on the origin of the nonlinearity, it is now generally agreed that the quadratic term in the Forchheimer equation is associated with the inertia effect in the laminar regime and is fundamentally different from the quadratic velocity dependence for turbulent flow. These theoretical efforts demonstrated the existence of the Forchheimer flow regime. Furthermore, it was shown that this flow regime can be uniquely characterized by the inertia resistance factor β, which, like the permeability, is also an intrinsic property of the porous media (Huang and Ayoub, 2008).

The other deficiency of Darcy's law is the assumption that all of the stress in the flow field is carried by the porous medium and the fluid is not subjected to any strain because of the viscous stresses.

This assumption cannot be regarded to be physically realistic for high permeability porous media where at least part of the viscous stress is borne by the fluid itself (Parvazinia et al, 2006). The *Brinkman model* considers the transition from Darcy flow to viscous free flow and is used for high permeability porous medium. The kinematic and dynamic boundary conditions are the two boundary conditions which should be satisfied at the pore/permeable medium interface. These are indicated with the continuity of the fluid velocity and the shear stresses at the boundaries. Darcy's law is insufficient to satisfy these boundary conditions. The Brinkman model facilitates the matching of boundary conditions at an interface between the larger pores and the permeable medium. The Brinkman equation for an isotropic medium where the permeability is constant in all direction at a point is:

$$\nabla p - \rho g = -\frac{\mu}{k}V + \mu_e (\nabla \cdot \nabla)V \qquad (4.62)$$

where μ_e is called effective viscosity. This theoretically takes into account the stress borne by the fluid and is a parameter to allow for matching boundary conditions across the free-fluid/porous medium interface. In many cases, it is set to be equal to the fluid viscosity.

The fluid flow equation can also be modified to include the effect of viscous free flow and the inertia nonlinear effect. The resulting equation is the *Darcy-Brinkman-Forchheimer model* or *Brinkman-Forchheimer model* for the fluid flow inside the porous medium.

$$\frac{\partial p}{\partial x} - \rho g_x = -\frac{\mu}{k}u - \rho \beta u^2 + \mu_e (\nabla \cdot \nabla)u \qquad (4.63)$$

The equation (4.64) is for 1-D single phase flow.

4.2.7 The Compositional System of Equations and Model Variables

The model consists of a set of basic equations in the form of differential equations derived from the conservation laws. These equations

Mathematical Formulation

for the case of $T_f = T_r = T$ are rewritten in the form of a set of differential equations in (4.65). We use Darcy's law as the equation of motion in this set of partial differential equations. These sets of equations consists of N_C mass balance equation for all hydrocarbon and water components, one partial differential equation obtained from the energy balance equation and an equation that was called the volume conservation equation.

$$\begin{cases} \dfrac{\partial}{\partial t}\left(\phi \sum_j \rho_j y_j S_j\right) - \nabla \cdot \sum_j \left[\rho_j y_{i,j} \dfrac{K_j}{\mu_j}(\nabla p_j - \rho_j g) + \rho_j S_j \underline{D}_{i,j} \nabla y_{i,j}\right] + q_i = 0 \\[2mm] \dfrac{\partial}{\partial t}\left[\phi \sum_j \rho_j S_j (\hat{u}_j + g \cdot R_j)\right] + \dfrac{\partial}{\partial t}[(1-\phi)\rho_r C_r (T - T_0)] - \nabla \cdot (k\nabla T) \\[2mm] \quad - \phi\left\{\nabla \cdot \left[\sum_j \rho_j S_j (h_j + g \cdot R_j) V_j\right]\right\} + e_s = 0 \\[2mm] (\phi \hat{c}_f + \hat{c}_r)\dfrac{\partial p}{\partial t} - \hat{c}r \dfrac{\partial T}{\partial t} = \sum_{i=1}^{N_c} \gamma_i \dfrac{\partial}{\partial t}\left(\phi \sum_j \rho_j y_{i,j} S_j\right) \end{cases}$$

(4.64)

The subscripts $i = 1, \ldots, N_C$ and $j = 1, \ldots, N_p$ are for the fluid components and phases, respectively. In other words, It is considered a system of N_C components existing in N_p phases. The phases are typically gas, oil and water or $N_p = 3$. The unknowns are $N_C N_p$ concentrations $y_{i,j}$, N_p saturations S_j, N_p pressures p_j and the reservoir temperature T for every point or every gridblock. These should be added by one to be able to find the fluid flow properties in extensive form. The total number of unknowns are $N_p (N_C + 2) + 2$. There are N_C mass balance equations, one energy conservation equation and one volume conservation equation. The total number of partial differential equations are $N_C + 2$.

Therefore, we must have $[N_p (N_C + 2) + 2 - (N_C + 2)]$ more equations. The additional equations are in the form of algebraic equations. There are N_p phase constraints.

$$\sum_i y_{i,j} = 1, \quad i = 1, \cdots, N_C \tag{4.65}$$

The capillary pressure relationships are the other ($N_p - 1$) algebraic equations. These equations relate the pressure in one phase to a reference phase pressure. The oil phase is normally selected as the reference phase.

$$\begin{cases} P_{cgo}(p,m_i,T) = p_g(p,m_i,T) - p_o(p,m_i,T) \\ P_{cwo}(p,m_i,T) = p_w(p,m_i,T) - p_o(p,m_i,T) \end{cases} \quad (4.66)$$

The phase saturation equation is the next equation.

$$\sum_j S_j = 1, \qquad j = 1, \cdots, N_p \quad (4.67)$$

For the case that phases are gas, oil and water, $S_g + S_o + S_w = 1$.

Since, it is assumed that there is a local thermodynamic equilibrium, there are N_c ($N_p - 1$) equilibrium relationships describing how each component partitions between phases due to the pressure and temperature variation.

$$y_{ij} = y_{ij}(p_j, m_i, T) \quad (4.68)$$

$$m_i = \sum_j y_{ij} m_j \quad (4.69)$$

It may be written in an expanded form for a case of gas, oil and water phases:

$$m_i = y_{ig} m_g + y_{io} m_o + y_{iw} m_w \quad (4.70)$$

There are N_c ($N_p - 1$) relationships for the separation of components between phases as expressed in (4.69) and (4.70).

The total number of algebraic equations that are obtained due to the local thermodynamic equilibrium for each grid block are N_c ($N_p - 1$) + 2 N_p. Therefore, there are only $N_c + 2$ primary unknowns that can be found by using the system of partial differential equations (4.65). As mentioned before, a reference pressure such is p_o and the temperature T and N_c components mass mi can be selected

Mathematical Formulation

as the unknown primary parameters at each grid block. However, there are many possible sets of primary variables that can be used for compositional problems.

4.3 Simplification of the General Compositional Model

We present some simplification of the general compositional model, namely:

- The black oil model; and
- The water oil model.

4.3.1 The Black Oil Model

The black-oil model is a simplified model to describe multiphase flow with mass interchange between phases in a porous medium. In the black oil model, it is assumed that the fluid system consists of a water component and two hydrocarbon components namely gas and oil. The fluid system flows in three phases that are gas, oil and water phases. In other words, the number of components and the number of phases are equal ($N_C = N_p$). The model is generally used for an isothermal reservoir where the temperature variations are neglected. The model consists of:

- Three equations from mass conservation of each component;
- Darcy's law for the volumetric flow rates;
- Saturations, capillary pressures, and phase constraints;
- Thermodynamic equilibrium condition; and
- An equation states the balance between the fluid volume and rock void.

The water components assumed to be immiscible with the other two hydrocarbon components. The gas components are assumed to be soluble in oil but usually not in water. It may be described as follows:

Components:	oil ≡ o	gas ≡ g	water ≡ w
Phases:	oil ≡ o	oil & gas ≡ o & g	water ≡ w
Concentration of oil, o:	$y_{oo} = \dfrac{m_{oo}}{m_o}$	$y_{og} = \dfrac{m_{og}}{m_g} = 0$	$y_{ow} = \dfrac{m_{ow}}{m_w} = 0$
Concentration of gas, g:	$y_{go} = \dfrac{m_{go}}{m_g}$	$y_{gg} = \dfrac{m_{gg}}{m_g} = 1$	$y_{gw} = \dfrac{m_{gw}}{m_w} = 0$
Concentration of water, w:	$y_{wo} = \dfrac{m_{wo}}{m_o} = 0$	$y_{wg} = \dfrac{m_{wg}}{m_g} = 0$	$y_{ww} = \dfrac{m_{ww}}{m_w} = 1$
$\sum_i y_{ij}$	1	1	1

It is also assumed that there is no diffusive and the dispersion mass flux and therefore, the mass balance equations are written in the form:

- Oil component o;

$$\frac{\partial}{\partial t}\left(\phi \rho_o y_{oo} S_o\right) - \nabla \cdot \left[\rho_o y_{oo} \frac{K_o}{\mu_o}\left(\nabla p_o - \rho_o g\right)\right] + q_o = 0 \quad (4.71)$$

- Gas component g; and

$$\frac{\partial}{\partial t}\begin{pmatrix}\phi \rho_o y_{go} S_o \\ + \phi \rho_g S_g\end{pmatrix} - \nabla \cdot \left[\begin{array}{c}\rho_o y_{go} \dfrac{K_o}{\mu_o}\left(\nabla p_o - \rho_o g\right) \\ + \rho_g y_{gg} \dfrac{K_g}{\mu_g}\left(\nabla p_g - \rho_g g\right)\end{array}\right] + q_g = 0 \quad (4.72)$$

- Water component w;

$$\frac{\partial}{\partial t}\left(\phi \rho_w S_w\right) - \nabla \cdot \left[\rho_w y_{ww}\frac{K_w}{\mu_w}\left(\nabla p_w - \rho_w g\right)\right] + q_w = 0 \quad (4.73)$$

The unknown variables are $y_{oo}, y_{go}, S_o, S_g, S_w, P_o, P_g, P_w$. There are a set of Constraints relationships as:

- Saturation constraint;

$$S_o + S_g + S_w = 1 \quad (4.74)$$

Mathematical Formulation

- Capillary pressures relationships; and

$$\begin{cases} p_{cwo} = p_o - p_w \\ p_{cog} = p_g - p_o \end{cases} \qquad (4.75)$$

- Phase constraints.

$$y_{oo} + y_{go} = 1 \qquad (4.76)$$

The total number of unknown variables is eight while the equations consisting three partial differential equations and four algebraic equations. The other equation is obtained from the local thermodynamic equilibrium at the grid points. It depends on the pressure distribution and the mass of the gas and oil components. The mass and oil components may be expressed by using the gas oil ratio (GOR) coefficient.

4.3.2 The Water Oil Model

The compositional model may be more simplified for the case of two-phase immiscible displacement of oil and water. It may be described as follows:

Components:	oil ≡ o	water ≡ w
Phases:	oil ≡ o	water ≡ w
Concentration of oil, o:	$y_{oo} = \dfrac{m_{oo}}{m_o} = 1$	$y_{ow} = \dfrac{m_{ow}}{m_w} = 0$
Concentration of water, w:	$y_{wo} = \dfrac{m_{wo}}{m_o} = 0$	$y_{ww} = \dfrac{m_{ww}}{m_w} = 1$
$\sum_i y_{ij}$	1	1

The temperature variation is assumed to be very small. The mass balance equations may be given in the form:

- Oil component o;

$$\frac{\partial}{\partial t}(\phi \rho_o S_o) - \nabla \cdot \left[\rho_o \frac{K_o}{\mu_o} (\nabla p_o - \rho_o g) \right] + q_o = 0 \qquad (4.77)$$

- Water component w;

$$\frac{\partial}{\partial t}(\phi \rho_w S_w) - \nabla \cdot \left[\rho_w \frac{K_w}{\mu_w}(\nabla p_w - \rho_w g) \right] + q_w = 0 \qquad (4.78)$$

By neglecting the local and convective variation of oil and water densities, these equations may be written in the form:

$$\begin{cases} \phi \dfrac{\partial S_o}{\partial t} - \nabla \cdot \left[\dfrac{K_o}{\mu_o}(\nabla p_o - \rho_o g) \right] + \dfrac{q_o}{\rho_o} = 0 \\ \phi \dfrac{\partial S_w}{\partial t} - \nabla \cdot \left[\dfrac{K_w}{\mu_w}(\nabla p_w - \rho_w g) \right] + \dfrac{q_w}{\rho_w} = 0 \end{cases} \qquad (4.79)$$

Taking into account the saturation constraint $S_o + S_w = 1$, it may be written that $\partial S_w/\partial t = -\partial S_o/\partial t$. If we add both mass balance equations we get

$$\nabla \cdot \left[\frac{K_o}{\mu_o}(\nabla p_o - \rho_o g) + \frac{K_w}{\mu_w}(\nabla p_w - \rho_w g) \right] = \frac{q_o}{\rho_o} + \frac{q_w}{\rho_w} \qquad (4.80)$$

It is defined that $q_o/\rho_o + q_w/\rho_w = q_t$ and taking into account the Darcy law, the equation (4.81) may be written in the following form.

$$\nabla \cdot [V_o + V_w] = q_t \qquad (4.81)$$

It is also defined that $V_t = V_o + V_w$. It is defines that $V_w = f_w V_t$. The function f_w is called as the fractional water flow rate and can be found as follows.

$$\frac{\mu_o V_o}{K_o} - \frac{\mu_w V_w}{K_w} = -\nabla(p_o - p_w) + (\rho_o - \rho_w)y_{go}$$

$$\frac{\mu_o(1-f_w)V_t}{K_o} - \frac{\mu_w f_w V_t}{K_w} = -\nabla p_{cow} + (\rho_o - \rho_w)g$$

$$\left(1 + \frac{\mu_w K_o}{\mu_o K_w}\right) f_w = 1 + \frac{K_o}{\mu_o V_t}\left[\nabla p_{cow} - (\rho_o - \rho_w)g\right]$$

$$f_w = \left(\frac{1}{1+\dfrac{\mu_w k_{ro}}{\mu_o k_{rw}}}\right) \left\{1 + \frac{K_o}{\mu_o V_t}\left[\nabla p_{cow} - (\rho_o - \rho_w)g\right]\right\} \qquad (4.82)$$

Mathematical Formulation

The mobility of a fluid may be defined as the ratio of relative permeability to the viscosity. The mobility of water and oil are:

$$\lambda_w = \frac{k_{rw}}{\mu_w}, \quad \lambda_o = \frac{k_{ro}}{\mu_o} \tag{4.83}$$

and the mobility ratio of oil to water is denoted by $M = \frac{\mu_w k_{ro}}{\mu_o k_{rw}}$. The fractional water flow rate may be written using the mobility of water and oil in the form:

$$f_w = \left(\frac{1}{1+M}\right)\left\{1 + \frac{K\lambda_o}{V_t}\left[\nabla p_{cow} - (\rho_o - \rho_w)g\right]\right\} \tag{4.84}$$

For the case where the capillary pressure between the oil and water is negligible and the gravity force is also neglected, the rate of the water flow to the total fluid rate is:

$$f_w = \frac{1}{1+M}. \tag{4.85}$$

The equation (4.82) may be given as:

$$\nabla \cdot \left[\underline{K}\lambda_t \nabla p\right] = q_t \tag{4.86}$$

where the notation λ_t is called the total mobility and is defined as $\lambda_t = \lambda_o + \lambda_w$.

The mass conservation equation for water flow can be modified by using the notation of water rate ratio f_w as:

$$\phi\frac{\partial}{\partial t}(S_w) - \nabla \cdot \left[f_w V_t\right] + \frac{q_w}{\rho_w} = 0 \tag{4.87}$$

The water is considered to be incompressible, $\rho_w = constant$. It is also assumed that the porosity does not vary with time at a given point. Following Buckley and Leverett (1942) and assuming that fractional water flow rate depends only on the water saturation $f_w = f_w(S_w)$, the mass conservation equation for the water is

$$\phi\frac{\partial}{\partial t}(S_w) - (\nabla \cdot V_t)f_w(S_w) - \frac{\partial f_w}{\partial S_w}(\nabla S_w \cdot V_t) + \frac{q_w}{\rho_w} = 0 \tag{4.88}$$

The equation (4.89) is a general form of the Buckley-Leverett equation. It describes the displacement of the saturation fronts in the reservoir and is called the frontal advance equation. The capillary pressure between the oil and water is a strong function of the water saturation. The frontal advance equation is a non-linear differential equation via of the fractional water flow rate equation (4.85).

4.4 Some Examples in Application of the General Compositional Model

We present some application of this general compositional model in simulation of real cases in petroleum engineering. The examples are:

- Isothermal volatile oil reservoir.
- Steam injection inside a dead oil reservoir.
- Steam injection inside an oil and gas reservoir.
- CO_2 injection inside an oil reservoir.

4.4.1 Isothermal Volatile Oil Reservoir

A volatile oil is defined as a high shrinkage crude oil near its critical point, (Moses, 1986). It is recognized as a type between black oil and gas-condensate fluid. As the reservoir pressure drops below the bubble point, the reservoir flow stream becomes in gas phase and the effective permeability to oil can exhibit a rapid decline. The thermodynamic behavior of a volatile oil is very sensitive to pressure and temperature changes, and hence the treatment of compositional alterations is important, (Syahrial, 1998).

There are two components: volatile oil and water. The volatile oil may exist in two phases of gas and oil. The water component is assumed to be only in liquid form as the water phase. The reservoir is assumed to be isothermal and therefore, the temperature is constant and there is no need for the energy balance equation. The dispersion and diffusive fluxes are neglected and only the mass is transfer due to the convective term.

Mathematical Formulation

Components:	volitile oil ≡ 1		water ≡ 2
Phases:	oil & gas ≡ o & g		water ≡ w
Concentration of 1	$y_{1o} = \dfrac{m_{1o}}{m_o} = 1$	$y_{1g} = \dfrac{m_{1g}}{m_g} = 1$	$y_{1w} = \dfrac{m_{1w}}{m_w} = 0$
Concentration of 2	$y_{2o} = \dfrac{m_{2o}}{m_o} = 0$	$y_{2g} = \dfrac{m_{2g}}{m_g} = 0$	$y_{2w} = \dfrac{m_{2w}}{m_w} = 1$
$\sum_i y_{ij}$	1	1	1

The governing equations are the mass balance for each component. It is assumed that the flow velocity is small and it is in the range of Darcy's flow. Therefore, the Darcy law can consider the equation of motion. The mass balance equation, considering the Darcy's law, may be given according to (4.56) as follows.

Mass balance equation:

Oil component 1

$$\frac{\partial}{\partial t}\left[\phi\begin{pmatrix}\rho_o y_{1o} S_o \\ +\rho_g y_{1g} S_g \\ +\rho_w y_{1w} S_w\end{pmatrix}\right] - \nabla \cdot \begin{bmatrix}\rho_o y_{1o}\dfrac{K_o}{\mu_o}(\nabla p_o - \rho_o g) \\ +\rho_g y_{1g}\dfrac{K_g}{\mu_g}(\nabla p_g - \rho_g g) \\ +\rho_w y_{1w}\dfrac{K_w}{\mu_{gw}}(\nabla p_w - \rho_w g)\end{bmatrix} + q_1 = 0 \quad (4.89)$$

Water component 2

$$\frac{\partial}{\partial t}\left[\phi\begin{pmatrix}\rho_o y_{2o} S_o \\ +\rho_g y_{2g} S_g \\ +\rho_w y_{2w} S_w\end{pmatrix}\right] - \nabla \cdot \begin{bmatrix}\rho_o y_{2o}\dfrac{K_o}{\mu_o}(\nabla p_o - \rho_o g) \\ +\rho_g y_{2g}\dfrac{K_g}{\mu_g}(\nabla p_g - \rho_g g) \\ +\rho_w y_{2w}\dfrac{K_w}{\mu_w}(\nabla p_w - \rho_w g)\end{bmatrix} + q_2 = 0 \quad (4.90)$$

The total unknowns in these two differential equations are six values for y_{ij}, three values for p_j, and three values for S_j at each point.

Therefore, we need 10 more equations that are obtained through the constraints and the local thermodynamic equilibrium equations.

Constraints:

1. Saturation

$$S_o + S_g + S_w = 1$$

2. Capillary pressures

$$\begin{cases} p_{cwo} = p_o - p_w \\ p_{cog} = p_g - p_o \end{cases}$$

3. Phase constraints

$$\begin{cases} y_{1o} + y_{2o} = 1 \\ y_{1g} + y_{2g} = 1 \\ y_{1w} + y_{2w} = 1 \end{cases}$$

Thermodynamic equilibrium equations:
The number of equations from the local thermodynamic equilibrium is $N_C (N_P - 1) = 2 \times 2 = 4$. These are:

$$y_{1w} = 0, \quad y_{2o} = y_{2g} = 0.$$

If the oil is at he bubble point pressure, then

$$p_g = p_{sat}.$$

If the oil is a pressure higher than the bubble point, then

$$S_g = 0.$$

If the reservoir pressure is under the oil bubble point pressure, then

$$S_o = 0.$$

The number of unknown and the number of equations are identical and using the algebraic equation, the differential equation can be solved by a proper numerical method for two intensive variables. The volume balance equation should be taken into account to get the extensive properties.

4.4.2 Steam Injection Inside a Dead Oil Reservoir

We consider a case of dead oil reservoir that in injected by the steam to recover the oil. The steam condenses and forms two phases of gas and water. Therefore, there are two components and three phases exist in such a problem. It is assumed that there is no dispersion flux.

Components:	oil $\equiv 1$	water $\equiv 2$	
Phases:	oil $\equiv o$	gas $\equiv g$	water $\equiv w$
Concentration of 1	$y_{1o} = \dfrac{m_{1o}}{m_o} = 1$	$y_{1g} = \dfrac{m_{1g}}{m_g} = 0$	$y_{1w} = \dfrac{m_{1w}}{m_w} = 0$
Concentration of 2	$y_{2o} = \dfrac{m_{2o}}{m_o} = 0$	$y_{2g} = \dfrac{m_{2g}}{m_g} = 1$	$y_{2w} = \dfrac{m_{2w}}{m_w} = 1$
$\sum_i y_{ij}$	1	1	1

Mass balance equation:

Oil component **1**

$$\frac{\partial}{\partial t}(\phi \rho_o y_{1o} S_o) - \nabla \cdot \rho_o y_{1o} \frac{K_o}{\mu_o}(\nabla p_o - \rho_o g) + q_1 = 0 \tag{4.91}$$

Water component **2**

$$\frac{\partial}{\partial t}(\phi \rho_g y_{2g} S_g + \phi \rho_w y_{2w} S_w)$$

$$-\nabla \cdot \left[\rho_g y_{2g} \frac{K_g}{\mu_g}(\nabla p_g - \rho_g g) + \rho_w y_{2w} \frac{K_w}{\mu w}(\nabla p_w - \rho_w g) \right] + q_2 = 0 \tag{4.92}$$

Energy balance equation:

$$\frac{\partial}{\partial t}\begin{bmatrix} \phi \rho_o S_o [C_o(T-T_0) + g \cdot R_o] \\ +\phi \rho_g S_g [\hat{u}_g + g \cdot R_g] \\ +\phi \rho_w S_w [C_w(T-T_0) + g \cdot R_w] \end{bmatrix} + \frac{\partial}{\partial t}\left[(1-\phi)\rho_r C_r(T-T_0)\right] - \nabla \cdot (k \nabla T)$$

$$-\phi \begin{Bmatrix} \nabla \cdot \left\{ \rho_o S_o \left[C_o(T-T_0) + g \cdot R_o \right] \frac{K_o}{\mu_o} (\nabla p_o - \rho_o g) \right\} \\ + \nabla \cdot \left\{ \rho_g S_g \left[h_g + g \cdot R_g \right] \frac{K_g}{\mu_g} (\nabla p_g - \rho_g g) \right\} \\ \nabla \cdot \left\{ \rho_w S_w \left[C_w(T-T_0) + g \cdot R_w \right] \frac{K_w}{\mu_w} (\nabla p_w - \rho_w g) \right\} \end{Bmatrix} + e_s = 0 \quad (4.93)$$

Constraints:

1. Saturation

$$S_o + S_g + S_w = 1$$

2. Capillary pressures

$$\begin{cases} p_{cwo} = p_o - p_w \\ p_{cog} = p_g - p_o \end{cases}$$

The unknowns are p_o, p_g, p_w, S_o, S_g, S_w and T while there are only six equations. The next set of equations is related to the local thermodynamic equilibrium.

Thermodynamic equilibrium relationships:
The last set of equations is related to the steam condition. If the steam is saturated and both the water and steam are present, the gas pressure is equal to the saturation pressure,

$$p_g = p_{sat}.$$

If there is superheated steam, then

$$Sw = 0.$$

In the case that there is no steam

$$Sg = 0.$$

This example is taken from Coats et al (1974).

4.4.3 Steam Injection in Presence of Distillation and Solution Gas

This example is for a case of three hydrocarbons: A hydrocarbon solution gas; a distillable oil potion; and a nonvolatile hydrocarbon

Mathematical Formulation

oil. The first two hydrocarbon components may exist in both oil and gas phases and the third one is only in oil phase. The steam is also injected to enhance the hydrocarbon recovery. The steam may exist in gas and water phases.

Components:	distillable oil ≡ 1	solution gas ≡ 2	nonvolatile oil ≡ 2	water ≡ 2
Phases:	oil & gas ≡ o & g	oil & gas ≡ o & g	gas ≡ g	water ≡ w
Concentration of 1	$y_{1o} = \dfrac{m_{1o}}{m_o}$	$y_{1g} = \dfrac{m_{1g}}{m_g}$		$y_{1w} = \dfrac{m_{1w}}{m_w} = 0$
Concentration of 2:	$y_{2o} = \dfrac{m_{2o}}{m_o}$	$y_{2g} = \dfrac{m_{2g}}{m_g}$		$y_{2w} = \dfrac{m_{2w}}{m_w} = 0$
Concentration of 3:	$y_{3o} = \dfrac{m_{3o}}{m_o}$	$y_{3g} = \dfrac{m_{3g}}{m_g} = 0$		$y_{3w} = \dfrac{m_{3w}}{m_w} = 0$
Concentration of 4:	$y_{4o} = \dfrac{m_{4o}}{m_o} = 0$	$y_{4g} = \dfrac{m_{4g}}{m_g}$		$y_{4w} = \dfrac{m_{4w}}{m_w} = 1$
$\sum_i y_{ij}$	1	1		1

Mass balance equation:

  Distillable oil component 1

$$\frac{\partial}{\partial t}\left[\phi(\rho_o y_{1o} S_o + \rho_g y_{1g} S_g)\right] - \nabla \cdot \left[\begin{array}{l}\rho_o y_{1o}\dfrac{K_o}{\mu_o}(\nabla p_o - \rho_o g) \\ + \rho_g y_{1g}\dfrac{K_g}{\mu_g}(\nabla p_g - \rho_g g)\end{array}\right] + q_1 = 0$$

(4.94)

  Solution gas component 2

$$\frac{\partial}{\partial t}\left[\phi(\rho_o y_{2o} S_o + \rho_g y_{2g} S_g)\right] - \nabla \cdot \left[\begin{array}{l}\rho_o y_{2o}\dfrac{K_o}{\mu_o}(\nabla p_o - \rho_o g) \\ + \rho_g y_{2g}\dfrac{K_g}{\mu_g}(\nabla p_g - \rho_g g)\end{array}\right] + q_2 = 0$$

(4.95)

Distillable oil component **3**

$$\frac{\partial}{\partial t}(\phi \rho_o y_{3o} S_o) - \nabla \cdot \left[\rho_o y_{3o} \frac{K_o}{\mu_o} (\nabla p_o - \rho_o g) \right] + q_3 = 0 \qquad (4.96)$$

Water component **4**

$$\frac{\partial}{\partial t}\left[\phi(\rho_g y_{2g} S_g + \rho_w y_{2w} S_w)\right] - \nabla \cdot \begin{bmatrix} \rho_g y_{4g} \dfrac{K_g}{\mu_g}(\nabla p_g - \rho_g g) \\ \rho_w y_{4w} \dfrac{K_w}{\mu_w}(\nabla p_w - \rho_w g) \end{bmatrix} + q_4 = 0$$

(4.97)

Energy balance equation:

$$\frac{\partial}{\partial t}\begin{Bmatrix} \phi \rho_o S_o [C_o(T-T_0) + g \cdot R_o] \\ + \phi \rho_g S_g [\hat{u}_g + g \cdot R_g] \\ + \phi \rho_w S_w [C_w(T-T_0) + g \cdot R_w] \end{Bmatrix} + \frac{\partial}{\partial t}\left[(1-\phi)\rho_r C_r(T-T_0)\right] - \nabla \cdot (k \nabla T)$$

$$-\phi \begin{Bmatrix} \nabla \cdot \left\{ \rho_o S_o [C_o(T-T_0) + g \cdot R_o] \dfrac{K_o}{\mu_o}(\nabla p_o - \rho_o g) \right\} \\ + \nabla \cdot \left\{ \rho_g S_g [h_g + g \cdot R_g] \dfrac{K_g}{\mu_g}(\nabla p_g - \rho_g g) \right\} \\ \nabla \cdot \left\{ \rho_w S_w [C_w(T-T_0) + g \cdot R_w] \dfrac{K_w}{\mu_w}(\nabla p_w - \rho_w g) \right\} \end{Bmatrix} + e_s = 0$$

(4.98)

Constraints:

4. Saturation

$$S_o + S_g + S_w = 1$$

5. Capillary pressures

$$\begin{cases} p_{cwo} = p_o - p_w \\ p_{cog} = p_g - p_o \end{cases}$$

Mathematical Formulation

6. *Phase constraints*

$$\begin{cases} y_{1o} + y_{2o} + y_{3o} + y_{4o} = 1 \\ y_{1g} + y_{2g} + y_{3g} + y_{4g} = 1 \\ y_{1w} + y_{2w} + y_{3w} + y_{4w} = 1 \end{cases} \quad (4.99)$$

The total unknowns are y_{ij}, p_j, S_j, T. The total number of them are 19. There are 4 mass balance equations, 1 energy balance equation and 6 constraints. We must have (19 − 11 = 8) more equations. This is equal to $(N_c (N_p - 1) = 4.3 - 1) = 8$. These equations are obtained from the local thermodynamic equilibrium. The algebraic equations are:

Thermodynamic equilibrium equations:

$$\begin{cases} y_{4o} = 0, \quad y_{3g} = 0, \quad y_{1w} = y_{2w} = y_{3w} = 0 \\ y_{4g} = y_{4g}(p_w, p_g, m_4) \\ y_{1g} = y_{1g}(p_g, p_o, m_1), \quad y_{2g} = y_{2g}(p_g, p_o, m_2) \end{cases} \quad (4.100)$$

Therefore, the number of unknown and the equations are equal to provide the solution for the variables. The volume balance equation should be taken into account to get the extensive properties. This example is taken from Coats (1976).

5
The Compositional Simulator Using the Engineering Approach

The basic formulations of the compositional model are described for an infinitesimal elementary volume in Chapter 5. These are partial differential equations that are nonlinear. Due to the complexity of the reservoir properties and geometry, the non-homogeneity of the reservoir fluid, and the nonlinear behavior of the partial differential equations governing the flow inside the porous matrix, analytical solutions can be obtained for very simplified cases and geometries, such as the Buckley-Leverett solution for oil/water production without the effect of capillary pressure. However, the analytical solution for the case of oil/water production without the effect of capillary pressure provides multiple solutions that need special attention to check the validity of them and the possibilities of the existence of different solutions for a certain problem.

In general, a numerical solution of the flow equations in a porous medium is imperative. The mathematical model should be converted into numerical formulations that are solved for a set of discrete points in the reservoir. These numerical formulations are normally a set of algebraic equations that are called discretized equations. The algebraic equations involving the unknown values of dependent variables are derived from the governing partial differential equations. The values of all the dependent variables at these grid points are considered as basic unknowns. Different algorithms may be applied to solve the discretized equations. Generally the method should be fast and provide a set of valid solution with an acceptable degree of accuracy. The region that attributed to a certain grid point is called the grid blocks. The number of grid blocks depends on the complexity of the reservoir properties and geometry, applied algorithm and the capabilities of the computational hardware facilities. It should be selected in a range that provides the solutions in a reasonable time frame.

It is necessary to apply proper numerical methods to handle both heterogeneity and anisotropy. Large discontinuities in medium properties require construction of numerical schemes with a proper definition of the effective conductivity across cell interfaces. To satisfy local continuity in flux between grid cells with strong discontinuities in permeability, control-volume methods are especially well suited. For time-dependent problems with large solution gradients, it is important that the methods can be combined with a fully implicit time stepping (Aavatsmark, 1996).

The engineering approach presented by Abou-Kassem (2006) is based on the control-volume method. We extended that for a general case inhomogeneous fluid in different phases through heterogeneous, anisotropic media. The general form of the conservation laws are given in Chapter 4. Discretized forms of those equation are given and several examples of them are presented.

5.1 Finite Control Volume Method

There may be several different ways to derive discretized equations for a given differential equation such as: finite difference; finite control volume; and finite element methods. The control volume methods ensure integral conservation of mass, momentum and energy over any group of control volumes and, of course, over the whole solution domain. This characteristic exists for any number of grid points (and not only for the limiting case of a large number of grid points) and is the most attractive feature from the reservoir engineer's point of view. Therefore, we discuss details of finite control volume method to solve the partial differential equations derived in Chapter 4.

Integrating the governing partial differential equations over a finite control volume (CV) gives discretized equations in the finite control volume method. The finite control volume may be called a computational cell or a grid block. The first step is the division of the solution domain into a finite number of control volumes or grid blocks. There may be two types of grid blocks, namely, structured and unstructured grid blocks. For the sake of discussing the finite control volume method, we considered a simple, structured grid block. It is important to note that grid blocks should not overlap and each control volume face should be unique to the two grid blocks which lie on either side of it. A typical finite volume grid blocks in 1-, 2- and 3-dimensional reservoir representation are shown in Fig. 5.2 using Cartesian coordinate system. Different methods can be employed to select positions of computational nodes and boundaries of computational cells or in other words to generate a structured computational grid. The boundaries of the grid blocks are usually decided as a first step and then a computational node is assigned at the center of each grid block.

5.1.1 Reservoir Discretization in Rectangular Coordinates

Reservoir discretization means that the reservoir is described by a set of blocks (or points) whose properties, dimensions, boundaries, and locations in the reservoir are well defined. Fig. 5.1 shows reservoir discretization in the x-direction, using block-centered grid, as one focuses on block i. The figure shows how the blocks are related to each other –block i and its neighboring blocks (blocks $i-1$ and $i+1$) –, block dimensions (Δx_i, $\Delta x_i - 1$, Δx_{i+1}), block boundaries ($x_{i-1/2}$, $x_{i+1/2}$), distances between the point that represents the block and block boundaries (δx_{i-}, δx_{i+}), and distances between the points representing the blocks ($\Delta x_{i-1/2}$, $\Delta x_{i+1/2}$). Reservoir discretization uses similar terminology in the y- and z-directions. Thus, reservoir discretization involves dividing the reservoir into n_x blocks in the x-direction, n_y blocks in the y-direction, and n_z blocks in the z-direction. Any reservoir block in a discretized reservoir is identified as block (i,j,k), where i, j and k are respectively the orders of the block in x-, y-, and z-directions with $1 \leq i \leq n_x$, $1 \leq j \leq n_y$, $1 \leq k \leq n_z$, see Fig. 5.2. In addition, each block is assigned elevation and rock properties such as porosity and permeability in the x-, y-, and z-directions. The transfer of fluids from one block to the rest of the

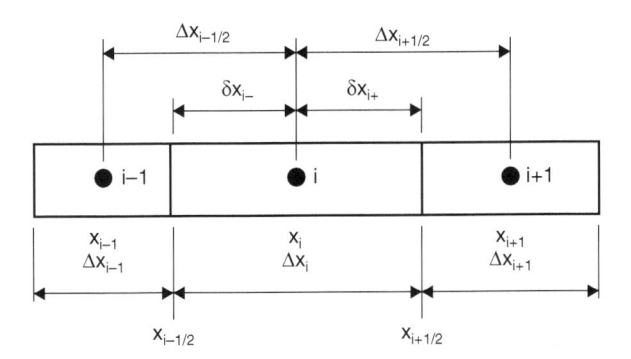

Figure 5.1 Relationships between block and its neighboring blocks in 1D flow.

(a) – 3-D reservoir representation

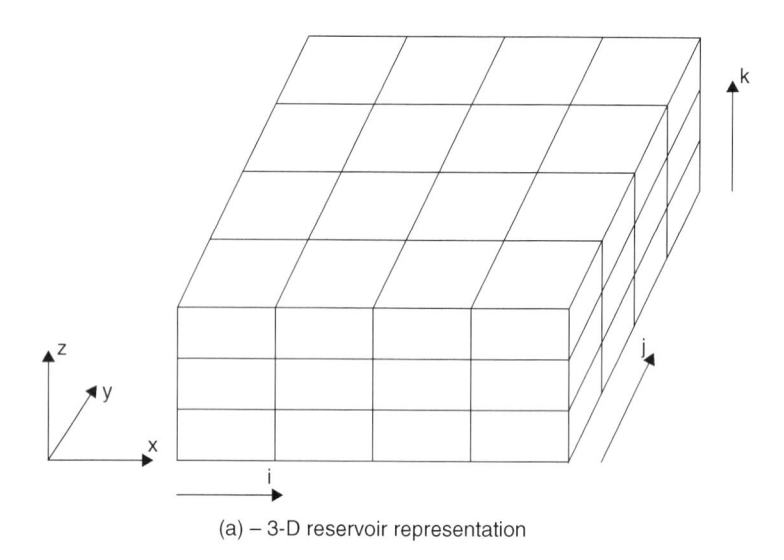

Figure 5.2 Representation of reservoir in x-, y- and z-directions. (a) – 3-D reservoir representation.

reservoir takes place through the immediate neighboring blocks. When the whole reservoir is discretized, each block is surrounded by a set (group) of neighboring blocks.

A control volume and the neighboring blocks to it are shown in Fig. 5.3a in 1D flow along the x-axis, Fig. 5.3b in 2D flow in the x-y plane, and Fig. 5.3c in 3D flow in x-y-z space.

Fig 5.3c shows that block (i, j, k) surrounded by blocks $(i-1, j, k)$ and $(i+1, j, k)$ in the x-direction, by blocks $(i, j-1, k)$ and $(i, j+1, k)$ in the y-direction, and by blocks $(i, j, k-1)$ and $(i, j, k+1)$ in the z-direction. The boundaries between block (i, j, k) and its neighboring (or surrounding) blocks are termed $(i-1/2, j, k)$ and $(i+1/2, j, k)$ in the x-direction, $(i, j-1/2, k)$ and $(i, j+1/2, k)$ in the y-direction, and $(i, j, k+1/2)$ and in the z-direction.

5.1.2 Discretization of Governing Equations

The governing equations are the mass and energy balance equations along with a motion equation such as the Darcy law for a general case of a non-isothermal reservoir. These governing equations should be integrated over each computational cell and over

The Compositional Simulator

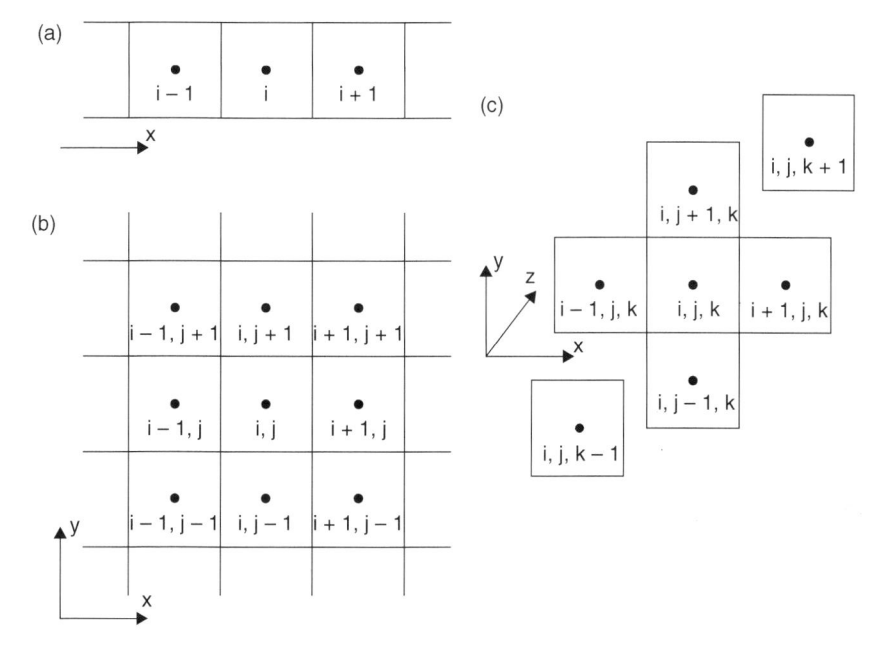

Figure 5.3 A block and its neighboring blocks: (a) 1D flow, (b) 2D flow; and (c) 3D flow.

a desired time interval. At first the mass balance equation for each component of the reservoir fluid is described and then the energy balance equation will be dealt with.

5.1.2.1 Components Mass Conservation Equation

The mass conservation equation for the component i is integrated over the control volume $C.V$ and the time interval Δt.

$$\int_{t}^{t+\Delta t} \left\{ \int_{C.V} \left[\frac{\partial}{\partial t}\left(\phi \sum_{j} r_j y_{i,j} S_j \right) + \nabla \cdot \sum_{j} \left(r_j y_{i,j} V_j - r_j S_j \underline{D}_{i,j} \nabla y_{i,j} \right) + q_i = 0 \right] dU \right\} dt \quad (5.1)$$

Using the divergence theorem, the integral of the divergence of convective and dispersion part of the mass flux over the control volume can convert into a surface integral over the boundaries of the control volume S. Therefore, it may be written that:

$$\int_t^{t+\Delta t} \left\{ \int_{C.V.} \left[\frac{\partial}{\partial t} \left(\phi \sum_j r_j y_{i,j} S_j \right) \right] dU \right. \\ \left. + \int_S \left[\sum_j \left(\rho_j y_{i,j} V_j - \rho_j S_j \underline{D}_{i,j} \nabla y_{i,j} \right) \right] \cdot n dS + \int_{C.V.} q_i dU = 0 \right\} dt \quad (5.2)$$

a. The accumulation term

The first term shows that variation of the mass of component i in a time interval Δt over a control volume say l. It is assumed that each grid block has fixed dimensions and the volume of the grid block l is U_l. It is also assumed that the properties of each fluid component is homogeneous along each control volume. The rock properties are also fixed throughout a given grid block. Thus, the integration of the first term may be given:

$$\int_t^{t+\Delta t} \left[\int_{C.V.} \frac{\partial}{\partial t} \left(\phi \sum_j r_j y_{i,j} S_j \right) dU \right] dt$$

$$= \int_{C.V.} \left[\int_t^{t+\Delta t} \frac{\partial}{\partial t} \left(\phi \sum_j r_j y_{i,j} S_j \right) dt \right] dU$$

$$= \int_{C.V.} \left(f \sum_j r_j y_{i,j} S_j \right) \bigg|_t^{t+\Delta t} dU$$

$$= \left[\left(\phi \sum_j r_j y_{i,j} S_j \right)_{t+\Delta t} - \left(\phi \sum_j r_j y_{i,j} S_j \right)_t \right] U_l \quad (5.3)$$

It is assumed $t = n \bullet \Delta t$, or in other words the time t corresponds to the time step n and the time $t + \Delta t$ corresponds to time step $n+1$. Using (5.3), the first term in (5.2) may be given as:

The accumulation
of component i $= \left[\left(\phi \sum_j \rho_j y_{i,j} S_j \right)_l^{n+1} - \left(\phi \sum_j \rho_j y_{i,j} S_j \right)_l^n \right] U_l$
in grid block l

(5.4)

The Compositional Simulator

b. The mass flux through the boundaries:
The second bracket in (5.2) shows the net flux of component i through the boundaries of control volume l. This surface integral is approximated as the summation of products of the integrand at the center of cell faces and cell faces area.

$$\int_S \left[\sum_j \left(\rho_j y_{i,j} V_j - \rho_j S_j \underline{D}_{i,j} \nabla y_{i,j} \right) \right] \cdot n \, dS$$

$$= \sum_l \left\{ \left[\sum_j \left(\rho_j y_{i,j} V_j - \rho_j S_j \underline{D}_{i,j} \nabla y_{i,j} \right) \right] \cdot n \right\} S_k \quad (5.5)$$

The notation S_k is the area of the face k of a given control volume. For a cubic grid block with the faces S_x, S_y, S_z in the direction x, y and z of a Cartesian coordinate system, the convective part of the mass flux of component i may be given according to (5.5) in the form:

The net convective mass flux of the component i $= \sum_l \left(\sum_j \rho_j y_{i,j} V_j \right) \cdot n \, S_k$

$$= \left[\left(\sum_j \rho_j y_{i,j} u_j \right)_{x_{i+\frac{1}{2}}} - \left(\sum_j \rho_j y_{i,j} u_j \right)_{x_{i-\frac{1}{2}}} \right] S_x$$

$$+ \left[\left(\sum_j \rho_j y_{i,j} v_j \right)_{y_{i+\frac{1}{2}}} - \left(\sum_j \rho_j y_{i,j} v_j \right)_{y_{i-\frac{1}{2}}} \right] S_y \quad (5.6)$$

$$+ \left[\left(\sum_j \rho_j y_{i,j} w_j \right)_{z_{i+\frac{1}{2}}} - \left(\sum_j \rho_j y_{i,j} w_j \right)_{z_{i-\frac{1}{2}}} \right] S_z$$

ADVANCED PETROLEUM RESERVOIR SIMULATION

Using (Eq. 5.5), the mass flux due to dispersion for the cubic control volume is given as:

The net dispersive mass flux of the component i $= -\sum_{l}\left(\sum_{j} \rho_j S_j \underline{D}_{i,j} \nabla y_{i,j}\right) \cdot n\, S_k$

$$= \begin{bmatrix} \left(\sum_j \rho_j S_j \left(D^{i,j}_{x,x} + D^{i,j}_{y,x} + D^{i,j}_{z,x}\right)\dfrac{\partial y_{i,j}}{\partial x}\right)_{x_{i-\frac{1}{2}}} \\ -\left(\sum_j \rho_j S_j \left(D^{i,j}_{x,x} + D^{i,j}_{y,x} + D^{i,j}_{z,x}\right)\dfrac{\partial y_{i,j}}{\partial x}\right)_{x_{i+\frac{1}{2}}} \end{bmatrix} S_x$$

$$+ \begin{bmatrix} \left(\sum_j \rho_j S_j \left(D^{i,j}_{x,y} + D^{i,j}_{y,y} + D^{i,j}_{z,y}\right)\dfrac{\partial y_{i,j}}{\partial y}\right)_{y_{i-\frac{1}{2}}} \\ -\left(\sum_j \rho_j S_j \left(D^{i,j}_{x,y} + D^{i,j}_{y,y} + D^{i,j}_{z,y}\right)\dfrac{\partial y_{i,j}}{\partial y}\right)_{y_{i+\frac{1}{2}}} \end{bmatrix} S_y$$

$$+ \begin{bmatrix} \left(\sum_j \rho_j S_j \left(D^{i,j}_{x,z} + D^{i,j}_{y,z} + D^{i,j}_{z,z}\right)\dfrac{\partial y_{i,j}}{\partial z}\right)_{z_{i-\frac{1}{2}}} \\ -\left(\sum_j \rho_j S_j \left(D^{i,j}_{x,z} + D^{i,j}_{y,z} + D^{i,j}_{z,z}\right)\dfrac{\partial y_{i,j}}{\partial z}\right)_{z_{i+\frac{1}{2}}} \end{bmatrix} S_z \qquad (5.7)$$

The summation of equations (5.6) and (5.7) gives the total flux at time t.

The total net mass flux of the component i $= \sum_{l}\left\{\sum_{j}\left(\rho_j y_{i,j} V_j - \rho_j S_j \underline{D}_{i,j} \nabla y_{i,j}\right)\right\} \cdot n\, S_k$

$$= \begin{bmatrix} \left(\sum_j \rho_j \left[y_{i,j} u_j - S_j \left(D^{i,j}_{x,x} + D^{i,j}_{y,x} + D^{i,j}_{z,x}\right)\dfrac{\partial y_{i,j}}{\partial x}\right]\right)_{x_{i+\frac{1}{2}}} \\ -\left(\sum_j \rho_j \left[y_{i,j} u_j - S_j \left(D^{i,j}_{x,x} + D^{i,j}_{y,x} + D^{i,j}_{z,x}\right)\dfrac{\partial y_{i,j}}{\partial x}\right]\right)_{x_{i-\frac{1}{2}}} \end{bmatrix} S_x$$

The Compositional Simulator

$$+ \begin{bmatrix} \left(\sum_j \rho_j \left[y_{i,j} v_j - S_j \left(D^{i,j}_{x,y} + D^{i,j}_{y,y} + D^{i,j}_{z,y}\right) \frac{\partial y_{i,j}}{\partial y}\right]\right)_{y_{i+\frac{1}{2}}} \\ -\left(\sum_j \rho_j \left[y_{i,j} v_j - S_j \left(D^{i,j}_{x,y} + D^{i,j}_{y,y} + D^{i,j}_{z,y}\right) \frac{\partial y_{i,j}}{\partial y}\right]\right)_{y_{i-\frac{1}{2}}} \end{bmatrix} S_y$$

$$+ \begin{bmatrix} \left(\sum_j \rho_j \left[y_{i,j} w_j - S_j \left(D^{i,j}_{x,z} + D^{i,j}_{y,z} + D^{i,j}_{z,z}\right) \frac{\partial y_{i,j}}{\partial z}\right]\right)_{z_{i+\frac{1}{2}}} \\ -\left(\sum_j \rho_j \left[y_{i,j} w_j - S_j \left(D^{i,j}_{x,z} + D^{i,j}_{y,z} + D^{i,j}_{z,z}\right) \frac{\partial y_{i,j}}{\partial z}\right]\right)_{z_{i-\frac{1}{2}}} \end{bmatrix} S_z \quad (5.8)$$

The equation (5.8) gives the integrand of the time integral between the time t and $t+\Delta t$. The integrand is replaced by the function $F(t)$. The integral of $F(t)$ in the interval Δt is equal to the area under the curve $F(t)$ between t and $t+\Delta t$. This area is equal to the area of a rectangle as shown in the Fig. 5.4. $F(t^m)$ is an average value of $F(t)$ in the interval of Δt. It is set that $F^m = F(t^m)$

$$\int_t^{t+\Delta t} F(t)dt = F^m \times \Delta t \quad (5.9)$$

The value of this integral can be calculated using the above equation provided that the value of F^m is known. In reality,

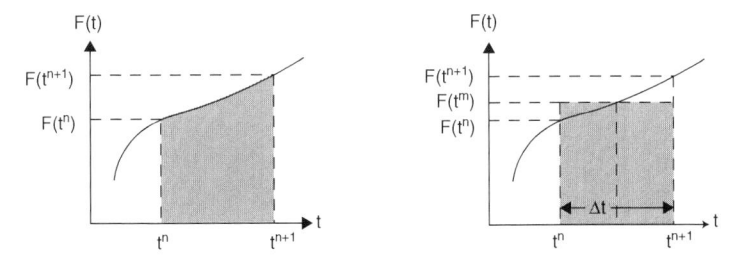

Figure 5.4 Representation of the time integral of $F(t)$ as the area under the curve (left) and the equivalent of it as the area under a rectangle with a dimension equal to an average value $F(t^m)$.

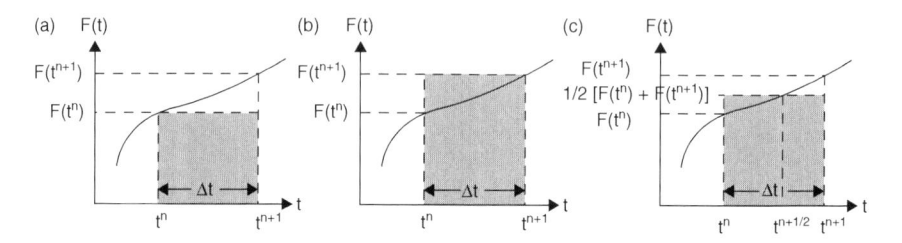

Figure 5.5 Some types of approximation of time integral for a function F.

however, F^m is not known and, therefore, it needs to be approximated. The approximation can be obtained by using different types of numerical integration. Some of them are illustrated in Fig. 5.5.

a.

Figure 5.5(a) shows a fully explicit approximation of the integral $\int_{t}^{t+\Delta t} F(t)dt$. The value of $F(t)$ at time step n is consider to approximate the integration.

$$\int_{t}^{t+\Delta t} F(t)dt = F(t^n) \times \Delta t \qquad (5.10)$$

b.

It may apply a fully implicit scheme and apply the value of the function $F(t)$ at a time step $(n+1)$. It is illustrated in Fig. 5.5(b).

$$\int_{t}^{t+\Delta t} F(t)dt = F(t^{n+1}) \times \Delta t \qquad (5.11)$$

c.

The trapezoidal integration rule may also be applied and approximate the function $F(t)$ with a linear function. This is shown in Fig. 5.5(c).

$$\int_{t}^{t+\Delta t} F(t)dt = \frac{1}{2}\left[F(t^n) + F(t^{n+1})\right] \times \Delta t \qquad (5.12)$$

The Compositional Simulator

Using (5.9), the total net mass flux through the boundaries of the control volume l in a time interval Δt can be given in the following form:

$$\left\{\sum_k\left[\sum_j\left(\rho_j y_{i,j} V_j - \rho_j S_j \underline{D}_{i,j} \nabla y_{i,j}\right)\right]\cdot \mathbf{n}\, S_k\right\}^m \times \Delta t \qquad (5.13)$$

The total mass flux in an extended form can be given by using (5.8).

The total net mass flux of the component i

$$= \left\{ \begin{array}{l} \left[\begin{array}{l} \left(\sum_j \rho_j\left[y_{i,j} u_j - S_j\left(D^{i,j}_{x,x} + D^{i,j}_{y,x} + D^{i,j}_{z,x}\right)\dfrac{\partial y_{i,j}}{\partial x}\right]\right)_{x_{i+\frac{1}{2}}} \\ -\left(\sum_j \rho_j\left[y_{i,j} u_j - S_j\left(D^{i,j}_{x,x} + D^{i,j}_{y,x} + D^{i,j}_{z,x}\right)\dfrac{\partial y_{i,j}}{\partial x}\right]\right)_{x_{i-\frac{1}{2}}} \end{array}\right] S_x \\[2ex] +\left[\begin{array}{l} \left(\sum_j \rho_j\left[y_{i,j} v_j - S_j\left(D^{i,j}_{x,y} + D^{i,j}_{y,y} + D^{i,j}_{z,y}\right)\dfrac{\partial y_{i,j}}{\partial y}\right]\right)_{y_{i+\frac{1}{2}}} \\ -\left(\sum_j \rho_j\left[y_{i,j} v_j - S_j\left(D^{i,j}_{x,y} + D^{i,j}_{y,y} + D^{i,j}_{z,y}\right)\dfrac{\partial y_{i,j}}{\partial y}\right]\right)_{y_{i-\frac{1}{2}}} \end{array}\right] S_y \\[2ex] +\left[\begin{array}{l} \left(\sum_j \rho_j\left[y_{i,j} w_j - S_j\left(D^{i,j}_{x,z} + D^{i,j}_{y,z} + D^{i,j}_{z,z}\right)\dfrac{\partial y_{i,j}}{\partial z}\right]\right)_{z_{i+\frac{1}{2}}} \\ -\left(\sum_j \rho_j\left[y_{i,j} w_j - S_j\left(D^{i,j}_{x,z} + D^{i,j}_{y,z} + D^{i,j}_{z,z}\right)\dfrac{\partial y_{i,j}}{\partial z}\right]\right)_{z_{i-\frac{1}{2}}} \end{array}\right] S_z \end{array} \right\}^m \times \Delta t$$

(5.14)

The average value of the integrand may be obtained by using the methods that have already been described for a function $F(t)$.

c. The source term:

The source term is associated with the production rate of component i or the injection rate of it into the reservoir in a time interval

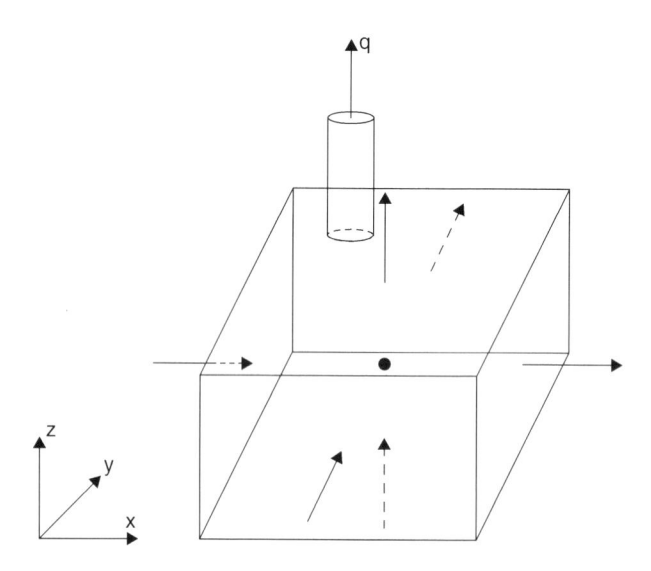

Figure 5.6 A grid block *l* with a production well with a rate $Q = \sum_j Q_j$.

Δt. The component *i* may produce in different phases. It can be associated with the phases volumetric rates from a single well located in block *l* as shown in Fig. 5.6.

$$\int_{C.V.} q_i dU = \sum_j y_{ij} \rho_j Q_j \qquad (5.15)$$

Where Q_j is volumetric rate of phase *j* at a given time *t*. The total production rate between *t* and $t+\Delta t$ is obtained by integrating over that time interval. Using (5.15), it may be given by an average value multiply by the time interval Δt.

$$\int_t^{t+\Delta t} \left(\int_{C.V.} q_i dU \right) dt = \left(\sum_j y_{ij} \rho_j Q_j \right)^m \times \Delta t \qquad (5.16)$$

Fluid production rates in multiphase flow are dependent on each other through at least relative permeability of different phases. In other words, the specification of the production rate of any phase implicitly dictates the production rates of the other phases. The concern in single-block wells is to estimate the production rate of phase *j* from wellblock *l* under different well operating conditions. (Abou-Kassem, 2006) are derived relations for the

The Compositional Simulator

production rates at different well operating conditions. It will be discussed in a follow up section.

The mass balance equation in discretized form can be obtained by substituting (5.4), (5.13) and (5.16) in (5.3).

$$\frac{U}{\Delta t}\left[\left(\phi \sum_j \rho_j y_{i,j} S_j\right)^{n+1} - \left(\phi \sum_j \rho_j y_{i,j} S_j\right)^n\right]$$

$$+\left\{\sum_k \left[\sum_j \left(\rho_j y_{i,j} V_j - \rho_j S_j \underline{D}_{i,j} \nabla y_{i,j}\right)\right] \cdot \mathrm{n}\right\} S_k\right\}^m$$

$$+\left(\sum_j y_{i,j} \rho_j Q_j\right)^m = 0 \qquad (5.17)$$

The equation (5.18) is for a grid block l of the discretized reservoir. If there is n_x grid blocks in the x-direction, n_y grid blocks in y-direction and n_z grid blocks in z-direction, there will be a total of $n_x \times n_y \times n_z$ discretized mass balance equations for component i. If there exist N_c components in the reservoir fluid, the total number of discretized mass balance equations is $N_c \times (n_x \times n_y \times n_z)$. The solution of these sets of algebraic equations provides the unknown variables.

5.1.2.2 Energy Balance Equation

The energy balance equation is given in (4.39) for the case of $T_f = T_r = T$ and in (4.40) and (4.41) for the case when $T_f \neq T_r$. These equations are integrated over the control volume C.V and the time interval Δt to obtain the control volume formulations.

Case a: Equal temperature $T_f = T_r = T$

We consider the case of equal temperature for the rock and fluid and the discretized formulation is derived for a grid block l. The partial differential equation for the energy balance equation for the case of $T_f = T_r = T$ is integrated over the finite control volume C.V and the time interval Δt.

$$\int_t^{t+\Delta t}\left\{\int_{C.V.}\frac{\partial}{\partial t}\left[\phi\sum_j \rho_j S_j\left(\hat{u}_j + g \cdot R_j\right)\right] + \frac{\partial}{\partial t}\left[(1-\phi)\rho_r C_r(T - T_0)\right] - \nabla \cdot (k\nabla T)\right.$$

$$\left. -\phi\left\{\nabla \cdot \left[\sum_j \rho_j S_j\left(h_j + g \cdot R_j\right)V_j\right]\right\} + e_s\right\} dU\right\} dt = 0$$

(5.18)

Using the divergence theorem, the integral of the divergent of the energy flux due to the conduction and convection over the control volume can be converted into a surface integral over the boundaries of the control volume S. Therefore, it may be written that:

$$\int_{t}^{t+\Delta t} \left\{ \int_{C.V.} \left[\frac{\partial}{\partial t} \left[\phi \sum_j \rho_j S_j \left(\hat{u}_j + g \cdot R_j \right) \right] + \frac{\partial}{\partial t} \left[(1-\phi) \rho_r C_r (T - T_0) \right] \right] dU - \int_S \left[(k\nabla T) + \phi \sum_j r_j S_j \left(h_j + g \cdot R_j \right) V_j \right] \cdot n \, ds + \int_{C.V.} e_s dU \right\} dt = 0$$

(5.19)

The energy balance equation in discretized form can be obtained with the same procedure as given for mass balance equation in 5.1.2.1. The final equation for the case of equal temperature is:

$$\left\{ \left[\phi \sum_j \rho_j S_j \left(\hat{u}_j + g \cdot R_j \right) + (1-\phi) \rho_r C_r (T - T_0) \right]^{n+1} \right. \\ \left. \left[\phi \sum_j \rho_j S_j \left(\hat{u}_j + g \cdot R_j \right) + (1-\phi) \rho_r C_r (T - T_0) \right]^n \right\} \\ + \left\{ \sum_k \left[(k\nabla T) + \phi \sum_j \rho_j S_j \left(h_j + g \cdot R_j \right) V_j \right] \cdot n \right\} S_k \right\}^m + (E_s)^m = 0$$

(5.20)

where $E_s = \int_{C.V.} e_s dU$.

Case b: Different temperature $T_f \neq T_r$

The equations (4.40) and (4.41) are the energy balance equations for the case where the rock and fluid temperatures are not equal. These partial differential equations are integrated over the finite control volume C.V and the time interval, Δt.

The Compositional Simulator

$$\begin{cases} \int\limits_{t}^{t+\Delta t} \left\{ \int\limits_{C.V.} \left\{ \begin{array}{l} \dfrac{\partial}{\partial t}\left[\phi \sum_j \rho_j S_j \left(\hat{u}_j + g \cdot R_j\right)\right] - \nabla \cdot \left(k_f \nabla T_f\right) \\ -\phi\left[\nabla \cdot \left(e_f V\right)\right] + h\left(T_r - T_f\right) + e_s \end{array} \right\} dU \right\} dt = 0 \\[2em] \int\limits_{t}^{t+\Delta t} \left\{ \int\limits_{C.V.} \left\{ \dfrac{\partial}{\partial t}\left[(1-\phi)\rho_r C_r \left(T_r - T_0\right)\right] - \nabla \cdot \left(k_f \nabla T_r\right) - h\left(T_r - T_f\right) \right\} dU \right\} dt = 0 \end{cases}$$

(5.21)

The notation $e_f = \Sigma_j \rho_j S_j (h_j + g \cdot R_j)$ is the total energy of a unit volume of the fluid system. These set of equations can be written by using the divergence theorem as follows:

$$\begin{cases} \int\limits_{t}^{t+\Delta t} \left\{ \int\limits_{C.V.} \left\{ \dfrac{\partial}{\partial t}\left[\phi \sum_j r_j S_j \left(\hat{u}_j + g \cdot R_j\right)\right] \right\} dU - \int\limits_{S} \left[k_f \nabla T_f + \phi e_f V\right] \cdot n \, ds \right\} dt = 0 \\[2em] \int\limits_{t}^{t+\Delta t} \left\{ \int\limits_{C.V.} \left\{ \begin{array}{l} \dfrac{\partial}{\partial t}\left[(1-\phi)r_r C_r \left(T_r - T_0\right)\right] \\ -h\left(T_r - T_f\right) \end{array} \right\} dU - \int\limits_{S} k_r \nabla T_r \cdot n \, ds \right\} dt = 0 \end{cases}$$

(5.22)

The discretized form of energy balance equation for the general case of $T_f \neq T_r$ is:

$$\begin{cases} \dfrac{U}{\Delta t}\left[\begin{array}{l} \left[\phi \sum_j r_j S_j \left(\hat{u}_j + g \cdot R_j\right)\right]^{n+1} \\ -\left[\phi \sum_j r_j S_j \left(\hat{u}_j + g \cdot R_j\right)\right]^{n} \end{array} \right] + \left\{ \begin{array}{l} \sum_k \left\{\left[k_f \nabla T_f + \phi e_f V\right] \cdot n\right\} S_k \\ + \int\limits_{C.V.} h\left(T_r - T_f\right) dU + E_s \end{array} \right\}^{m} = 0 \\[2em] \dfrac{U}{\Delta t}\left[\begin{array}{l} \left[(1-\phi)r_r C_r \left(T_r - T_0\right)\right]^{n+1} \\ -\left[(1-\phi)r_r C_r \left(T_r - T_0\right)\right]^{n} \end{array} \right] + \left\{ \begin{array}{l} \sum_k k_r \nabla T_r \cdot n S_k \\ -\int\limits_{C.V.} h\left(T_r - T_f\right) dU \end{array} \right\}^{m} = 0 \end{cases}$$

(5.23)

5.1.3 Discretization of Motion Equation

The motion equations are described in Chapter 4 section 4.2.6. They describe the relationship for fluid volumetric velocity. A petroleum reservoir is generally a heterogeneous anisotropic medium. In the most simplified form, the volumetric flux of a component α of the reservoir fluid is approximated by the Darcy law, $V_j = -\frac{K_j}{\mu_j}\nabla\Phi_j$ as given in (5.56). Using the notation of relative permeability, the Darcy law can be given in the following form.

$$V_j = -\underline{K}\frac{k_{rj}}{\mu_j}\nabla\Phi_j = -\underline{K}\lambda_j\nabla\Phi_j \qquad (5.24)$$

The notation λ_j is the mobility of phase j. The notation $\underline{K}$ is the space-dependent permeability tensor. Off-diagonal elements in the permeability tensor may exist if the coordinate directions are not aligned with the principal directions for $\underline{K}$.

If a one dimensional fluid flow is taken into account in x-direction, the discretized form of flow rate per unit cross-sectional area (the volumetric velocity) for a phase j from a block $i-1$ to block i as shown in Fig. 5.1 according to Darcy's law (5.24) is:

$$V_{jx}\big|_{x-\frac{1}{2}} = \frac{k_x k_{rj}}{\mu_j}\frac{\Phi_{j_{i-1}} - \Phi_{j_i}}{\Delta x_{i-\frac{1}{2}}} \qquad (5.25)$$

The potential difference between block $i-1$ and block i is

$$\Phi_{j_{i-1}} - \Phi_{j_i} = p_{j_{i-1}} - p_{j_i} - p_{j_{i-1}}g\left(z_{j_{i-1}} - z_{j_i}\right) \qquad (5.26)$$

for phase j. Substituting (5.26) in (5.25) yields

$$V_{jx}\big|_{x-\frac{1}{2}} = \frac{k_x k_{rj}}{\mu_j}\frac{\left(p_{j_{i-1}} - p_{j_i}\right) - p_{j_{i-1}}g\left(z_{j_{i-1}} - z_{j_i}\right)}{\Delta x_{i-\frac{1}{2}}} \qquad (5.27)$$

The equation (5.27) can be rewritten as:

$$V_{jx}\big|_{x-\frac{1}{2}} = \frac{k_x\big|_{x-\frac{1}{2}}}{\Delta x_{i-\frac{1}{2}}}\left(\frac{k_{rj}}{\mu_j}\right)\bigg|_{x-\frac{1}{2}}\left[\left(p_{j_{i-1}} - p_{j_i}\right) - p_{j_{i-1}}g\left(z_{j_{i-1}} - z_{j_i}\right)\right] \qquad (5.28)$$

The Compositional Simulator

Likewise, the fluid volumetric velocity of phase j from block $i+1$ to block i in Fig. 5.1 is expressed as:

$$V_{jx}\big|_{x-\frac{1}{2}} = \frac{k_x\big|_{x_{i+\frac{1}{2}}}}{\Delta x_{i+\frac{1}{2}}} \left(\frac{k_{rj}}{\mu_j}\right)\bigg|_{x_{i+\frac{1}{2}}} \left[(p_{j_{i+1}} - p_{j_i}) - p_{j_{i-1}} g(z_{j_{i+1}} - z_{j_i})\right] \quad (5.29)$$

Note that fluid volumetric velocity of phase j from block i to block $i+1$ is the negative of the value given by (5.28). It may be written that:

$$V_{jx}\big|_{x\mp\frac{1}{2}} = \frac{k_x\big|_{x_{i\mp\frac{1}{2}}}}{\Delta x_{i\mp\frac{1}{2}}} \left(\frac{k_{rj}}{\mu_j}\right)\bigg|_{x_{i\mp\frac{1}{2}}} \left[(p_{j_{i\pm1}} - p_{j_i}) - p_{j_{i-1}} g(z_{j_{i\pm1}} - z_{j_i})\right] \quad (5.30)$$

For multidimensional flow in rectangular coordinates, the fluid volumetric velocity of phase j from neighboring blocks to block (i, j, k) in the x, y, and z directions are:

$$V_{jx}\big|_{x\mp\frac{1}{2}} = \frac{k_x\big|_{x_{i\mp\frac{1}{2},j,k}}}{\Delta x_{i\mp\frac{1}{2},j,k}} \left(\frac{k_{rj}}{\mu_j}\right)\bigg|_{x_{i\mp\frac{1}{2},j,k}} \left[(p_{j_{i\pm1,j,k}} - p_{j_{i,j,k}}) - p_{j_{i-1}} g(z_{j_{i\pm1,j,k}} - z_{j_{i,j,k}})\right]$$

$$V_{jy}\big|_{y\mp\frac{1}{2}} = \frac{k_y\big|_{y_{i,j\mp\frac{1}{2},k}}}{\Delta} \left(\frac{k_{rj}}{\mu_j}\right)\bigg|_{x_{i,j\mp\frac{1}{2},k}} \left[(p_{j_{i,j\pm1,k}} - p_{j_{i,j,k}}) - p_{j_{i-1}} g(z_{j_{i,j\pm1,k}} - z_{j_{i,j,k}})\right]$$

$$V_{jz}\big|_{z\mp\frac{1}{2}} = \frac{k_z\big|_{z_{i,j,k\mp\frac{1}{2}}}}{\Delta z_{i,j,k\mp\frac{1}{2}}} \left(\frac{k_{rj}}{\mu_j}\right)\bigg|_{z_{i,j,k\mp\frac{1}{2}}} \left[(p_{j_{i,j,k\pm1}} - p_{j_{i,j,k}}) - p_{j_{i-1}} g(z_{j_{i,j,k\pm1}} - z_{j_{i,j,k}})\right]$$

$$(5.31)$$

The fluid volumetric velocity is $V_j = V_{jx}\hat{i} + V_{jy}\hat{j} + V_{jz}\hat{k}$.

5.2 Uniform Temperature Reservoir Compositional Flow Equations in a 1-D Domain

The numerical formulation for component mass balance equation and the energy balance equation with the Darcy law were explained in previous sections. The numerical model is based on the finite control volume method. The formulations should be combined to obtain a set of equations for practical applications. At first we consider a constant temperature reservoir and developed the compositional formulation for single and multidimensional reservoirs. We will combine the component mass balance equation with the Darcy law to obtain the flow equation. We start with the case of 1-D domain and then developed the formulation for a multidimensional domain.

The mass balance for component i for a block l in 1-D flow over a time step $\Delta t = t^{n+1} - t^n$ is derived in this chapter using the Darcy law as the motion equation which describes the volumetric velocity relationship. The general form of discretized form of mass balance equation for a 3-D flow is given in (5.17). It is assumed that there exists no mass transfer due to thermal and molecular diffusion. The mass balance equation for 1-D flow x-direction is simplified as, Fig. 5.1:

$$\sum_j \left\{ \frac{(U)_l}{\Delta t} \left[\left(\phi \rho_j y_{i,j} S_j\right)^{n+1} - \left(\phi \rho_j y_{i,j} S_j\right)^n \right] + \left[\begin{array}{l} \left(\rho_j y_{i,j}\right)_{i+1/2} V_{jx_{i+1/2}} A_{x_{i+1/2}} - \\ \left(\rho_j y_{i,j}\right)_{i-1/2} V_{jx_{i-1/2}} A_{x_{i-1/2}} \end{array} \right]^m + \left(y_{ij} \rho_j Q_j\right)^m \right\} = 0$$

(5.32)

Substituting (5.30) in (5.32), it can be written that:

$$\sum_j \left\{ \begin{array}{l} \left[\left(\rho_j y_{i,j}\right)_{i-1/2} \left(\frac{A_x k_x}{\Delta x}\right)_{x_{i-1/2}} \left(\frac{k_{rj}}{\mu_j}\right)\bigg|_{x_{i-1/2}} \left[\left(p_{j_{i-1}} - p_{j_i}\right) - \rho_{j_{i-1}} g \left(z_{j_{i-1}} - z_{j_i}\right) \right] \\ + \left(\rho_j y_{i,j}\right)_{i+1/2} \left(\frac{A_x k_x}{\Delta x}\right)_{x_{i+1/2}} \left(\frac{k_{rj}}{\mu_j}\right)\bigg|_{x_{i+1/2}} \left[\left(p_{j_{i+1}} - p_{j_i}\right) - \rho_{j_{i+1}} g \left(z_{j_{i+1}} - z_{j_i}\right) \right] \end{array} \right]^m \\ - \left(y_{ij} \rho_j Q_j\right)^m = \frac{(U\phi)_l}{\Delta t} \left[\left(\rho_j y_{i,j} S_j\right)^{n+1} - \left(\rho_j y_{i,j} S_j\right)^n \right]_l \end{array} \right\}$$

(5.33)

The Compositional Simulator

There may exist three phases of gas, oil and water, ($j = g$, o & w) in the reservoir fluid. To consider a simplified and practical sense, the oil and water phases are assumed immiscible; i.e., the hydrocarbon oil components do not dissolve in the water phase and the water component does not dissolve in the oil phase. Furthermore, the water phase does not transport any component other than the water component. Water component may evaporate into the gas phase. Therefore, two-phase oil and gas physical equilibrium describes mass transfer between the oil and gas phases, and the Dalton and Raoult laws describe the mass transfer of the water component between the water and gas phases. Let the oil phase contain n_c hydrocarbon components, the water phase consists of the water component only. Therefore, the gas phase consists of $n_c + 1$ components. For hydrocarbon component i, where $i = 1, \ldots, n_c$,

$$(\rho_g y_{i,g})^m_{l-1/2} \left(\frac{A_x k_x}{\Delta x}\right)_{x_{l-1/2}} \left(\frac{k_{rg}}{\mu_g}\right)^m_{x_{l-1/2}} \left[(p_{g_{l-1}} - p_{g_l}) - \rho_{g_{l-1}} g(z_{g_{l-1}} - z_{g_l})\right]$$

$$+ (\rho_g y_{i,g})^m_{l+1/2} \left(\frac{A_x k_x}{\Delta x}\right)_{x_{l+1/2}} \left(\frac{k_{rg}}{\mu_g}\right)^m_{x_{l+1/2}} \left[(p_{g_{l+1}} - p_{g_l}) - \rho_{g_{l+1}} g(z_{g_{l+1}} - z_{g_l})\right]$$

$$+ (\rho_o y_{i,o})^m_{l-1/2} \left(\frac{A_x k_x}{\Delta x}\right)_{x_{l-1/2}} \left(\frac{k_{ro}}{\mu_o}\right)^m_{x_{l-1/2}} \left[(p_{o_{l-1}} - p_{o_l}) - \rho_{o_{l-1}} g(z_{o_{l-1}} - z_{o_l})\right]$$

$$+ (\rho_o y_{i,o})^m_{l+1/2} \left(\frac{A_x k_x}{\Delta x}\right)_{x_{l+1/2}} \left(\frac{k_{ro}}{\mu_o}\right)^m_{x_{l+1/2}} \left[(p_{o_{l+1}} - p_{o_l}) - \rho_{o_{l+1}} g(z_{o_{l+1}} - z_{o_l})\right]$$

$$- (y_{ij} \rho_j Q_j + y_{io} \rho_o Q_o)^m = \frac{(U)_l}{\Delta t} \left[\begin{array}{c}(\phi \rho_g y_{i,g} S_g + \phi \rho_o y_{i,o} S_o)^{n+1} \\ -(\phi \rho_g y_{i,g} S_g + \phi \rho_o y_{i,o} S_o)^n\end{array}\right]_l.$$

(5.32a)

For the water component, where $i = n_c + 1$,

$$(\rho_g y_{i,g})^m_{l-1/2} \left(\frac{A_x k_x}{\Delta x}\right)_{x_{l-1/2}} \left(\frac{k_{rg}}{\mu_g}\right)^m_{x_{l-1/2}} \left[(p_{g_{l-1}} - p_{g_l}) - \rho_{g_{l-1}} g(z_{g_{l-1}} - z_{g_l})\right]$$

$$+ (\rho_g y_{i,g})^m_{l+1/2} \left(\frac{A_x k_x}{\Delta x}\right)_{x_{l+1/2}} \left(\frac{k_{rg}}{\mu_g}\right)^m_{x_{l+1/2}} \left[(p_{g_{l+1}} - p_{g_l}) - \rho_{g_{l+1}} g(z_{g_{l+1}} - z_{g_l})\right]$$

$$+\left(\rho_w\right)_{l-1/2}^m \left(\frac{A_x k_x}{\Delta x}\right)_{x_{l-1/2}} \left(\frac{k_{rw}}{\mu_w}\right)_{x_{l-1/2}}^m \left[\left(p_{w_{l-1}} - p_{w_l}\right) - \rho_{w_{l-1}} g\left(z_{w_{l-1}} - z_{w_l}\right)\right]$$

$$+\left(\rho_w\right)_{l+1/2}^m \left(\frac{A_x k_x}{\Delta x}\right)_{x_{l+1/2}} \left(\frac{k_{rw}}{\mu_w}\right)_{x_{l+1/2}}^m \left[\left(p_{w_{l+1}} - p_l\right) - \rho_{w_{l+1}} g\left(z_{w_{l+1}} - z_{w_l}\right)\right]$$

$$-\left(y_{ig}\rho_g Q_g + \rho_w Q_w\right)^m = \frac{(U)_l}{\Delta t}\left[\begin{array}{l}\left(\phi\rho_g y_{i,g} S_g + \phi\rho_w S_w\right)^{n+1} \\ -\left(\phi\rho_g y_{i,g} S_g + \phi\rho_w S_w\right)^n\end{array}\right]_l.$$

(5.33a)

The derivation (5.32a) and (5.33a) are based on the assumption of validity of Darcy's law to estimate the phase volumetric velocities between block i and its neighboring blocks $i-1$ and $i+1$. Such validity is widely accepted by petroleum engineers.

To simplify presenting the compositional mass balance equations in a multi dimensional domain, it is defined:

$$\left(\mathbb{T}_{i,o}^m\right)_{l\mp 1/2} = \left(\rho_o y_{i,o}\right)_{l\mp 1/2}^m \left(\frac{A_x k_x}{\Delta x}\right)_{x_{l\mp 1/2}} \left(\frac{k_{ro}}{\mu_o}\right)_{x_{l\mp 1/2}}^m = \left(\rho_o y_{i,o}\right)_{l\mp 1/2}^m \left(\mathbb{T}_o^m\right)_{l\mp 1/2}$$

(5.34)

where $i = 1, \ldots, n_c$;

$$\left(\mathbb{T}_{i,g}^m\right)_{l\mp 1/2} = \left(\rho_g y_{i,g}\right)_{l\mp 1/2}^m \left(\frac{A_x k_x}{\Delta x}\right)_{x_{l\mp 1/2}} \left(\frac{k_{rg}}{\mu_g}\right)_{x_{l\mp 1/2}}^m = \left(\rho_g y_{i,g}\right)_{l\mp 1/2}^m \left(\mathbb{T}_g^m\right)_{l\mp 1/2}$$

(5.35)

where $i = 1, \ldots, n_c$; and

$$\left(\mathbb{T}_{i,w}^m\right)_{l\mp 1/2} = \left(\rho_w\right)_{l\mp 1/2}^m \left(\frac{A_x k_x}{\Delta x}\right)_{x_{l\mp 1/2}} \left(\frac{k_{rw}}{\mu_w}\right)_{x_{l\mp 1/2}}^m = \left(\rho_w\right)_{l\mp 1/2}^m \left(\mathbb{T}_w^m\right)_{l\mp 1/2} \quad (5.36)$$

where $i = n_c + 1$. According to (Abou-Kassem, 2006), it may be defined that:

$$\left(\mathbb{T}_{i,j}^m\right)_{l\mp 1,l} = \left(\mathbb{T}_{i,j}^m\right)_{l\mp 1/2}^m = G_{l\mp 1/2}\left(\frac{k_{rj}}{\mu_j}\right)_{x_{l\mp 1/2}}^m = G_{l\mp 1,l}\left(\frac{k_{rj}}{\mu_j}\right)_{x_{l\mp 1,l}}^m \quad (5.37)$$

The Compositional Simulator

where $G_{l\mp 1,l} = G_{l\mp 1,l} = \left(\frac{A_x k_x}{\Delta x}\right)_{x_{l\mp 1/2}}$ are a geometric function. The notation $(T_{i,j}^m)_{l\mp 1,l}$ is called the transmissibility between the blocks $l \mp 1, l$ for the phase j. Using the definition (5.34), (5.35), (5.36) and (5.37), the mass balance equation for a hydrocarbon component takes a compact form as:

$$\left(\mathbb{T}_{i,g}^m\right)_{l-1,l}\begin{bmatrix}(p_{g_{l-1}}-p_{g_l}) \\ -\rho_{g_{l-1}}g(z_{g_{l-1}}-z_{g_l})\end{bmatrix} + \left(\mathbb{T}_{i,g}^m\right)_{l+1,l}\begin{bmatrix}(p_{g_{l+1}}-p_{g_l}) \\ -\rho_{g_{l+1}}g(z_{g_{l-1}}-z_{g_l})\end{bmatrix}$$

$$+ \left(\mathbb{T}_{i,o}^m\right)_{l-1,l}\begin{bmatrix}(p_{o_{l-1}}-p_{o_l}) \\ -\rho_{o_{l-1}}g(z_{o_{l-1}}-z_{o_l})\end{bmatrix} + \left(\mathbb{T}_{i,o}^m\right)_{l-1,l}\begin{bmatrix}(p_{o_{l+1}}-p_{o_l}) \\ -\rho_{o_{l-1}}g(z_{o_{l+1}}-z_{o_l})\end{bmatrix}$$

$$- \left(y_{ig}\rho_g Q_g + y_{io}\rho_o Q_o\right)^n = \frac{(U)_l}{\Delta t}\begin{bmatrix}\left(\phi\rho_g y_{i,g}S_g + \phi\rho_o y_{i,o}S_o\right)^{n+1} \\ -\left(\phi\rho_g y_{i,g}S_g + \phi\rho_o y_{i,o}S_o\right)^n\end{bmatrix}_l.$$

(5.38)

For the water component, where $i = n_c + 1$,

$$\left(\mathbb{T}_{i,g}^m\right)_{l-1,l}\begin{bmatrix}(p_{g_{l-1}}-p_{g_l}) \\ -\rho_{g_{l-1}}g(z_{g_{l-1}}-z_{g_l})\end{bmatrix} + \left(\mathbb{T}_{i,g}^m\right)_{l+1,l}\begin{bmatrix}(p_{g_{l+1}}-p_{g_l}) \\ -\rho_{g_{l+1}}g(z_{g_{l-1}}-z_{g_l})\end{bmatrix}$$

$$+ \left(\mathbb{T}_{i,w}^m\right)_{l-1,l}\begin{bmatrix}(p_{w_{l-1}}-p_{w_l}) \\ -\rho_{w_{l-1}}g(z_{w_{l-1}}-z_{wl})\end{bmatrix} + \left(\mathbb{T}_{i,w}^m\right)_{l-1,l}\begin{bmatrix}(p_{w_{l+1}}-p_{o_l}) \\ -\rho_{w_{l-1}}g(z_{w_{l+1}}-z_{wl})\end{bmatrix}$$

$$- \left(y_{ig}\rho_g Q_g + \rho_w Q_w\right)^n = \frac{(U)_l}{\Delta t}\begin{bmatrix}\left(\phi\rho_g y_{i,g}S_g + \phi\rho_w S_w\right)^{n+1} \\ -\left(\phi\rho_g y_{i,g}S_g + \phi\rho_w S_w\right)^n\end{bmatrix}_l.$$

(5.39)

The transmissibility between blocks l, n for a phase j can be given as:

$$\left(\mathbb{T}_{i,j}^m\right)_{l,n} = G_{l,n}\left(\frac{k_{rj}}{\mu_j}\right)_{l,n}^m = \left(\mathbb{T}_{j}^m\right)_{l,n}\left(\rho_j y_{i,j}\right)_{l,n}^m \quad (5.40)$$

176 ADVANCED PETROLEUM RESERVOIR SIMULATION

for $j = g, o$ & w. The directional geometric factors for the direction x, y and z according to Abou-Kassem (2006) are:

Table 5.1 Directional geometric factors in a rectangular grid (Abou-Kassem, 2006)

Direction	Geometric Factor
x	$G_{x_{i\mp 1/2,j,k}} = \dfrac{2}{\Delta x_{i,j,k}/\left(A_{x_{i,j,k}} k_{x_{i,j,k}}\right) + \Delta x_{i\mp 1,j,k}/\left(A_{x_{i\mp 1,j,k}} k_{x_{i\mp 1,j,k}}\right)}$
y	$G_{y_{i,j\mp 1/2,k}} = \dfrac{2}{\Delta y_{i,j,k}/\left(A_{x_{i,j,k}} k_{x_{i,j,k}}\right) + \Delta y_{i,j\mp 1,k}/\left(A_{x_{i,j\mp 1,k}} k_{x_{i,j\mp 1,k}}\right)}$
z	$G_{z_{i,j,k\mp 1/2}} = \dfrac{2}{\Delta z_{i,j,k}/\left(A_{x_{i,j,k}} k_{x_{i,j,k}}\right) + \Delta z_{i,j,k\mp 1}/\left(A_{x_{i,j,k\mp 1}} k_{x_{i,j,k\mp 1}}\right)}$

Following the works of Abou-Kassem (2006), the mass balance equation can be written for a block l in a compact form as:

$$\sum_{n=y_n}\left\{\left(\mathbb{T}_{i,g}^m\right)_{l-1,l}\left[\begin{array}{c}\left(p_{g_n}-p_{g_l}\right)\\-\rho_{g_n}g\left(z_{g_n}-z_{g_l}\right)\end{array}\right]+\left(\mathbb{T}_{i,o}^m\right)_{l,n}\left[\begin{array}{c}\left(p_{o_n}-p_{o_l}\right)\\-\rho_{o_n}g\left(z_{o_n}-z_{o_l}\right)\end{array}\right]\right\}$$
$$-\left(y_{ig}\rho_g Q_g + y_{io}\rho_o Q_o\right)^m = \frac{(U)_l}{\Delta t}\left[\begin{array}{c}\left(\phi\rho_g y_{i,g}S_g + \phi\rho_o y_{i,o}S_o\right)^{n+1}\\-\left(\phi\rho_g y_{i,g}S_g + \phi\rho_o y_{i,o}S_o\right)^n\end{array}\right]_l. \quad (5.41)$$

for a hydrocarbon component $i = 1, \ldots, n_c$. For the water component $i = n_c + 1$, the mass balance equation in a compact form is:

$$\sum_{n=y_n}\left\{\left(\mathbb{T}_{i,g}^m\right)_{l-1,l}\left[\begin{array}{c}\left(p_{g_n}-p_{g_l}\right)\\-\rho_{g_n}g\left(z_{g_n}-z_{g_l}\right)\end{array}\right]+\left(\mathbb{T}_{i,w}^m\right)_{l,n}\left[\begin{array}{c}\left(p_{o_n}-p_{o_l}\right)\\-\rho_{o_n}g\left(z_{o_n}-z_{o_l}\right)\end{array}\right]\right\}$$
$$-\left(y_{ig}\rho_g Q_g + \rho_w Q_w\right)^m = \frac{(U)_l}{\Delta t}\left[\begin{array}{c}\left(\phi\rho_g y_{i,g}S_g + \phi\rho_w S_w\right)^{n+1}\\-\left(\phi\rho_g y_{i,g}S_g + \phi\rho_w S_w\right)^n\end{array}\right]_l. \quad (5.42)$$

The Compositional Simulator

Where $\psi_n = \{l-1, l+1\}$ is the neighboring block to the block l. The equations (5.41) and (5.42) are for an interior grid block or in other words, are for blocks $l = 2, 3, \ldots, n_x - 1$. To include the boundary blocks, it may be possible to use a fictitious well instead of the boundary conditions and extend a general form of the formulation that can be applied for all grid blocks either interior blocks or boundary blocks. The general form of formulation is:

$$\sum_{n=y_n} \left\{ (\mathbb{T}_{i,g}^m)_{l,n} \begin{bmatrix} (p_{g_n} - p_{g_l}) \\ -\rho_{g_n} g(z_{g_n} - z_{g_l}) \end{bmatrix} + (\mathbb{T}_{i,o}^m)_{l,n} \begin{bmatrix} (p_{o_n} - p_{o_l}) \\ -\rho_{o_n} g(z_{o_n} - z_{o_l}) \end{bmatrix} \right\}$$

$$+ \sum_{n=x_n} (q_i^m)_{l,n} - \left(y_{ig} \rho_g Q_g + y_{io} \rho_o Q_o \right)_l^n = \frac{(U)_l}{\Delta t} \left[\begin{array}{c} \left(\phi \rho_g y_{i,g} S_g + \phi \rho_o y_{i,o} S_o \right)^{n+1} \\ - \left(\phi \rho_g y_{i,g} S_g + \phi \rho_o y_{i,o} S_o \right)^n \end{array} \right]_l$$

(5.43)

for a hydrocarbon component $i = 1, \ldots, n_c$. The notation $(q_i^m)_{l,n}$ is a fictitious well rate to model the boundary. It is equal to:

$$(q_i^m)_{l,n} = \left(y_{ig} \rho_g Q_g + y_{io} \rho_o Q_o \right)_{l,n}^m \qquad (5.44)$$

For the water component $i = n_c + 1$, the mass balance equation in a compact form for all grid blocks is:

$$\sum_{n=y_m} \left\{ (\mathbb{T}_{i,g}^m)_{l,n} \begin{bmatrix} (p_{g_n} - p_{g_l}) \\ -\rho_{g_n} g(z_{g_n} - z_{g_l}) \end{bmatrix} + (\mathbb{T}_{i,w}^m)_{l,n} \begin{bmatrix} (p_{o_n} - p_{o_l}) \\ -\rho_{o_n} g(z_{o_n} - z_{o_l}) \end{bmatrix} \right\}$$

$$+ \sum_{n=x_n} (q_i^m)_{l,n} - \left(y_{ig} \rho_g Q_g + \rho_w Q_w \right)_l^m = \frac{(U)_l}{\Delta t} \left[\begin{array}{c} \left(\phi \rho_g y_{i,g} S_g + \phi \rho_w S_w \right)^{n+1} \\ - \left(\phi \rho_g y_{i,g} S_g + \phi \rho_w y_w S_w \right)^n \end{array} \right]_l$$

(5.45)

$$(q_i^m)_{l,n} = \left(y_{ig} \rho_g Q_g + \rho_w Q_w \right)_{l,n}^m. \qquad (5.46)$$

5.3 Compositional Mass Balance Equation in a Multidimensional Domain

Reservoir blocks have a three-dimensional shape whether fluid flow is 1D, 2D, or 3D. The number of existing neighboring blocks and the number of reservoir boundaries shared by a reservoir block add up to six, as is the case in 3D flow. Existing neighboring blocks contribute to flow to or from the block, whereas reservoir boundaries may or may not contribute to flow depending on the dimensionality of flow and the prevailing boundary conditions. The dimensionality of flow implicitly defines those reservoir boundaries that do not contribute to flow at all. In 1D-flow problems, all reservoir blocks have four reservoir boundaries that do not contribute to flow. In 1D flow in the x-direction, the reservoir south (b_s), north (b_n), lower (b_l), and upper (b_u) boundaries do not contribute to flow to any reservoir block, including boundary blocks. These four reservoir boundaries (b_s, b_n, b_l, b_u) in this case are discarded as if they did not exist. As a result, an interior reservoir block has two neighboring blocks (one to the left and another to right of the block) and no reservoir boundaries, whereas a boundary reservoir block has one neighboring block (Block 2 for $n=1$ and Block n_x-1 for $n=n_x$) and one reservoir boundary (b_w for $n=1$ and b_E for $n=n_x$). In 2D flow problems, all reservoir blocks have two reservoir boundaries that do not contribute to flow at all. For example, in 2D flow in the xy plane the reservoir lower and upper boundaries do not contribute to flow to any reservoir block, including boundary blocks. These two reservoir boundaries (b_l, b_u) are discarded as if they did not exist. As a result, an interior reservoir block has four neighboring blocks and no reservoir boundaries; a reservoir block that falls on one reservoir boundary has three neighboring blocks and one reservoir boundary; and a reservoir block that falls on two reservoir boundaries has two neighboring blocks and two reservoir boundaries. In 3D flow problems, any of the six reservoir boundaries may contribute to flow depending on the specified boundary condition. An interior block has six neighboring blocks. It does not share any of its boundaries with any of the reservoir boundaries. A boundary block may fall on one, two, or three of the reservoir boundaries. Therefore, a boundary block that falls on one, two, or three reservoir boundaries has five, four, or three neighboring blocks, respectively. If the notation Ψ_n is a set showing the neighboring blocks and ξ_n denotes a set

The Compositional Simulator

showing the number of boundaries, the above discussion leads to a few conclusions related to the number of elements contained in sets ψ_n and ξ_n:

1) For an interior reservoir block, set ψ_n contains two, four, or six elements for a 1D, 2D, or 3D flow problem, respectively, and set ξ_n contains no elements or, in other words, is empty.
2) For a boundary reservoir block, set ψ_n contains less than two, four, or six elements for a 1D, 2D, or 3D flow problem, respectively, and set ξ_n is not empty.
3) The sum of the number of elements in sets ψ_n and ξ_n for any reservoir block is a constant that depends on the dimensionality of flow. This sum is two, four, or six for a 1D, 2D, or 3D flow problem, respectively.

Therefore, equations (5.43) and (5.45) are applicable for 1D, 2D, or 3D flow problems. In 1D flow in the x-direction (see Fig. 5.7a), ψ_n for a block $n \equiv i$ has a maximum of two elements of $[(i-1), (i+1)]$, and ξ_n is empty or has a maximum of one element of (b_w, b_E). In 2D flow in the x-y plane (Fig. 5.7b), ψ_n for $n \equiv (i,j)$ has a maximum of four elements of $[(i-1,j), (i, j-1), (i, (j+1), (i+1,j)]$, and ξ_n is empty or has a maximum of two element of (b_S, b_w, b_E, b_N). In 3D flow in the x-y-z plane (Fig. 5.7c), Ψ_n for $n \equiv (i, j, k)$ has a maximum of six elements of $[(i-1, j, k), (i, j-1, k), (i, j, (k-1), (i+1, j, k), (i, j+1, k), (i, j, k+1)]$, and ξ_n is empty or has a maximum of three element of $(b_L, b_S, b_w, b_E, b_N, b_U)$. The relations for the fictitious well, (5.44) and (5.45), are not only valid for the (i, j, k) notation but also valid for any block-ordering scheme. As for example, if blocks are ordered using natural

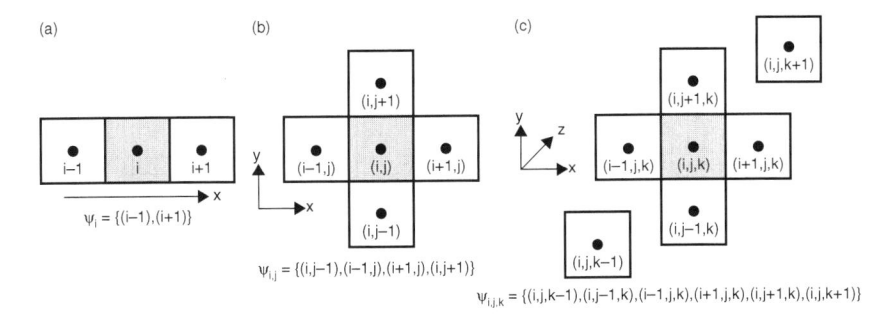

Figure 5.7 A Block and its neighboring blocks in 1D, 2D, and 3D using engineering notation.

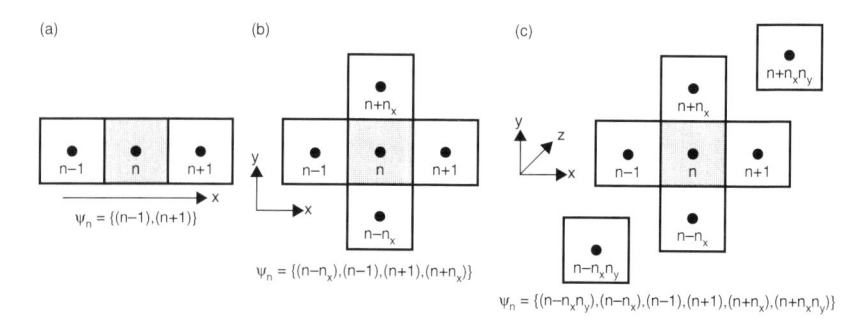

Figure 5.8 A block and its neighboring blocks in 1D, 2D, and 3D using natural ordering.

ordering with blocks being ordered along the x-axis, followed the y-axis, and finally along the z-axis, then the set of neighboring blocks are defined as shown in Fig. 5.8. That is to say, for block n, $\psi_n = 0\ [(n-1), (n+1)]$ for 1D flow along the x-axis (Fig. 5.8a), $\psi_n = 0\ [(n-n_x), (n-1), (n+1), (n+n_x)]$ for 2D flow in the x-y plane (Fig. 5.8b), and $\Psi_n = 0\ [(n-n_x n_y), (n-n_x), (n-1), (n+1)\ (n+n_x), (n+n_x n_y)]$ for 3D flow in the xyz space (Fig. 5.8c). In summary, $\psi_n = 0$ a set whose elements are the existing neighboring blocks to block n in the reservoir, $\xi_n = 0$ a set whose elements are the reservoir boundaries (b_L, b_S, b_w, b_E, b_N, b_u) that are shared by block n. It is either an empty set for interior blocks or a set that contains one element for boundary blocks that fall on one reservoir boundary, two elements for boundary blocks that fall on two reservoir boundaries, or three elements for blocks that fall on three reservoir boundaries. An empty set implies that the block does not fall on any reservoir boundary; i.e., block n is an interior block, and hence $\dot{\Sigma}_{l=\xi_n}(q_i^m)_{l,n}$ for $i = 1, 2, \cdots, n_c + 1$. The notation $(q_i^m)_{l,n}$ is component i flow rate of the fictitious well that represents transfer of component i between reservoir boundary l and block n because of a boundary condition.

5.3.1 Implicit Formulation of Compositional Model in Multi-Dimensional Domain

The explicit, implicit, and Crank-Nicolson formulations are derived from (5.43) and (5.45) by specifying the approximation of time t^m as t^n, $t^{n+1/2}$, or t^{n+1}, which are illustrated in Fig. 5.5. The explicit formulation, however, is not used in multiphase flow because of time step limitations, and the Crank-Nicolson formulation is not commonly

THE COMPOSITIONAL SIMULATOR

used. Consequently, we limit our presentation to the implicit formulation. In the following equations, fluid gravity is dated at old time level n instead of new time level $n+1$, as this approximation does not introduce any noticeable errors (Coats, George, and Marcum, 1974). The implicit model is presented below for a hydrocarbon component $i = 1, \ldots, n_c$.

$$\sum_{n=y_n} \left\{ (\mathbb{T}_{i,g}^{n+1})_{l,n} \begin{bmatrix} (p_{g_n} - p_{g_l})^{n+1} \\ -\rho_{g_n}^{n+1} g(z_{g_n} - z_{g_l}) \end{bmatrix} + (\mathbb{T}_{i,o}^{n+1})_{l,n} \begin{bmatrix} (p_{o_n} - p_{o_l})^{n+1} \\ -\rho_{o_n}^{n+1} g(z_{o_n} - z_{o_l}) \end{bmatrix} \right\}$$

$$+ \sum_{n=x_n} (q_i^{n+1})_{l,n} - (y_{ig}\rho_g Q_g + y_{io}\rho_o Q_o)_l^{n+1} = \frac{(U)_l}{\Delta t} \begin{bmatrix} (\phi\rho_g y_{i,g} S_g + \phi\rho_o y_{i,o} S_o)^{n+1} \\ -(\phi\rho_g y_{i,g} S_g + \phi\rho_o y_{i,o} S_o)^n \end{bmatrix}_l$$

(5.47)

Where (q_i^{n+1}) is a fictitious well rate to model the boundary.

$$(q_i^{n+1})_{l,n} = (y_{ig}\rho_g Q_g + y_{io}\rho_o Q_o)_{l,n}^{n+1} \tag{5.48}$$

The implicit model for the water component $i = n_c + 1$ is in the form:

$$\sum_{n=y_m} \left\{ (\mathbb{T}_{i,g}^{n+1})_{l,n} \begin{bmatrix} (p_{g_n} - p_{g_l})^{n+1} \\ -\rho_{g_n}^{n+1} g(z_{g_n} - z_{g_l}) \end{bmatrix} + (\mathbb{T}_{i,w}^{n+1})_{l,n} \begin{bmatrix} (p_{o_n} - p_{o_l})^{n+1} \\ -\rho_{o_n}^{n+1} g(z_{o_n} - z_{o_l}) \end{bmatrix} \right\}$$

$$+ \sum_{n=x_n} (q_i^{n+1})_{l,n} - (y_{ig}\rho_g Q_g + \rho_w Q_w)_l^{n+1} = \frac{(U)_l}{\Delta t} \begin{bmatrix} (\phi\rho_g y_{i,g} S_g + \phi\rho_w S_w)^{n+1} \\ -(\phi\rho_g y_{i,g} S_g + \phi\rho_w S_w)^n \end{bmatrix}_l$$

(5.49)

where:

$$(q_i^{n+1})_{l,n} = (y_{ig}\rho_g Q_g + \rho_w Q_w)_{l,n}^{m} \tag{5.50}$$

$$(\mathbb{T}_{i,o}^{n+1})_{l,n} = (\mathbb{T}_o^{n+1})_{l,n} (y_{io}\rho_o) \quad \text{for} \quad i = 1, 2, \cdots, n_c \tag{5.51}$$

$$\left(\mathbb{T}_{i,g}^{n+1}\right)_{l,n} = \left(\mathbb{T}_{g}^{n+1}\right)_{l,n} \left(y_{ig}\rho_g\right) \quad \text{for} \quad i = 1,2,\cdots,n_c+1 \qquad (5.52)$$

$$\left(\mathbb{T}_{i,w}^{n+1}\right)_{l,n} = \left(\mathbb{T}_{w}^{n+1}\right)_{l,n} \left(y_{iw}\rho_w\right) \quad \text{for} \quad i = n_c+1 \qquad (5.53)$$

$$\left(\mathbb{T}_{j}^{n+1}\right)_{l,n} = G_{l,n}\left(\frac{k_{rj}}{\mu_j}\right)_{l,n}^{n+1} \quad \text{for} \quad i = g,o,w \qquad (5.54)$$

Each block contributes $n_c + 6$ equations:

- $n_c + 1$ component flow equations;
- two capillary pressure relationships, and
- three constraint equations.

There are $n_c + 6$ unknowns in each block:

- n_c-mass fractions in the oil phase $y_{i,o}$ for $i = 1,2,\ldots,n_c$;
- three-phase pressures (p_g, p_o, p_w); and
- three-phase saturations (S_g, S_o, S_w).

The component flow equation for one of the oil components, say the lightest oil component ($c = 1$), can be replaced with a pressure equation. The pressure equation is obtained by adding all $n_c + 1$ component flow equations. Thee mass fraction constraint for oil and gas phases are used when necessary to replace the sum of mass fractions in the oil and gas phases with the value of one. The resulting pressure equation in this case is

$$\sum_{n=y_m} \left\{ \begin{aligned} &\left(\mathbb{T}_g^{n+1}\right)_{l,n} \rho_g^{n+1} \begin{bmatrix} \left(p_{g_n} - p_{g_l}\right)^{n+1} \\ -\rho_{g_n}^n g\left(z_{g_n} - z_{g_l}\right) \end{bmatrix} \\ &+ \left(\mathbb{T}_o^{n+1}\right)_{l,n} \rho_o^{n+1} \begin{bmatrix} \left(p_{o_n} - p_{o_l}\right)^{n+1} \\ -\rho_{o_n}^n g\left(z_{o_n} - z_{o_l}\right) \end{bmatrix} \\ &+ \left(\mathbb{T}_w^{n+1}\right)_{l,n} \rho_w^{n+1} \begin{bmatrix} \left(p_{w_n} - p_{w_l}\right)^{n+1} \\ -\rho_{w_n}^n g\left(z_{o_n} - z_{o_l}\right) \end{bmatrix} \end{aligned} \right\} + \sum_{n=x_n}\left(\rho_g Q_g + \rho_o Q_o + \rho_w Q_w\right)_{l,n}^{n+1}$$

$$-\left(\rho_g Q_g + \rho_o Q_o + \rho_w Q_w\right)^{n+1} = \frac{(U)_l}{\Delta t}\left[\begin{array}{c}\left(\phi\rho_g y_{i,g} S_g + \phi\rho_o y_{i,o} S_o\right)^{n+1} \\ -\left(\phi\rho_g y_{i,g} S_g + \phi\rho_o y_{i,o} S_o\right)^n\end{array}\right]_l \quad (5.55)$$

5.3.2 Reduced Equations of Implicit Compositional Model in Multidimensional Domain

The compositional model reduced equations are obtained by eliminating the equations, which do not contain flow terms from the set of equations comprising the model. We limit our presentation to the $p_o - S_w - S_g$ formulation, i.e., the formulation that uses p_o, S_w and S_g as the primary unknowns for phase flow in the reservoir. The secondary unknowns in this formulation are p_g, p_w, and S_o. The primary unknowns in composition are the n_c mass fractions in oil phase, ($y_{i,o}$ for $i = 1,2,\ldots,n_c$). Other formulations using $p_o - p_w - p_g$, $p_o - p_{cow} - p_{cgo}$, $p_o - p_{cow} - S_g$ or $p_o - S_w - p_{cgo}$ break down for negligible or zero capillary pressures. To obtain the reduced set of equations of the model for each block, we express the secondary unknowns in the component equations in terms of the primary unknowns and eliminate the secondary unknowns from the flow equations. In the reduced set of equations, each block contributes $n_c + 3$ equations:

- the oil-phase composition constraint;
- the gas-phase composition constraint;
- $n_c + 1$ oil component flow equations;
- the water component flow equation; and
- the pressure equation.

The $n_c + 3$ primary unknowns in each block are:

- n_c mass fractions in the oil phase $y_{i,o}$ for $i = 1, 2, \ldots, n_c$;
- the saturation of the gas and water phases S_g, S_w; and
- the pressure of the oil-phase p_o.

The equations used to eliminate the secondary unknowns (p_w, p_g, S_o) are the capillary pressure relationships:

$$p_w = p_o - P_{cow}(S_w) \quad \text{and} \quad p_g = p_o - P_{cgo}(S_g); \quad (5.56)$$

and the phase saturation constraint equation.

$$S_o = 1 - S_w - S_g \qquad (5.57)$$

Once the primary unknowns are solved for, the phase saturation constraint and capillary pressure relationships are used to solve for the secondary unknowns for each reservoir block.

The equations and primary unknowns for each reservoir block are aligned in order to obtain a diagonally dominant matrix and to alleviate the need for pivoting during forward Gaussian elimination. The model-reduced equations have the following order:

- the gas-phase composition constraint equation;

$$\sum_{i=1}^{n_c+1} y_{i,g}^{n+1} = 1 \qquad (5.58)$$

- the oil-phase composition constraint equation;

$$\sum_{i=1}^{n_c} y_{i,o}^{n+1} = 1 \qquad (5.59)$$

- the second lightest oil component flow equation ($c = 2$);

$$\sum_{n=y_n} \left\{ \begin{array}{c} \left(\mathbb{T}_{2,g}^{n+1}\right)_{l,n} \left[\begin{array}{c} \left(p_{g_n} - p_{g_l}\right)^{n+1} \\ -\rho_{g_n}^{n+1} g\left(z_{g_n} - z_{g_l}\right) \end{array} \right] \\ + \left(\mathbb{T}_{2,o}^{n+1}\right)_{l,n} \left[\begin{array}{c} \left(p_{o_n} - p_{o_l}\right)^{n+1} \\ -\rho_{o_n}^{n+1} g\left(z_{o_n} - z_{o_l}\right) \end{array} \right] \end{array} \right\} + \sum_{n=x_n} \left(q_2^{n+1}\right)_{l,n} - \left(\begin{array}{c} y_{2,g} \rho_g Q_g \\ + y_{2,o} \rho_o Q_o \end{array} \right)_l^{n+1}$$

$$= \frac{(U)_l}{\Delta t} \left[\left(\phi \rho_g y_{2,g} S_g + \phi \rho_o y_{2,o} S_o\right)_l^{n+1} - \left(\phi \rho_g y_{2,g} S_g + \phi \rho_o y_{2,o} S_o\right)_l^n \right]$$

$$(5.60)$$

- the third lightest oil component flow equation ($c = 3$);

The Compositional Simulator

$$\sum_{n=y_n}\left\{\begin{array}{c}\left(\mathbb{T}_{3,g}^{n+1}\right)_{l,n}\left[\begin{array}{c}\left(p_{g_n}-p_{g_l}\right)^{n+1}\\-\rho_{g_n}^{n+1}g\left(z_{g_n}-z_{g_l}\right)\end{array}\right]\\+\left(\mathbb{T}_{3,o}^{n+1}\right)_{l,n}\left[\begin{array}{c}\left(p_{o_n}-p_{o_l}\right)^{n+1}\\-\rho_{o_n}^{n+1}g\left(z_{o_n}-z_{o_l}\right)\end{array}\right]\end{array}\right\}+\sum_{n=x_n}\left(q_3^{n+1}\right)_{l,n}-\left(\begin{array}{c}y_{3,g}\rho_g Q_g\\+y_{3,o}\rho_o Q_o\end{array}\right)_l^{n+1}$$

$$=\frac{(U)_l}{\Delta t}\left[\begin{array}{c}\left(\phi\rho_g y_{3,g}S_g+\phi\rho_o y_{3,o}S_o\right)^{n+1}\\-\left(\phi\rho_g y_{3,g}S_g+\phi\rho_o y_{3,o}S_o\right)^n\end{array}\right]_l \qquad (5.61)$$

- the heaviest oil component flow equation ($c = n_c$)

$$\sum_{n=y_n}\left\{\begin{array}{c}\left(\mathbb{T}_{n_c,g}^{n+1}\right)_{l,n}\left[\begin{array}{c}\left(p_{g_n}-p_{g_l}\right)^{n+1}\\-\rho_{g_n}^{n+1}g\left(z_{g_n}-z_{g_l}\right)\end{array}\right]\\+\left(\mathbb{T}_{n_c,o}^{n+1}\right)_{l,n}\left[\begin{array}{c}\left(p_{o_n}-p_{o_l}\right)^{n+1}\\-\rho_{o_n}^{n+1}g\left(z_{o_n}-z_{o_l}\right)\end{array}\right]\end{array}\right\}+\sum_{n=x_n}\left(q_{n_c}^{n+1}\right)_{l,n}-\left(\begin{array}{c}y_{n_c,g}\rho_g Q_g\\+y_{n_c,o}\rho_o Q_o\end{array}\right)_l^{n+1}$$

$$=\frac{(U)_l}{\Delta t}\left[\begin{array}{c}\left(\phi\rho_g y_{n_c,g}S_g+\phi\rho_o y_{n_c,o}S_o\right)^{n+1}\\-\left(\phi\rho_g y_{n_c,g}S_g+\phi\rho_o y_{n_c,o}S_o\right)^n\end{array}\right]_l \qquad (5.62)$$

- the water component flow equation ($c = n_c + 1$)

$$\sum_{n=y_m}\left\{\begin{array}{c}\left(\mathbb{T}_{n_c+1,g}^{n+1}\right)_{l,n}\left[\begin{array}{c}\left(p_{g_n}-p_{g_l}\right)^{n+1}\\-r_{g_n}^{n+1}g\left(z_{g_n}-z_{g_l}\right)\end{array}\right]\\+\left(\mathbb{T}_{n_c+1,w}^{n+1}\right)_{l,n}\left[\begin{array}{c}\left(p_{o_n}-p_{o_l}\right)^{n+1}\\-r_{o_n}^{n+1}g\left(z_{o_n}-z_{o_l}\right)\end{array}\right]\end{array}\right\}+\sum_{n=x_n}\left(q_{n_c+1}^{n+1}\right)_{l,n}-\left(\begin{array}{c}y_{n_c+1,g}r_g Q_g\\+r_w Q_w\end{array}\right)_l^{n+1}$$

$$=\frac{(U)_l}{\Delta t}\left[\begin{array}{c}\left(\phi r_g y_{n_c+1,g}S_g+\phi r_w S_w\right)^{n+1}\\-\left(\phi r_g y_{n_c+1,g}S_g+\phi r_w S_w\right)^n\end{array}\right]_l \qquad (5.63)$$

- and the pressure equation.

$$\sum_{n=y_n}\left\{\begin{array}{c}\left(\mathbb{T}_g^{n+1}\right)_{l,n}\rho_g^{n+1}\left[\begin{array}{c}\left(p_{g_n}-p_{g_l}\right)^{n+1}\\-\rho_{g_n}^n g\left(z_{g_n}-z_{g_l}\right)\end{array}\right]\\+\left(\mathbb{T}_o^{n+1}\right)_{l,n}\rho_o^{n+1}\left[\begin{array}{c}\left(p_{o_n}-p_{o_l}\right)^{n+1}\\-\rho_{o_n}^n g\left(z_{o_n}-z_{o_l}\right)\end{array}\right]\\+\left(\mathbb{T}_w^{n+1}\right)_{l,n}\rho_w^{n+1}\left[\begin{array}{c}\left(p_{w_n}-p_{w_l}\right)^{n+1}\\-\rho_{w_n}^n g\left(z_{o_n}-z_{o_l}\right)\end{array}\right]\end{array}\right\}+\sum_{n=x_n}\left(\begin{array}{c}\rho_g Q_g\\+\rho_o Q_o\\+\rho_w Q_w\end{array}\right)_{l,n}^{n+1}-\left(\begin{array}{c}\rho_g Q_g\\+\rho_o Q_o\\+\rho_w Q_w\end{array}\right)_l^{n+1}$$

$$=\frac{(U)_l}{\Delta t}\left[\begin{array}{c}\left(\phi\rho_g y_{i,g}S_g+\phi\rho_o y_{i,o}S_o\right)^{n+1}\\-\left(\phi\rho_g y_{i,g}S_g+\phi\rho_o y_{i,o}S_o\right)^n\end{array}\right]_l$$

(5.64)

With this choice of equation ordering, the primary unknowns have the order of $S_g, y_{1,o}, y_{2,o}, y_{3,o}, \cdots y_{n_c-1,o}, y_{n_c,o}, S_w$ and p_o. Therefore, the vector of primary unknowns for block l is:

$$X_n = \left[S_g\ y_{1,o}\ y_{2,o}\ y_{3,o}\ \cdots\ y_{n_c-1,o}\ y_{n_c,o}\ S_w\ p_o\right]_l^T \quad (5.65)$$

5.3.3 Well Production and Injection Rate Terms

5.3.3.1 Production Wells

The production rates of component i associated with phases $j = g, o, w$ from a single well located in block l are expressed in the following form:

$$q_{i,l}^{n+1} = \left(y_{ig}\rho_g Q_g + y_{io}\rho_o Q_o\right)_l^{n+1} \quad \text{for} \quad i=1,2,\cdots,n_c \quad (5.66)$$

$$q_{i,l}^{n+1} = \left(y_{ig}\rho_g Q_g + \rho_w Q_w\right)_l^{n+1} \quad \text{for} \quad i = n_c + 1 \quad (5.67)$$

The Compositional Simulator

Fluid production rates in multiphase flow are dependent on each other through at least relative permeabilities. In other words, the specification of the production rate of any phase implicitly dictates the production rates of the other phases. The concern in single-block wells is to estimate the production rate of phase $j = g, o, w$ from well-block l under different well operating conditions. The equations presented in this section are derived from the recent work of Abou-Kassem (2006).

a. Shut-in well

$$Q_{j,l} = 0 \quad \text{for} \quad j = g, o, w \tag{5.68}$$

b. Specified well flow rate

$$Q_{j,l} = G_{w_l} \left(\frac{k_{rj}}{\mu_j} \right)_l \left(p_o - p_{wf} \right)_l \tag{5.69}$$

Where G_{w_l} is estimated as suggested by Peaceman (1983) and p_{wf_l} is estimated from the well rate specification Q_{s_j} using:

$$p_{wf_l} = p_{o_l} + \frac{Q_{sj}}{G_{w_l} \sum_{j \in \eta_{prd}} M_{j_l}} \tag{5.70}$$

Where η_{prd} and M_j depend on the type of well rate specification as listed in Table 5.2.

Table 5.2 Well rate specification and definitions of set η_{prd} and M_j

Well Rate Specification, Q_{s_j}	Set of Specified Phases, η_{prd}	Phase Relative Mobility, M_j
Q_{os_j}	{o}	k_{rj}/μ_j
Q_{Ls_j}	{o, w}	k_{rj}/μ_j
Q_{Ts_j}	{o, w, g}	k_{rj}/μ_j

Substitution (6.62) into (6.61) yields:

$$Q_{j,l} = \left(\frac{k_{rj}}{\mu_j}\right)_l \frac{Q_{sj}}{G_{w_l} \sum_{j \in n_{prd}} M_{jl}} \quad \text{for} \quad j = g, o, w. \tag{5.71}$$

c. Specified well pressure gradient

For a specified well pressure gradient, the production rate of phase $j = g, o, w$ from well-block l is given by:

$$Q_{j,l} = 2\pi r_w k_{H_l} h_l \left(\frac{k_{rj}}{\mu_j}\right)_l \left.\frac{\partial p}{\partial r}\right|_{r_w}. \tag{5.72}$$

d. Specified well FBHP

If the FBHP of a well $p_{wf_{ref}}$ is specified, then the production rate of phase $j = g, o, w$ from well-block l can be estimated using (5.69) using $p_{wf_n} = p_{wf_{ref}}$.

5.3.3.2 Injection Wells

For injection wells, one phase (usually water, gas of known composition, or oil of known composition) is injected. The mobility of the injected fluid at reservoir conditions in a well-block is equal to the sum of the mobilities of all phases present in the well-block.

$$M_{inj} = \sum_{j \in \eta_{inj}} M_{jl} \tag{5.73}$$

Where the notations η_{inj} and M_j are:

$$\eta_{inj} = \{o, w, g\}, \quad M_j = k_{rj}/\mu_j \tag{5.74}$$

The concern here is to estimate the injection rate of the injected phase (usually water or gas) into well-block l under different well operating conditions. Of course, the rates of injection of the remaining phases are set to zero.

a. Shut-in well

$$Q_{j,l} = 0 \quad \text{for} \quad j = g, o, w \tag{5.75}$$

The Compositional Simulator

b. Specified well flow rate

The injection rate of the injected fluid $j = w$ or g into well-block l is given by

$$Q_{j,l} = -G_{w_l} M_{inj_l} \left(p_o - p_{wf} \right) \quad (5.76)$$

For a single-block well $Q_{j,l} = Q_{s_j}$ and (5.76) is used to estimate p_{wf_l}.

$$p_{wf_l} = p_l + \frac{Q_{sj}}{G_{w_l} \left(\dfrac{M_{inj}}{B_j} \right)_l} \quad (5.77)$$

Then, the injection rate of the injected fluid $j = w$ or g into well-block l ($Q_{j,l}$) is estimated using (5.76). The use of (5.76) requires solving for p_{wf_l} implicitly along with the reservoir block pressures. An explicit treatment, however, uses (5.77) at old time level n to estimate $p_{wf_l}^n$, which is subsequently substituted into (5.76) to estimate the injection rate of the injected phase $j = w$ or g into well-block l.

c. Specified well pressure gradient

For a specified well pressure gradient, the injection rate of fluid $j = w$ or g into well-block l is given by:

$$Q_{j,l} = -2\pi r_w k_{H_l} h_l M_{inj_l} \left. \frac{\partial p}{\partial r} \right|_{r_w} . \quad (5.78)$$

d. Specified well FBHP

If the FBHP of a well p_{wf_l} is specified, then the injection rate of the injected fluid $j = w$ or g into well-block l can be estimated using (5.76).

5.3.4 Fictitious Well Rate Terms (Treatment of Boundary Conditions)

Component i production rates from fictitious wells are:

$$\left(q_i^{n+1} \right)_{l,n} = \left(y_{ig} \rho_g Q_g + y_{io} \rho_o Q_o \right)_{l,n}^{n+1} \quad \text{for} \quad i = 1, 2, \cdots, n_c \quad (5.79)$$

$$\left(q_i^{n+1} \right)_{l,n} = \left(y_{ig} \rho_g Q_g + \rho_w Q_w \right)_{l,n}^{n+1} \quad \text{for} \quad i = n_c + 1. \quad (5.80)$$

A reservoir boundary can be subject to one of four conditions:

1. no-flow boundary;
2. constant flow boundary;
3. constant pressure gradient boundary;
4. constant pressure boundary.

The fictitious well rate equations presented are derived from those presented by Abou-Kassem (2006). The fictitious well rate of phase j, $(q_j^{n+1}{}_{b,bB})$ reflects fluid transfer of phase j between the boundary block, (bB), and the reservoir boundary itself, (b), or the block next to the reservoir boundary that falls outside the reservoir. In multiphase flow, a reservoir boundary may (1) separate two segments of one reservoir that has the same fluids, (2) separate an oil reservoir from a water aquifer, or (3) seal off the reservoir from a neighboring reservoir. If the neighboring reservoir segment is an aquifer, then either water invades the reservoir across the reservoir boundary Water Oil Contact (WOC) or reservoir fluids leave the reservoir block to the aquifer.

a. Specified Pressure Gradient Boundary Condition

For a specified pressure gradient at the reservoir left (west) boundary,

$$q_j^{n+1}{}_{b,bB} = -\left(\frac{k_l k_{rj} A_l}{\mu_j}\right)^{n+1}_{bB}\left[\left.\frac{\partial p_j}{\partial l}\right|^{n+1}_b - (\rho_j g)^n_{bB} \left.\frac{\partial z}{\partial l}\right|_b\right] \quad \text{for} \quad j = g,o,w \quad (5.81)$$

and at the reservoir right (east) boundary, and at the reservoir right (east) boundary,

$$q_j^{n+1}{}_{b,bB} = \left(\frac{k_l k_{rj} A_l}{\mu_j}\right)^{n+1}_{bB}\left[\left.\frac{\partial p_j}{\partial l}\right|^{n+1}_b - (\rho_j g)^n_{bB} \left.\frac{\partial z}{\partial l}\right|_b\right] \quad \text{for} \quad j = g,o,w. \quad (5.82)$$

In (5.73) and (5.74), the specified pressure gradient may replace the phase pressure gradient at the boundary. These two equations apply to fluid flow across a reservoir boundary that separates two segments of the same reservoir or across a reservoir boundary that represents WOC with fluids being lost to the water aquifer.

The Compositional Simulator

If the reservoir boundary represents WOC and water invades the reservoir, then

$$q^{n+1}_{j\ b,bB} = \left(\frac{k_l A_l}{\mu_j}\right)^{n+1}_{bB} (k_{rw})^{n+1}_{aq} \left[\left.\frac{\partial p_j}{\partial l}\right|_b^{n+1} - (\rho_w g)^n_{bB} \left.\frac{\partial z}{\partial l}\right|_b\right] \quad (5.83)$$

for the reservoir left (west) boundary, and

$$q^{n+1}_{w\ b,bB} = \left(\frac{k_l A_l}{\mu_j}\right)^{n+1}_{bB} (k_{rw})^{n+1}_{aq} \left[\left.\frac{\partial p_j}{\partial l}\right|_b^{n+1} - (\rho_w g)^n_{bB} \left.\frac{\partial z}{\partial l}\right|_b\right] \quad (5.84)$$

for the reservoir right (east) boundary. Moreover,

$$q^{n+1}_{o\ b,bB} = q^{n+1}_{g\ b,bB} = 0 \quad (5.85)$$

note that, in (5.75) and (5.76), the rock and fluid properties in the aquifer are approximated by those of the boundary block properties because of the lack of geologic control in aquifers and because the effect of oil/water capillary pressure is neglected. In addition, $(k_{rw})^{n+1}_{aq} = 1$ because $S_w = 1$ in the aquifer.

b. Specified Flow Rate Boundary Condition

If the specified flow rate stands for water influx across a reservoir boundary, then

$$q^{n+1}_{w\ b,bB} = q_{sp} \quad (5.86)$$

In addition,

$$q^{n+1}_{o\ b,bB} = q^{n+1}_{g\ b,bB} = 0. \quad (5.87)$$

If, however, the specified flow rate stands for fluid transfer between two segments of the same reservoir or fluid loss to an aquifer across WOC, then

$$q^{n+1}_{j\ b,bB} = \frac{\left(\dfrac{k_{rj}}{\mu_j}\right)^{n+1}_{bB}}{\sum_{l\in\{o,w,f,g\}} \left(\dfrac{k_{rl}}{\mu_l}\right)^{n+1}_{bB}} q_{sp} \quad \text{for } j = g, o, w. \quad (5.88)$$

The effects of gravity forces and capillary pressures are neglected in (5.80).

c. No-flow Boundary Condition

This condition results from vanishing permeability at a reservoir boundary or because of symmetry about a reservoir boundary. In either case, for a reservoir no-flow boundary,

$$q_{j\ b,bB}^{n+1} = 0 \quad \text{for} \quad j = g, o, w. \tag{5.89}$$

d. Specified Boundary Pressure Condition

This condition arises due to the presence of wells on the other side of a reservoir boundary that operate to maintain voidage replacement and as a result keep the boundary pressure (p_b) constant. The flow rate of phase j across a reservoir boundary that separates two segments of the same reservoir or across a reservoir boundary that represents WOC with fluid loss to an aquifer is estimated using

$$q_{j\ b,bB}^{n+1} = \left(\frac{k_l k_{rj} A_l}{\mu_j (\Delta l/2)}\right)_{bB}^{n+1} \left[\left(p_b - p_{bB}^{n+1}\right) - \left(\rho_w g\right)_{bB}^{n}\left(z_b - z_{bB}\right)\right] \quad \text{for} \quad j = g, o, w.$$

(5.90)

If the reservoir boundary represents WOC with water influx, then

$$q_{j\ b,bB}^{n+1} = \left(\frac{k_l A_l}{\mu_w (\Delta l/2)}\right)_{bB}^{n+1} \left(k_{rw}\right)_{bB}^{n+1} \left[\left(p_b - p_{bB}^{n+1}\right) - \left(\rho_w g\right)_{bB}^{n}\left(z_b - z_{bB}\right)\right] \tag{5.91}$$

In addition,

$$q_{o\ b,bB}^{n+1} = q_{g\ b,bB}^{n+1} = 0. \tag{5.92}$$

Note that, in Eq. 5.91, the rock and fluid properties in the aquifer are approximated by those of the boundary block properties because of the lack of geologic control in aquifers. It should also take into account that $(k_{rw})_{aq}^{n+1} = 1$ because $S_w = 1$ in the aquifer.

The Compositional Simulator

It is worth mentioning that when reservoir boundary b stands for WOC, the flow rate of phase j across the reservoir boundary is determined from the knowledge of the upstream point between reservoir boundary b and boundary block bB. If b is upstream to bB (i.e., when $\Delta\Phi_w > 0$), the flow is from the aquifer to the reservoir boundary block and (5.83) applies for water and $q_o^{n+1}{}_{b,bB} = q_g^{n+1}{}_{b,bB} = 0$. If b is downstream to bB (i.e., when $\Delta\Phi_w < 0$), the flow is from the reservoir boundary block to the aquifer and (5.82) applies for all phases. The water potential between the reservoir boundary and the reservoir boundary block is defined as $\Delta\Phi_w = (p_b - p_{bB}) - \rho_w g(z_b - z_{bB})$.

5.4 Variable Temperature Reservoir Compositional Flow Equations

5.4.1 Energy Balance Equation

The numerical formulation for component mass balance equation for a uniform reservoir has been developed in sections 5.2 and 5.3. When the reservoir temperature is not constant, it is necessary to consider the energy balance equation. The numerical model was introduced in subsection 5.1.2.2 for the energy balance equation for two different cases based on the finite control volume method. The formulations should be combined with the Darcy law to obtain a set of equations for practical applications. The relationships for the enthalpy and internal energy of the fluid j using the heat capacity at constant temperature C_p and the heat capacity at constant volume C_v are also applied to express the energy balance equation a function of reservoir temperature.

$$h_j = C_p(T - T_o) \quad \text{and} \quad \hat{u}_j = C_v(T - T_o) \tag{5.93}$$

Substituting (5.85) in (5.20) and taking into account a 1-D domain as given in Fig. 5.1, the energy balance equation takes the form:

$$\frac{U}{\Delta t}\left\{\left[(T-T_o)\left(\sum_j \phi \rho_j S_j C_{vj} + (1-\phi)\rho_r C_r\right) - g\phi\sum_j \rho_j S_j Z_j\right]^{n+1} - \left[(T-T_o)\left(\sum_j \phi \rho_j S_j C_{vj} + (1-\phi)\rho_r C_r\right) - g\phi\sum_j \rho_j S_j Z_j\right]^n\right\}$$

$$+ \begin{bmatrix} \left(\dfrac{kA_{x_{i+1/2}}}{\Delta x}\right)_{x_{i+1/2}} (T_{i+1} - T_i) \\ + \left(\dfrac{kA_{x_{i-1/2}}}{\Delta x}\right)_{x_{i-1/2}} (T_{i-1} - T_i) \end{bmatrix}^m$$

$$+ \begin{bmatrix} \sum_j \left[\phi \rho_j S_j C_{pj} (T - T_0)\right]_{x_{i+1/2}} V_{jx_{i+1/2}} A_{x_{i+1/2}} \\ - \sum_j \left[\phi \rho_j S_j C_{pj} (T - T_0)\right]_{x_{i-1/2}} V_{jx_{i-1/2}} A_{x_{i-1/2}} \end{bmatrix}^m + (E_s)^m = 0$$

(5.94)

Using the discretized form of Darcy's law and substituting in 5.86, it can be written that

$$\dfrac{U}{\Delta t} \left\{ \begin{bmatrix} (T - T_0)\left(\sum_j \phi \rho_j S_j C_{vj} + (1-\phi)\rho_r C_r\right) - g\phi \sum_j \rho_j S_j Z_j \end{bmatrix}^{n+1} \\ - \begin{bmatrix} (T - T_0)\left(\sum_j \phi \rho_j S_j C_{vj} + (1-\phi)\rho_r C_r\right) - g\phi \sum_j \rho_j S_j Z_j \end{bmatrix}^n \right\}$$

$$+ \begin{bmatrix} \sum_j \left[\phi \rho_j S_j C_{pj} (T - T_0)\right]_{x_{i+1/2}} A_{x_{i+1/2}} \dfrac{k_x|_{x_{i+1/2}}}{\Delta x_{i+1/2}} \left(\dfrac{k_{rj}}{m_j}\right)\bigg|_{x_{i-1/2}} \\ \times \begin{bmatrix} (p_{j_{i+1}} - p_{j_i}) - \\ \rho_{j_{i-1}} g(z_{j_{i+1}} - z_{j_i}) \end{bmatrix} - \sum_j \left[\phi \rho_j S_j C_{pj} (T - T_0)\right]_{x_{i-1/2}} \\ \times A_{x_{i-1/2}} \dfrac{k_x|_{x_{i+1/2}}}{\Delta x_{i-1/2}} \left(\dfrac{k_{rj}}{m_j}\right)\bigg|_{x_{i-1/2}} \begin{bmatrix} (p_{j_{i-1}} - p_{j_i}) - \\ \rho_{j_{i-1}} g(z_{j_{i-1}} - z_{j_i}) \end{bmatrix} \end{bmatrix}^m$$

The Compositional Simulator

$$+\left[\begin{array}{c}\left(\dfrac{kA_{x_{i+1/2}}}{\Delta x}\right)_{x_{i+1/2}}(T_{i+1}-T_i)\\+\left(\dfrac{kA_{x_{i-1/2}}}{\Delta x}\right)_{x_{i-1/2}}(T_{i-1}-T_i)\end{array}\right]^m+(E_s)^m=0 \quad (5.95)$$

Taking into account the definition of phase transmissibility for a grid-block i as:

$$(\mathbb{T}_j^m)_{i\mp 1/2}=\left(\dfrac{A_x k_x}{\Delta x}\right)_{x_{i\mp 1/2}}\left(\dfrac{k_{ro}}{\mu_o}\right)^m_{x_{i\mp 1/2}} \quad (5.96)$$

The energy balance equation may be given as:

$$\dfrac{U}{\Delta t}\left\{\left[(T-T_0)\left(\sum_j \phi\rho_j S_j C_{vj}+(1-\phi)\rho_r C_r\right)-g\phi\sum_j \rho_j S_j Z_j\right]^{n+1}\right.$$
$$\left.-\left[(T-T_0)\left(\sum_j \phi\rho_j S_j C_{vj}+(1-\phi)\rho_r C_r\right)-g\phi\sum_j \rho_j S_j Z_j\right]^n\right\}$$

$$+\left[\begin{array}{c}\sum_j [\phi\rho_j S_j C_{pj}(T-T_0)]_{x_{i+1/2}}(\mathbb{T}_j^m)_{i+1/2}\left[\begin{array}{c}(p_{j_{i+1}}-p_{j_i})\\-\rho_{j_{i-1}}g(z_{j_{i+1}}-z_{j_i})\end{array}\right]\\-\sum_j [\phi\rho_j S_j C_{pj}(T-T_0)]_{x_{i-1/2}}(\mathbb{T}_j^m)_{i-1/2}\left[\begin{array}{c}(p_{j_{i-1}}-p_{j_i})\\-\rho_{j_{i-1}}g(z_{j_{i-1}}-z_{j_i})\end{array}\right]\end{array}\right]^m$$

$$+\left[\begin{array}{c}\left(\dfrac{kA_{x_{i+1/2}}}{\Delta x}\right)_{x_{i+1/2}}(T_{i+1}-T_i)\\+\left(\dfrac{kA_{x_{i-1/2}}}{\Delta x}\right)_{x_{i-1/2}}(T_{i-1}-T_i)\end{array}\right]^m+(E_s)^m=0$$

(5.97)

for $j=g,o,w$. To obtain a generalized form of the equation and provide the ability to extend the formulation for a multidimensional domain, we apply the procedure adopted by Abou-Kassem (2006) and write the energy balance equation in the following form for a grid block *l* in a compact form as:

$$\frac{U}{\Delta t}\left\{ \begin{array}{l} \left[(T-T_0)\left(\sum_j \phi \rho_j S_j C_{vj} + (1-\phi)\rho_r C_r\right) - g\phi \sum_j \rho_j S_j Z_j\right]^{n+1} \\ -\left[(T-T_0)\left(\sum_j \phi \rho_j S_j C_{vj} + (1-\phi)\rho_r C_r\right) - g\phi \sum_j \rho_j S_j Z_j\right]^n \end{array} \right\}$$

$$\sum_{n=y_n}\left\{ \begin{array}{l} \sum_j \left[\phi \rho_j S_j C_{pj}(T-T_0)\right]_n (\mathbb{T}_j) \begin{bmatrix} (p_{j_n} - p_{j_l}) \\ -\rho_{j_n} g(z_{j_n} - z_{j_l}) \end{bmatrix} \\ +\left(\frac{kA}{\Delta x}\right)_n (T_n - T_l) \end{array} \right\}^m + (E_s)^m = 0$$

(5.98)

for *j* = *g*, *o*, *w*. Where $\Psi_N = \{l-1, l+1\}$ is the neighboring block to the block *l*. The equations (6.98) is for a interior grid block or in other words, are for blocks *l* = 2, 3, ..., $n_x - 1$. To include the boundary blocks, it may be possible to use a fictitious well instead of the boundary conditions and extend a general form of the formulation that can be applied for all grid blocks either interior blocks or boundary blocks. The general form of formulation is:

$$\frac{U}{\Delta t}\left\{ \begin{array}{l} \left[(T-T_0)\left(\sum_j \phi \rho_j S_j C_{vj} + (1-\phi)\rho_r C_r\right) - g\phi \sum_j \rho_j S_j Z_j\right]^{n+1} \\ -\left[(T-T_0)\left(\sum_j \phi \rho_j S_j C_{vj} + (1-\phi)\rho_r C_r\right) - g\phi \sum_j \rho_j S_j Z_j\right]^n \end{array} \right\}$$

$$\sum_{n=y_n}\left\{ \begin{array}{l} \sum_j \left[\phi \rho_j S_j C_{pj}(T-T_0)\right]_{x_n} (\mathbb{T}_j) \begin{bmatrix} (p_{j_n} - p_{j_l}) \\ -\rho_{j_n} g(z_{j_n} - z_{j_l}) \end{bmatrix} \\ +\left(\frac{kA}{\Delta x}\right)_n (T_n - T_l) \end{array} \right\}^m + (E_s)^m + \sum_{n=x_n} E_{l,n}^m = 0$$

(5.99)

The Compositional Simulator

The notations ψ_n and ξ_n are the sets showing the neighboring blocks and boundaries, respectively, as described in derivation of component mass balance equations in previous sections.

The equation (5.91) is applicable for 1D, 2D, or 3D flow problems. In 1D, ψ_n for a block n has a maximum of two elements of $[(i - 1), (i + 1)]$, and ξ_n is empty or has a maximum of one element of (b_w, b_E). In 2D, ψ_n for n has a maximum of four elements of $[(i - 1, j), (i, j - 1), (i, (j + 1), (i + 1, j)]$, and ξ_n is empty or has a maximum of two elements of (b_S, b_W, b_E, b_N). In 3D flow, ξ_n for n has a maximum of six elements of $[(i - 1, j, k), (i, j - 1, k), (i, j, (k - 1), (i + 1, j, k), (i, j + 1, k), (i, j, k + 1)]$, and ξ_n is empty or has a maximum of three elements of $(b_L, b_S, b_W, b_E, b_N, b_U)$. These are illustrated in Figs. 5.7. In summary, ψ_n = a set whose elements are the existing neighboring blocks to block n in the reservoir, ξ_n = a set whose elements are the reservoir boundaries $(b_L, b_S, b_W, b_E, b_N, b_U)$ that are shared by block n. It is either an empty set for interior blocks or a set that contains one element for boundary blocks that fall on one reservoir boundary, two elements for boundary blocks that fall on two reservoir boundaries, or three elements for blocks that fall on three reservoir boundaries.

5.4.2 Implicit Formulation of Variable Temperature Reservoir Compositional Flow Equations

A fully implicit formulation for a uniform temperature reservoir was introduced in section 5.3.1. If the reservoir temperature is not constant in different directions, we should also consider the energy balance equation to obtain the temperature distribution along the reservoir. The implicit form of energy balance equation is in the form:

$$\frac{U}{\Delta t}\left\{\left[(T-T_0)\left(\sum_j \phi\rho_j S_j C_{vj} + (1-\phi)\rho_r C_r\right) - g\phi \sum_j \rho_j S_j Z_j\right]^{n+1} - \left[(T-T_0)\left(\sum_j \phi\rho_j S_j C_{vj} + (1-\phi)\rho_r C_r\right) - g\phi \sum_j \rho_j S_j Z_j\right]^n\right\}$$

$$\sum_{n=y_n} \left\{ \begin{bmatrix} \sum_j \left[\phi \rho_j S_j C_{pj}(T-T_0)\right]_n (\mathbb{T}_j) \begin{bmatrix} (p_{j_n}-p_{j_l}) \\ -\rho_{j_n} g(z_{j_n}-z_{j_l}) \end{bmatrix} \end{bmatrix}^{n+1} \\ + \left(\frac{kA}{\Delta x}\right)_n (T_n - T_l) \end{bmatrix} \right.$$
$$+ (E_s)^{n+1} + \sum_{n=x_n} E_{l,n}^{n+1} = 0 \qquad (5.100)$$

This equation with the component mass balance equation and the constraint equations provide a set of algebraic equations to find all primary unknown including the temperature at gridblocks. Following the description in section 6.3.2 and taking into account the energy balance equation, each block contributes $n_c + 4$ equations:

- the oil-phase composition constraint;
- the gas-phase composition constraint;
- $n_c + 1$ oil component flow equations;
- the water component flow equation;
- the pressure equation; and
- the energy balance equation.

The $n_c + 4$ primary unknowns in each block are:

- n_c mass fractions in the oil phase for $y_{i,o}$, $i = 1, 2, \ldots, n_c$;
- the saturation of the gas and water phases S_g, S_w;
- the pressure of the oil-phase p_o; and
- the temperature of the gridblock.

The equations and primary unknowns for each reservoir block are aligned in order to obtain a diagonally dominant matrix and to alleviate the need for pivoting during forward Gaussian elimination. The model-reduced equations have the following order:

- the gas-phase composition constraint equation;

$$\sum_{i=1}^{n_c+1} y_{i,g}^{n+1} = 1 \qquad (5.101)$$

The Compositional Simulator

- the oil-phase composition constraint equation;

$$\sum_{i=1}^{n_c} y_{i,o}^{n+1} = 1 \qquad (5.102)$$

- the second lightest oil component flow equation ($i = 2$);

$$\sum_{n=y_n} \left\{ \left(\mathbb{T}_{2,g}^{n+1}\right)_{l,n} \begin{bmatrix} \left(p_{g_n} - p_{g_l}\right)^{n+1} \\ -\rho_{g_n}^{n+1} g\left(z_{g_n} - z_{g_l}\right) \end{bmatrix} + \left(\mathbb{T}_{2,o}^{n+1}\right)_{l,n} \begin{bmatrix} \left(p_{o_n} - p_{o_l}\right)^{n+1} \\ -\rho_{o_n}^{n+1} g\left(z_{o_n} - z_{o_l}\right) \end{bmatrix} \right\} + \sum_{n=x_n} \left(q_2^{n+1}\right)_{l,n} - \begin{pmatrix} y_{2,g} \rho_g Q_g \\ +y_{2,o} \rho_o Q_o \end{pmatrix}_l^{n+1}$$

$$= \frac{(U)_l}{\Delta t} \begin{bmatrix} \left(\phi \rho_g y_{2,g} S_g + \phi \rho_o y_{2,o} S_o\right)^{n+1} \\ -\left(\phi \rho_g y_{2,g} S_g + \phi \rho_o y_{2,o} S_o\right)^n \end{bmatrix}_l$$

(5.103)

- the third the third lightest oil component flow equation ($i = 3$);

$$\sum_{n=y_n} \left\{ \left(\mathbb{T}_{3,g}^{n+1}\right)_{l,n} \begin{bmatrix} \left(p_{g_n} - p_{g_l}\right)^{n+1} \\ -\rho_{g_n}^{n+1} g\left(z_{g_n} - z_{g_l}\right) \end{bmatrix} + \left(\mathbb{T}_{3,o}^{n+1}\right)_{l,n} \begin{bmatrix} \left(p_{o_n} - p_{o_l}\right)^{n+1} \\ -\rho_{o_n}^{n+1} g\left(z_{o_n} - z_{o_l}\right) \end{bmatrix} \right\} + \sum_{n=x_n} \left(q_3^{n+1}\right)_{l,n} - \begin{pmatrix} y_{3,g} \rho_g Q_g \\ +y_{3,o} \rho_o Q_o \end{pmatrix}_l^{n+1}$$

$$= \frac{(U)_l}{\Delta t} \begin{bmatrix} \left(\phi \rho_g y_{3,g} S_g + \phi \rho_o y_{3,o} S_o\right)^{n+1} \\ -\left(\phi \rho_g y_{3,g} S_g + \phi \rho_o y_{3,o} S_o\right)^n \end{bmatrix}_l$$

(5.104)

- the heaviest oil component flow equation ($i = n_c$)

$$\sum_{n=y_n}\left\{\left(\mathbb{T}^{n+1}_{n_c,g}\right)_{l,n}\left[\begin{array}{c}(p_{g_n}-p_{g_l})^{n+1}\\-\rho^{n+1}_{g_n}g(z_{g_n}-z_{g_l})\end{array}\right]+\left(\mathbb{T}^{n+1}_{n_c,o}\right)_{l,n}\left[\begin{array}{c}(p_{o_n}-p_{o_l})^{n+1}\\-\rho^{n+1}_{o_n}g(z_{o_n}-z_{o_l})\end{array}\right]\right\}+\sum_{n=x_n}\left(q^{n+1}_{n_c}\right)_{l,n}-\left(\begin{array}{c}y_{n_c,g}\rho_g Q_g\\+y_{n_c,o}\rho_o Q_o\end{array}\right)^{n+1}_l$$

$$=\frac{(U)_l}{\Delta t}\left[\begin{array}{c}(\phi\rho_g y_{n_c,g}S_g+\phi\rho_o y_{n_c,o}S_o)^{n+1}\\-(\phi\rho_g y_{n_c,g}S_g+\phi\rho_o y_{n_c,o}S_o)^n\end{array}\right]_l$$

(5.105)

- the water component flow equation ($i = n_c + 1$);

$$\frac{U}{\Delta t}\left\{\left[(T-T_0)\left(\sum_j\phi\rho_j S_j C_{vj}+(1-\phi)\rho_r C_r\right)-g\phi\sum_j r_j S_j Z_j\right]^{n+1}\right.$$
$$\left.-\left[(T-T_0)\left(\sum_j\phi\rho_j S_j C_{vj}+(1-\phi)r_r C_r\right)-g\phi\sum_j \rho_j S_j Z_j\right]^n\right\}$$

$$+\sum_{n=y_n}\left\{\sum_j\left[\phi\rho_j S_j C_{pj}(T-T_0)\right]_n(\mathbb{T}_j)\left[\begin{array}{c}(p_{j_n}-p_{j_l})\\-\rho_{j_n}g(z_{j_n}-z_{j_l})\end{array}\right]^{n+1}\right.$$
$$\left.+\left(\frac{kA}{\Delta x}\right)_n(T_n-T_l)\right\}$$

$$+(E_s)^{n+1}+\sum_{n=x_n}E^{n+1}_{l,n}=0 \qquad (5.108)$$

- the pressure equation; and

$$X_n=\left[S_g\ y_{1,o}\ y_{2,o}\ y_{3,o}\ \cdots\ y_{n_c-1,o}\ y_{n_c,o}\ S_w\ p_o\ T\right]^T_l \qquad (5.109)$$

The Compositional Simulator

- the energy balance equation.

$$\frac{U}{\Delta t}\left\{\left[(T-T_0)\left(\sum_j \phi\rho_j S_j C_{vj} + (1-\phi)\rho_r C_r\right) - g\phi\sum_j \rho_j S_j Z_j\right]^{n+1} \right.$$
$$\left. -\left[(T-T_0)\left(\sum_j \phi\rho_j S_j C_{vj} + (1-\phi)\rho_r C_r\right) - g\phi\sum_j \rho_j S_j Z_j\right]^n\right\}$$

$$+\sum_{n=y_n}\left\{\sum_j\left[\phi\rho_j S_j C_{pj}(T-T_0)\right]_n (\mathbb{T}_j)\left[\begin{array}{c}(p_{j_n}-p_{j_l})\\-\rho_{j_n}g(z_{j_n}-z_{j_l})\end{array}\right]\right.$$
$$\left. +\left(\frac{kA}{\Delta x}\right)_n (T_n-T_l)\right\}$$

$$+(E_s)^{n+1} + \sum_{n=x_n} E_{l,n}^{n+1} = 0 \qquad (5.110)$$

With this choice of equation ordering, the primary unknowns have the order of $S_g, y_{1,o}, y_{2,o}, y_{3,o} \cdots, y_{n_c-1,o}, S_w, p_o$ and T. Therefore, the vector of primary unknowns for block l is:

$$X_n = \begin{bmatrix} S_g & y_{1,o} & y_{2,o} & y_{3,o} & \cdots & y_{n_c-1,o} & y_{n_c,o} & S_w & p_o & T \end{bmatrix}_l^T \qquad (5.101)$$

5.5 Solution Method

A complete formulation for a non-uniform temperature reservoir is taken into account. As described already a fully implicit method is adopted to find the solution for the n_c+4 primary unknowns in each block. In the fully implicit method, transmissibilities, well production rates, and fictitious well rates if present are dated at the current time level (time level n_c+1). Gravities are dated at the old time level as mentioned earlier. The model equations (constraint equations and flow equations for all components) for all blocks are solved simultaneously in a fully implicit simulator using Newton's Iteration. However, because the constraint equations do not have inter-block terms, they can be eliminated at the matrix level without influencing the solution or the stability of the fully implicit scheme. This elimination results in considerable savings both in CPU time and memory storage requirements.

5.5.1 Solution of Model Equations Using Newton's Iteration

The fully implicit iterative equations for the model are derived using the Coats, Ramesh, and Winestock (1977) procedure. Each equation of the model is written in a residual from at time level $n+1$, i.e., all terms are placed on one side of an equation and the other side is zero. Each term at time level $n+1$ in the resulting equation is approximated by its value at the current iteration level $v+1$, which in turn can be approximated by its value at the last iteration level v, plus a linear combination of the unknowns arising from partial differentiation with respect to all primary unknowns. The unknown quantities in the resulting equation are the changes over an iteration of all the primary unknowns in the original equation. This approach does not use conservative expansions of accumulation terms and, therefore, the resulting equations do not conserve material balance during iterations; but they do preserve it at convergence. The resulting fully implicit iterative equations for block l are derived as follows.

$$\vec{R}_l^{n+1} = 0 \tag{5.110}$$

Where:

$$\vec{R}_l^{n+1} = \begin{Bmatrix} R_1 \\ R_2 \\ R_3 \\ R_4 \\ R_5 \\ \vdots \\ R_{n_c+1} \\ R_{n_c+2} \\ R_{n_c+3} \\ R_{n_c+4} \end{Bmatrix}^{n+1} = \begin{Bmatrix} R_{(y_{i,g})_l}^{n+1} \\ R_{(y_{i,o})_l}^{n+1} \\ R_{(i=2)_l}^{n+1} \\ R_{(i=3)_l}^{n+1} \\ R_{(i=4)_l}^{n+1} \\ \vdots \\ R_{(i=n_c)_l}^{n+1} \\ R_{(i=n_c+1)_l}^{n+1} \\ R_{(p)_l}^{n+1} \\ R_{(T)_l}^{n+1} \end{Bmatrix} \tag{5.111}$$

The Compositional Simulator

$$R^{n+1}_{(y_{i,g})_l} = \sum_{i=1}^{n_c+1} y_{i,g}^{n+1} - 1 \tag{5.112}$$

$$R^{n+1}_{(y_{i,o})_l} = \sum_{i=1}^{n_c} y_{i,o}^{n+1} - 1 \tag{5.113}$$

$$R^{n+1}_{(i=2,3,\cdots,n_c)_l} = \sum_{n=y_n} \left\{ \left(\mathbb{T}^{n+1}_{i,g}\right)_{l,n} \begin{bmatrix} \left(p_{g_n} - p_{g_l}\right)^{n+1} \\ -\rho_{g_n}^{n+1} g\left(z_{g_n} - z_{g_l}\right) \end{bmatrix} + \left(\mathbb{T}^{n+1}_{i,o}\right)_{l,n} \begin{bmatrix} \left(p_{o_n} - p_{o_l}\right)^{n+1} \\ -\rho_{o_n}^{n+1} g\left(z_{o_n} - z_{o_l}\right) \end{bmatrix} \right\}$$

$$+ \sum_{n=x_n} \left(q_i^{n+1}\right)_{l,n} - \left(\begin{matrix} y_{i,g} \rho_g Q_g \\ +y_{i,o} \rho_o Q_o \end{matrix}\right)_l^{n+1}$$

$$- \frac{(U)_l}{\Delta t} \left[\left(\phi \rho_g y_{i,g} S_g + \phi \rho_o y_{i,o} S_o\right)^{n+1} - \left(\phi \rho_g y_{i,g} S_g + \phi \rho_o y_{i,o} S_o\right)^{n} \right]_l \tag{5.114}$$

$$R^{n+1}_{(i=n_c+1)_l} = \sum_{n=y_n} \left\{ \left(\mathbb{T}^{n+1}_{n_c+1,g}\right)_{l,n} \begin{bmatrix} \left(p_{g_n} - p_{g_l}\right)^{n+1} \\ -\rho_{g_n}^{n+1} g\left(z_{g_n} - z_{g_l}\right) \end{bmatrix} + \left(\mathbb{T}^{n+1}_{n_c+1,w}\right)_{l,n} \begin{bmatrix} \left(p_{o_n} - p_{o_l}\right)^{n+1} \\ -\rho_{o_n}^{n+1} g\left(z_{o_n} - z_{o_l}\right) \end{bmatrix} \right\}$$

$$+ \sum_{n=x_n} \left(q_{n_c+1}^{n+1}\right)_{l,n} - \left(y_{n_c+1,g} \rho_g Q_g + \rho_w Q_w\right)_l^{n+1}$$

$$- \frac{(U)_l}{\Delta t} \left[\left(\phi \rho_g y_{n_c+1,g} S_g + \phi \rho_w S_w\right)^{n+1} - \left(\phi \rho_g y_{n_c+1,g} S_g + \phi \rho_w S_w\right)^{n} \right]_l \tag{5.115}$$

$$R_{(p)_l}^{n+1} = \sum_{n=y_n} \left\{ \begin{array}{l} \left(\mathbb{T}_g^{n+1}\right)_{l,n} \rho_g^{n+1} \left[\begin{array}{l} \left(p_{g_n} - p_{g_l}\right)^{n+1} \\ -\rho_{g_n}^{n+1} g\left(z_{g_n} - z_{g_l}\right) \end{array} \right] \\ + \left(\mathbb{T}_o^{n+1}\right)_{l,n} \rho_o^{n+1} \left[\begin{array}{l} \left(p_{o_n} - p_{o_l}\right)^{n+1} \\ -\rho_{o_n}^{n+1} g\left(z_{o_n} - z_{o_l}\right) \end{array} \right] \\ + \left(\mathbb{T}_w^{n+1}\right)_{l,n} \rho_w^{n+1} \left[\begin{array}{l} \left(p_{w_n} - p_{w_l}\right)^{n+1} \\ -\rho_{w_n}^{n+1} g\left(z_{o_n} - z_{o_l}\right) \end{array} \right] \end{array} \right\} + \sum_{n=x_n} \left(\begin{array}{l} \rho_g Q_g \\ +\rho_o Q_o \\ +\rho_w Q_w \end{array} \right)_{l,n}^{n+1}$$

$$- \left(\begin{array}{l} \rho_g Q_g \\ +\rho_o Q_o \\ +\rho_w Q_w \end{array} \right)_l^{n+1} - \frac{(U)_l}{\Delta t} \left[\begin{array}{l} \left(\phi \rho_g y_{n_c+1,g} S_g + \phi \rho_w S_w\right)^{n+1} \\ - \left(\phi \rho_g y_{n_c+1,g} S_g + \phi \rho_w S_w\right)^n \end{array} \right]_l$$

(5.116)

$$R_{(T)_l}^{n+1} = \frac{U}{\Delta t} \left\{ \begin{array}{l} \left[(T-T_0)\left(\sum_j \phi \rho_j S_j C_{vj} + (1-\phi) r_r C_r\right) - g\phi \sum_j \rho_j S_j Z_j \right]^{n+1} \\ - \left[(T-T_0)\left(\sum_j \phi \rho_j S_j C_{vj} + (1-\phi) \rho_r C_r\right) - g\phi \sum_j \rho_j S_j Z_j \right]^n \end{array} \right\}$$

$$+ \sum_{n=y_n} \left\{ \begin{array}{l} \sum_j \left[\phi \rho_j S_j C_{pj} (T-T_0)\right]_n \left(\mathbb{T}_j\right)_l \left[\begin{array}{l} \left(p_{j_n} - p_{j_l}\right) \\ -\rho_{j_n} g\left(z_{j_n} - z_{j_l}\right) \end{array} \right] \right\}^{n+1} \\ + \left(\frac{kA}{\Delta x}\right)_n (T_n - T_l) \end{array} \right\}$$

$$+ (E_s)^{n+1} + \sum_{n=x_n} E_{l,n}^{n+1}$$

(5.117)

The Compositional Simulator

Let the number of gridblocks is N, it can be written that:

$$\vec{R}^{n+1} = \begin{Bmatrix} \vec{R}_1^{n+1} \\ \vec{R}_2^{n+1} \\ \vec{R}_3^{n+1} \\ \vec{R}_4^{n+1} \\ \vdots \\ \vec{R}_{N-2}^{n+1} \\ \vec{R}_{N-1}^{n+1} \\ \vec{R}_N^{n+1} \end{Bmatrix} \qquad (5.118)$$

$$\vec{R}^{n+1} \cong \vec{R}_{n+1}^{v+1} \cong \vec{R}_{n+1}^{v} + [J]_{n+1}^{v} \delta \vec{X}_{n+1}^{v+1} = 0 \qquad (5.119)$$

$$\vec{R}_{n+1}^{v} + [J]_{n+1}^{v} \delta \vec{X}_{n+1}^{v+1} = 0 \qquad (5.120)$$

$$[J]_{n+1}^{v} \delta \vec{X}_{n+1}^{v+1} = -\vec{R}_{n+1}^{v} \qquad (5.121)$$

Where:

$$d\vec{X}_{n+1}^{v+1} = \left\{ \delta \vec{X}_1^{v+1} \; \delta \vec{X}_2^{v+1} \; \delta \vec{X}_3^{v+1} \; \cdots \cdots \; \delta \vec{X}_{N-2}^{v+1} \; \delta \vec{X}_{N-1}^{v+1} \; \delta \vec{X}_N^{v+1} \right\}^T \qquad (5.122)$$

$$\delta \vec{X}_I^{v+1} = \delta \vec{X}_I^{v+1} - \delta \vec{X}_I^{v} \qquad (5.123)$$

$$[J] = \left\{ [J_1] \; [J_2] \; [J_3] \; [J_4] \; \cdots \cdots \; [J_{N-2}] \; [J_{N-1}] \; [J_N] \right\}^T \qquad (5.124)$$

$$[J_1] = \left\{ [J_{1,1}] \; [J_{1,2}] \; \cdots \; [J_{1,1+n_x}] \; \cdots \; [J_{1,1+n_x n_y}] \; \cdots \right\} \quad (5.125)$$

$$[J_2] = \left\{ [J_{2,1}] \; [J_{2,2}] \; \cdots \; [J_{2,2+n_x}] \; \cdots \; [J_{2,2+n_x n_y}] \; \cdots \right\} \quad (5.126)$$

$$[J_3] = \left\{ [J_{1,1}] \; [J_{1,2}] \; \cdots \; [J_{1,3+n_x}] \; \cdots \; [J_{1,3+n_x n_y}] \; \cdots \right\} \quad (5.127)$$
... ...

$$[J_l] = \left\{ [J_{1,1}] \; [J_{1,2}] \; \cdots \; [J_{1,l+n_x}] \; \cdots \; [J_{1,l+n_x n_y}] \; \cdots \right\} \quad (5.128)$$
... ...

$$[J_{N-2}] = \left\{ [J_{N-2,1}] \; [J_{N-1,2}] \; \cdots \; [J_{N-1,N-1+n_x}] \; \cdots \; [J_{N-2,N-2+n_x n_y}] \; \cdots \right\} \quad (5.129)$$

$$[J_{N-1}] = \left\{ [J_{N-1,1}] \; [J_{N-1,2}] \; \cdots \; [J_{N-1,N-1+n_x}] \; \cdots \; [J_{N-1,N-1+n_x n_y}] \; \cdots \right\} \quad (5.130)$$

$$[J_{N-2}] = \left\{ [J_{N,1}] \; [J_{N,2}] \; \cdots \; [J_{N,N+n_x}] \; \cdots \; [J_{N,N+n_x n_y}] \; \cdots \right\} \quad (5.131)$$

In general, $[J_{n,l}]$ for block n, is defined as:

$$[J_{n,l}] = \begin{bmatrix} \frac{\partial R_{(y_{i,g})_n}}{(\partial S_g)_l} & \frac{\partial R_{(y_{i,g})_n}}{(\partial y_{1,o})_l} & \frac{\partial R_{(y_{i,g})_n}}{(\partial y_{2,o})_l} & \cdots & \frac{\partial R_{(y_{i,g})_n}}{(\partial S_w)_l} & \frac{\partial R_{(y_{i,g})_n}}{(\partial p_o)_l} & \frac{\partial R_{(y_{i,g})_n}}{(\partial T)_l} \\ \frac{\partial R_{(y_{i,o})_n}}{(\partial S_g)_l} & \frac{\partial R_{(y_{i,o})_n}}{(\partial y_{1,o})_l} & \frac{\partial R_{(y_{i,o})_n}}{(\partial y_{2,o})_l} & \cdots & \frac{\partial R_{(y_{i,o})_n}}{(\partial S_w)_l} & \frac{\partial R_{(y_{i,o})_n}}{(\partial p_o)_l} & \frac{\partial R_{(y_{i,o})_n}}{(\partial T)_l} \\ \frac{\partial R_{(i=2)_n}}{(\partial S_g)_l} & \frac{\partial R_{(i=2)_n}}{(\partial y_{1,o})_l} & \frac{\partial R_{(i=2)_n}}{(\partial y_{2,o})_l} & \cdots & \frac{\partial R_{(i=2)_n}}{(\partial S_w)_l} & \frac{\partial R_{(i=2)_n}}{(\partial p_o)_l} & \frac{\partial R_{(i=2)_n}}{(\partial T)_l} \\ \frac{\partial R_{(i=3)_n}}{(\partial S_g)_l} & \frac{\partial R_{(i=3)_n}}{(\partial y_{1,o})_l} & \frac{\partial R_{(i=3)_n}}{(\partial y_{2,o})_l} & \cdots & \frac{\partial R_{(i=3)_n}}{(\partial S_w)_l} & \frac{\partial R_{(i=3)_n}}{(\partial p_o)_l} & \frac{\partial R_{(i=3)_n}}{(\partial T)_l} \\ & & & \vdots & & & \\ \frac{\partial R_{(i=n_c+1)_n}}{(\partial S_g)_l} & \frac{\partial R_{(i=n_c+1)_n}}{(\partial y_{1,o})_l} & \frac{\partial R_{(i=n_c+1)_n}}{(\partial y_{2,o})_l} & \cdots & \frac{\partial R_{(i=n_c+1)_n}}{(\partial S_w)_l} & \frac{\partial R_{(i=n_c+1)_n}}{(\partial p_o)_l} & \frac{\partial R_{(i=n_c+1)_n}}{(\partial T)_l} \\ \frac{\partial R_{(p)_n}}{(\partial S_g)_l} & \frac{\partial R_{(p)_n}}{(\partial y_{1,o})_l} & \frac{\partial R_{(p)_n}}{(\partial y_{2,o})_l} & \cdots & \frac{\partial R_{(p)_n}}{(\partial S_w)_l} & \frac{\partial R_{(p)_n}}{(\partial p_o)_l} & \frac{\partial R_{(p)_n}}{(\partial T)_l} \\ \frac{\partial R_{(T)_n}}{(\partial S_g)_l} & \frac{\partial R_{(T)_n}}{(\partial y_{1,o})_l} & \frac{\partial R_{(T)_n}}{(\partial y_{2,o})_l} & \cdots & \frac{\partial R_{(T)_n}}{(\partial S_w)_l} & \frac{\partial R_{(T)_n}}{(\partial p_o)_l} & \frac{\partial R_{(T)_n}}{(\partial T)_l} \end{bmatrix}$$

(5.132)

A complete description of the method and evaluation of the partial differentiation of the elements of $\vec{R}_i^{n+1}$ can be obtained in (Abou-Kassem, 2008).

5.6 The Effects of Linearization

Nonlinearities include phase compositions, phase transmissibilities, well production (injection) rates, fictitious well rates, and accumulation terms. The flow equation is obtained by combining the mass balance equation and Darcy's law in an integral form for a discrete reservoir. The variation of the integrand is neglected for a selected time interval to recast the flow equation in an algebraic form. This algebraic form of flow equation is nonlinear due to the dependency of involved parameters to each other. The algebraic equations for individual reservoir blocks form a set of nonlinear simultaneous algebraic equations. It is necessary to impose some simplification and linearization in order to obtain numerical description for the flow characteristics. There is a need to observe the role of linearizations. The effects of these simplifications and linearizations during the solving process are investigated for two different cases:

- A single phase flow of natural gas through a reservoir of 20 acre spacing and 30 ft net thickness. The reservoir is horizontal and described by four gridblocks in the radial direction.
- A multiphase flow of water and oil through a horizontal reservoir of length L = 1200 ft, width W = 350 ft and height H = 40 ft. The effect of the dimension of the reservoir blocks, the time interval, the pressure dependent parameters and the water saturation related parameters are studied.

5.6.1 Case I: Single Phase Flow of a Natural Gas

An example (Abou-Kassem et al, 2006a) on a natural gas reservoir of 20 acre spacing and 30 ft net thickness is taken as the test case in this chapter. The reservoir is horizontal and described by four gridblocks in the radial direction. It is also assumed that the reservoir has homogeneous and isotropic rock properties with $k = 15$ md and $\varphi = 0.13$. A vertical well ($d = 0.5$ ft) produces from the reservoir at a rate of 1 MMscf/D. The initial reservoir pressure is 4015 psia. The pressure distribution at different time intervals needs to be calculated. The flow is considered to be in radial direction without any variation in $z-$ and $\theta-$directions.

The governing equations of fluid flow through porous media are obtained by combining two basic engineering concepts including the principle of mass conservation and the constitutive equation. In reservoir simulation, the constitutive property, i.e., the rate of fluid movement into (or out of) the reservoir volume element is described by Darcy's law and is related to potential gradient. Therefore, the

The Compositional Simulator

combination of Darcy's law with the conservation of mass results in the flow equation. The resulting differential form of the flow equation is nonlinear. The discrete form of the flow equation may be obtained directly using the finite control volume method described in section 5.2. The reservoir is divided into gridblocks in different directions and the flow equation is written for each of these gridblocks. Finally, the resulting equations are a system of nonlinear algebraic equations that give the pressure distribution along the reservoir at any time.

The general form of the flow equation for Gridblock n can be written as:

$$\sum_{l=\psi_n} \mathbb{T}_{l,n}^{n+1}\left[\left(p_l^m - p_n^m\right) - \gamma_{l,n}^m\left(Z_l^m - Z_n^m\right)\right] + \sum_{l=\xi_n} q_{SC_{l,n}}^m + q_{SC_n}^m$$

$$= \frac{U}{a_c \Delta t}\left[\left(\frac{\phi}{B}\right)_n^{v+1} - \left(\frac{\phi}{B}\right)_n^v\right] \qquad (5.133)$$

The transmissibility along the r-direction in cylindrical coordinates is defined as:

$$\mathbb{T}_{i\mp\frac{1}{2},j,k} = G_{r_{i\mp\frac{1}{2},j,k}}\left(\frac{1}{\mu B}\right)_{i\mp\frac{1}{2},j,k} \qquad (5.134)$$

Logarithmic spacing constant, α_{lg}, geometric factor (G) in the r-direction and bulk volume (U) are calculated using the recently reported simple and explicit equations (Abou-Kassem, 2006)

$$\alpha_{lg} = \left(\frac{r_e}{r_w}\right)^{1/n_r}$$

$$G_{r_{i-1/2,j,k}} = \frac{\beta_c \Delta\theta_j}{\left\{\log_e\left[\alpha_{lg}\log_e\left(\alpha_{lg}\right)/\left(\alpha_{lg}-1\right)\right]/\left(\Delta z_{i,j,k} k_{r_{i,j,k}}\right)\right.}$$
$$\left. + \log_e\left[\left(\alpha_{lg}-1\right)/\log_e\left(\alpha_{lg}\right)\right]/\left(\Delta z_{i-1,j,k} k_{r_{i-1,j,k}}\right)\right\}}$$

$$G_{r_{i+1/2,j,k}} = \frac{\beta_c \Delta\theta_j}{\left\{\log_e\left[\left(\alpha_{lg}-1\right)/\log_e\left(\alpha_{lg}\right)\right]/\left(\Delta z_{i,j,k} k_{r_{i,j,k}}\right)\right.}$$
$$\left. + \log_e\left[\alpha_{lg}\log_e\left(\alpha_{lg}\right)/\left(\alpha_{lg}-1\right)\right]/\left(\Delta z_{i,j,k} k_{r_{i,j,k}}\right)\right\}}$$

$$V_{b_{i,j,k}} = \left\{ \left(\alpha_{lg}^2 - 1\right)^2 / \left[\alpha_{lg}^2 \log_e\left(\alpha_{lg}^2\right)\right] \right\} r_i^2 \left(\frac{1}{2}\Delta\theta_j\right)\Delta z_{i,j,}$$

for $i = 1, 2, 3, \cdots n_r - 1;$ $j = 1, 2, 3, \cdots n_\theta;$ and $k = 1, 2, 3, \cdots n_z;$

$$V_{b_{i,j,k}} = \left\{ 1 - \left[\log_e\left(\alpha_{lg}\right)/\left(\alpha_{lg}-1\right)\right]^2 \left(\alpha_{lg}^2 - 1\right)^2 / \left[\alpha_{lg}^2 \log_e\left(\alpha_{lg}^2\right)\right] \right\}$$
$$\times r_e^2 \left(\frac{1}{2}\Delta\theta_j\right)\Delta z_{n_r,j,k}$$

for $i = n;$ $j = 1, 2, 3, \cdots n_\theta;$ and $k = 1, 2, 3, \cdots n_z.$

The flow equation can be written as

$$\sum_{l=\psi_n} \mathbb{T}_{l,n}^{n+1}\left(p_l^m - p_n^m\right) + q_{SC_n}^m = \frac{U}{a_c \Delta t}\left[\left(\frac{\phi}{B}\right)_n^{v+1} - \left(\frac{\phi}{B}\right)_n^v\right] \quad (5.135)$$

Equation (5.126) can be simplified to (5.127) if $\frac{f}{B}$ is considered to be a linear function of pressure.

$$\sum_{l=\psi_n} \mathbb{T}_{l,n}^{n+1}\left(p_l^m - p_n^m\right) + q_{SC_n}^m = \frac{U}{a_c \Delta t}\left(\frac{\phi}{B}\right)_n' \left[p_n^{v+1} - p_n^v\right] \quad (5.136)$$

where, $\left(\frac{\phi}{B}\right)_n'$ is the chord slope of $\left(\frac{\phi}{B}\right)_n$ between p_n^{v+1} and p_n^v.

Although, μ and B are functions of the fluid temperature and pressure, it is assumed that temperature remains constant throughout the reservoir during the production period. The gas formation volume factor, $GFVF$ or B_g, and viscosity, μ as functions of reservoir pressure are shown in Table 5.3.

They are expressed in mathematical form using the polynomial of fourth order for μ and the power function for B as shown in Fig. 5.9a. These two fluid properties are also fitted with the spline functions of different degrees as shown in Fig. 5.9b. It is found that the quadratic and cubic splines give a very good approximation to the variation of μ and B with P.

The Compositional Simulator

Table 5.3 The variation of the gas formation volume factor and viscosity with reservoir pressure (Abou-Kassem et al, 2006a)

pressue (psia)	GFVF (RB/scf)	Viscosity (cp)	pressue (psia)	GFVF (RB/scf)	Viscosity (cp)
215	0.016654	0.0126	2215	0.001318	0.0167
415	0.008141	0.0129	2415	0.001201	0.0173
615	0.005371	0.0132	2615	0.001109	0.0180
815	0.003956	0.0135	2815	0.001032	0.0186
1015	0.003114	0.0138	3015	0.000972	0.0192
1215	0.002544	0.0143	3215	0.000922	0.0198
1415	0.0002149	0.0147	3415	0.000878	0.0204
1615	0.001857	0.0152	3615	0.000840	0.0211
1815	0.001630	0.0156	3815	0.000808	0.0217
2015	0.001459	0.0161	4015	0.000779	0.0223

The flow equation for the gridblocks as specified in (5.126) can also be expressed in the following form

$$\begin{cases} G\left(\dfrac{1}{\mu B}\right)_j^m (p_2^m - p_1^m) - 10^6 = \beta_1 \left[\left(\dfrac{\phi}{B}\right)_1^{v+1} - \left(\dfrac{\phi}{B}\right)_1^v\right] \\ G\left(\dfrac{1}{\mu B}\right)_j^m (p_1^m - p_2^m) + G\left(\dfrac{1}{\mu B}\right)_j^m (p_3^m - p_2^m) = \beta_2 \left[\left(\dfrac{\phi}{B}\right)_2^{v+1} - \left(\dfrac{\phi}{B}\right)_2^v\right] \\ G\left(\dfrac{1}{\mu B}\right)_j^m (p_2^m - p_3^m) + G\left(\dfrac{1}{\mu B}\right)_j^m (p_4^m - p_3^m) = \beta_3 \left[\left(\dfrac{\phi}{B}\right)_3^{v+1} - \left(\dfrac{\phi}{B}\right)_3^v\right] \\ G\left(\dfrac{1}{\mu B}\right)_j^m (p_3^m - p_4^m) = \beta_4 \left[\left(\dfrac{\phi}{B}\right)_4^{v+1} - \left(\dfrac{\phi}{B}\right)_4^v\right] \end{cases}$$

(5.128)

212   Advanced Petroleum Reservoir Simulation

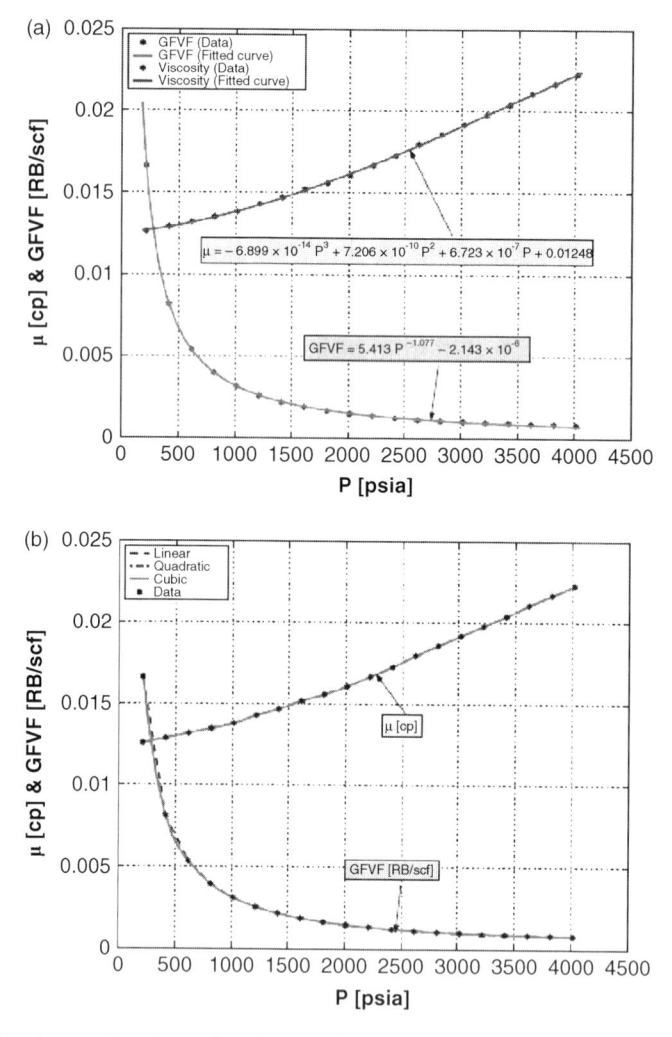

Figure 5.9 Approximation of variation of μ and *GFVF* with *P* (a) using continuous functions (b) using spline functions of different order.

The Compositional Simulator

and according to (5.127) as

$$\begin{cases} G\left(\dfrac{1}{\mu B}\right)_j^m (p_2^m - p_1^m) - 10^6 = \beta_1 \left(\dfrac{\phi}{B}\right)_1' [p_1^{v+1} - p_1^v] \\ G\left(\dfrac{1}{\mu B}\right)_j^m (p_1^m - p_2^m) + G\left(\dfrac{1}{\mu B}\right)_j^m (p_3^m - p_2^m) = \beta_2 \left(\dfrac{\phi}{B}\right)_2' [p_2^{v+1} - p_2^v] \\ G\left(\dfrac{1}{\mu B}\right)_j^m (p_2^m - p_3^m) + G\left(\dfrac{1}{\mu B}\right)_j^m (p_4^m - p_3^m) = \beta_3 \left(\dfrac{\phi}{B}\right)_3' [p_3^{v+1} - p_3^v] \\ G\left(\dfrac{1}{\mu B}\right)_j^m (p_3^m - p_4^m) = \beta_4 \left(\dfrac{\phi}{B}\right)_4' [p_4^{v+1} - p_4^v] \end{cases}$$

(5.129)

where, $\beta_1 = \dfrac{U}{a_c \Delta t}$ is constant for each block. The function $\left(\dfrac{1}{\mu B}\right)_j^m$ depends on pressure distribution in the reservoir. Two cases regarding subscript j are considered

I) pressure at the upstream, and
II) pressure at the i-th Gridblock.

The superscript m is related to the time step, which is taken as $m = v + 1$

For Case (I), (5.128) is written as

$$\begin{cases} G\left(\dfrac{1}{\mu B}\right)_j^{v+1} (p_2^{v+1} - p_1^{v+1}) - 10^6 = \beta_1 \left[\left(\dfrac{\phi}{B}\right)_1^{v+1} - \left(\dfrac{\phi}{B}\right)_1^v\right] \\ G\left(\dfrac{1}{\mu B}\right)_j^{v+1} (p_1^{v+1} - p_2^{v+1}) + G\left(\dfrac{1}{\mu B}\right)_j^{v+1} (p_3^{v+1} - p_2^{v+1}) = \beta_2 \left[\left(\dfrac{\phi}{B}\right)_2^{v+1} - \left(\dfrac{\phi}{B}\right)_2^v\right] \\ G\left(\dfrac{1}{\mu B}\right)_j^{v+1} (p_2^{v+1} - p_3^{v+1}) + G\left(\dfrac{1}{\mu B}\right)_j^{v+1} (p_4^{v+1} - p_3^{v+1}) = \beta_3 \left[\left(\dfrac{\phi}{B}\right)_3^{v+1} - \left(\dfrac{\phi}{B}\right)_3^v\right] \\ G\left(\dfrac{1}{\mu B}\right)_j^{v+1} (p_3^{v+1} - p_4^{v+1}) = \beta_4 \left[\left(\dfrac{\phi}{B}\right)_4^{v+1} - \left(\dfrac{\phi}{B}\right)_4^v\right] \end{cases}$$

(5.130)

and (5.129) is given in the form

$$\begin{cases} G\left(\dfrac{1}{\mu B}\right)_j^{v+1} \left(p_2^{v+1}-p_1^{v+1}\right)-10^6 = \beta_1 \left(\dfrac{\phi}{B}\right)_1' \left[p_1^{v+1}-p_1^v\right] \\[6pt] G\left(\dfrac{1}{\mu B}\right)_j^{v+1} \left(p_1^{v+1}-p_2^{v+1}\right)+G\left(\dfrac{1}{\mu B}\right)_j^{v+1} \left(p_3^{v+1}-p_2^{v+1}\right) = \beta_2 \left(\dfrac{\phi}{B}\right)_2' \left[p_2^{v+1}-p_2^v\right] \\[6pt] G\left(\dfrac{1}{\mu B}\right)_j^{v+1} \left(p_2^{v+1}-p_3^{v+1}\right)+G\left(\dfrac{1}{\mu B}\right)_j^{v+1} \left(p_4^{v+1}-p_3^{v+1}\right) = \beta_3 \left(\dfrac{\phi}{B}\right)_3' \left[p_3^{v+1}-p_3^v\right] \\[6pt] G\left(\dfrac{1}{\mu B}\right)_j^{v+1} \left(p_3^{v+1}-p_4^{v+1}\right) = \beta_4 \left(\dfrac{\phi}{B}\right)_4' \left[p_4^{v+1}-p_4^v\right] \end{cases}$$

(5.131)

The pressure distribution at the center of gridblocks is computed using (5.130) and (5.131). Here μ and B vary with pressure, however, k is assumed constant. The time step is chosen as $\Delta t = 1$ *month*. Figure 5.10a shows pressure at Gridblock 1 as a function of time when the variation of μ and B with pressure is approximated using continuous functions. It is observed that there is no substantial discrepancy between the pressure values obtained through original formulation and approximate formulation at different time steps. However, the computational results suggest that, with increasing time, the difference between the pressures from (5.130) and (5.131) increases. The effect of the simplification of the formulation is more evident when the cubic spline is applied to approximate the variation of μ and B with pressure as shown in Fig. 5.10b.

5.6.2 Effect of Interpolation Functions and Formulation

The pressure distribution at the center of gridblocks is also obtained using different interpolation functions and various formulation. The results for Gridblock 1 is shown in Figs. 5.11a and 5.11b. The nonlinear continuous functions as given in Fig. 5.9, linear interpolation and cubic spline are applied to approximate μ and B in both figures. The transient pressure results using cubic spline and linear interpolation are very close to each other for both linear formulations (Fig. 5.10a) and original formulations (Fig. 5.10b). It is also noticed that the continuous functional interpolation shows higher values for pressure at different time steps. Such observation is found to be

THE COMPOSITIONAL SIMULATOR 215

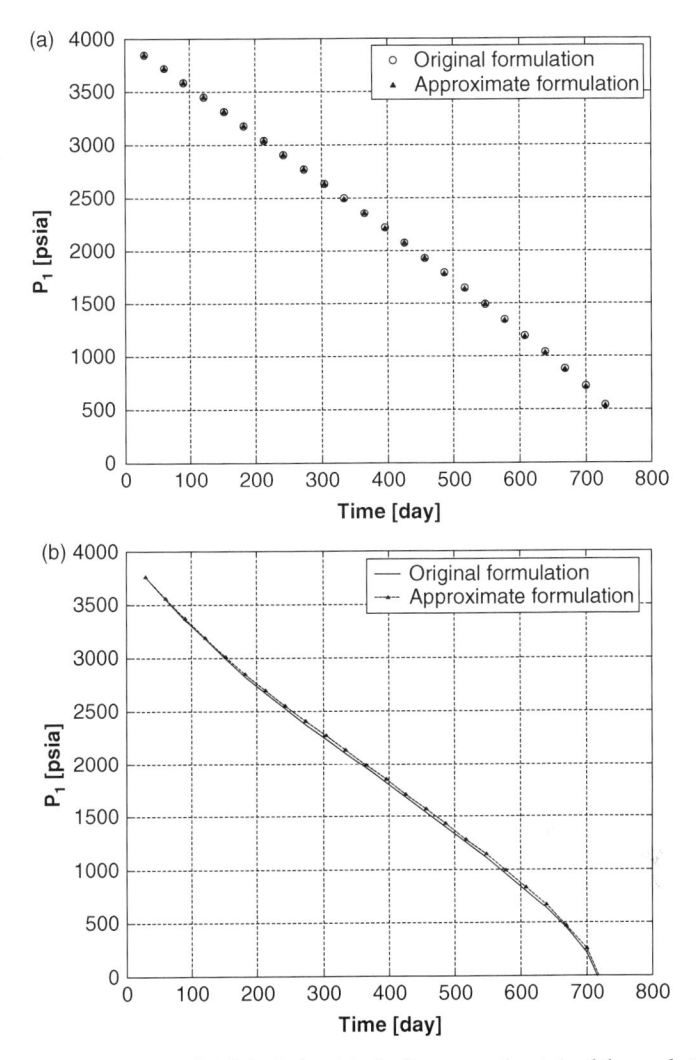

Figure 5.10 Pressure at Gridblock 1 with the linear and original formulation a) using nonlinear continuous functions b) using cubic spline

true for other gridblocks. The solution with the cubic spline is faster and more accurate.

5.6.3 Effect of Time Interval

The effect of the time interval on the accuracy of the computation is obtained by using different time steps (Δt), which was varied from 1 day to 4 months. Results are based on formulation following (5.130)

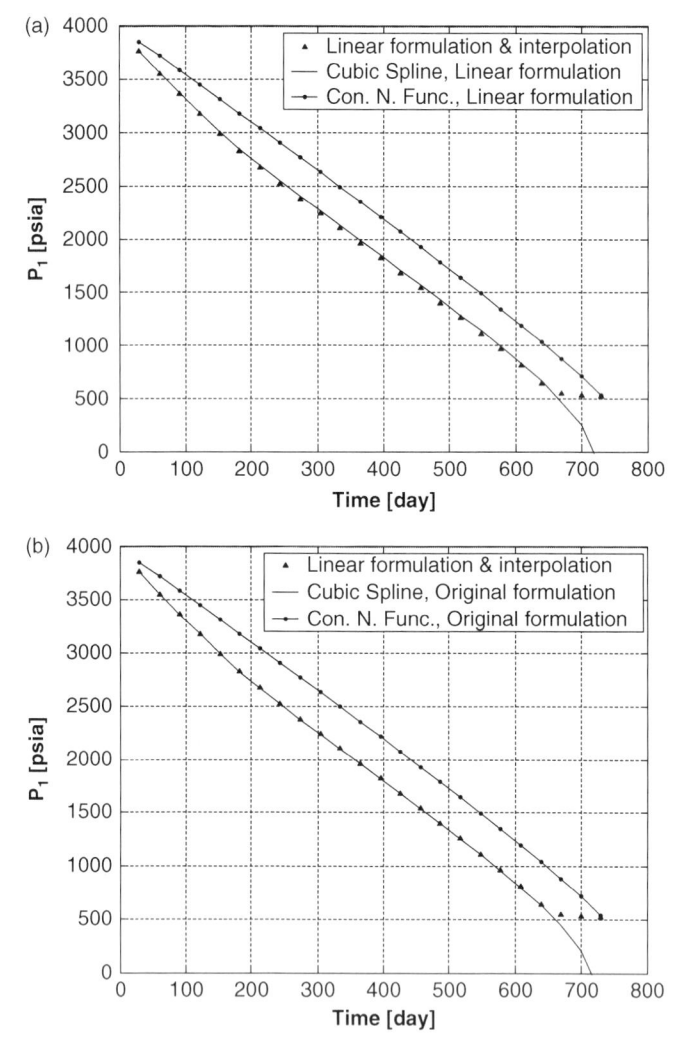

Figure 5.11 Pressure at Gridblock 1 using the linear interpolation, cubic spline and nonlinear continuous functions a) with linear formulation b) with original formulation

and (5.131) while the interpolation function for μ and B follows the cubic spline. Fig. 5.12, which is the solution with original formulation at different time steps, shows that the accuracy is not affected by the time interval. In most severe case, the mean relative error is

$$\text{Relative error} = \sum_{i=1}^{n} \left(\frac{p_{i_{\Delta t=4\,\text{months}}} - p_{i_{\Delta t=1\,\text{months}}}}{p_{i_{\Delta t=4\,\text{months}}}} \right) / n = 0.014\% \quad (5.132)$$

The Compositional Simulator

Table 5.4 Relative error at different time steps using linear formulation and cubic spline interpolation

Time step (Δt)	Relative error
2 days	0.09%
0.5 month	0.68%
1 month	1.46%
2 months	3.19%
4 months	7.82%

Relative errors are also calculated for the case of linear formulation and the results of it at different time steps are shown in Table 5.4. The table clearly suggests that the linearized formulation is more sensitive to the value of Δt.

5.6.4 Effect of Permeability

To investigate the effect of permeability, computation is also carried out with the original formulation. As a case study, a 10% variation in permeability between the boundary blocks is assumed. Such variation can be described by a linear relationship between permeability and pressure through the following equation

$$k = 0.004286 \; p + 15.22 \tag{5.133}$$

During this computation, permeability, k, in each time step and iteration is renewed for the gridblocks. The variation of pressure with time for Gridblock 1 with constant and variable permeability is shown in Fig. 5.13a. For the sake of clarity, the results are repeated in a tabular form 5.13b. It is noticed that there is a small difference between the constant permeability result and the variable permeability result. It is also observed that the margin of difference is more evident in initial time than later time.

5.6.5 Effect of Number of Gridblocks

All of the previous computations are carried out with 4 gridblocks. In order to investigate the effect of the number of gridblocks, the number of gridblocks is varied from 4 to 64. For this case, the

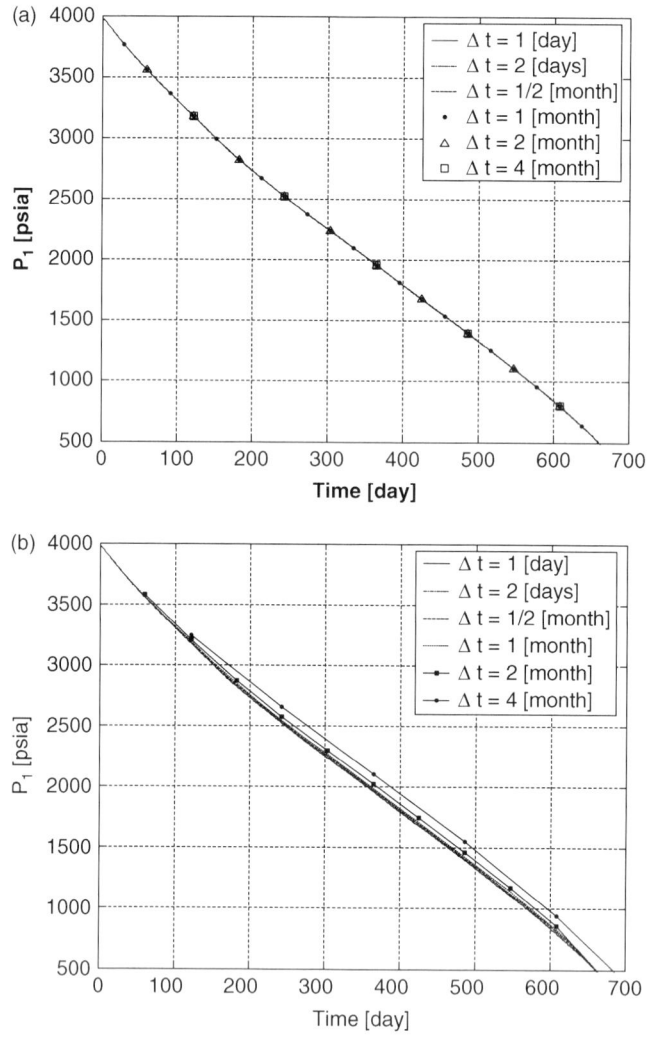

Figure 5.12 The effect of the time step in computation a) Cubic spline interpolation using original formulation b) Cubic spline interpolation using linear formulation

original formulation as given in (5.126) is applied and the systems of algebraic equations are written for each number of gridblocks. The cubic spline is taken as the interpolation function to include the variation of μ and B; and k is considered to vary according to (5.133). The Newton method is followed to solve the systems of algebraic equations. Figure 5.15a through 5.15d illustrate the effect of a number of gridblocks at various time steps. It is observed that when the number of gridblocks is increased from 4 to 8, there is

The Compositional Simulator

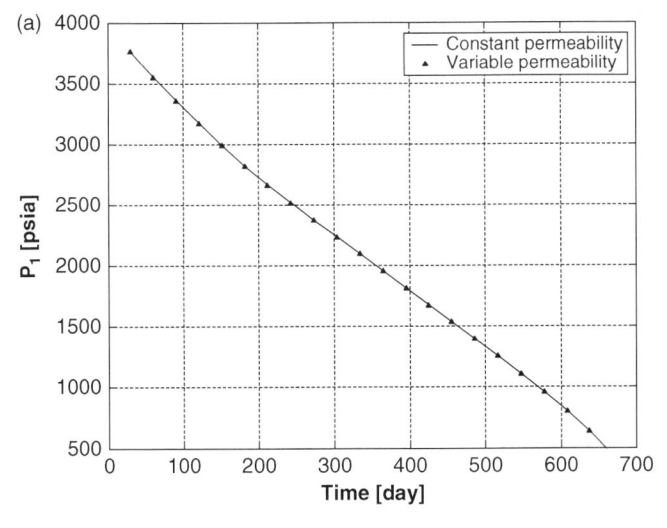

Figure 5.13 a) The effect of permeability variation on pressure variation with time b) tabular presentation

(b)

Time [months]	P_1 (k = const.) [psia]	P_1 (k = f(p)) [psia]
1	3765.668915	3762.476196
2	3554.151242	3551.149301
6	2822.768896	2820.363056
12	1959.327022	1957.484815
15	1540.271151	1538.676903
18	1108.891024	1107.635304
20	798.990998	798.110175

a difference in the pressure values predicted. However, increasing from 8 to 16 or more than that provides a smooth curve at all four time steps.

5.6.6 Spatial and Transient Pressure Distribution Using Different Interpolation Functions

Figure 5.16a through 5.16d show pressure distribution based on cubic spline and linear interpolation along the reservoir at different

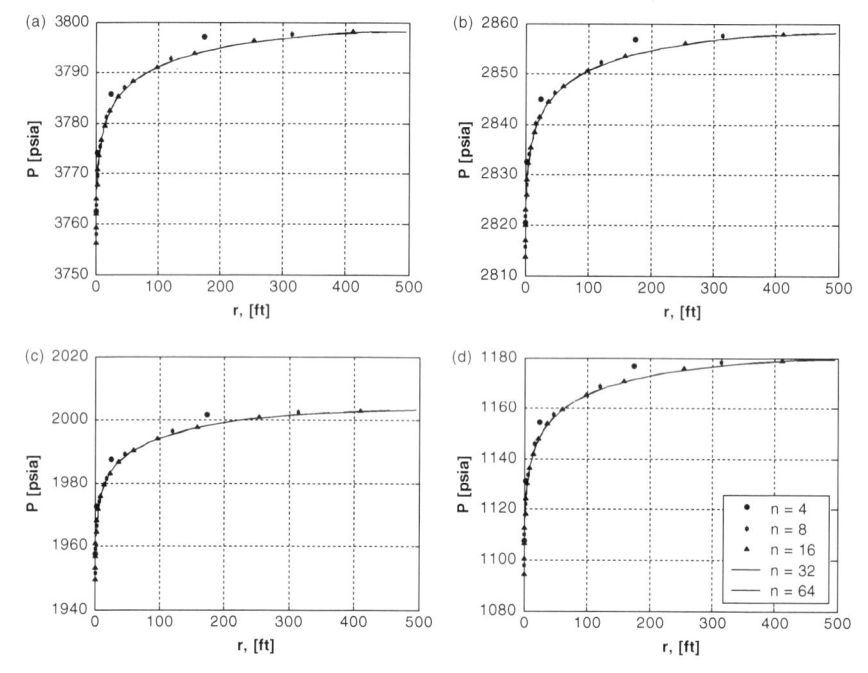

Figure 5.14 The pressure distribution using different number of gridblocks
a) $t = 1$ month, b) $t = 6$ months, c) $t = 12$ months and d) $t = 18$ months

times. These figures show that the linear interpolations of μ and B give higher values of pressure than those predicted by the cubic spline technique for all time period along the reservoir radius. However, the difference is very small and is only about 5 psia in the severe case.

The original formulation is taken into account and the cubic spline is used to approximate the variation of μ and B. The system of nonlinear algebraic equations is obtained using (5.128) and is given as

$$\begin{cases} G\left(\dfrac{1}{\mu B}\right)_j^{v+1} \left(p_2^{v+1}-p_1^{v+1}\right)-10^6 = \beta_1\left[\left(\dfrac{\phi}{B}\right)_1^{v+1}-\left(\dfrac{\phi}{B}\right)_1^{v}\right] \\ G\left(\dfrac{1}{\mu B}\right)_j^{v+1}\left(p_1^{v+1}-p_2^{v+1}\right)+G\left(\dfrac{1}{\mu B}\right)_j^{v+1}\left(p_3^{v+1}-p_2^{v+1}\right)=\beta_2\left[\left(\dfrac{\phi}{B}\right)_2^{v+1}-\left(\dfrac{\phi}{B}\right)_2^{v}\right] \end{cases}$$

The Compositional Simulator

$$\begin{cases} G\left(\dfrac{1}{\mu B}\right)_j^{v+1}\left(p_2^{v+1}-p_3^{v+1}\right)+G\left(\dfrac{1}{\mu B}\right)_j^{v+1}\left(p_4^{v+1}-p_3^{v+1}\right)=\beta_3\left[\left(\dfrac{\phi}{B}\right)_3^{v+1}-\left(\dfrac{\phi}{B}\right)_3^v\right] \\ G\left(\dfrac{1}{\mu B}\right)_j^{v+1}\left(p_3^{v+1}-p_4^{v+1}\right)=\beta_4\left[\left(\dfrac{\phi}{B}\right)_4^{v+1}-\left(\dfrac{\phi}{B}\right)_4^v\right] \end{cases}$$

(5.134)

A new method is used to formulate the problem. The first equation of (5.134) is nonlinear, for which the unknown is p_1^{v+1}. As a nonlinear equation, equation one of (5.134) has the potential to give more than one solution. Using p_1^{v+1} from the solution of The first and the second equations of (5.134) can be solved for p_2^{v+1} that may give more than one solution for each of p_1^{v+1}. The procedure is continued the similar way, i.e., applied to the third and forth equations of (5.134). Therefore, theoretically, multiple solutions can be expected for p_3^{v+1} and consequently, for p_4^{v+1}.

To examine the feasibility of the technique mentioned above, we start with the first equation of (5.134). It is found that there are two solutions for p_1 as obtained in the first iteration. The solutions are: p_1 = 4004.566 *psia* and 9312.639 *psia*.

The second equation is solved for p_2 using the result of p_1 and similarly the remaining equations of (5.134) are solved for p_3 and p_4. Table 5.5 shows the results after the first iteration:

The first and second sets give results, which are not only unexpected, but also unrealistic. During the computation, the second iteration resulted in break down of the solutions. However, the last set provides with satisfactory results following several iterations at desired time interval (the table only shows pressure at t=30.42 days).

The pressure values for the gridblocks are also obtained by utilizing (5.134). In this set, the system of algebraic equations is set up

Table 5.5 Pressure solution at t=30.42 days for four gridblocks

Time [days]	P_1 [psia]	P_2 [psia]	P_3 [psia]	P_4 [psia]
30.420	9312.639809	−5404.929667	−384.716801	577.929652
30.420	4004.566162	−5147.603731	−310.418879	606.315484
30.420	4004.566162	4009.783158	4012.393337	4012.512134

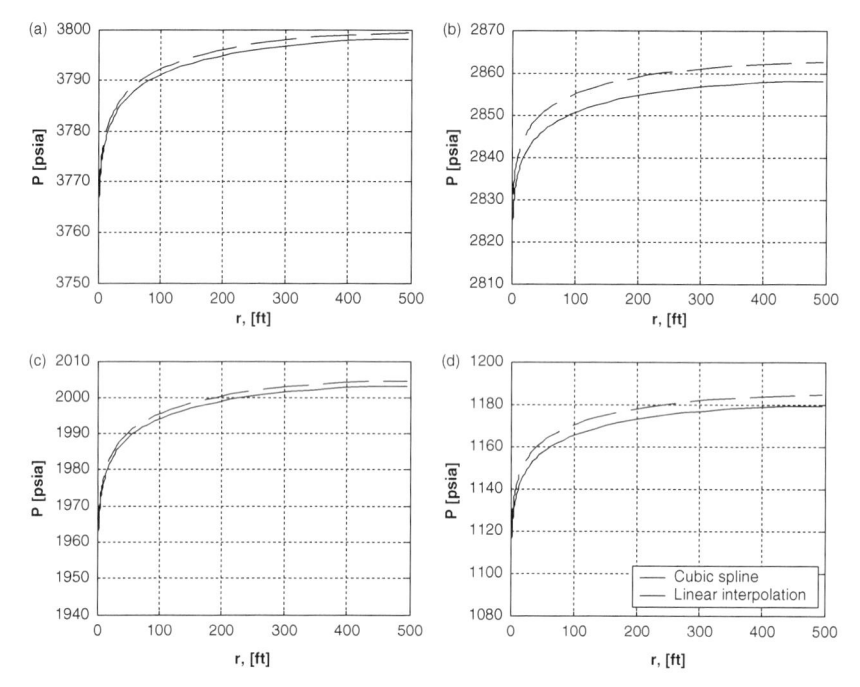

Figure 5.15 The pressure distribution along the reservoir radius using cubic spline and linear interpolation for variation of μ and B when $n = 64$ a) $t = 1$ month, b) $t = 6$ months, c) $t = 12$ months and d) $t = 18$ months

with the properties of the gridblock itself. During the computation, μ and B are updated using cubic spline and linear interpolation. Both techniques show almost same results as evident in Fig. 5.16a. The utilization of gridblock and upstream flow properties are also examined in Fig. 5.16b. The system of algebraic equations (5.130) is based on the upstream flow properties. The figure shows that the results based on these two formulations are also very close to each other. However, the difference increases as time increases (>500 days).

5.6.7 CPU Time

In continuing discussion of the previous section, it is important to note the CPU time required for computation. In fact, the main constraint during computation with (5.134) is the computing time. The CPU time required to compute pressure for all four gridblocks for a period of 1.5 years with $\Delta t = 1$ month and using the formulation of (5.134), is approximately 510 seconds. On the contrary, the same problem when utilizing the formulation of (5.130), takes only 1.156 seconds, which is significantly lower than the previous ones.

The Compositional Simulator

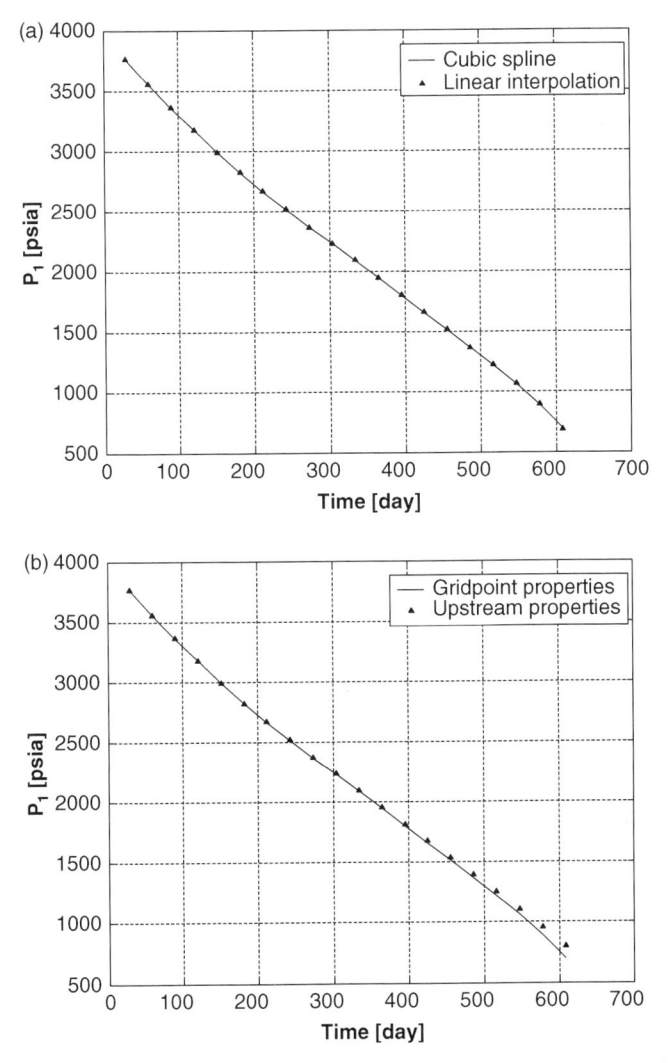

Figure 5.16 The pressure at Gridblock 1 using different sets of system of algebraic equations and various interpolations a) Cubic spline and linear interpolation using (5.130) b) Cubic spline using (5.130) and (5.134)

It is important when the time step and the number of gridblocks are increased. The computation of the pressure distribution takes t = 1.672[sec] with formulation (5.130) for 64 gridblocks and $\Delta t = 1$ months. The time to compute the pressure with formulation (6) is t = 31.172[sec] with $\Delta t = 1$ months for a period of 1.5 years. It indicates that the formulation (6) is more efficient in time than the formulation (9).

The effects of the nonlinear behavior of some fluid and formation properties and the simplification of the governing equations and the possibility of having multiple solutions are investigated in this example. The pressure distribution along the reservoir is computed with different types of the formulations while the viscosity and the fluid formation volume factor are approximated with different types of the interpolating functions.

The effect of the linearization in the governing equations is significant. Linearization of the coefficient of the formulation may lead to wrong prediction of the pressure distribution along the reservoir. The variation of the fluid and formation properties are obtained more properly if they are approximated with spline functions. The order of the piecewise polynomials has a very minimal effect if the original formulation is applied. However, the computation shows that the pressure dependent properties have a very weak nonlinearity effect and they may be neglected during the computations.

The problem is also formulated in such a way to produce multiple solutions for the pressure distribution in the gridblocks. This formulation is promising and shows the mathematical potential of having multiple solutions. More investigation is required to confirm if multiple-valued solutions of physical significance exist.

5.6.8 Case II: An Oil/Water Reservoir

A horizontal reservoir of length $L = 1200\,ft$, width $W = 350\,ft$ and height $H = 40\,ft$ is taken into account, as shown in Fig. 5.17. The flow is in the direction of length to a 7 inches production well located in the end of the reservoir. The pressure at the left side boundary of the reservoir is kept constantly the same as the initial reservoir pressure during the production process. The initial pressure of the reservoir is $p_r = 4000$ psia. The gas oil ratio is $GOR = 400\,SCF/STB$ and the bubble pressure is $p_B = 2000$ psia. The specific gravity of oil and gas are $\gamma_o = 0.876$ and

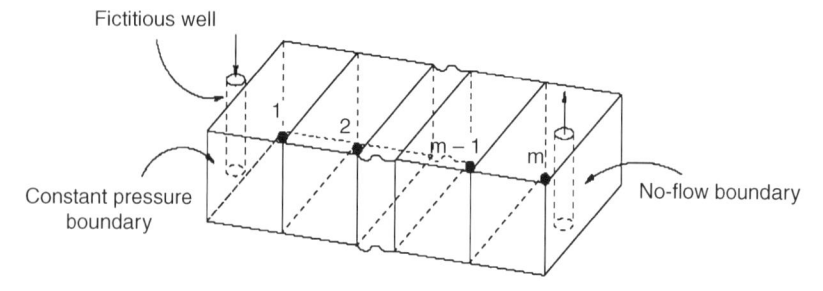

Figure 5.17 A schematic diagram of the reservoir

The Compositional Simulator

$\gamma_g = 0.75$, respectively. The reservoir temperature is $T_R = 150°F$ initially and is assumed to be constant during the production process. It is also assumed that the reservoir has homogeneous rock properties initially with porosity $\varphi = 0.27$ and permeability $k = 270$ md.

Two cases are taken into account:
  a) There is flow of oil into the reservoir in the left hand side boundary to keep this side a constant pressure boundary. Therefore, there is a single phase flow reservoir.
  b) The pressure is kept constant by the presence of a strong aquifer. The flow of the water into the reservoir caused a multi-phase flow of oil and water.

Single phase flow: The pressure is keep constant in the left hand side by an oil flow into the reservoir in that boundary. The flow equation can be written as:

$$\sum_{l=y_n} \mathbb{T}^{n+1}_{ol,n}\left[\left(p^{n+1}_{ol}-p^{n+1}_{on}\right)-\gamma^n_{ol,n}\left(Z_l-Z_n\right)\right] + \sum_{l=x_n} q^{n+1}_{osc_{l,n}} + q^{n+1}_{osc_n}$$

$$= \frac{U}{\alpha_c \Delta t}\left[\left(\frac{\phi}{B_o}\right)^{n+1}_n - \left(\frac{\phi}{B_o}\right)^n_n\right] \tag{5.135}$$

for each grid point. The notation $\mathbb{T}^{n+1}_{ol,n}$ is the oil transmissibility between l and n gridblocks and may be given in the form

$$\mathbb{T}^{n+1}_{ol,n} = G_{l,n}\left(\frac{1}{\mu_o B_o}\right)_{l,n} \tag{5.136}$$

Where

$$G_{l,n} = G = \frac{\beta_c A_x k_x}{\Delta x} \tag{5.137}$$

The rate of water flow into the reservoir to keep a constant pressure is obtained by

$$q^{n+1}_{wsc_b,BB} = \left[\frac{\beta_c A_x k_x}{\mu_o B_o \left(\frac{\Delta x}{2}\right)}\right]^{n+1}_{bB}\left[\left(p_b - p^{n+1}_{bB}\right) - \left(\gamma_w\right)^{n+1}_{bB}\left(Z_b - Z_{bB}\right)\right] \tag{5.138}$$

where p_b is the boundary pressure ($b \equiv$ boundary) and p_{bB} is the pressure of the boundary block ($B \equiv$ boundary block).

It is taken into account that the reservoir is divided into m gridblocks of the same length as shown in Fig. 5.18. Since the gridblocks have the same length and same cross-sectional area and the flow is 1-D, the geometrical factors and bulk volumes are equal for all gridblocks.

$$G_{l,n} = G = \frac{\beta_c A_x k_x}{\Delta x} = \frac{0.001127 \times (350 \times 40) \times 270}{300} = 14.2002$$

$$U = 300 \times 350 \times 40 = 4200000 \ ft^3$$

The set of flow equations for the gridblocks is:

$$\begin{cases} G\left(\frac{1}{\mu_o B_o}\right)^{n+1}_j \left(p^{n+1}_{o_2} - p^{n+1}_{o_1}\right) + 2G\left(\frac{1}{\mu_o B_o}\right)^{n+1}_1 \left(4000 - p^{n+1}_{o_2}\right) \\ \qquad = \frac{U}{\alpha_c \Delta t}\left[\left(\frac{\phi}{B_o}\right)^{n+1}_1 - \left(\frac{\phi}{B_o}\right)^n_1\right] \\ \ldots \ldots \ldots \ldots \ldots \\ \ldots \ldots \ldots \ldots \ldots \\ G\left(\frac{1}{\mu_o B_o}\right)^{n+1}_1 \left(p^{n+1}_{o_{i-1}} - p^{n+1}_{o_i}\right) + G\left(\frac{1}{\mu_o B_o}\right)^{n+1}_1 \left(p^{n+1}_{o_{i+1}} - p^{n+1}_{o_i}\right) \\ \qquad = \frac{U}{\alpha_c \Delta t}\left[\left(\frac{\phi}{B_o}\right)^{n+1}_i - \left(\frac{\phi}{B_o}\right)^n_i\right] \\ \ldots \ldots \ldots \ldots \ldots \\ \ldots \ldots \ldots \ldots \ldots \\ G\left(\frac{1}{\mu_o B_o}\right)^{n+1}_{m-1} \left(p^{n+1}_{o_{m-1}} - p^{n+1}_{o_m}\right) - q_o = \frac{U}{\alpha_c \Delta t}\left[\left(\frac{\phi}{B_o}\right)^{n+1}_m - \left(\frac{\phi}{B_o}\right)^n_m\right] \end{cases} \quad (5.139)$$

where, $1 < i < m$. The distribution of the pressure along the reservoir is obtained by solving (5.139). This set of algebraic equations is nonlinear. The nonlinearity is due to the fact that the viscosity and

The Compositional Simulator

the oil formation volume factor are functions of oil pressure. They change with the variation of pressure. The porosity of the formation is also a function of the fluid pressure inside it.

$$\mu_o = 1 + c_\mu (p - p_B), \quad c_\mu = 5.7 \times 10^{-5} \frac{cp}{psia}$$

$$B_o = B_{oB} e^{c_o(p-p_B)}, \quad B_{oB} = 1.22 \frac{RB}{bbl}, \quad c_o = 5.666 \times 10^{-6} \ psia^{-1} \quad (5.140)$$

$$\phi = \phi^o \left[c_\phi (p - p_R) \right], \quad c_\phi = 1 \times 10^{-6}$$

The Newton method is applied to find the solution for the system of algebraic equations. The solutions are obtained iteratively by updating the coefficients to reach a convergent solution based on an assigned minimum error between two consecutive results. The minimum error is taken into account to be less than 10^{-6}.

The pressure distribution takes several days to reach a steady state condition and it does not vary any more as shown in Fig. 5.18. The pressure distributions are shown for different periods of time. The pressure distribution after t = 4 [days] is very close to the steady state condition when the pressure distribution does not change with

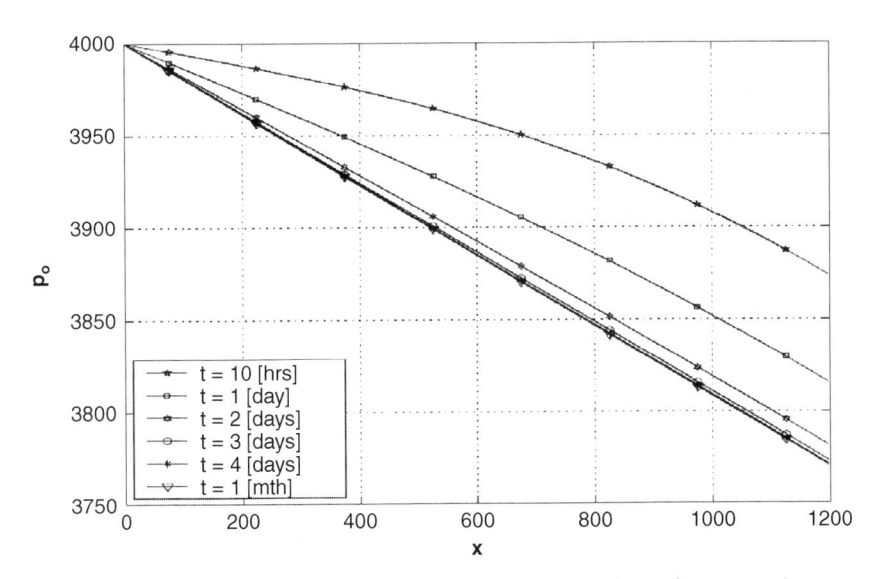

Figure 5.18 The pressure distribution at a different time along the reservoir

228   Advanced Petroleum Reservoir Simulation

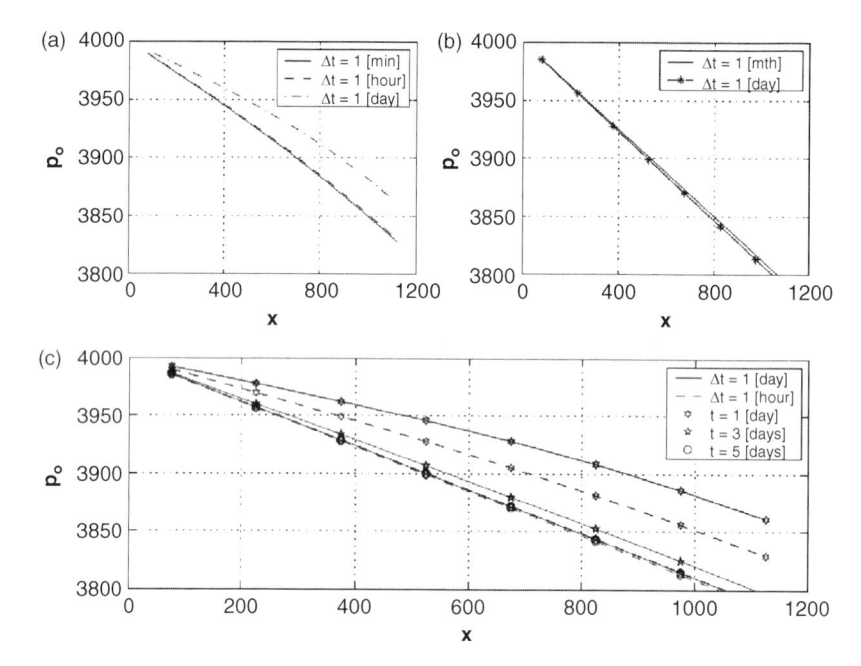

Figure 5.19 The pressure distribution at different time along the reservoir

the time anymore. The result are computed with m = 8 and 256. The solution with m = 8 is given with a different marker at various periods of time. The solution with m = 256 is given with the solid line at a different time. The results are the same with both numbers of gridblocks.

The solution with different values of Δt is given in Fig. 5.19. The solution is obtained at different time with Δt = 1 [min], 1 [hour], 1 [day] and 1 [month]. The time step has no effect on the final steady state condition. However, it is necessary to use small values of Δt to predict the pressure distribution and the time to reach a steady state condition. The solution with Δt = 1 [min] is carried out for only a period of one day with n = 8. The results with Δt = 1 [hour] is obtained with the same number of gridblocks for a period of one month. The pressure distribution is given for a period of one day with Δt = 1 [min], 1 [hour], 1 [day] in Fig. 5.19a. It is necessary to have small values of Δt to predict the variation of pressure in the period of unsteady process. The pressure distribution with Δt = 1 [day] is more than the ones obtained with smaller values of Δt. The pressure distribution after one day of operation is almost the same with $\Delta t = 1$ [min] and 1 [hour]. This indicates that it is not necessary to adopt very small values of time interval. The pressure distribution with $\Delta t = 1$ [month] is compared with the solution obtained with $\Delta t = 1$ [hour] in Fig. 5.19b. The solution with $\Delta t = 1$ [month] shows some unsteady behavior after a month of operation that is not correct. It is better to have a smaller time interval to predict more precisely the behavior of fluid flow inside the reservoir. The pressure distributions for the first few days of production process are given in Fig. 5.19c with $\Delta t = 1$ [hour] and 1 [day]. The solutions are different for the first three days but after five days of operation the results are almost the same with both time intervals. It seems that a time interval of $\Delta t = 1$ [day] is sufficient to provide a precise results for pressure distribution in unsteady state and the final steady state conditions.

Multi-phase flow, Oil/Water: The pressure on the left hand side boundary is kept constant by a strong aquifer that replace the production oil with water. This causes the change in water saturation during the production process. The pressure distribution in oil phase is also affected by the capillary pressure in the interface of oil and water. The flow equations for oil/water flow model for each gridblock are:

- for the oil component;

$$\sum_{l=\psi_n} \mathbb{T}_{ol,n}^{n+1}\left[\left(p_{ol}^{n+1} - p_{on}^{n+1}\right) - \gamma_{ol,n}^n (Z_l - Z_n)\right] + \sum_{l=\xi_n} q_{osc_{l,n}}^{n+1} + q_{osc_n}^{n+1}$$
$$= \frac{U}{a_c \Delta t}\left[\left(\frac{\phi(1-S_w)}{B_o}\right)_b^{n+1} - \left(\frac{\phi(1-S_w)}{B_o}\right)_n^n\right] \qquad (5.141)$$

- for the water component,

$$\sum_{l=\psi_n} \mathbb{T}_{wl,n}^{n+1}\left[\left(p_{o_l}^{n+1} - p_{o_n}^n\right) - \left(p_{cow_l}^{n+1} - p_{cow_n}^n\right) - \gamma_{w_{l,n}}^n (Z_l - Z_n)\right]$$
$$+ \sum_{l=\xi_n} q_{wsc_{l,n}}^{n+1} + q_{wsc_n}^{n+1} = \frac{U}{a_c \Delta t}\left[\left[\frac{\phi S_w}{B_w}\right]_b^{n+1} - \left[\frac{\phi S_w}{B_w}\right]_n^n\right] \qquad (5.142)$$

Where

$$\mathbb{T}_{ol,n}^{n+1} = G_{l,n}\left(\frac{1}{\mu_p B_p}\right) k_{rp_{l,n}} \qquad (5.143)$$

is the transmissibility of oil ($p \equiv o$) or water ($p \equiv w$). The notation $G_{l,n}$ is the geometric factor between blocks n and l and is obtained by

$$G_{l,n} = \left(\frac{\beta_c A_x k_x}{\Delta x}\right)_{l,n} \qquad (5.144)$$

The specific pressure, which is assigned for the boundary of the reservoir, is modeled by the amount of the water that is replaced from a strong aquifer. The amount of water to produce a constant pressure at the boundary is obtained by

$$q_{wsc_b,BB}^{n+1} = \left[\frac{\beta_c A_x k_x}{\mu_w B_w \left(\frac{\Delta x}{2}\right)}\right]_{bB}^{n+1} (k_{rw})_{aq}^{n+1} \left[\left(p_b - p_{bB}^{n+1}\right) - (\gamma_w)_{bB}^{n+1}(Z_b - Z_{bB})\right]$$

(5.145)

where the relative permeability at the aquifer is equal one $((k_{rw})_{aq}^{n+1} = 1)$. If the reservoir is assumed to consist of m gridblocks, Fig. 5.17, the flow equations may be written for the oil phase in the form,

$$\begin{cases} G\left(\frac{k_{ro}}{\mu_o B_o}\right)_1^{n+1}\left(p_{o_2}^{n+1} - p_{o_1}^{n+1}\right) = \frac{U}{a_c \Delta t}\left\{\left[\frac{\phi(1-S_w)}{B_o}\right]_1^{n+1} - \left[\frac{\phi(1-S_w)}{B_o}\right]_1^n\right\} \\ \cdots\cdots\cdots\cdots\cdots \\ \cdots\cdots\cdots\cdots\cdots \\ G\left(\frac{k_{ro}}{\mu_o B_o}\right)_1^{n+1}\left(p_{o_{i-1}}^{n+1} - p_{o_i}^{n+1}\right) + G\left(\frac{k_{ro}}{\mu_o B_o}\right)_1^{n+1}\left(p_{o_{i+1}}^{n+1} - p_{o_i}^{n+1}\right) \\ = \frac{U}{a_c \Delta t}\left\{\left[\frac{\phi(1-S_w)}{B_o}\right]_i^{n+1} - \left[\frac{\phi(1-S_w)}{B_o}\right]_i^n\right\} \end{cases}$$

The Compositional Simulator

$$\begin{cases} \cdots\cdots\cdots\cdots\cdots\cdots \\ \cdots\cdots\cdots\cdots\cdots\cdots \\ G\left(\dfrac{k_{ro}}{\mu_o B_o}\right)_{m-1}^{n+1}\left(p_{o_{m-1}}^{n+1}-p_{o_m}^{n+1}\right)-q_o = \dfrac{U}{\alpha_c \Delta t}\left\{\left[\dfrac{\phi(1-S_w)}{B_o}\right]_m^{n+1}-\left[\dfrac{\phi(1-S_w)}{B_o}\right]_m^n\right\} \end{cases}$$

(5.146)

and for the water phase

$$\begin{cases} G\left(\dfrac{k_{rw}}{\mu_w B_w}\right)_1^{n+1}\left[\left(p_{o_2}^{n+1}-p_{o_1}^{n+1}\right)-\left(p_{cow_2}^{n+1}-p_{cow_1}^n\right)\right]+2G\left(\dfrac{1}{\mu_w B_w}\right)_1^{n+1} \\ \times\left[4000-\left(p_{o_1}^{n+1}-p_{cow_1}^n\right)\right] = \dfrac{U}{\alpha_c \Delta t}\left[\left(\dfrac{\phi S_w}{B_w}\right)_1^{n+1}-\left(\dfrac{\phi S_w}{B_w}\right)_1^n\right] \\ \cdots\cdots\cdots\cdots\cdots\cdots \\ \cdots\cdots\cdots\cdots\cdots\cdots \\ G\left(\dfrac{k_{rw}}{\mu_w B_w}\right)_{i-1}^{n+1}\left[\left(p_{o_{i-2}}^{n+1}-p_{o_i}^{n+1}\right)-\left(p_{cow_{i-1}}^{n+1}-p_{cow_i}^n\right)\right]+G\left(\dfrac{k_{rw}}{\mu_w B_w}\right)_i^{n+1} \\ \left[\left(p_{o_{i+1}}^{n+1}-p_{o_i}^{n+1}\right)-\left(p_{cow_{i+1}}^{n+1}-p_{cow_i}^n\right)\right] = \dfrac{U}{\alpha_c \Delta t}\left[\left(\dfrac{\phi S_w}{B_w}\right)_i^{n+1}-\left(\dfrac{\phi S_w}{B_w}\right)_i^n\right] \\ \cdots\cdots\cdots\cdots\cdots\cdots \\ \cdots\cdots\cdots\cdots\cdots\cdots \\ G\left(\dfrac{k_{rw}}{\mu_w B_w}\right)_{m-1}^{n+1}\left[\left(p_{o_{m-1}}^{n+1}-p_{o_m}^{n+1}\right)-\left(p_{cow_{m-1}}^{n+1}-p_{cow_m}^n\right)\right]-q_m \\ = \dfrac{U}{\alpha_c \Delta t}\left[\left(\dfrac{\phi S_w}{B_w}\right)_m^{n+1}-\left(\dfrac{\phi S_w}{B_w}\right)_m^n\right] \end{cases}$$

(5.147)

where $1<i<m$. The unknowns are $p_{o_1}^{n+1}, p_{o_2}^{n+1}, \cdots, p_{o_m}^{n+1}$, $S_{w_1}^{n+1}, S_{w_2}^{n+1}, \cdots, S_{w_m}^{n+1}$. These two systems of simultaneous algebraic equation (SAE) (5.146) and (5.147) are nonlinear due to the dependency of the fluids and formation properties to the variation of the pressure and water saturation along the reservoir.

The water can be taken into account as an incompressible fluid without loss of accuracy. It is also assumed that the reservoir temperature is constant and therefore, the water phase viscosity is also constant, $\mu_w = 0.52\ cp$, $B_w = 1$. The oil viscosity and the oil formation volume factor are functions of reservoir pressure and temperature (p_R & T_R), bubble point pressure (p_b), and the amount of the gas/oil ratio (GOR), $\mu_o = f_1\ (T_R, P_R, P_b, GOR)$ and $B_o = f_2\ (T_R, p_R, p_b, GOR)$. The variation of μ_o, B_o and also the media porosity A ϕ are given in (6.140). The relative permeability of oil and water and also the capillary pressure depend on the water saturation of the reservoir, $k_{rw} = f(S_w)$, $k_{ro} = f(S_w)$ & $p_{cow} = f(S_w)$.

Since the pressure is assumed to be kept constant by a high pressure aquifer, the water saturation is increased and owing to that both the relative permeabilities of oil and water and the capillary pressure are changed during the production process. It is assumed that the connate water saturation and the minimum possible oil saturation to produce oil are $S_{wc} = 0.18$ and $S_{oc} = 0.1$, respectively. The variation of the water and oil relative permeability are expressed by

$$k_{rw} = 0.59439\ S_{wn}^4, \quad k_{ro} = \left(1 - S_{wn}\right)^2 \tag{5.148}$$

where $S_{wn} = \frac{S_w - S_{iw}}{1 - S_{wc} - S_{oc}}$ is called the normalized water saturation. The capillary pressure variation is given in Fig. 5.20 in tabular and graphical form as a function of the water saturation.

The oil phase pressures at the grid points may be obtained by combining the equations for the oil phase (5.146) and the water

S_{wn}	P_c [atm]	S_{wn}	P_c [atm]
0.00	3.9944	0.30	0.3600
0.01	3.5846	0.36	0.2700
0.02	3.1975	0.42	0.1980
0.05	2.2576	0.48	0.1450
0.08	1.6209	0.56	0.0920
0.11	1.2035	0.65	0.0550
0.15	0.8747	0.72	0.0350
0.18	0.7010	0.87	0.0100
0.21	0.5709	0.95	0.0020
0.25	0.4592	1.00	0.0000

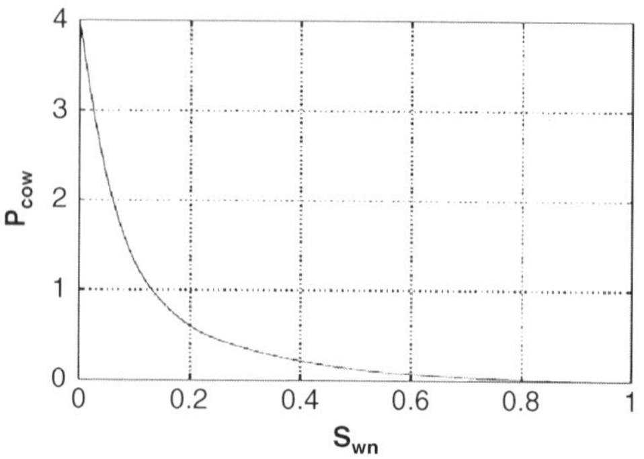

Figure 5.20 Capillary pressure variation with water saturation.

The Compositional Simulator

phase (5.147). The equations (5.146) are multiplied by B_w and (5.147) are multiplied by $B_{o_n}^{n+1}$ and then added together. It gives the following system of simultaneous equations for the oil pressure at

$$\begin{cases} G\left(\dfrac{k_{ro}}{\mu_o B_o B_w}\right)_1^{n+1}\left(p_{o_2}^{n+1}-p_{o_1}^{n+1}\right)+G\left(\dfrac{k_{rw}}{\mu_w B_o B_w}\right)_1^{n+1} \\ \times\left[\left(p_{o_2}^{n+1}-p_{o_1}^{n+1}\right)-\left(p_{cow_2}^{n+1}-p_{cow_1}^{n}\right)\right]+2G\left(\dfrac{1}{\mu_w B_o B_w}\right)_1^{n+1} \\ \times\left[4000-\left(p_{o_1}^{n+1}-p_{cow_1}^{n}\right)\right]=\dfrac{U}{a_o \Delta t}\left[\left(\dfrac{\phi}{B_o B_w}\right)_1^{n+1}-\left(\dfrac{\phi}{B_o B_w}\right)_1^{n}\right] \end{cases}$$

$$\begin{cases} \cdots\cdots\cdots\cdots\cdots\cdots \\ \cdots\cdots\cdots\cdots\cdots\cdots \\ G\left(\dfrac{k_{ro}}{\mu_o B_o B_w}\right)_{i-1}^{n+1}\left(p_{o_{i-1}}^{n+1}-p_{o_i}^{n+1}\right)+G\left(\dfrac{k_{rw}}{\mu_w B_w}\right)_{i-1}^{n+1}\left(\dfrac{1}{B_o}\right)_i^{n+1} \\ \left[\left(p_{o_{i-1}}^{n+1}-p_{o_i}^{n+1}\right)-\left(p_{cow_{i-1}}^{n+1}-p_{cow_i}^{n}\right)\right] \\ G\left(\dfrac{k_{ro}}{\mu_o B_o B_w}\right)_{i-1}^{n+1}\left(p_{o_{i+1}}^{n+1}-p_{o_i}^{n+1}\right)+G\left(\dfrac{k_{rw}}{\mu_w B_o B_w}\right)_i^{n+1} \\ \left[\left(p_{o_{i+1}}^{n+1}-p_{o_i}^{n+1}\right)-\left(p_{cow_{i+1}}^{n+1}-p_{cow_i}^{n}\right)\right]=\dfrac{U}{\alpha_c \Delta t}\left[\left(\dfrac{\phi}{B_o B_w}\right)_i^{n+1}-\left(\dfrac{\phi}{B_o B_w}\right)_i^{n}\right] \\ \cdots\cdots\cdots\cdots\cdots\cdots \\ \cdots\cdots\cdots\cdots\cdots\cdots \\ G\left(\dfrac{k_{ro}}{\mu_o B_o B_w}\right)_{m-1}^{n+1}\left(p_{o_{m-1}}^{n+1}-p_{o_m}^{n+1}\right)+G\left(\dfrac{k_{rw}}{\mu_w B_w}\right)_{m-1}^{n+1}\left(\dfrac{1}{B_o}\right)_m^{n+1} \\ \left[\left(p_{o_{m-1}}^{n+1}-p_{o_m}^{n+1}\right)-\left(p_{cow_{m-1}}^{n+1}-p_{cow_m}^{n}\right)\right] \\ -\dfrac{q_w}{B_o}-\dfrac{q_o}{B_w}=\dfrac{U}{a_c \Delta t}\left[\left(\dfrac{\phi}{B_o B_w}\right)_m^{n+1}-\left(\dfrac{\phi}{B_o B_w}\right)_m^{n}\right] \end{cases}$$

(5.149)

The system of equations (5.149) is solved for oil pressure distribution when the water saturation is assumed to be unvaried. The computed oil pressure is applied to compute the new water saturation by using the set of algebraic equations (5.147). The procedure is continued for another time step to find the oil pressure and water saturation during the production process.

The systems of algebraic equations (5.147) and (5.149) are nonlinear due to the dependency of the coefficient to the pressure and water saturation distribution. The solution of these systems of algebraic equations are obtained iteratively in each time step by renewing the coefficients to reach a minimum error. The minimum error is taken into account to be less than 10^{-6}. The Newton method is applied to solve these systems of algebraic equations (5.147) and (5.149). Fig. 20 shows the capillary pressure data used in this case. We investigate:

- the effect of the number of gridblocks;
- the effect of the value of time interval Δt; and
- the effect of the variation of the fluid and formation properties.

The effect of gridblocks number: It was shown that the number of gridblocks does not affect the accuracy of the result in the single phase flow in the previous section. The nonlinearity produced is due to the variation of fluid pressure in single phase flow. In the multi phase flow, the nonlinearity is due to the variation of pressure and the saturations of the fluid during the production process. We are going to show the effectiveness of the variation of pressure and fluid saturation in the final solution. At this part, we study the effect of the number of gridblocks on the accuracy of the results. The computation is carried on with a time interval of $\Delta t = 1$[day]. The reservoir is divided into $n = 32, 64, 128$, and 256 and the pressure and water saturation distributions are obtained at $t = 1$ [day], 10[days], 200 [days] and 1[year]. The pressure and water saturation distributions are shown in Figs. 5.5 and 5.6 respectively for different time intervals with various number of gridblocks.

The results of computations are shown in Fig. 5.21a for $t = 1$[day] of the production process. There is not a significant change on S_w and therefore, the fluid flow is almost one phase. The number of gridblocks has no effect on the pressure distribution. However, the

The Compositional Simulator

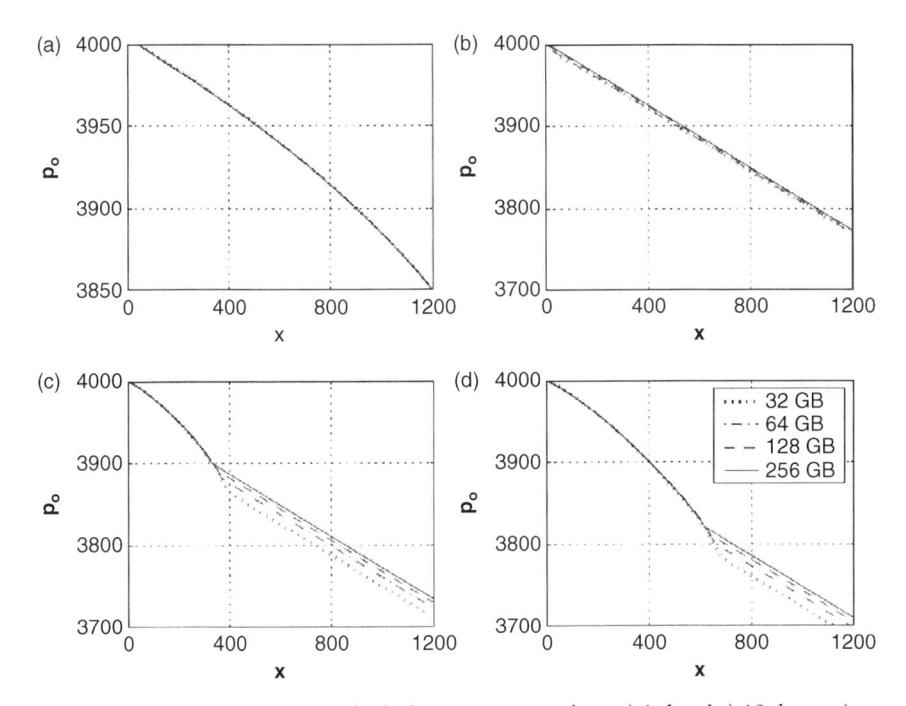

Figure 5.21 The distribution of oil phase pressure after a-) 1 day, b-) 10 *days*, c-) 200 *days*, and d-) 1 *year* with different number of Gridblocks (GB).

water saturation is started to increase along the reservoir and the number of gridblocks has significant effects on the water saturation distribution. Therefore, it is necessary to have as many gridblocks as possible to predict the water saturation distribution exactly.

The water saturation is increased just near the constant pressure boundary for $t = 10$[days] as shown in Fig. 5.22b. The number of gridblocks is important to predict the distance of the affected area and the amount of the water saturation. This is shown that it is necessary to increase the number of gridblocks to get a more precise solution. The effect on the pressure distribution is not so significant as the influence on the water saturation. However, the effect of two phase flow show that it is also good to have more gridblocks to get more precise pressure distribution.

The pressure distribution for $t = 200$[days] and 1[year] is depicted in Figs. 5.21c and 5.21d. The pattern of pressure distributions is different in the area of two phase flow with the area of single phase

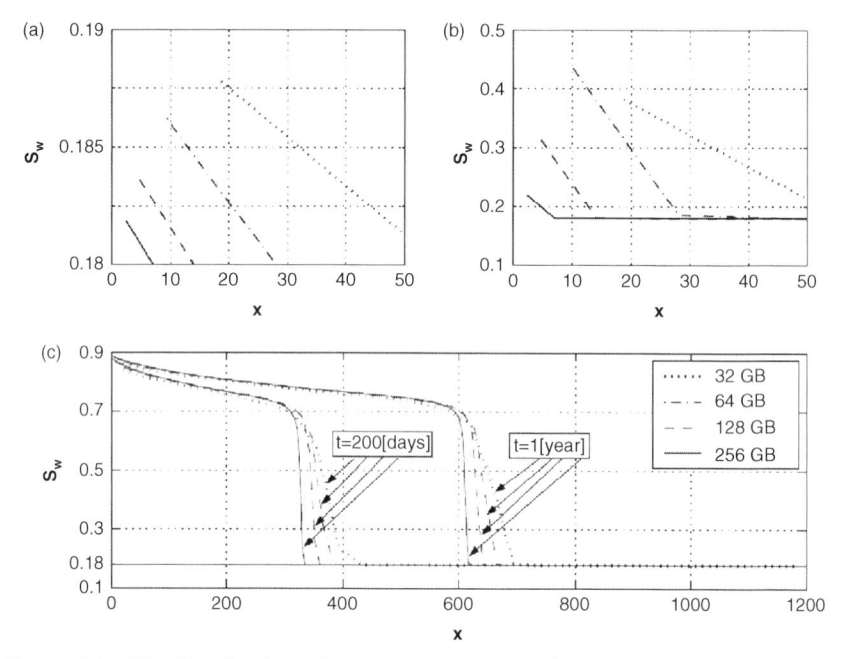

Figure 5.22 The distribution of water saturation at a-) 1 day, b-) 10 *days*, and c-) 200 *days* and 1 *year* with various number of Gridblocks (GB).

flow. There is a point of discontinuity at the different fluid flow regime. The number of gridblocks has influence in the exact prediction of the point of discontinuity in the pressure distribution. Consequently, this affects the pressure distribution in the area of single phase flow of oil. The water saturation at $t = 200$[days] and 1[year] is shown in Fig. 5.22c with different numbers of gridblocks. The exact distribution of water saturation is dependent on the number of Gridblocks. It is necessary to have as many gridblocks as possible to find the exact distribution of water saturation along the reservoir.

The effect of time interval, Δt: The computation in the single phase flow showed that the value of Δt is important in the unsteady state flow condition and in prediction of the time that the flow become independent to the time variation. It can be realized from the computation in the previous subsection that the multi phase flow of oil and water does not reach a steady state condition. The fluid quality and consequently the flow characteristics change with time and a steady state condition is never reached. The choice of Δt depends on the computer facilities and the speed and capacity of them but

The Compositional Simulator

it should be kept as small as possible. We use a PC with an Intel Pentium(R) M processor of 1.73 GHZ and 504 MB of RAM. It is taken into account that $\Delta t = 1$ [hr],1 [day] and 15 [days]. The computations are carried out with $n = 256$ gridblocks.

The pressure and water saturation distributions are shown in Figs. 5.23 and 5.24, respectively, at $t = 3,6,9$ and 12 [months] of production. The pressure distributions with $\Delta t = 1$ [hr], 1 [day] show a very small difference. The effects on the water saturation are more pronounced with $\Delta t = 1$ [hr], 1 [day]. The results with $\Delta t = 15$ [days] show a significant difference in both pressure and water saturation distribution with the others. This shows that it is necessary to have small value of Δt to produce accurate and reliable solutions. The less the Δt the more accurate are the results for the pressure and water saturation. However, a time interval of $\Delta t = 1$ day gives relatively accurate solutions that can be relied on in engineering problem and can be obtained pretty fast with a normal computer facility.

The effect of variation of parameters: The flow depends on two types of parameters. The first type are those that are changed with

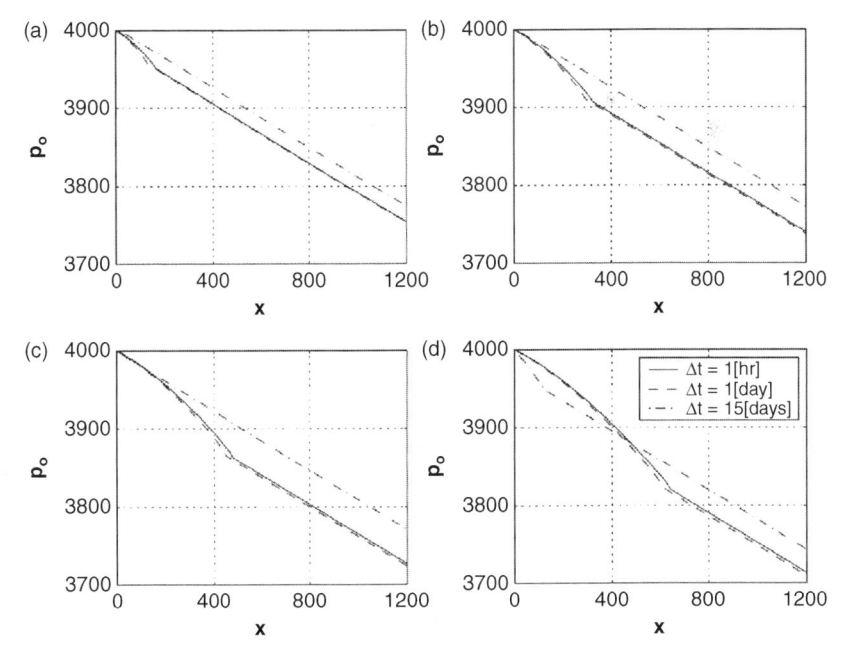

Figure 5.23 The distribution of oil phase pressure after a) $t = 3$ [mth], b) $t = 6$ [mth], c) t = 9 [mth], and d)$t = 12$ [mth] with different values of Δt.

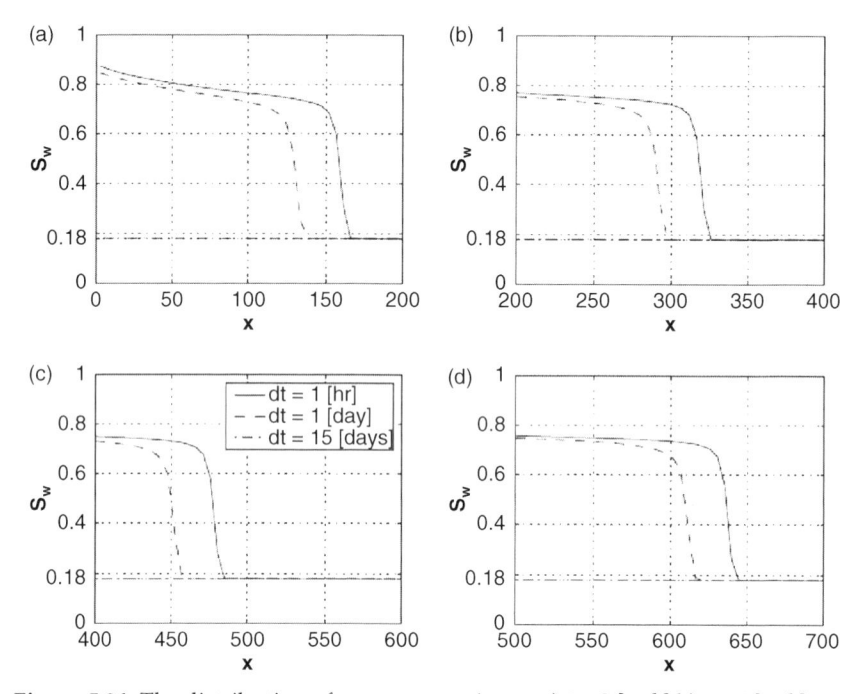

Figure 5.24 The distribution of water saturation at a) $t = 3$ [mth] b) $t = 6$ [mth], c) $t = 9$ [mth], and d) $t = 12$ [mth] with different values of t.

the variation of pressure, such as the fluid formation volume factor, the viscosity of the oil and the porosity of the formation. The second type depends on the variation of water saturation. This is the relative permeability of oil and water and the capillary pressure. The parameters are categorized based on their dependency on the pressure or water saturation and the effects of each group are studied separately.

a- The pressure dependent parameters: The effect of the variation of the pressure dependent parameters is first taken into account separately and then they are considered all together.

a-1) The oil formation volume factor, B_o
The value of the fluid formation volume factor in the process of oil production is varied between 1.2063 to 1.22 from the reservoir pressure to the bubble point pressure. We neglect the compressibility of the fluid and take that $B_o = 1.22$ during the production process. The pressure and water saturation distribution are computed for a year with $\Delta t = 1$[day] and $n = 256$. The result for $t = 3, 6, 9$ and 12 [*months*]

The Compositional Simulator

are given in Fig. 5.25. The obtained results are then compared with the solutions when the fluid formation volume factor is changed with change in pressure. There is not a significant difference at early months of production. But, the differences are increased with time and with the increase in production time. However, it shows that the variation of B_o has a minor effect on the distribution of p_o and S_w.

a-2) The oil viscosity, μ_o

The variation of the viscosity with the pressure is neglected and it is assumed that $\mu_o = 1.1340\ cp$ during the production process. This corresponds to the viscosity at the initial reservoir pressure, $p_R = 4000$ psia. The computations are carried out for a year with $\Delta t = 1\ [day]$ and $m = 256$. The pressure and water saturation distributions are shown in Fig. 5.26 for $t = 3, 6, 9$ and $12\ [months]$. The results are compared with the case when $\mu_o = f(p_o)$ as given in (5.27). There are not significant differences between the results when $\mu_o = const.$ and $\mu_o = f(p_o)$ for various depicted times. It indicates that the variation of viscosity has not a major effect on the pressure and water saturation distributions.

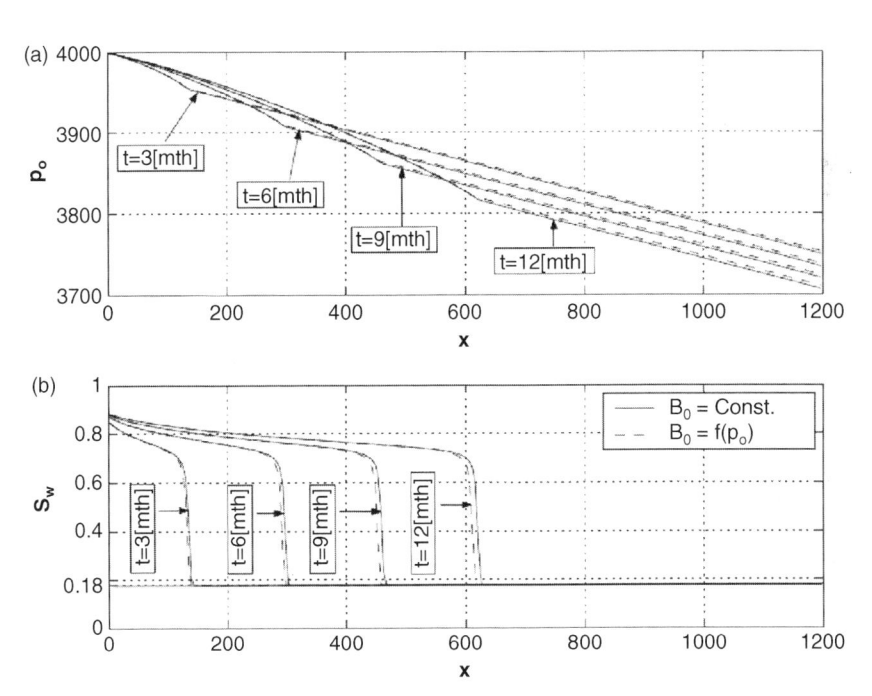

Figure 5.25 The distribution of oil phase pressure and water saturation for constant and variable fluid formation volume factors.

a-3) The porosity of the formation, ϕ

The porosity variation due to pressure change is neglected in this part. It is assumed that $\phi = \phi^o = 0.27$ during the production process. The computations are carried out for a year with $\Delta t = 1[day]$ and $m = 256$. The pressure and water saturation distributions do not demonstrate any difference between two cases of $\mu_o = const.$ and $\mu_o = f(p_o)$. The pressure and water distributions are shown in Fig. 5.27. There is not any difference between the graphs at a certain time with constant and variable ϕ, respectively, and the diagrams coincide completely.

a-4) The combined effect of variation of B_o, μ_o and ϕ: In this part, the simultaneous effect of all pressure dependent parameters are taken into account. It is assumed that $B_o = 1.22$, $\mu_o = 1.1340\ cp$ and $\phi = \phi_o = 0.27$ during the production process. The computations are carried out for a year with $\Delta t = 1[day]$ and $n = 256$. The pressure and water saturation distributions are given in Fig. 5.28 for $t = 1[year]$. The results are compared with the cases of:

- $B_o = const.$ when μ_o and ϕ are function of p_o;
- $\mu_o = const.$ when B_o and ϕ are function of p_o;
- $\phi = const.$ when μ_o and ϕ are function of p_o;
- B_o, μ_o and ϕ are function of p_o;

The oil formation volume factor B_o has the major effect while the porosity variation has the minor effect on the oil pressure and water saturation distributions. Neglecting the variation of the pressure dependent parameters gives maximum relative errors in the pressure distribution as follows with respect to the case when the variation of all pressure dependent parameters are taken into account.

- $E_{po} = 0.1623\%$ when B_o, μ_o & $\phi = const.$;
- $E_{po} = 0.1050\%$ when $B_o = const. = const.$ and μ_o & $\phi = f(p_o)$;
- $E_{po} = 0.0545\%$ when $\mu_o = const.$ and B_o & $\phi = f(p_o)$; and
- $E_{po} = 0.0014\%$ when $\phi = const.$ and B_o & $\mu_o = f(p_o)$

These show that the variation of pressure dependent parameters do not produce significant effects and they may be assumed constant.

The Compositional Simulator

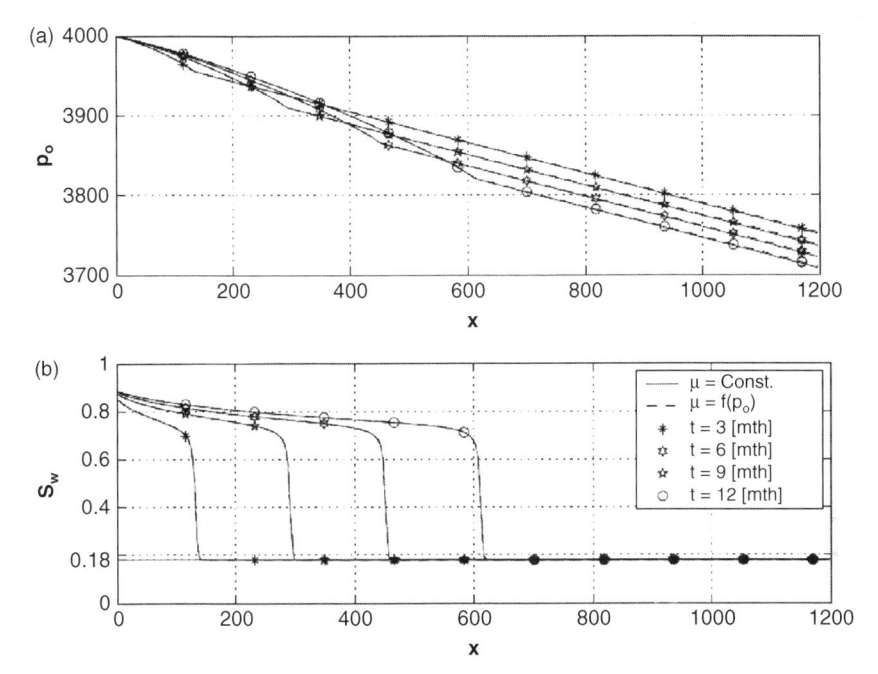

Figure 5.26 The distribution of oil phase pressure and water saturation for constant and variable oil viscosity.

b- The water saturation dependent parameters: The water permeability of water and oil and the capillary pressure are dependent on the variation of water saturation. If the variation of these parameters with the water saturation is neglected, the problem is reduced to a single phase flow problem that is addressed in previous section. The pressure distribution of the single and two-phase flows are given for $t = 1[year]$ in Fig. 5.29. The pressure distribution with neglecting the variation of the pressure dependent parameters is also depicted to provide a comparison and to find out the effect of different parameters on the final results. The effect of the pressure dependent parameters is very small compared with the influence of the water saturation variation.

Conclusion: The flow inside a porous media is a nonlinear phenomenon. The derived mathematical models are also nonlinear for such a flow regardless of the fact that many assumptions are made during the formulation of the flow. The nonlinearity of the equations is due to the dependency of the fluid and formation properties on the unknown variables that should be calculated. The effects of the variation of fluid and formation properties on the oil pressure and

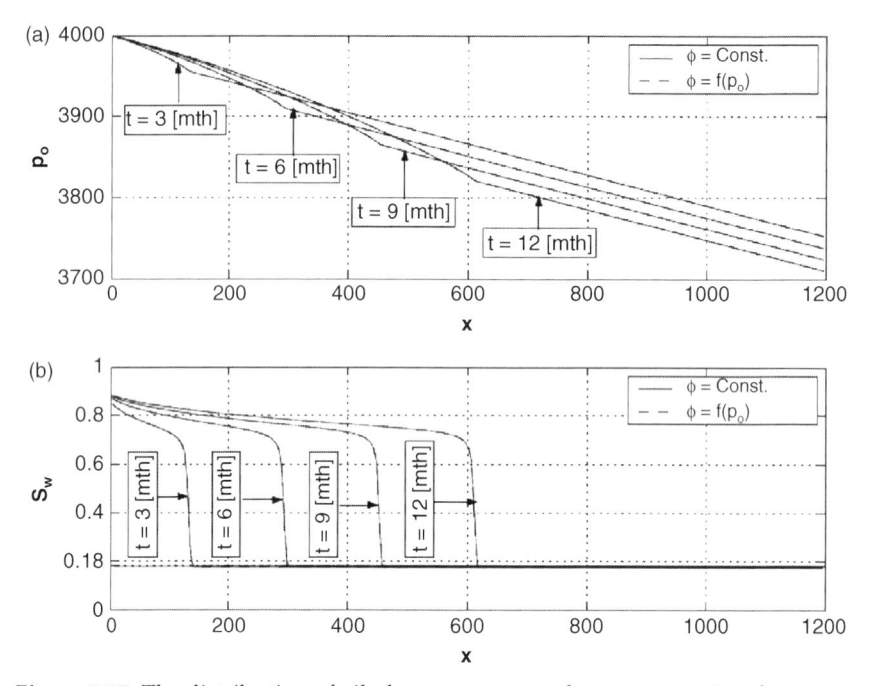

Figure 5.27 The distribution of oil phase pressure and water saturation for constant and variable porosity.

water-saturation distributions were studied for a sample reservoir in this chapter. The parameters are categorized into two groups and the effects of each of them on the final solution are sought. The effect of the number of gridblocks and that of the time interval are also studied for two cases of single and double phase flow (oil/water) problems.

In the single phase flow, the number of grid point creates little or no loss of accuracy. Increasing the number of gridblocks gives better distribution of pressure but does not affect the accuracy of the results. It is true for both the steady and unsteady state conditions. The time interval value is important to find the time that the regime of the fluid flow changes from an unsteady state to a steady state condition. The smaller the time interval the better description of the unsteady state flow conditions is obtained. A time interval of $\Delta t = 1[day]$ is sufficient to predict the behavior of the flow in unsteady state conditions for this particular case study.

The quality of the fluid in two phase (oil/water) system changes continuously, as such, no steady state condition is attainable. The flow properties are functions of the pressure variation and the water

The Compositional Simulator

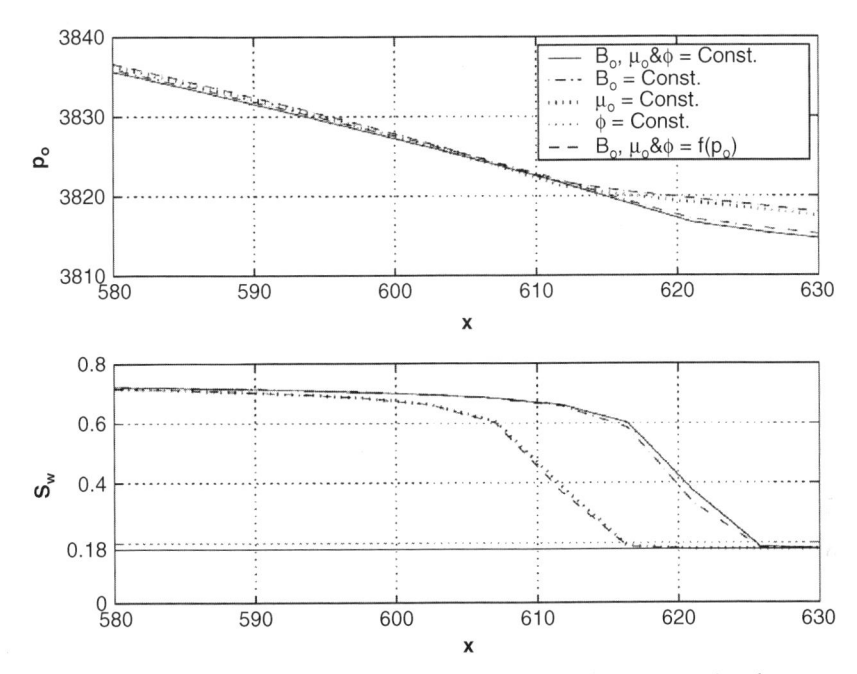

Figure 5.28 The distribution of oil phase pressure and water saturation for constant and variable pressure dependent parameters.

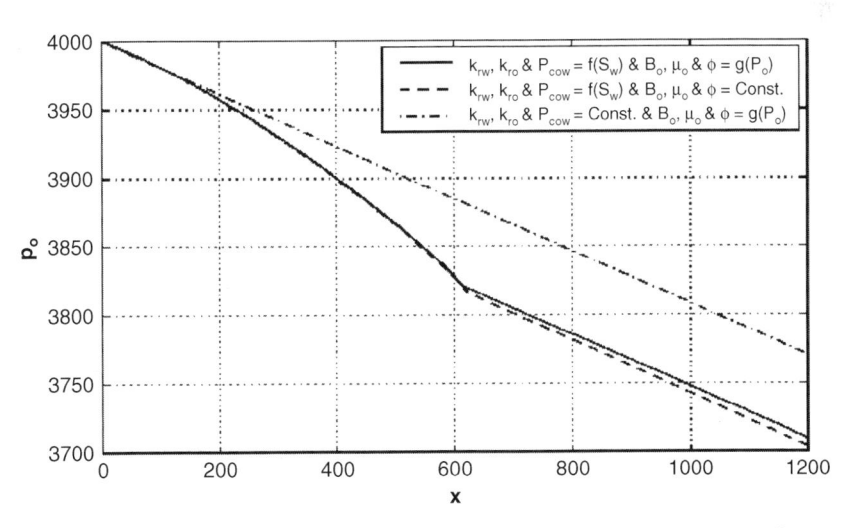

Figure 5.29 The distribution of oil phase saturation and water saturation for constant and variable pressure dependent parameters.

saturation. The effect of the variation of the pressure-dependent properties is very small in comparison with the effect of the saturation-dependent parameters variation on the final solution. The computation shows that the effect of the pressure-dependent parameters can be neglected with little loss of accuracy. However, neglecting the saturation-dependent parameters reduces the two-phase flow problem to a single phase flow problem.

6

A Comprehensive Material Balance Equation for Oil Recovery

The material balance equation (MBE) forms the core of the petroleum engineering problems. Conventional MBE involves a number of simplifications that arise from many assumptions. The most important assumption includes time-invariant rock and fluid properties. Because, real fluids and real rocks are time-dependent, it is important to remove this assumption. This chapter introduces a flexible option to have more complex phenomena included in describing a petroleum reservoir. The time-dependence is included through variable compressibilities that are dependent on pressure as well as time. The resulting formulation, however, is not free from uncertainties. This new version of MBE is applicable in the case for which rock and fluid compressibilities are available as a function of pressure and time.

6.1 Background

Nowadays, a high powered modern computing technique is minimizing the challenges of a very high accuracy and efficiency in complex calculations in reservoir simulation. Therefore, it is not necessary to have approximate solutions in reservoir engineering formulations. The MBE, one of the most widely used techniques in reservoir engineering, is an excellent example of this. MBE is used for estimating the original hydrocarbon-in-place. It is also used for calculating the decline in average reservoir pressure with depletion. In the majority of cases, the conventional formulation of the material balance is satisfactory. However, certain circumstances, which are sometimes unpredictable, demand formulations with greater accuracy. Proper understanding of reservoir behavior and predicting future performance is necessary to have knowledge of the driving mechanisms that control the behavior of fluids within reservoirs. The overall performance of an oil reservoir is mainly determined by the nature of energy (i.e. driving mechanism) available for moving oil toward the wellbore. There are basically six driving mechanisms that provide the natural energy necessary for oil recovery. These mechanisms are rock and liquid expansion drive, depletion drive, gas cap drive, water drive, gravity drainage drive and combination drive (Dake, 1978; Ahmed, 2002).

When an oil reservoir initially exists at a pressure higher than its bubble-point pressure, the reservoir is called an undersaturated oil reservoir (Ahmed, 2000). At pressures above the bubble-point, crude oil, connate water, and rock are the only materials present in the formation. There may also be dissolved gas in oil and water. As the reservoir pressure declines, rock and fluids expand due to their individual compressibilities. The reservoir rock compressibility is the result of two factors; expansion of the individual rock grains and formation compaction. Both of the above two factors are the results of a decrease of fluid pressure within the pore spaces, and both tend to reduce the pore volume through the reduction of the porosity. As the expansion of fluids and reduction in pore volume occur with decreasing reservoir pressure, crude oil and water will be forced out of the pore space toward the wellbore. In such a situation, the reservoir will experience a rapid pressure decline because liquids and rocks are only slightly compressible. If there is an existence of some dissolved gas, the oil reservoir under this driving mechanism is characterized by a constant gas-oil ratio that is equal to the gas solubility at the bubble point pressure. This driving mechanism is considered the least efficient driving force and usually results in the recovery of only a small percentage of the total initial oil- in-place (IOIP).

The dependency of formation compressibility is well known in the literature (Hall, 1953; Dobrynin, 1961; Ramagost and Farshad, 1981; Fetkovich et al, 1991; Fetkovich et al, 1998; Rahman et al, 2006a; Rahman et al, 2006b). The porosity dependent correlation for formation compressibility was first developed by Hall (1953). The Hall correlation is good for normal pressure because it was introduced based on sandstones laboratory data. It has a tendency to underpredict formation compressibility at high pressure situations. The formation compressibility correlation by Dobrynin (1961) can predict a relative change over a range of net overburden pressures. His correlation is valid for over-pressured reservoirs. However, it does not predict the initial formation compressibility. There are very limited published correlations for high-pressured reservoirs (Rahman et al, 2006a).

In this case, expansion of oil, water, rock, and dissolved gas in oil and water are incorporated. In addition, total water associated with the oil reservoir volume is taken care of. "Associated" water comprises connate water, water within interbedded shales and nonpay reservoir rock, and any limited aquifer volume (Fetkovich et al, 1998). Fetkovich and co-authors (Fetkovich et al, 1991; Fetkovich et al, 1998) introduced a MBE for high-pressured gas reservoir. They used a cumulative effective compressibility term to characterize the expansion drive mechanism. They account for pressure-dependent rock and water compressibility of the formation, all water and rock volumes associated with the reservoir and available for expansion including a limited aquifer volume. Ramagost and Farshad (1981) modified the traditional MBE (Dake, 1978; Ahmed, 2000; Craft and Hawkins, 1959; Havlena and Odeh, 1963; Havlena and Odeh, 1964). They proposed a new plotting function for gas-in-place based on an improved version of the conventional MBE. In some situations with certain combinations of compressibility and saturation values, it has been observed that the Ramagost and Farshad MBE should have been more accurate in predicting the average reservoir pressure. This is also reported and examined by Rahman and co-authors (Rahman et al, 2006a). However, the MBE developed by Ramagost and Farshad is the simplified version of the MBE of other researchers (Fetkovich et al, 1991; Fetkovich et al, 1998; Rahman et al, 2006a). All these former researchers use the expansion drive mechanism in developing MBE for a gas reservoir.

The available literature shows that very few researchers consider the variable compressibility concept in generating MBE. Normally, the inclusion of the rock and liquid expansion drive are neglected due to insignificant effects in MBE which is explained earlier. However, pore compressibility can sometimes be very large in shallow unconsolidated reservoirs. As an example, in the Boliver Coast field in Venezuela pore compressibility was measured over $100 \times 10^{-6}\, psi^{-1}$. In such a situation, it is not wise to neglect this drive mechanism (Dake, 1978).

Therefore, this case is an attempt to investigate the effects of expansion drive mechanism in a typical oil field using the PVT data available in the literature. A new, rigorous MBE for oil flow in a compressible formation with residual fluid saturations is presented in this study. A new dimensionless parameter, C_{epm} in MBE is identified to represent the whole expansion drive mechanism. It is also explained how this parameter can predict the behavior of MBE.

All water and rock volumes associated with the reservoir and available for expansion, including a limited aquifer volume with formation fluid and rock volume expansion are added in this dimensionless parameter (Fetkovich et al, 1991; Fetkovich et al, 1998).

6.2 Permeability Alteration

For fractured porous media, it is assumed that the Boussinesq approximation (Ghorayeb and Firoozabadi, 2000; Islam and Nandakumar, 1990) is valid in the range of temperature and pressure so that the density, ρ is constant except in the buoyancy term ($\rho g z$) where it varies linearly with the temperature, T_f. So, this approximation is invoked to model the permeability variation with respect to heat transfer in the formation which is (Hossain and Islam, 2009):

$$\rho_r = \rho_f \left[1 - \beta \left(T_r - T_f \right) \right] \quad (6.1)$$

The fluid velocity in the media can be described by using Darcy's law (Hossain and Islam, 2009):

$$u = -\frac{K_{(x,t)}}{\phi_{(x,t)} \mu} \left\{ \frac{\partial p}{\partial x} + \rho_f \left[1 - \beta \left(T_r - T_f \right) \right] g \right\} \quad (6.2)$$

Substituting $T_s = T_f = T$ in equation (6.2) becomes:

$$K_{(x,t)} = -\frac{u \phi_{(x,t)} \mu}{\frac{\partial p}{\partial x} + \rho_f \left[1 - \beta \left(T_r - T \right) \right] g} \quad (6.3)$$

The momentum equation for a single phase 1D can be approximated by (Hossain and Islam, 2009):

$$\frac{\partial p}{\partial x} = \rho_f g - \frac{\mu}{K_i} u - \frac{\rho_f}{\phi_{(x,t)}} \frac{\partial u}{\partial t} \quad (6.4)$$

A Material Balance Equation for Oil Recovery

Substituting equation (6.4) in to equation (6.3) becomes:

$$K_{(x,t)} = -\frac{u\phi\mu}{\rho_f g - \frac{\mu}{K_i}u - \frac{\rho_f}{\phi_{(x,t)}}\frac{\partial u}{\partial t} + \rho_f\left[1-\beta(T_r - T)\right]g}$$

$$K_{(x,t)} = -\frac{u\phi\mu}{2\rho_f g - \frac{\mu}{K_i}u - \frac{\rho_f}{\phi_{(x,t)}}\frac{\partial u}{\partial t} - \rho_f\beta gT_r + \rho_f\beta gT} \quad (6.5)$$

6.3 Porosity Alteration

Kühn et al (2002) used the Kozeny–Carman equation (Carman, 1956) which links permeability to the effective pore radius and the formation factor to produce a general permeability-porosity relationship (Pape et al, 1999) as a three term power series of porosity:

$$K_{(x,t)} = A\phi + B\phi^p + C(10\phi)^q \quad (6.6)$$

where, the exponents p and q depend on the fractal dimension. The coefficients A, B, and C need to be calibrated for each type of reservoir sandstone or chemical pore-space modification. For average sandstone (porosity range 2–40%), Pape et al (1999) found coefficients and exponents:

$$K_{(x,t)} = 31\phi + 7463\phi^2 + 191(10\phi)^{10} \quad (6.7)$$

To investigate the effects of temperature on porosity, a steady state situation is considered where velocity does not change for a particular time interval. So, for a constant space permeability and porosity, equation (6.6) can be differentiated with respect to time and can be written as:

$$\frac{\partial K}{\partial t} = 31\frac{\partial \phi}{\partial t} + 14926\phi\frac{\partial \phi}{\partial t} + 191 \times 10^{11}\phi^9\frac{\partial \phi}{\partial t}$$

$$\frac{\partial K}{\partial t} = \left(31 + 14926\phi + 191 \times 10^{11}\phi^9\right)\frac{\partial \phi}{\partial t} \quad (6.8)$$

For a specific locality where permeability and porosity are constant, permeability can be differentiated with respect to time, t and equation (6.5) becomes:

$$\frac{\partial K}{\partial t} = -\frac{\left(2\rho_f g - \frac{\mu}{K_i} u - \frac{\rho_f}{\phi} \frac{\partial u}{\partial t} - \rho_f \beta g T_r + \rho_f \beta g T\right)\left(u\mu \frac{\partial \phi}{\partial t} + \phi\mu \frac{\partial u}{\partial t}\right)}{\left(2\rho_f g - \frac{\mu}{K_i} u - \frac{\rho_f}{\phi} \frac{\partial u}{\partial t} - \rho_f \beta g T_r + \rho_f \beta g T\right)^2}$$

$$-\frac{\phi u \mu \left(-\frac{\mu}{K_i} \frac{\partial u}{\partial t} + \frac{\rho_f}{\phi^2} \frac{\partial \phi}{\partial t} \frac{\partial u}{\partial t} - \frac{\rho_f}{\phi} \frac{\partial^2 u}{\partial t^2}\right)}{\left(2\rho_f g - \frac{\mu}{K_i} u - \frac{\rho_f}{\phi} \frac{\partial u}{\partial t} - \rho_f \beta g T_r + \rho_f \beta g T\right)^2} \quad (6.5a)$$

Neglecting the second derivative of velocity, the above equation becomes:

$$\frac{\partial K}{\partial t} = -\frac{\left(2\rho_f g - \frac{\mu}{K_i} u - \frac{\rho_f}{\phi} \frac{\partial u}{\partial t} - \rho_f \beta g T_r + \rho_f \beta g T\right)}{\left(2\rho_f g - \frac{\mu}{K_i} u - \frac{\rho_f}{\phi} \frac{\partial u}{\partial t} - \rho_f \beta g T_r + \rho_f \beta g T\right)^2}$$

$$\frac{\left(u\mu \frac{\partial \phi}{\partial t} + \phi\mu \frac{\partial u}{\partial t}\right) - \phi u \mu \left(-\frac{\mu}{K_i} \frac{\partial u}{\partial t} + \frac{\rho_f}{\phi^2} \frac{\partial \phi}{\partial t} \frac{\partial u}{\partial t}\right)}{\left(2\rho_f g - \frac{\mu}{K_i} u - \frac{\rho_f}{\phi} \frac{\partial u}{\partial t} - \rho_f \beta g T_r + \rho_f \beta g T\right)^2} \quad (6.5b)$$

Let us assume

$$A_1 = 2\rho_f g - \frac{\mu}{K_i} u - \frac{\rho_f}{\phi} \frac{\partial u}{\partial t} - \rho_f \beta g T_r + \rho_f \beta g T \quad (6.5c)$$

The above equation becomes:

$$\frac{\partial K}{\partial t} = \frac{\phi u \mu \left(-\frac{\mu}{K_i} \frac{\partial u}{\partial t} + \frac{\rho_f}{\phi^2} \frac{\partial \phi}{\partial t} \frac{\partial u}{\partial t}\right) + A_1 \left(u\mu \frac{\partial \phi}{\partial t} + \phi\mu \frac{\partial u}{\partial t}\right)}{A_1^2} \quad (6.9)$$

A Material Balance Equation for Oil Recovery

Equating equation (6.8) and equation (6.9) becomes:

$$\left(31 + 14926\phi + 191 \times 10^{11}\phi^9\right)\frac{\partial \phi}{\partial t}$$

$$= \frac{\phi u \mu \left(-\frac{\mu}{K_i}\frac{\partial u}{\partial t} + \frac{p_f}{\phi^2}\frac{\partial \phi}{\partial t}\frac{\partial u}{\partial t}\right) + A_1\left(u\mu\frac{\partial \phi}{\partial t} + \phi\mu\frac{\partial u}{\partial t}\right)}{A_1^2}$$

$$\left(31 + 14926\phi + 191 \times 10^{11}\phi^9\right)A_1^2\frac{\partial \phi}{\partial t}$$

$$= u\mu\left(\frac{p_f}{\phi}\frac{\partial u}{\partial t} + A_1\right)\frac{\partial \phi}{\partial t} + \phi\mu\left(A_1 - \frac{\mu u}{K_i}\right)\frac{\partial u}{\partial t}$$

$$\left[\left(31 + 14926\phi + 191 \times 10^{11}\phi^9\right)A_1^2 - u\mu\left(\frac{p_f}{\phi}\frac{\partial u}{\partial t} + A_1\right)\right]\frac{\partial \phi}{\partial t}$$

$$= \phi\mu\left(A_1 - \frac{\mu u}{K_i}\right)\frac{\partial u}{\partial t} \quad (6.9a)$$

Again let

$$A_2 = \left(31 + 14926\phi + 191 \times 10^{11}\phi^9\right)A_1^2 - u\mu\left(\frac{p_f}{\phi}\frac{\partial u}{\partial t} + A_1\right) \quad (6.9b)$$

Putting the value of A_2 in the above equation,

$$\frac{\partial \phi}{\partial t} = \frac{\phi\mu\left(A_1 - \frac{\mu u}{K_i}\right)}{A_2}\frac{\partial u}{\partial t} \quad (6.10)$$

6.4 Pore Volume Change

Equation (6.10) represents the variation of porosity with time. Now, in order to develop the change of pore volume with time due to pressure depletion, temperature, heat transfer etc. through porous medium, one can determine this from the definition of porosity. So, the relation between porosity and pore volume can be written as:

$$V_{pore} = \phi V_{bulk}$$

$$\frac{\partial V_{pore}}{\partial t} = V_{bulk} \frac{\partial \phi}{\partial t} \quad (6.11)$$

6.5 A Comprehensive MBE with Memory for Cumulative Oil Recovery

The equation (6.40) of Appendix 6A shows the derivation of the new material balance equation (MBE) that can be written using the expression of equation (6.49) as:

$$N_p B_o - (W_e - W_p B_w) = N(B_o - B_{oi} + B_{oi} c_{epm}) \quad (6.12)$$

where,

$$c'_{epm} = \frac{\left[S_{oi} c_o + S_{wi} c_w + S_{gi} c_g \left(\frac{R_{soi}}{B_{oi}} + \frac{R_{swi}}{B_{wi}} \right) B_{gi} + c_s + M(c_w + c_s) \right]}{1 - S_{wi}}$$

$$\times \frac{[p_i - p(t)]}{1 - S_{wi}} \quad (6.13)$$

In equation (6.13), the average pressure decline for a particular time t from the start of production may be calculated using the time-dependent rock/fluid properties with stress-strain model. The mathematical explanation and the derivation of the stress-strain formulation are described in Hossain et al, 2007. They gave the stress-strain relationship as follows:

$$\tau_T = (-1)^{0.5} \times \left(\frac{\partial \sigma}{\partial T} \frac{\Delta T}{a_D M_a} \right) \times \left[\frac{\int_0^t (t - \xi)^{-\alpha} \left(\frac{\partial^2 p}{\partial \xi \partial x} \right) d\xi}{\Gamma(1-\alpha)} \right]^{0.5}$$

$$\times \left(\frac{6 K \mu_0 \eta}{\frac{\partial p}{\partial x}} \right)^{0.5} \times e^{\left(\frac{E}{RT_T} \right)} \frac{du_x}{dy} \quad (6.14)$$

A Material Balance Equation for Oil Recovery

$$\left(\frac{6K\mu_0\eta}{\frac{\partial p}{\partial x}}\right)^{0.5} = \frac{\tau_T}{(-1)^{0.5} \times \left(\frac{\partial \sigma}{\partial T}\frac{\Delta T}{a_D M_a}\right) \times \left[\frac{\int_0^t (t-\xi)^{-a}\left(\frac{\partial^2 p}{\partial \xi \partial x}\right)d\xi}{\Gamma(1-a)}\right]^{0.5} \times e^{\left(\frac{E}{RT}\right)}\frac{du_x}{dy}}$$

$$\left(\frac{6K\mu_0\eta}{\frac{\partial p}{\partial x}}\right) = -\frac{\tau_T^2}{\left(\frac{\partial \sigma}{\partial T}\frac{\Delta T}{a_D M_a}\right)^2 \times \left[\frac{\int_0^t (t-\xi)^{-a}\left(\frac{\partial^2 p}{\partial \xi \partial x}\right)d\xi}{\Gamma(1-a)}\right]^{0.5} \times e^{2\left(\frac{E}{RT}\right)}\left(\frac{du_x}{dy}\right)^2}$$

$$\frac{\partial p}{\partial x} = -\frac{6K\mu_0\eta\left(\frac{\partial \sigma}{\partial T}\frac{\Delta T}{a_D M_a}\right)^2 \times \left[\frac{\int_0^t (t-\xi)^{-a}\left(\frac{\partial^2 p}{\partial \xi \partial x}\right)d\xi}{\Gamma(1-a)}\right] \times e^{2\left(\frac{E}{RT}\right)} \times \left(\frac{du_x}{dy}\right)^2}{\tau_T^2}$$

$$\frac{\partial p/\partial t}{\partial x/\partial t} = -\frac{6K\mu_0\eta\left(\frac{\partial \sigma}{\partial T}\frac{\Delta T}{a_D M_a}\right)^2 \times \left[\frac{\int_0^t (t-\xi)^{-a}\left(\frac{\partial^2 p}{\partial \xi \partial x}\right)d\xi}{\Gamma(1-a)}\right] \times e^{2\left(\frac{E}{RT}\right)} \times \left(\frac{du_x}{dy}\right)^2}{\tau_T^2}$$

$$\frac{\partial p}{\partial t} = -\frac{6K\mu_0\eta\left(\frac{\partial \sigma}{\partial T}\frac{\Delta T}{a_D M_a}\right)^2 \times \left[\frac{\int_0^t (t-\xi)^{-a}\left(\frac{\partial^2 p}{\partial \xi \partial x}\right)d\xi}{\Gamma(1-a)}\right]^{0.5} \times e^{2\left(\frac{E}{RT}\right)} \times \left(\frac{du_x}{dy}\right)^2}{\tau_T^2}\frac{\partial x}{\partial t}$$

For a limit of $\Delta p, \Delta x, \Delta t \to 0$ the above equation can be written as:

$$\frac{\Delta p}{\Delta t} = \frac{6K\mu_0\eta\left(\frac{\partial \sigma}{\partial T}\frac{\Delta T}{a_D M_a}\right)^2 \times \left[\frac{\int_0^t (t-\xi)^{-a}\left(\frac{\partial^2 p}{\partial \xi \partial x}\right)d\xi}{\Gamma(1-a)}\right]^{0.5} \times e^{2\left(\frac{E}{RT}\right)} \times \left(\frac{du_x}{dy}\right)^2}{\tau_T^2} u_x$$

$$\Delta p = \frac{6K\mu_0\eta\left(\frac{\Delta T}{a_D M_a}\frac{\partial \sigma}{\partial T}\right)^2 \times \left[\frac{\int_0^t (t-\xi)^{-a}\left(\frac{\partial^2 p}{\partial \xi \partial x}\right)d\xi}{\Gamma(1-a)}\right]^{0.5} \times e^{2\left(\frac{E}{RT}\right)} \times \left(\frac{du_x}{dy}\right)^2}{\tau_T^2} u_x \Delta t$$

$$p_i - p_{(t)} = \frac{6K\mu_0\eta\left(\frac{\Delta T}{a_D M_a}\frac{\partial\sigma}{\partial T}\right)^2 \times \left[\dfrac{\int_0^t (t-\xi)^{-\alpha}\left(\frac{\partial^2 p}{\partial \xi \partial x}\right)d\xi}{\Gamma(1-\alpha)}\right] \times e^{2\left(\frac{E}{RT_T}\right)} \times \left(\frac{du_x}{dy}\right)^2}{\tau_T^2} - u_x \Delta t$$

(6.15)

The change of pressure with time and space can be calculated using the stress-strain equation (6.15). This change of pressure is directly related to oil production performance of a well. Therefore, substituting equation (6.15) into equation (6.13):

$$C'_{epm} = \frac{\left[S_{oi}c_o + S_{wi}c_w + S_{gi}c_g\left(\frac{R_{soi}}{B_{oi}} + \frac{R_{swi}}{B_{wi}}\right)B_{gi} + c_s + M(c_w + c_s)\right]}{1 - S_{wi}}$$

$$\times \frac{\left[\dfrac{6K\mu_0\eta(L-x)\left(\frac{\partial\sigma}{\partial T}\frac{\Delta T}{a_D M_a}\right)^2\left[\int_0^t (t-\xi)^{-\alpha}\left(\frac{\partial^2 p}{\partial \xi \partial x}\right)d\xi\right]e^{2\left(\frac{E}{RT_T}\right)}\left(\frac{du_x}{dy}\right)^2}{\tau_T^2\{\Gamma(1-\alpha)\}} - u_x \Delta t\right]}{1 - S_{wi}}$$

(6.16)

where C'_{epm} is the modified dimensionless parameter which depends on rock/fluid memory and other related fluid and rock properties and Δt is the time difference between the start of production and a particular time which is actually time, t.

The C'_{epm} of equation (6.16) can be used in equation (6.12) to represent the time dependent rock/fluid properties and other properties which are related to the formation fluid and formation itself. Therefore, equation (6.12) can be written as:

$$N_p = \frac{N}{B_o}(B_o - B_{oi} + B_{oi}C'_{epm}) + \frac{1}{B_o}(W_e - W_p B_w)$$

We know that initial oil in place, N can be defined as;

$$N = \frac{V\phi(1-S_{wi})}{B_{oi}}$$

A MATERIAL BALANCE EQUATION FOR OIL RECOVERY 255

Substituting this relationship with the above MBE becomes;

$$N_p = \frac{V\phi(1-S_{wi})}{B_{oi}B_o}(B_o - B_{oi} + B_{oi}C'_{epm}) + \frac{1}{B_o}(W_e - W_p B_w)$$

Now, consider

$$A_3 = \frac{V(1-S_{wi})}{B_{oi}B_o}(B_o - B_{oi} + B_{oi}C'_{epm})$$

Substituting this into the above equation;

$$N_p = \phi A_3 + \frac{1}{B_o}(W_e - W_p B_w) \qquad (6.17)$$

The equation (6.17) represents the rigorous, new MBE with memory where every possibility of time dependent rock/fluid properties are considered.

6.6 Numerical Simulation

The numerical results of the dimensionless parameters, C_{epm} and C_{eHO} based on the models presented by equations (6.41), (6.44), (6.49), and (6.51) of Appendix 6A can be obtained by solving these equations. A volumetric undersaturated reservoir with no gascap gas is considered for the simulation. The reservoir initial pressure is $p_i = 4000$ *psi*. Table 6.1 presents the rock and fluid properties that have been used in solving the above mentioned equations. Trapezoidal method is used to solve the exponential integral. All computation is carried out by MATLAB 6.5.

To calculate IOIP using Havlena and Odeh (1963, 1964) straight line method, the Virginia Hills Beaverhill Lake field (Ahmed, 2002) data and additional data, listed in Table 6.1, are considered. The initial reservoir pressure is 3685 psi. The bubble-point pressure was calculated as 1500 psi. Table 6.2 shows the field production and PVT data.

Table 6.1 Reservoir rock and fluid properties for simulation

Rock and Fluid Properties (Hall, 1953; Dake, 1978; Ahmed, 2000)	
B_{gi} = 0.00087 rb/scf	c_w = 3.62 × 10^{-6} psi^{-1}
B_{oi} = 1.2417 rb/stb	R_{soi} = 510.0 scf/stb
B_{wi} = 1.0 rb/stb	R_{swi} = 67.5 scf/stb
c_g = 500.0 × 10^{-6} psi^{-1}	S_{gi} = 20%
c_o = 15.0 × 10^{-6} psi^{-1}	S_{oi} = 60%
c_s = 4.95 × 10^{-6} psi^{-1}	S_{wi} = 20%

Table 6.2 The field production and PVT data (Example 11-3: of Ahmed (2002))

Volumetric Average Pressure psi	No. of Producing Wells	B_o rb/stb	N_p mstb	W_p mstb
3685	1	1.3102	0	0
3680	2	1.3104	20.481	0
3676	2	1.3104	34.750	0
3667	3	1.3105	78.557	0
3664	4	1.3105	101.846	0
3640	19	1.3109	215.681	0
3605	25	1.3116	364.613	0
3567	36	1.3122	542.985	0.159
3515	48	1.3128	841.591	0.805
3448	59	1.3130	1273.530	2.579
3360	59	1.3150	1691.887	5.008
3275	61	1.3160	2127.077	6.500
3188	61	1.3170	2575.330	8.000

A Material Balance Equation for Oil Recovery

6.6.1 Effects of Compressibilities on Dimensionless Parameters

Figures 6.1(a) – (d) present the variation of dimensionless parameters with average reservoir pressures when associated volume fraction is not considered. The figures give a general idea of how the fluid and formation compressibilities play a role on MBE when pressure varies. The plotting of Fig. 6.1(a) is based on equation (6.41) where variable compressibilities of the fluids and formation are taken care of. The trend of the curve is a non-linear exponential type. When reservoir pressure starts to deplete, C_{epm} increases and it reaches its highest value at $p=0$. Figure 6.1(b) is plotted based on equation (6.44) where constant compressibilities of the fluids and formation are considered. The trend of the curve is still in the form of a non-linear exponential type. The numerical data and shape of the curve is almost

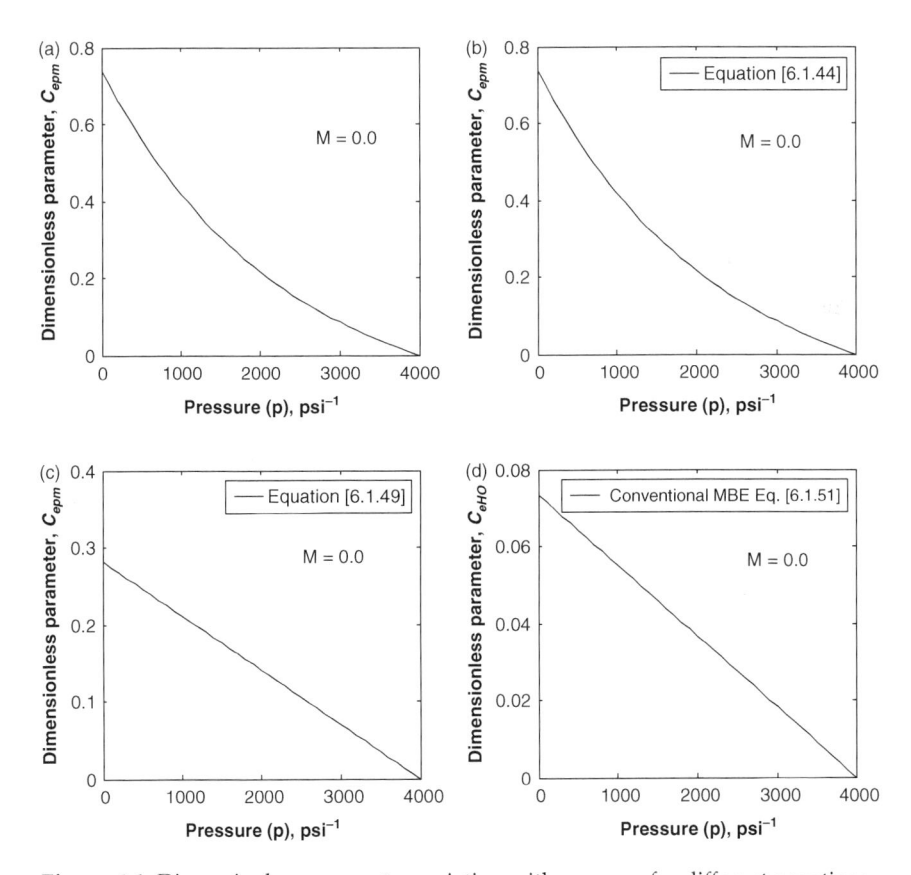

Figure 6.1 Dimensionless parameter variation with pressure for different equations (Equation no. refers to Hossain and Islam, 2009).

same as Fig. 6.1(a). Figure 6.1(c) is plotted using the equation (6.49) where constant compressibilities and an approximation of the exponential terms are considered. A straight line curve produces where the numerical values are less than that of Fig. 6.1(a) and 6.1(b). The dimensionless parameter from the use of conventional MBE shows a straight line curve where the numerical values are much less than that of the previous presentation (Fig. 6.1(d)).

6.6.2 Comparison of Dimensionless Parameters Based on Compressibility Factor

Figure 6.2 explains how the values of the dimensionless parameter can vary with reservoir pressure for a given set of compressibility and saturation. It compares between the dimensionless parameter, C_{epm} of the proposed model and the conventional dimensional parameter, C_{eHO}. The curves of the figure have been generated using equations (6.41), (6.44), (6.49) and (6.51). The depleted reservoir

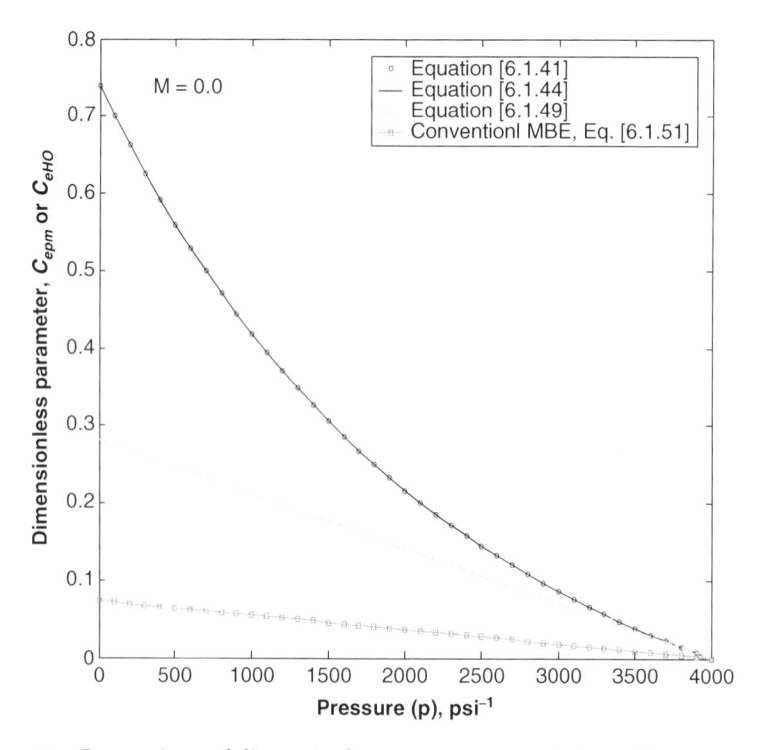

Figure 6.2 Comparison of dimensionless parameters variation with pressure for different equations (Equation no. refers to Hossain and Islam, 2009).

A Material Balance Equation for Oil Recovery 259

pressure of $p=0$ gives the maximum value of C_{epm} or C_{eHO}. The constant or variable compressibility (Equations (6.41), and (6.44)) does not make any significant difference in computation up to a certain level of accuracy, which is approximately 10^{-3} %. At low pressure, the magnitude of C_{epm} increases very fast compared with other two equations (6.49) and (6.51). The pattern and nature of the curves are already explained in Fig. 6.1. The change of dimensionless parameter is low for conventional MBE. However, when the expression of the proposed MBE's parameter is simplified, it turns to reduce the magnitude of the dimensionless parameter. This clearly means that the simplified version of the proposed model will be the same if it will be further simplified. Therefore, it may be concluded that the use of conventional MBE overestimates the IOIP. The proposed model is closer to reality. This issue will be discussed in a later section.

6.6.3 Effects of *M* on Dimensionless Parameter

If we consider the associated volume in the reservoir, all the available or probable pressure support from rock and water as well as from fluids are being accounted for the proposed MBE with dimensionless parameter. Figure 6.3 has been generated for a specific reservoir where several M values have been considered. The figure shows the variation of C_{epm} with average reservoir pressure for different M values. Figures 6.3(a)–6.3(c) present the C_{epm} variation for the proposed equations (6.41), (6.44) and (6.49) respectively. These curves have specific characteristics depending on the pressure dependence of rock and fluids (water, oil and dissolved gas) compressibilities. These curves have relatively less variant at high pressure, increase gradually as pressure decreases, and finally rise sharply at low pressure especially after 1,000 psi. All the curves in Fig. 6.3 have the same characteristics except the numerical values of the dimensionless parameter, C_{epm}. For every equation, if M increases, the curve shifts upward in the positive direction of C_{epm}. The difference in C_{epm} due to M is more dominant at low pressure. This trend of the curve indicates that the matured reservoir feels more contributions from associated volume of the reservoir.

6.6.4 Effects of Compressibility Factor with *M* Values

Figures 6.4(a)–6.4(c) illustrate C_{epm} verses pressure for different proposed equations (6.41), (6.44), and (6.49) at several M values

260　Advanced Petroleum Reservoir Simulation

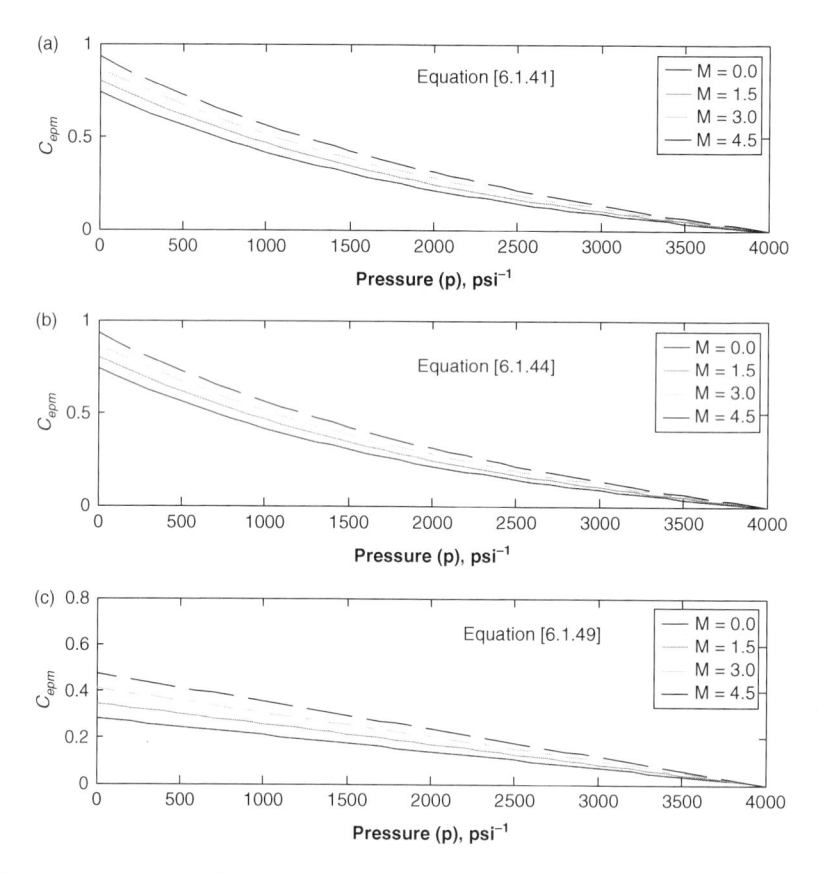

Figure 6.3 Dimensionless parameter variations with pressure at different M ratios (Equation no. refers to Hossain and Islam, 2009).

of 0.0, 1.5, 3.0, 4.5 respectively. The shape and characteristics of all curves are same as Fig. 6.3. When variable compressibilities are considered (Equation (6.41)) with pressure, there is a big difference at low pressure with constant compressibilities and exponential approximation equation (Equation (6.49)) for all M values. However, there is no significant change in equation (6.41) and (6.44) at different M values. It should be mentioned here that as M increases, C_{epm} increases. This is true for all the equations.

6.6.5 Comparison of Models Based on RF

Figure 6.5 illustrates the underground withdrawal, F verses the expansion term $E_0 + E_{cepm}$ for the proposed MBE (Equation (6.43)) with

A Material Balance Equation for Oil Recovery

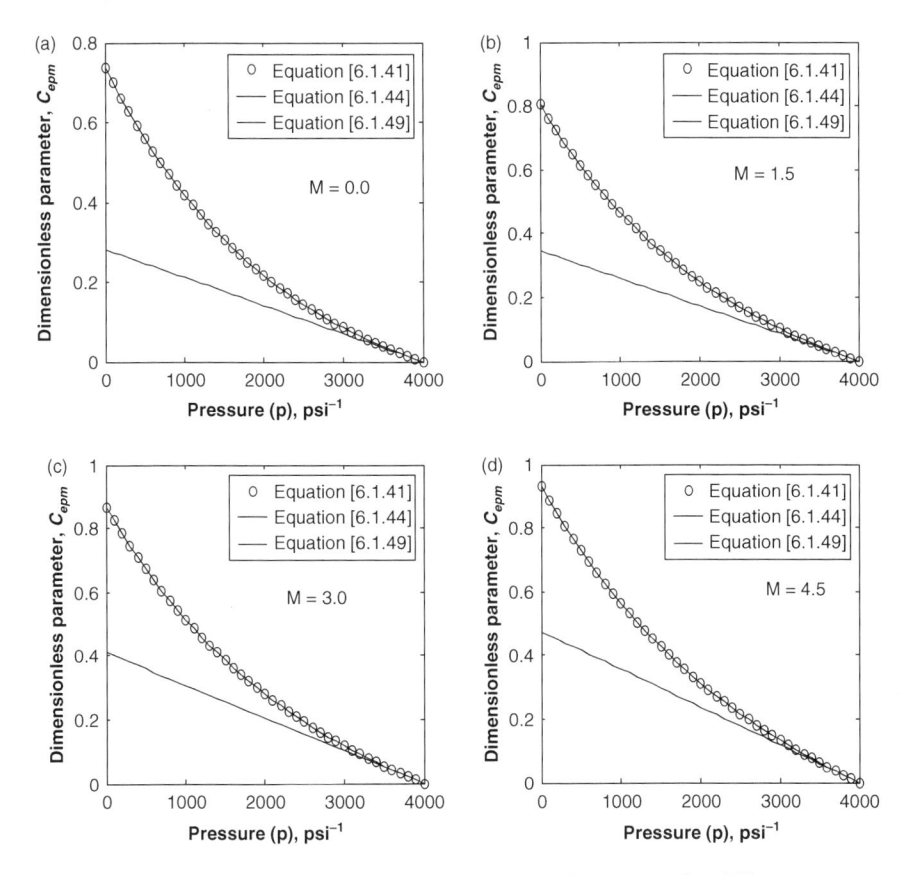

Figure 6.4 Dimensionless parameters variation with pressure for different equations (Equation no. refers to Hossain and Islam, 2009).

Equations (6.44) and (6.49) where associated volume ratio is ignored. The conventional MBE (Equation (6.52)) with equation (6.51) is also shown in the same graph. Using best fit curve fitting analysis, these plottings give a straight line passing through the origin with a slope of N. IOIP is identified as 73.41 mmstb, 68.54 mmstb and 175.75 mmstb for the MBE with equations (6.44), (6.49), and (6.51) respectively. The corresponding recovery factors are calculated as 3.76%, 3.51%, and 1.46%. Therefore, the inclusions of the probable parameters increase the ultimate oil recovery. Here, linear plot indicates that the field is producing under volumetric performance ($W_e = 0$) which is strictly by pressure depletion and fluid and rock expansion.

Figure 6.6 shows a plotting of ($F/E_0 + E_{cepm}$) verses cumulative production, N_p for the proposed MBE (Equation (6.42)) with

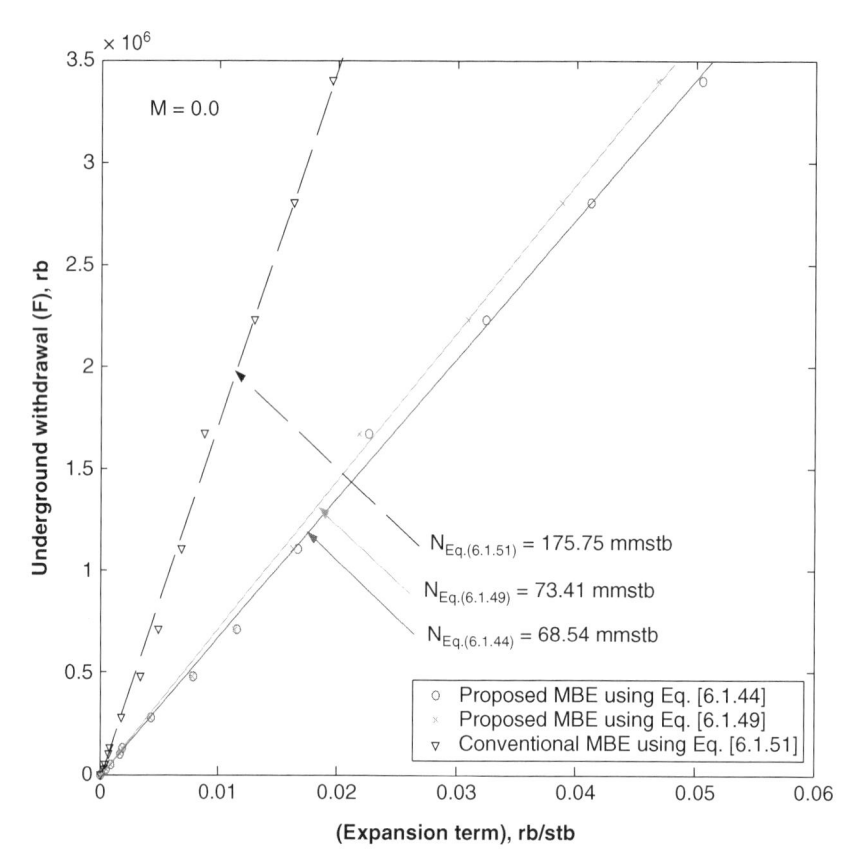

Figure 6.5 Underground withdrawal vs. Expansion term for N calculation (Equation no. refers to Hossain and Islam, 2009).

equations (6.44), (6.49), and $(F/E_0 + E_{ceHO})$ vs. N_p for the conventional MBE (Equation (6.50)). In this figure, associated volume ratio is also ignored. The best fit plot for all the equations indicate that the reservoir has been engaged by water influx, abnormal pore compaction or a combination of these two (Dake, 1978; Ahmed, 2000). In our situation we ignore the water influx($W_e = 0$). Therefore, we may conclude that the reservoir behavior is an indication of pore compactions and fluids and rocks expansion.

6.6.6 Effects of *M* on MBE

The recovery factor of the proposed model is higher than that of conventional MBE (Fig. 6.5). Moreover, the comprehensive proposed

A Material Balance Equation for Oil Recovery

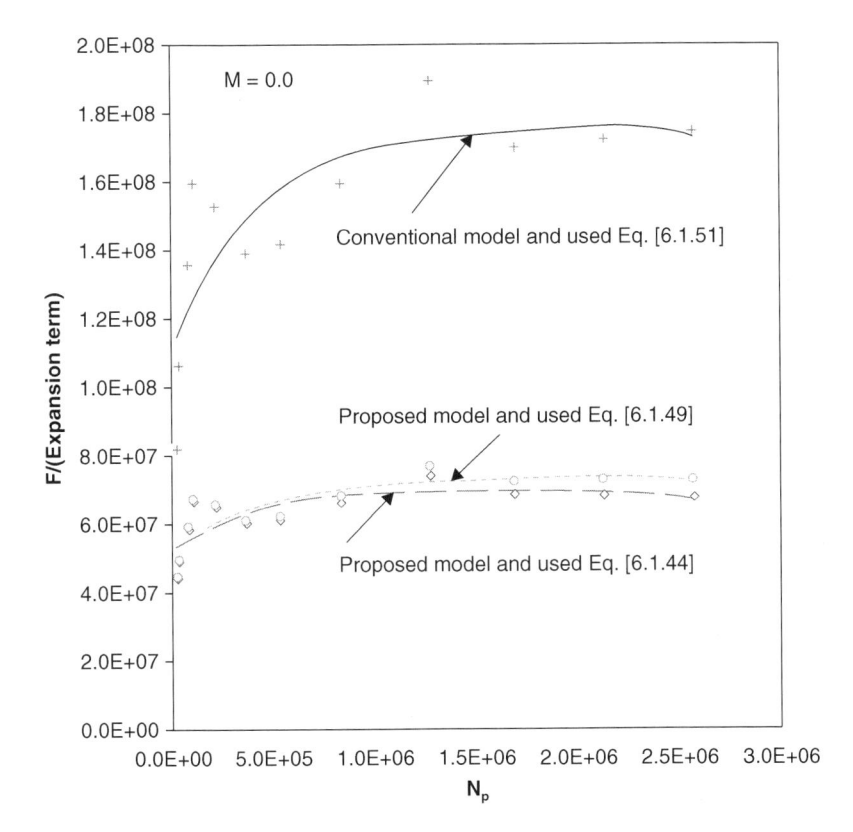

Figure 6.6 $F/(E_o + E_{cepm}$ or $E_{ceHO})$ vs. N_p (Equation no. refers to Hossain and Islam, 2009).

MBE, equation (6.43) with equation (6.41) has higher Recovery factor (RF) than that of using equation (6.44) with equation (6.43). Therefore, to show the effects of M values, Fig. 6.7 illustrates F vs. (E_0+E_{cepm}) for only the proposed MBE with equation (6.44). The straight line plotting passing through the origin of the figure gives 59.78 mmstb, 53.01 mmstb and 47.61 mmstb of IOIP for M = 1.0, 2.0, 3.0 respectively. The corresponding RF values are calculated as 4.31%, 4.86%, and 5.4%. So, RF increases with the increase of M values which correspond if there is an associated volume of a reservoir, it should be considered in the MBE calculations; otherwise there might be some error in getting the true production history of reservoir life.

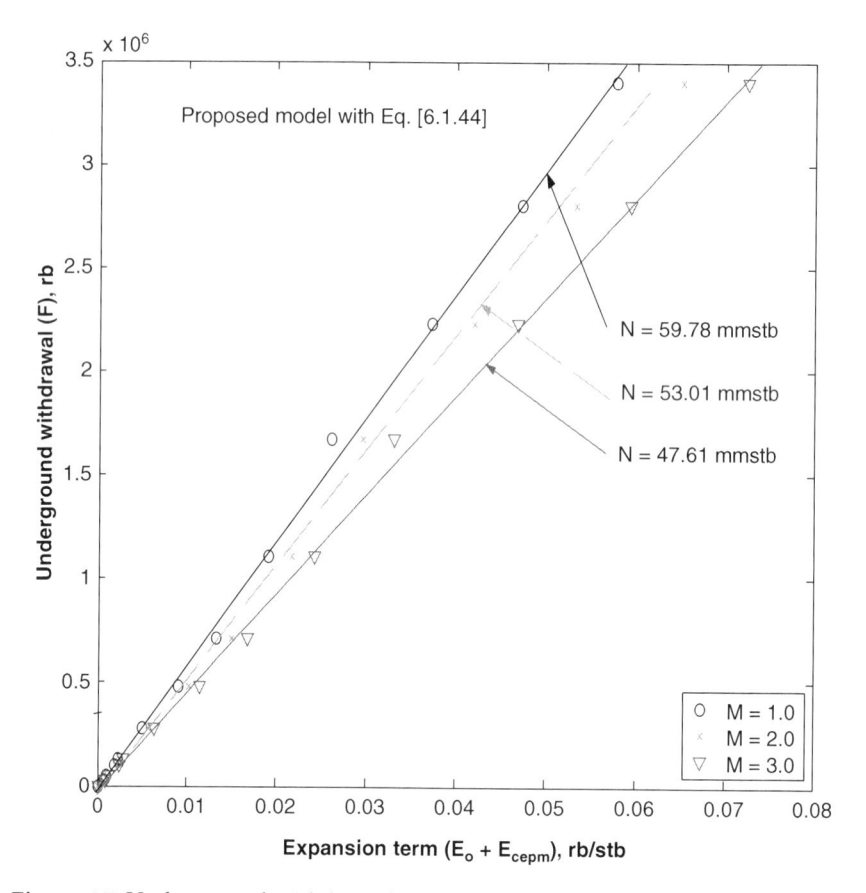

Figure 6.7 Underground withdrawal vs. Expansion term for different M values (Equation no. refers to Hossain and Islam, 2009).

6.7 Appendix 6A: Development of an MBE for a Compressible Undersaturated Oil Reservoir

The natural energy available to utilize in the primary recovery relies on the expansion of fluids and rock. The component describing the reduction in the hydrocarbon pore volume due to the expansion of initial (connate) water and the reservoir rock can not be neglected for an undersaturated oil reservoir. The water compressibility, c_w and rock compressibility, c_s are generally of the same order of magnitude as the compressibility of the oil. The effect of these two

A Material Balance Equation for Oil Recovery

components, however, can generally be neglected for a gascap drive reservoir or when the reservoir pressure drops below the bubble-point pressure. The compressibility coefficient, c which describes the changes in the volume (expansion) of the fluid or material with changing pressure for an isothermal system is given by $c = -\frac{1}{V}\frac{\partial V}{\partial p}$. To describe the reservoir depletion by the definition of compressibility, it is more illustrative to express it in the form, $dV = cV\Delta p$, where dV is an expansion and Δp is a pressure drop, both of which are positive. To produce oil, reservoir wells should be drilled into the oil zone. If reservoir is in contact with a gascap and an aquifer, the oil production due to a uniform pressure drop, Δp, in the entire system, will have components due to the separate expansion of the oil, gas and water, which can be treated as oil production due to separate inter molecular expansion of phases, thus,

$$\Delta V_{total} = \Delta V_o + \Delta V_w + \Delta V_g \qquad (6.18)$$

It is evident that the contribution of ΔV_{total} supplies by the oil, water and gas expansions will only be significant V_o, V_w, and V_g the initial volume of oil and water are large. So the total fluid expansion in the reservoir can play a role in the oil recovery mechanism.

6.7.1 Development of a New MBE

To develop a MBE for an undersaturated reservoir with no gascap gas, the reservoir pore is considered as an idealized container. MBE is derived by considering the whole reservoir as a homogeneous tank of uniform rock and fluid properties. We are considering oil and water as the only mobile phases in the compressible rock and the residual fluid saturation ($c_s \neq 0$, $c_o \neq 0$, $c_w \neq 0$, $S_{oi} \neq 0$, $S_{gi} \neq 0$, $S_{wi} \neq 0$). The derivation includes pressure-dependent rock and water compressibilities (with gas evolving solution). The inclusions of a limited aquifer volume, all water and rock volumes associated[6,7] with the reservoir available for expansion are recognized. The volumetric balance expressions can be derived to account for all volumetric changes which occur during the natural productive life of the reservoir. Therefore, MBE can be presented in terms of volume changes in reservoir barrels (rb) where all the probable fluids and media changes are taken care as:

Pore volume occupied by the oil initially in place + originally dissolved gas at time, t = 0 and d, rb

=

Pore volume occupied by the remaining oil at a given time, t and at p, rb
+
Change in oil volume due to oil expansion at a given time t and at p, $(-\Delta V_o)$, rb
+
Change in water volume due to connate water expansion at a given time t and at p, $(-\Delta V_w)$, rb
+
Change in dissolved gas volume due to gas expansion at a given time, t and at p, $(-\Delta V_g)$, rb
+
Change in pore volume due to reduction at a given time, t and at p, $(-\Delta V_s)$, rb
+
Change in associated volume due to expansion and reduction of water and pore volume at a given time, t and at p, $(-\Delta V_A)$, rb
+
Change in water volume due to water influx and water production at a given time, t and at p, rb

(6.19)

Now,

a. Pore volume occupied by the initial oil in place and originally dissolved gas = NB_{oi}.
b. Pore volume occupied by the remaining oil at p = $(N - N_p)B_o$

The isothermal fluid and formation compressibilities are defined according to the above discussion as (Dake, 1978; Ahmed, 2000; Rahman et al, 2006a):

Oil:

$$c_o = -\frac{1}{V_o}\frac{\partial V_o}{\partial p_p}\bigg]_T \qquad (6.20)$$

A Material Balance Equation for Oil Recovery

Water:

$$c_w = -\frac{1}{V_w}\frac{\partial V_w}{\partial p}\bigg]_T \tag{6.21}$$

Gas:

$$c_g = -\frac{1}{V_g}\frac{\partial V_g}{\partial p}\bigg]_T \tag{6.22}$$

Solid rock formation:

$$c_s = -\frac{1}{V_s}\frac{\partial V_s}{\partial p}\bigg]_T \tag{6.23}$$

If p_i is the initial reservoir pressure and p is the average reservoir pressure at current time t, one can write down the expressions for ΔV_o, ΔV_w, ΔV_g, and ΔV_s by integrating equation (6.20) through equation (6.23), assuming compressibilities to be pressure dependent, and by subsequent algebraic manipulations as (Rahman et al, 2006a):

$$\Delta V_o = -V_{oi}\left(1 - e^{\int_p^{p_i} c_o dp}\right) \tag{6.24}$$

$$\Delta V_w = -V_{wi}\left(1 - e^{\int_p^{p_i} c_w dp}\right) \tag{6.25}$$

$$\Delta V_g = -V_{gi}\left(1 - e^{\int_p^{p_i} c_g dp}\right) \tag{6.26}$$

$$\Delta V_s = V_{si}\left(1 - e^{\int_p^{p_i} c_s dp}\right) \tag{6.27}$$

In addition to above, associated volume is also included in developing a MBE with expansion drive mechanism. The "associated" volume is an additional reservoir part which is not active in oil/gas production. However, this part of the reservoir may accelerate the oil recovery by its water and rock expansion-contraction. Fetkovich et al (Fetkovich et al, 1991; Fetkovich et al, 1998) defined this term as "the non-net pay part of reservoir where interbedded shales and poor quality rock is assumed to be 100% water-saturated". The interbedded non-net pay volume and limited aquifer volumes are referred to as "associated" water volumes and both contribute to water influx during depletion. They used a volume fraction, M to represent the associated volume effects in the conventional MBE (Dake, 1978; Ahmed, 2000; Craft and Hawkins, 1959; Havlena and Odeh, 1963; Havlena and Odeh, 1964) for gas reservoir. M is defied as the ratio of associated pore volume to reservoir pore volume. If we consider this associated volume change, this part will be contributing as an additional expansion term in MBE. Therefore, the volume change would be:

$$\Delta V_A = -\left(-\Delta V_{Aw}\right) + \Delta V_{AS} = V_{Awi}\left(e^{\int_p^{p_i} c_w dp} - 1\right) + V_{Asi}\left(1 - e^{\int_p^{p_i} c_s dp}\right) \quad (6.28)$$

Note that ΔV_o, ΔV_w, and ΔV_g, have negative values due to expansion, and ΔV_s, has a positive value due to contraction. The initial fluid and pore volumes can be expressed as

$$V_{oi} = \frac{NB_{oi}}{1-S_{wi}} S_{oi} \quad (6.29)$$

$$V_{wi} = \frac{NB_{oi}}{1-S_{wi}} S_{wi} \quad (6.30)$$

$$V_{gi} = \frac{(NB_{oi}) \times (R_{soi}/B_{oi}) \times B_{gi} + (NB_{oi}) \times (R_{swi}/B_{wi}) \times B_{gi}}{1-S_{wi}} S_{gi} \quad (6.31)$$

$$V_{si} = \frac{NB_{oi}}{1-S_{wi}} \quad (6.32)$$

$$V_{Awi} = M \frac{NB_{oi}}{1-S_{wi}} \quad \text{(as 100\% water saturated e.g., } S_w = 1.0\text{)} \quad (6.33)$$

$$V_{Asi} = M \frac{NB_{oi}}{1-S_{wi}} \quad (6.34)$$

A Material Balance Equation for Oil Recovery

Substituting equation (6.29) through (6.29) in equation (6.24) to (6.24) respectively, one can write down the equations as:

$$\Delta V_o = -V_{oi}\left(1 - e^{\int_p^{p_i} c_o dp}\right) = -\frac{NB_{oi}}{1-S_{wi}} S_{oi}\left(1 - e^{\int_p^{p_i} c_o dp}\right) \tag{6.35}$$

$$\Delta V_w = -V_{wi}\left(1 - e^{\int_p^{p_i} c_w dp}\right) = -\frac{NB_{oi}}{1-S_{wi}} S_{wi}\left(1 - e^{\int_p^{p_i} c_w dp}\right) \tag{6.36}$$

$$\Delta V_g = -V_{gi}\left(1 - e^{\int_p^{p_i} c_g dp}\right)$$

$$= -\frac{(NB_{oi}) \times (R_{soi}/B_{oi}) \times B_{gi} + (NB_{oi}) \times (R_{swi}/B_{wi}) \times B_{gi}}{1-S_{wi}}$$

$$\times S_{gi}\left(1 - e^{\int_p^{p_i} c_g dp}\right) \tag{6.37}$$

$$\Delta V_s = V_{si}\left(1 - e^{-\int_p^{p_i} c_s dp}\right) = -\frac{NB_{oi}}{1-S_{wi}}\left(1 - e^{-\int_p^{p_i} c_s dp}\right) \tag{6.38}$$

$$\Delta V_A = \Delta V_{Aw} + \Delta V_{As}$$

$$= M\frac{NB_{oi}}{1-S_{wi}}\left(e^{\int_p^{p_i} c_w dp} - 1\right) + M\frac{NB_{oi}}{1-S_{wi}}\left(1 - e^{-\int_p^{p_i} c_s dp}\right) \tag{6.39}$$

Substituting Equation (6.35) through (6.39) in Equation (6.19), one can write down the MBE as:

$$NB_{oi} = (N - N_p)B_o$$

$$+ \frac{NB_{oi}}{1-S_{wi}} S_{oi}\left(e^{\int_p^{p_i} c_o dp} - 1\right) + \frac{NB_{oi}}{1-S_{wi}} S_{wi}\left(e^{\int_p^{p_i} c_w dp} - 1\right)$$

$$+ \frac{(NB_{oi}) \times (R_{soi}/B_{oi}) \times B_{gi} + (NB_{oi}) \times (R_{swi}/B_{wi}) \times B_{gi}}{1-S_{wi}}$$

$$\times S_{gi}\left(e^{\int_p^{p_i} c_g dp} - 1\right)$$

$$+\frac{NB_{oi}}{1-S_{wi}}S_{oi}\left(1-e^{-\int_{p}^{p_i}c_s dp}\right)+M\frac{NB_{oi}}{1-S_{wi}}\left(e^{\int_{p}^{p_i}c_w dp}-1\right)$$

$$+M\frac{NB_{oi}}{1-S_{wi}}\left(1-e^{-\int_{p}^{p_i}c_s dp}\right)+\left(W_e-W_p B_w\right)$$

$$NB_{oi}=\left(N-N_p\right)B_{oi}+\left(W_e-W_p B_w\right)$$

$$+\frac{NB_{oi}}{1-S_{wi}}\begin{bmatrix}S_{oi}\left(e^{\int_{p}^{p_i}c_o dp}-1\right)+S_{wi}\left(e^{\int_{p}^{p_i}c_w dp}-1\right)\\+S_{gi}\left(e^{\int_{p}^{p_i}c_g dp}-1\right)\left(\frac{R_{soi}}{B_{oi}}+\frac{R_{swi}}{B_{wi}}\right)\times B_{gi}\\+\left(1-e^{-\int_{p}^{p_i}c_s dp}\right)+M\left(e^{\int_{p}^{p_i}c_w dp}-1\right)\\+M\left(1-e^{-\int_{p}^{p_i}c_s dp}\right)\end{bmatrix}$$

$$N_p B_o - \left(W_e - W_p B_w\right) = NB_o - NB_{oi}$$

$$+NB_{oi}\left[\frac{\begin{aligned}&S_{oi}\left(e^{\int_{p}^{p_i}c_o dp}-1\right)+S_{wi}\left(e^{\int_{p}^{p_i}c_w dp}-1\right)+S_{gi}\left(e^{\int_{p}^{p_i}c_g dp}-1\right)\\&\left(\frac{R_{soi}}{B_{oi}}+\frac{R_{swi}}{B_{wi}}\right)B_{gi}+\left(1-e^{-\int_{p}^{p_i}c_s dp}\right)\\&+M\left\{\left(e^{\int_{p}^{p_i}c_w dp}-1\right)+\left(1-e^{-\int_{p}^{p_i}c_s dp}\right)\right\}\end{aligned}}{1-S_{wi}}\right]$$

$$N_p B_o - \left(W_e - W_p B_w\right) = N\left(B_o - B_{oi} + B_{oi}C_{epm}\right) \qquad (6.40)$$

where,

$$C_{epm}=\left[\frac{\begin{aligned}&S_{oi}\left(e^{\int_{p}^{p_i}c_o dp}-1\right)+S_{wi}\left(e^{\int_{p}^{p_i}c_w dp}-1\right)+S_{gi}\left(e^{\int_{p}^{p_i}c_g dp}-1\right)\left(\frac{R_{soi}}{B_{oi}}+\frac{R_{swi}}{B_{wi}}\right)\\&\times B_{gi}+\left(1-e^{-\int_{p}^{p_i}c_s dp}\right)+M\left[\left(e^{\int_{p}^{p_i}c_w dp}-1\right)+\left(1-e^{-\int_{p}^{p_i}c_s dp}\right)\right]\end{aligned}}{1-S_{wi}}\right]$$

$$(6.41)$$

A Material Balance Equation for Oil Recovery

The equation (6.40) is the new, rigorous and comprehensive MBE for an undersaturated reservoir with no gascap gas and above the bubble-point pressure. The above MBE is very rigorous because it considers the fluid and formation compressibilities as any function of pressure. It is deeming oil and water as the only mobile phases in the compressible rock. This rigorous MBE is applicable for a water-drive system with a history of water production in an undersaturated reservoir. Here, associated volume ratio, the residual and dissolved phase saturations are also considered. Equation (6.41) is an expression of the proposed dimensionless parameter, C_{epm} where all the probable and available expansions are illustrated. When the water influx term is not significant (i.e. $W_e = 0$), Equation (6.40) may be modified to:

$$N_p B_o + W_p B_w = N \left(B_o - B_{oi} + B_{oi} C_{epm} \right) \quad (6.42)$$

Equation (6.42) can be written in the form of straight line's MBE as (Havlena and Odeh, 1963; Havlena and Odeh, 1964):

$$N = \frac{F}{E_o + E_{cepm}} \quad (6.43)$$

where,

$$F = N_p B_o + W_p B_w$$
$$E_o = B_o - B_{oi}$$
$$E_{cepm} = B_{oi} C_{epm}$$

Now, if we consider a constant data of oil, water, gas and formation compressibilities (e.g. compressibilities are not functions of pressure), equation (6.40) remains unchanged. However, equation (6.41) can be modified by integrating the power of exponents as:

$$C_{epm} = \frac{\left\{ \begin{array}{l} S_{oi}\left(e^{c_o(p_i-p)}-1\right) + S_{wi}\left(e^{c_w(p_i-p)}-1\right) + S_{gi}\left(e^{c_g(p_i-p)}-1\right)\left(\dfrac{R_{soi}}{B_{oi}}+\dfrac{R_{swi}}{B_{wi}}\right) \\ \times B_{gi} + \left(1-e^{-c_s(p_i-p)}\right) + M\left[\left(e^{c_w(p_i-p)}-1\right)+\left(1-e^{-c_s(p_i-p)}\right)\right] \end{array} \right\}}{1-S_{wi}}$$

$$(6.44)$$

The equation (6.44) is still a rigorous expression for use in the case of constant compressibilities. This equation can be further approximated by the exponential terms for small values of the exponents as:

$$e^{c_o(p_i-p)} \approx 1 + c_o(p_i - p) \tag{6.45}$$

$$e^{c_w(p_i-p)} \approx 1 + c_w(p_i - p) \tag{6.46}$$

$$e^{c_g(p_i-p)} \approx 1 + c_g(p_i - p) \tag{6.47}$$

$$e^{-c_s(p_i-p)} \approx 1 + c_s(p_i - p) \tag{6.48}$$

Substituting equations (6.45) through (6.48) in equation (6.44)

$$C_{epm} = \frac{S_{oi}c_o + S_{wi}c_w + S_{gi}c_g\left(\frac{R_{soi}}{B_{oi}} + \frac{R_{swi}}{B_{wi}}\right)B_{gi} + c_s + M(c_w + c_s)}{1 - S_{wi}}(p_i - p)$$

If we consider the average reservoir pressure

$$C_{epm} = \frac{S_{oi}c_o + S_{wi}c_w + S_{gi}c_g\left(\frac{R_{soi}}{B_{oi}} + \frac{R_{swi}}{B_{wi}}\right)B_{gi} + c_s + M(c_w + c_s)}{1 - S_{wi}}\Delta p \tag{6.49}$$

The dimensionless parameter, C_{epm} expressed in terms of fluids (oil, water and gas) and formation compressibilities, and saturation is same to that of conventional MBE if we neglect the expansion of dissolved gas in oil and water, and the associated volume expansion. This issue will be discussed later.

6.7.2 Conventional MBE

The general form of conventional MBE can be written as (Dake, 1978; Ahmed, 2000; Craft and Hawkins, 1959; Havlena and Odeh, 1963; Havlena and Odeh, 1964):

A Material Balance Equation for Oil Recovery

$$N_p\left[B_o + (R_p - R_s)B_g\right] + W_pB_w = N\left[(B_o - B_{oi}) + (R_{si} - R_s)B_g\right]$$

$$+ mNB_{oi}\left(\frac{B_g}{B_{gi}} - 1\right) + NB_{oi}(1+m)$$

$$\times \frac{S_{wi}C_w + C_s}{1 - S_{wi}}\Delta p + W_e + W_{inj}B_w + G_{inj}B_{ginj}$$

If no water and gas injection well exists, the equation becomes as:

$$N_p\left[B_o + (R_p - R_s)B_g\right] + W_pB_w = N\left[(B_o - B_{oi}) + (R_{si} - R_s)B_g\right]$$

$$+ mNB_{oi}\left(\frac{B_g}{B_{gi}} - 1\right) + NB_{oi}(1+m)$$

$$\times \frac{S_{wi}C_w + C_s}{1 - S_{wi}}\Delta p + W_e$$

The equation may be further simplified for a volumetric and undersaturated reservoir. Here, if the residual oil compressibility is considered as well, the above equation turns to reduce the Havlena and Odeh equation (Havlena and Odeh, 1963; Havlena and Odeh, 1964):

$$N_pB_o + W_pB_w = N(B_o - B_{oi}) + NB_{oi}\frac{S_{oi}C_o + S_{wi}C_w + C_s}{1 - S_{wi}}\Delta p$$

$$N_pB_o + W_pB_w = N(B_o - B_{oi} + B_{oi}C_{eHO}) \tag{6.50}$$

where,

$$C_{eHO} = \frac{S_{oi}C_o + S_{wi}C_w + C_s}{1 - S_{wi}}\Delta p \tag{6.51}$$

The equation (6.42) resembles the MBE (Equation (6.50)) of Havlena and Odeh (1963, 1964) for a volumetric and undersaturated reservoir with no gascap except the pattern of dimensionless parameter, C_{epm} and Δp. Equation (6.50) can further be written as the straight line form of Havlena and Odeh MBE as:

$$N = \frac{F}{E_o + E_{ceHO}} \quad (6.52)$$

where,

$$F = N_p B_o + W_p B_w$$
$$E_o = B_o - B_{oi}$$
$$E_{ceHO} = B_{oi} C_{eHO}$$

Now, if we neglect dissolved gas saturation and associated volume expansion in equation (6.49) and define the pressure at time t as the average pressure, the equation becomes as:

$$C_{epm} = \frac{S_{oi} c_o + S_{wi} c_w + c_s}{1 - S_{wi}} \Delta p$$

The right hand side of above equation is same as stated in equation (6.51).

6.7.3 Significance of C_{epm}

The dimensionless parameter, C_{epm}, in the below equation (6.53) can be considered as the effective strength of the energy source for oil production in expansion drive oil recovery. This is only due to the compressible residual fluids (oil, water and dissolved gas) and rock expansions of the reservoir. This value does not account for the oil compressibility as stated by other researchers (Dake, 1978; Fetkovich et al, 1991; Fetkovich et al, 1998; Ahmed, 2000; Rahman et al, 2006a). If we see the expression of C_{epm} as presented in equation (6.41), it is a function of the current reservoir pressure, fluid compressibilities, initial saturations, dissolved gas properties engaged in water and oil, and associated volume fraction. C_{epm} in equation (6.44) is still a function all those parameters except a set of constant compressibilities instead of variable compressibilities. The final simplified form of C_{epm} presented in equation (6.49) is dependent on current average reservoir pressure drop and other related parameters as stated above. The

A Material Balance Equation for Oil Recovery

dimensionless parameter, C_{epm} is an important parameter in the proposed MBE (Equation (6.40)) because it can be used as an analytic tool to predict how the MBE will behave for the relevant input data.

Some specific significant properties may well be explained to elucidate the new version of C_{epm}, presented in equations (6.41) and (6.44). To explain these significance of C_{epm}, equation (6.40) can be rearranged as:

$$NB_{oi}\left(1 - C_{epm} - \frac{W_e - W_p B_w}{NB_{oi}}\right) = NB_o\left(1 - \frac{N_p}{N}\right)$$

$$\frac{1}{B_o}\left(1 - C_{epm} - \frac{W_e - W_p B_w}{NB_{oi}}\right) = \frac{1}{B_{oi}}\left(1 - \frac{N_p}{N}\right) \quad (6.53)$$

The equation (6.53) is used to explain the effects of C_{epm}, where two cases have been considered.

6.7.4 Water Drive Mechanism with Water Production

The equation (6.53) gives a limit of expansion plus water drives mechanism for initial fluid situations, water influx and water production. This limit may be expressed as:

$$0 \leq \left(C_{epm} - \frac{W_e - W_p B_w}{NB_{oi}}\right) < 1.0 \quad (6.54)$$

The equation (6.54) is true for any given average reservoir pressure. The lower limit in equation (6.54) is due to the reality of C_{epm}, W_e and W_p where all these parameters are zero at the initial reservoir pressure. The upper limit is characterized by the fact that the right-hand side term in equation (6.53) is zero, when all the original oil-in-place has been produced. However, practically it is not possible to reach the production level up to that mark. Therefore, the upper limit should be less than 1. Hence, it may be concluded that if the numerical values beyond this range comes out, there might be some problem in input data or there might be a problem in calculating or assigning the average reservoir pressure. So, it is a tool to diagnose or predict the reservoir behavior in the early stage of production.

6.7.5 Depletion Drive Mechanism with No Water Production

When depletion drive mechanism (with no water influx) with no water production is considered, equation (6.53) gives a limit for this drive mechanism as:

$$0 \leq C_{epm} < 1.0 \qquad (6.55)$$

The limits are identified using the same argument as the water drive mechanism, presented in the previous section. If the cited limits are violated at a given time, there is no chance of calculating any reasonable values of the average reservoir pressure. Therefore, the limits are the indications of decision tools about the reservoir and fluid properties and decline criteria.

7

Modeling Viscous Fingering During Miscible Displacement in a Reservoir

Viscous fingering is a phenomenon caused by the momentum difference between the displacing fluid of high viscosity and the displaced fluid of low viscosity. This phenomenon plays a vital role in numerous applications of enhanced oil recovery (EOR). This is particularly relevant for heavy oil and tar sand for which such fingering may occur in both isothermal and non-isothermal modes. The selection of the displacing fluid with particular characteristics becomes extremely significant in avoiding the growth and the propagation of its finger-patterned flow, which leads to extremely poor displacement efficiency in both miscible and immiscible displacement schemes for enhanced oil recovery. Currently the petroleum industry is faced with great discrepancy between prediction and field performance of these displacement schemes because of the presence of the phenomenon of viscous fingering, which ultimately renders the flow schemes unstable. Yet, no rigorous numerical model for predicting viscous fingering has been presented to-date. This chapter first introduces a new numerical scheme that can handle both the onset and propagation of viscous fingers. Then, it uses this numerical scheme to solve an array of viscous fingering problems. Finally, it shows good agreement with experimental data. As an example, only the case of isothermal viscous fingering is considered. However, the formulation is quite similar for non-isothermal cases that show double diffusive convention and it is expected that this numerical scheme will also apply to non-isothermal applications.

7.1 Improvement of the Numerical Scheme

The numerical solution of the convective dispersion equation, a key equation for the phenomenon of viscous fingering, has been an area of active research for many years within and outside the petroleum engineering community. The past and on-going research effort in this area reflects a great deal of difficulties in solving the convective dispersion equation. While advection and dispersion are simultaneous processes, they promote the transport in a different manner. Mathematically, this means the simultaneous treatment of the hyperbolic terms (associated with advection) and the parabolic terms (associated with dispersion), and therefore the solution to the problem is not easily tackled by any numerical method.

The convective dispersion equation has been solved earlier by using various numerical methods such as fully implicit, partially implicit, and fully explicit schemes. It has been perceived that the most accurate scheme is the *DuFort–Frankel* scheme with the limitation of its stability. The *Barakat–Clark* scheme has been proven to be unconditionally stable and more accurate than the fully implicit one. It is also interesting to note that the *Barakat–Clark* approximation has been found to be consistent with parabolic partial differential equations.

All of these partially implicit schemes work out perfectly with the central difference model as far as the accuracy in space is concerned. The partially implicit schemes are stable and practically insensitive to the discretization step-size in space. The accuracy in time, however, has not been targeted successfully to a higher order using these schemes. It is due to the convergence limitation of the central weighing scheme, which also has an inherent tendency to create artificial oscillations. Even most of the fully implicit schemes, which are unconditionally stable with forward (or backward) models, have the accuracy of the order of Δt in time.

Solving the convective-dispersion equation by the fully explicit method has the primary disadvantage of instability. Instability in the numerical solution of partial differential equations resembles what is encountered when solving ordinary differential equations in a way that results from an amplification of errors (of both the truncation and round-off variety). It is beyond the scope of this book to explore mathematically the details of this instability. When dispersion and advection terms are solved simultaneously using the explicit scheme, a combined stability criterion for 1-D has been proven and found earlier as:

$$\frac{2D\Delta t}{(\Delta x)^2} + \frac{u\Delta t}{\Delta x} \le 1 \tag{7.0}$$

where D is the dispersion term while u represents the velocity term.

A finite difference approximation may possess favorable stability properties in the sense that it will generate an approximate solution that converges to the exact solution, as the time or/and spatial intervals are refined. In doing so, undesirable error growth is also not admitted. At the same time, the errors associated with the generated solution on a specific interval can compare poorly with those associated with the solution generated on the same interval by a less stable approximation. An important discussion can be made on the topic of consistency, stability and convergence of finite difference formulae for partial differential equations. Successive refinement of the interval Δt may generate a finite difference solution that is stable, but that may converge to the solution of a different differential equation. Local truncation errors associated with any particular finite difference approximation can easily be obtained by Taylor's series. The finite difference equation is said to be consistent and compatible with the differential equation if the local truncation errors (the right hand side of the expanded series) tend to zero as the space or time interval approaches to zero ($h, k \to 0$).

In 2005, Bokhari and Islam extended the work of Aboudheir et al (1999) in 2-D and improved the accuracy in time to the order of Δt^4, by combining the *Barakat–Clark* scheme with the time term expressed in the central difference form. The results are compared with the classic combination of the *Barakat–Clark* scheme and the time term expressed in forward difference form. The comparison is also extended to the *DuFort–Frankel* (1953) scheme, which is claimed to be unconditionally stable.

7.1.1 The Governing Equation

A 2-D convective diffusive equation can be written as:

$$\frac{\partial C}{\partial t} = D_x \frac{\partial^2 C}{\partial x^2} + D_y \frac{\partial^2 C}{\partial x^2} - u \frac{\partial C}{\partial x} - v \frac{\partial C}{\partial x} \tag{7.1}$$

Where D_x and D_y denote lateral and transfer diffusion terms, u and v are the velocity terms in the X and Y directions and C denotes the concentration term. Equation 7.1 can be solved by numerous finite difference schemes, both explicitly and implicitly. In the current study, three methods have been employed namely *DuFort–Frankel*, *Barakat–Clark (FTD)* with the forward time difference scheme and the proposed scheme, *Barakat–Clark (CTD)* using the central time difference term. The following sections are comprised of basic consideration and the comparison of the results of these schemes.

The given initial and boundary conditions to solve equation 7.1 have been partially taken from the work done by Aboudheir et al (1999) and can be written for 2-D as:

$$C(x,y,0) = 0.0 \quad \text{for} \quad 0 \leq x \leq 1, \ 0 \leq y \leq 1$$
$$C(0,y,t) = t \quad \text{for} \quad t \leq 1, \ 0 \leq y \leq 1$$
$$C(0,y,t) = 2 - t \quad \text{for} \quad 1 < t \leq 2, 0 \leq y \leq 1$$
$$C(0,y,t) = 0.0 \quad \text{for} \quad t \leq 2, 0 \leq y \leq 1$$
$$C(1,y,t) = -C_x(x,y,t) \quad \text{for} \quad 0 \leq t \leq 3, 0 \leq y \leq 1$$
$$\left.\frac{\partial C}{\partial x}\right|_{y=0} = 0.0 \quad \text{for} \quad 0 \leq t \leq 3, 0 \leq x \leq 1$$
$$\left.\frac{\partial C}{\partial x}\right|_{y=1} = 0.0 \quad \text{for} \quad 0 \leq t \leq 3, 0 \leq x \leq 1$$

(7.2)

The boundary condition given by equation (7.2) can be explained by Fig. (7.1). According to this inlet boundary condition, the concentration increases linearly from a value of zero at time = 0 to unity at time = 1.

Afterwards, this inlet concentration starts to decrease linearly and approaches to a value of zero at time = 2. The concentration at time ≥ 2 has also been taken as zero. This phenomenon is known as *Graded Injection* and is commonly used for polymer injection during enhanced oil recovery.

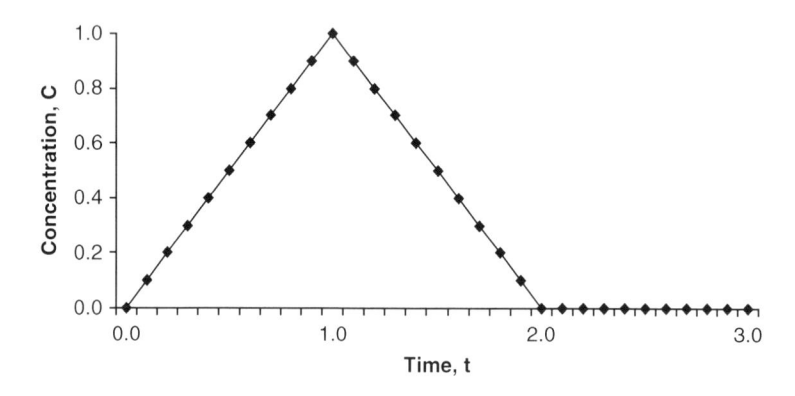

Figure 7.1 Inlet boundary condition – Graded Injection.

7.1.2 Finite Difference Approximations

The derivations of finite difference equations for *Barakat–Clark FTD* and *DuFort–Frankel* schemes are well known. In this present example, a mixed approach is used in approximating different terms of equation 7.1 by the finite difference scheme to target the consistent accuracy throughout the equation. In all these approximations, a partially implicit central difference in space scheme is used for the parabolic dispersion term while for approximating the hyperbolic convection term, an explicit in time two-point central weighing scheme is utilized.

7.1.2.1 Barakat–Clark FTD Scheme

The *Barakat–Clark* scheme with forward weighing for time term (FTD) is an explicit-finite difference approximation procedure, which has proven to be unconditionally stable for the solution of a general multidimensional non-homogeneous diffusion equation. Similar to *Fully Implicit* methods, there is no severe limitation of the time step size in this method. The beauty of this method is its simplicity in the formulation and solution technique. The solution is basically the arithmetic average of two multilevel finite difference equations, the solution of either of which can be used to approximate the solution of equation 7.1. These two equations calculate two intermediate concentration values at a time level of $n+1$ and therefore, their average gives the solution of equation 7.1. Although, the *Barakat–Clark* scheme overcomes the limitation of time step size (Equation 7.1) of explicit method, its real strength is its extraordinarily stability both in space and time. This stability will be discussed in detail in the next section.

Using the *Barakat–Clark FTD* scheme, equation 7.1 can be approximated by the following two multilevel finite difference equations:

$$\frac{Ca_{i,j}^{n+1} - Ca_{i,j}^{n}}{\Delta t} = Dx \frac{Ca_{i+1,j}^{n} - Ca_{i,j}^{n} - Ca_{i,j}^{n+1} + Ca_{i-1,j}^{n+1}}{\Delta x^2}$$

$$+ Dy \frac{Ca_{i+1,j}^{n} - Ca_{i,j}^{n} - Ca_{i,j}^{n+1} + Ca_{i,j-1}^{n+1}}{\Delta y^2}$$

$$- U \frac{Ca_{i+1,j}^{n} - Ca_{i-1,j}^{n}}{2\Delta x} - V \frac{Ca_{i,j+1}^{n} - Ca_{i,j-1}^{n}}{2\Delta y} \quad (7.4)$$

$$\frac{Cb_{i,j}^{n+1} - Cb_{i,j}^{n}}{\Delta t \, (8.4.2)} = Dx \frac{Cb_{i+1,j}^{n+1} - Cb_{i,j}^{n+1} - Cb_{i,j}^{n} + Cb_{i-1,j}^{n}}{\Delta x^2}$$

$$+ Dy \frac{Cb_{i,j+1}^{n+1} - Cb_{i,j}^{n+1} - Cb_{i,j}^{n} + Cb_{i,j-1}^{n}}{\Delta y^2}$$

$$- U \frac{Cb_{i+1,j}^{n} - Cb_{i-1,j}^{n}}{2\Delta x} - V \frac{Cb_{i,j+1}^{n} - Cb_{i,j-1}^{n}}{2\Delta y} \quad (7.5)$$

Where $Ca_{i,j}$ and $Cb_{i,j}$ are the solution of the above equations and they satisfy the initial and boundary conditions given above. For these circumstances the solution of equation (7.3) at any time interval $n+1$ can be written as:

$$C_{i,j}^{n+1} = \left(Ca_{i,j}^{n+1} + Cb_{i,j}^{n+1} \right) / 2$$

It is to be noted here that the left hand sides of equations (7.4) and (7.5) are the time terms written in the forward difference form. To calculate $Ca_{i,j}^{n+1}$ and $Cb_{i,j}^{n+1}$, equations 7.4 and 7.5 can be rearranged to give the simple representation of the solution as:

$$Ca_{i,j}^{n+1} = a Ca_{i,j}^{n} + b \left(Ca_{i+1,j}^{n} + Ca_{i-1,j}^{n+1} \right)$$
$$+ c \left(Ca_{i,j+1}^{n} + Ca_{i,j-1}^{n+1} \right) - d \left(Ca_{i+1,j}^{n} - Ca_{i-1,j}^{n} \right)$$
$$- e \left(Ca_{i,j+1}^{n} - Ca_{i,j-1}^{n} \right) \quad (7.4a)$$

$$Cb_{i,j}^{n+1\,n+1} = a Cb_{i,j}^{n} + b \left(Cb_{i+1,j}^{n+1} + Cb_{i-1,j}^{n} \right)$$
$$+ c \left(Cb_{i,j+1}^{n+1} + Cb_{i,j-1}^{n} \right) - d \left(Cb_{i+1,j}^{n} - Cb_{i-1,j}^{n} \right)$$
$$- e \left(Cb_{i,j+1}^{n} - Cb_{i,j-1}^{n} \right) \quad (7.5a)$$

where $\frac{D_x \Delta t}{\Delta x^2} = P$, $\frac{D_y \Delta t}{\Delta y^2} = Q$, $\frac{U \Delta t}{2\Delta x} = R$, $\frac{V \Delta t}{2\Delta y} = S$

While $a = \frac{(1-P-Q)}{(1+P+Q)}$, $b = \frac{P}{(1+P+Q)}$, $c = \frac{Q}{(1+P+Q)}$, $d = \frac{R}{(1+P+Q)}$ and $e = \frac{S}{(1+P+Q)}$ $Ca_{i,j}^{n+1}$ can be calculated from equation 7.4 explicitly. In this case, calculations proceed from the grid points nearest to the boundaries $x = 0$, and $y = 0$ in a sequence of increasing i and j. The needed values of the terms $Ca_{i-1,j}^{n+1}$ & $Ca_{i,j-1}^{n+1}$ are known from the boundary conditions. Similarly $Cb_{i,j}^{n+1}$ can be calculated explicitly from equation 7.5 beginning at the boundaries $x = 1$, $y = 1$ marching in a sequence of decreasing i and j. Here the required values of the terms $Ca_{i+1,j}^{n+1}$ & $Ca_{i,j+1}^{n+1}$ will be taken from the boundary conditions.

7.1.2.2 DuFort–Frankel Scheme

DuFort–Frankel, which is classified as an unconditionally stable method for solving partial differential equations explicitly has an error of $O(h^2, k^2)$. According to this scheme, equation 7.1 can be written in finite difference form as follows:

$$\frac{C_{i,j}^{n+1} - C_{i,j}^{n-1}}{2\Delta t} = Dx \frac{C_{i+1,j}^n - C_{i,j}^{n+1} - C_{i,j}^{n-1} + C_{i-1,j}^n}{\Delta x^2}$$
$$+ Dy \frac{C_{i,j+1}^n - C_{i,j}^{n+1} - C_{i,j}^{n-1} + C_{i,j-1}^n}{\Delta y^2}$$
$$- U \frac{C_{i+1,j}^n - C_{i-1,j}^n}{2\Delta x} - V \frac{C_{i,j+1}^n - C_{i,j-1}^n}{2\Delta y} \quad (7.6)$$

It is evident from equation 7.) that three time levels are involved, and if we compare it with the *Barakat–Clark* scheme, we will find that unlike that scheme, the left hand side of the equation is in central difference form. This renders the accuracy of the equation to the $O(k^2)$ in time. equation 7.6 can be rearranged to calculate $Cb_{i,j}^{n+1}$ explicitly as:

$$C_{i,j}^{n+1} = f C_{i,j}^{n-1} + g C_{i-1,j}^n + h C_{i+1,j}^n + l C_{i,j-1}^n + m C_{i,j+1}^n \quad (7.7)$$

where,

$$f = \frac{(1 - 2P - 2Q)}{(1 + 2P + 2Q)}, \quad g = \frac{(2P + 2R)}{(1 + 2P + 2Q)}, \quad h = \frac{(2P - 2R)}{(1 + 2P + 2Q)},$$
$$l = \frac{(2Q + 2S)}{(1 + 2P + 2Q)} \quad \text{and} \quad m = \frac{(2Q - 2S)}{(1 + 2P + 2Q)}$$

7.1.3 Proposed Barakat–Clark CTD Scheme

To achieve a higher accuracy in time with the *Barakat–Clark* method, central weighing scheme for time term is utilized in this proposed scheme. The advantage for using this modified scheme will be discussed in the next section. Using this scheme, Equation 7.1 can be approximated by the following two multi-level finite difference equations which having the form of:

$$\frac{Ca_{i,j}^{n+1} - Ca_{i,j}^{n-1}}{2\Delta t} = Dx \frac{Ca_{i+1,j}^{n} - Ca_{i,j}^{n} - Ca_{i,j}^{n+1} + Ca_{i-1,j}^{n+1}}{\Delta x^2}$$

$$+ Dy \frac{Ca_{i,j+1}^{n} - Ca_{i,j}^{n} - Ca_{i,j}^{n+1} + Ca_{i,j-1}^{n+1}}{\Delta y^2}$$

$$- U \frac{Ca_{i+1,j}^{n} - Ca_{i-1,j}^{n}}{2\Delta x} - V \frac{Ca_{i,j+1}^{n} - Ca_{i,j-1}^{n}}{2\Delta y} \quad (7.8)$$

$$\frac{Cb_{i,j}^{n+1} - Cb_{i,j}^{n-1}}{2\Delta t} = Dx \frac{Cb_{i+1,j}^{n+1} - Cb_{i,j}^{n+1} - Cb_{i,j}^{n} + Cb_{i-1,j}^{n}}{\Delta x^2}$$

$$+ Dy \frac{Cb_{i,j+1}^{n+1} - Cb_{i,j}^{n+1} - Cb_{i,j}^{n} + Cb_{i,j-1}^{n}}{\Delta y^2}$$

$$- U \frac{Cb_{i+1,j}^{n} - Cb_{i-1,j}^{n}}{2\Delta x} - V \frac{Cb_{i,j+1}^{n} - Cb_{i,j-1}^{n}}{2\Delta y} \quad (7.9)$$

Equations 7.8 and 7.9 can be arranged to have $Ca_{i,j}^{n+1}$ & $Cb_{i,j}^{n+1}$ as follows:

$$Ca_{i,j}^{n+1} = a Ca_{i,j}^{n-1} + q Ca_{i-1,j}^{n+1} + r Ca_{i,j-1}^{n+1} + p Ca_{i,j}^{n} + h Ca_{i+1,j}^{n} + m Ca_{i,j+1}^{n}$$
$$+ w Ca_{i-1,j}^{n} + x Ca_{i,j-1}^{n} \quad 7.10)$$

$$Cb_{i,j}^{n+1} = a Cb_{i,j}^{n-1} + q Cb_{i+1,j}^{n+1} + r Cb_{i,j+1}^{n+1} + g Cb_{i-1,j}^{n} - p Cb_{i,j}^{n}$$
$$+ l Cb_{i,j-1}^{n} - w Cb_{i+1,j}^{n} - x Cb_{i,j+1}^{n} \quad (7.11)$$

where

$$o = \frac{1}{(1+2P+2Q)}, \quad p = \frac{(2P+2Q)}{(1+2P+2Q)}, \quad q = \frac{2P}{(1+2P+2Q)},$$

$$r = \frac{2Q}{(1+2P+2Q)}, \quad w = \frac{2R}{(1+2P+2Q)} \quad \text{and} \quad x = \frac{2S}{(1+2P+2Q)}$$

7.1.3.1 Boundary Conditions

In the case study, the flow is considered to be in the y-direction. The inlet boundary condition across the x-direction can be given as a function of time and the outlet boundary condition can be written in backward weighing scheme utilizing three-point approach after discretizing for j = M as:

$$C_{i,M}^n = \left(4C_{i,M-1}^n - C_{i,M-2}^n\right)/\left(3 + 2\Delta y\right) \quad (7.12)$$

Whereas, the other two boundaries that are taken as "No Slip" boundaries and can be approximated in finite difference form using three-point forward and backward weighing schemes respectively as:

$$C_{0,j}^n = \left(4C_{1,j}^n - C_{2,j}^n\right)/3 \quad (7.13)$$

$$C_{L,j}^n = \left(4C_{L-1,j}^n - C_{L-2,j}^n\right)/3 \quad (7.14)$$

where L and M are the total lengths in the x and y directions respectively.

7.1.4 Accuracy and Truncation Errors

The central weighing scheme tends to create artificial oscillations (Mathews, 1992 referenced by Bokhari and Islam, 2005). Specifically, the numerical solution oscillates with respect to the true solution. Because of this problem, alternative spatial weighing has been developed for the advection term. A frequently used scheme is the forward/upstream (also called upwind) scheme. Although the upstream/forward difference scheme avoids the artificial oscillation with the central weighing scheme, it is accurate only to the first order and it introduces a second order truncation error. This error contributes to the numerical dispersion because of a similar effect to physical dispersion. To overcome this problem, the *Barakat–Clark* scheme is used due to its stability to include the central weighing scheme for time term in an attempt to see the impact of this new combination towards the accuracy in time.

From the expansion of Taylor series, truncation errors can easily be found for different schemes. The time term represented in forward difference has an inherent accuracy of the order of Δt, while the central weighing/difference scheme is accurate to the order of Δt^2. The solution of the equation by the *Barakat–Clark* scheme is the average of two intermediate equations (Equations 7.4. and 7.5) and it should be noted that either of these equations can be used to approximate the solution of the differential equation. If we apply Taylor expansion to both of these equations, due to the similarity in the magnitude of all terms, the terms having opposite signs tend to cancel out in the final solution. This consequence causes the truncation error to be of a higher order of magnitude. For the *Barakat–Clark FDT* scheme, the overall accuracy of the solution in time is approximately of the order of Δt^2 while with *Barakat–Clark CDT* schemes, this accuracy becomes of the order of Δt^4.

7.1.5 Some Results and Discussion

In the case study, the 2-D convective dispersion equation is solved using *Barakat–Clark FTD*, *DuFort–Frankel* and *Barakat–Clark CTD* schemes. Both the diffusion terms $(D_x$ and $D_y)$ are assumed to be equal to unity in all above-mentioned schemes. While the velocity term along the X-direction, U, is considered to be zero due to unidirectional flow consideration along the Y-direction. The velocity term, V, along the Y-axis is taken as constant having a value of unity. Results are analyzed using the outlet concentration. Consistent values of time (k = 0.001) and space intervals (h = 0.1) are used for general discussion of the results. It is to be noted here that all these schemes produce similar results using these time and space discretizations.

Figure 7.2 shows the inlet and outlet average concentration of the three schemes. It is clear from Fig. 7.2 that the inlet concentration

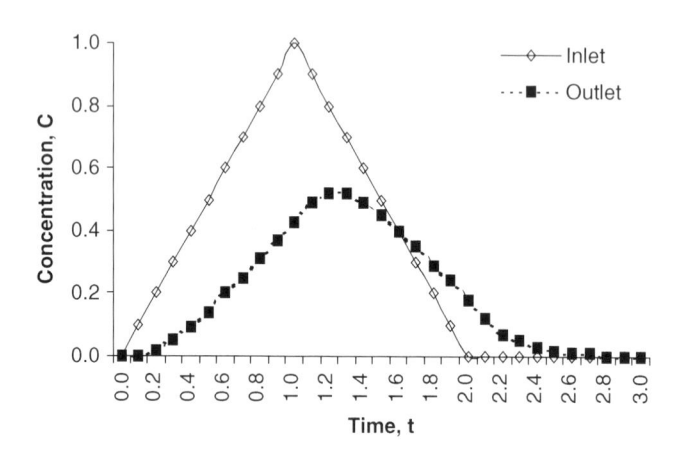

Figure 7.2 Inlet and outlet concentration (averaged of three schemes), h = 0.1 and k = 0.001.

Modeling Viscous Fingering

increases linearly with time according to the inlet boundary condition to reach its maximum value of unity at time = 1 and then it starts decreasing linearly to a value of zero at time = 2. The breakthrough in the outlet concentration curve takes place at time = 0.1 and its value increases smoothly with a lag in concentration in comparison with inlet concentration. The maximum value of outlet concentration observed is about 0.52, which lags slightly behind as compared to the peak value of the inlet concentration.

As mentioned earlier, three schemes, *Barakat–Clark FTD*, *DuFort–Frankel*, and *Barakat–Clark CTD*, gives the identical results and there is no discrepancy in the values observed along X-direction for these schemes. This identical behavior is shown in Figs. 7.3a, 7.3b, and 7.3c for the three different schemes, respectively. Comparison of these three schemes shows that they give similar concentration distribution in space and time. Figs. 7.4a and 7.4b show the average values of the outlet concentration along X-direction for these schemes. Some very interesting results have been observed while examining the relative accuracy of these three methods. First to check the sensitivity of these three schemes towards the spatial grid

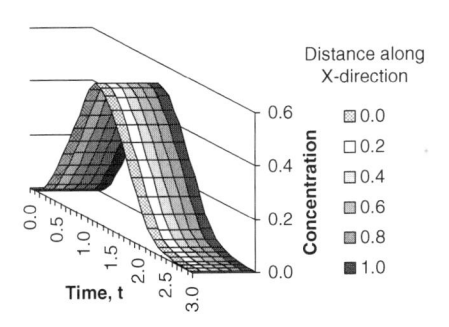

Figure 7.3a Outlet concentration (Barkat–Clark FTD), h = 0.1, k = 0.001.

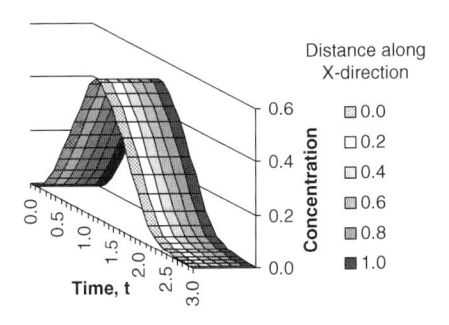

Figure 7.3b Outlet concentration (DuFort–Frankel), h = 0.1, k = 0.001.

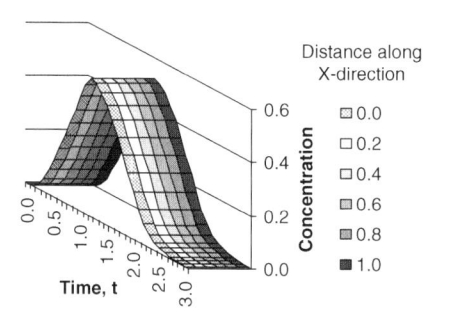

Figure 7.3c Outlet concentration (Barkat–Clark CTD) h = 0.1 and k = 0.001.

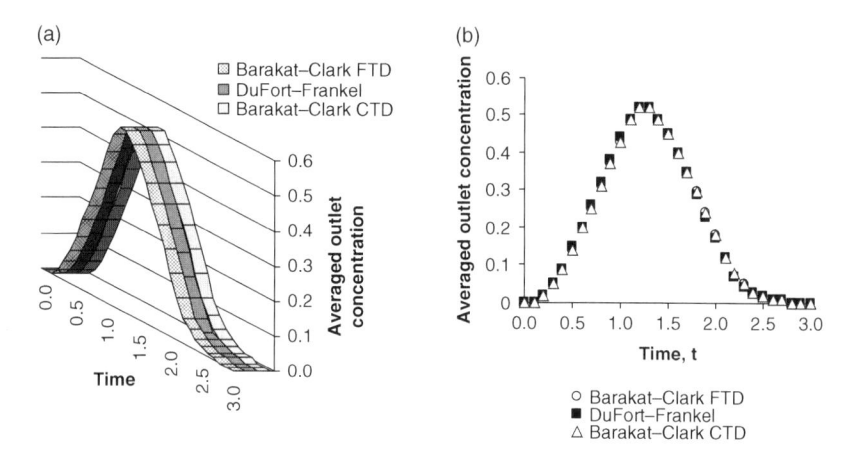

Figures 7.4a and b Comparison between outlet concentrations, calculated with different schemes, h = 0.1, k = 0.001.

size, three different sets of runs were conducted using time step size of 0.001, 0.01 and 0.1, respectively. In all these runs, the outlet concentration was observed at time = 1. In the first set, grid size in space, h, was varied from 0.03 to 0.5 and was equal for both X and Y directions ($\Delta x = \Delta y$).

The result of this set of runs is shown in Fig. 7.5. It can be inferred from the figure that the stability of all three schemes for time step of 0.001 is excellent. The proposed *Barakat–Clark CTD* scheme proves to be equally stable and accurate in space as compared to the *Barakat–Clark FTD* scheme. An important point is observed in this set of runs that the *DuFort–Frankel* scheme becomes unstable with a grid size greater than 0.38. As well, the grid size could not be decreased below 0.03 because of computer memory limitation.

Modeling Viscous Fingering

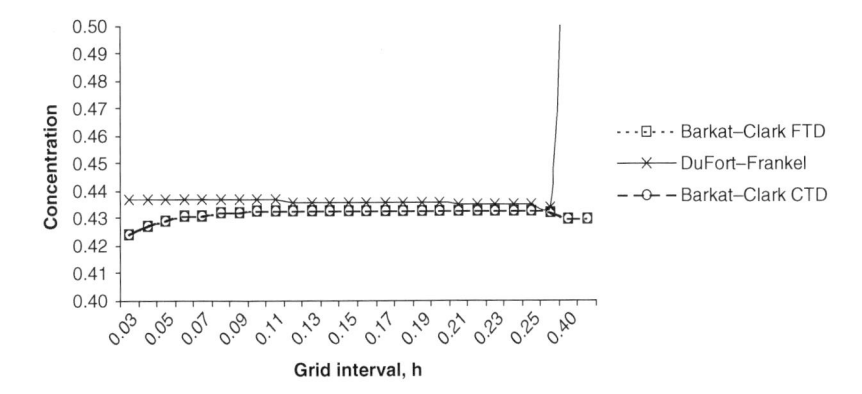

Figure 7.5a Effect of spatial grid size on outlet concentration for different schemes.

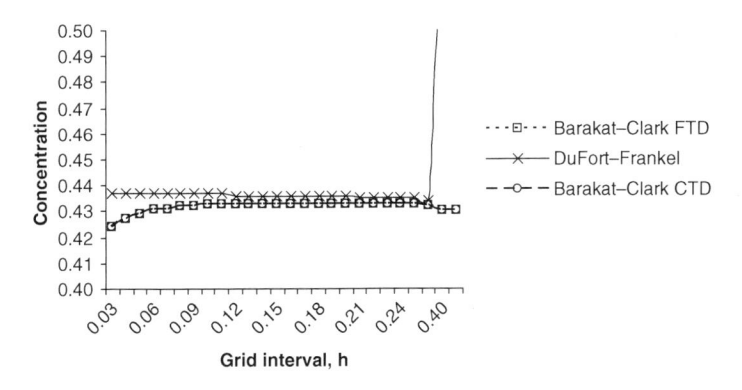

Figure 7.5b Effect of spatial grid size on outlet concentration for different schemes, $t = 1$ and $k = 0.001$.

Therefore, it is difficult to determine the stability of these schemes for the grid sizes below 0.03 and for a time step size of 0.001. In the range of 0.03 to 0.38, all three schemes give stable and almost identical results. To come to a conclusion for the stability and accuracy of these schemes two other sets of runs were conducted by increasing the time step size. Figure 7.6 shows the result of the second set of runs with k = 0.01. From the figure, it is evident that the *DuFort–Frankel* scheme, which is considered to be unconditionally stable, does show instability. For this case, the stable region ranges from 0.020 to 0.35. In this range, the *DuFort–Frankel* scheme shows excellent accuracy while both of the *Barakat–Clark* schemes give almost identical results with a stable behavior throughout the grid size domain. Even though the *Barakat–Clark* schemes do not describe

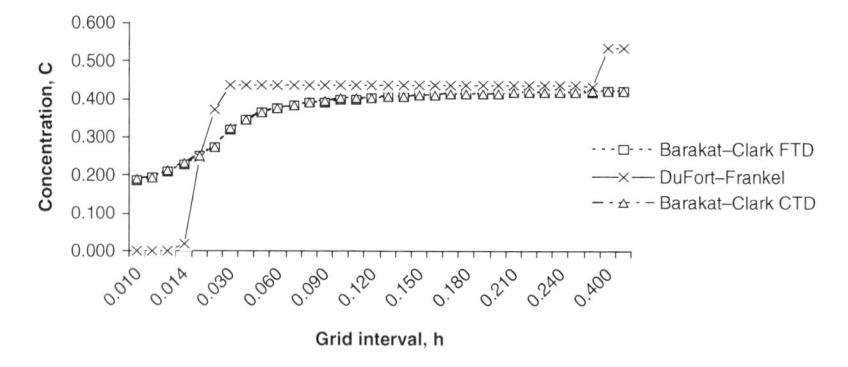

Figure 7.6 Effect of spatial grid size on outlet concentration for different schemes, $t = 1$ and $k = 0.01$.

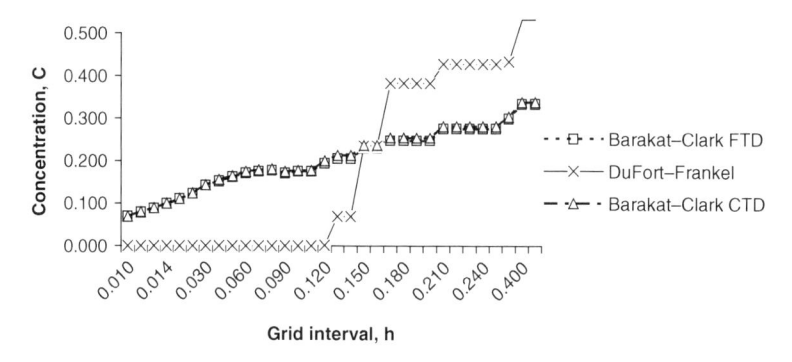

Figure 7.7 Effect of spatial grid size on outlet concentration for different schemes, $t = 1, k = 0.1$.

the actual physical problem quite well at grid intervals below 0.1 but for grid sizes above 0.1, these schemes show reasonable accuracy. The same applies to larger grid sizes, for which *DuFort–Frankel* scheme fails to converge.

Figure 7.7 also dictates the same underlying phenomenon of instability of the *DuFort–Frankel* scheme. In this set of runs time step size has been taken as 0.1. From the figure, one can conclude that the stability of both of the *Barakat–Clark* schemes are better than that of the *DuFort–Frankel* scheme, which in this case is quite unstable and demonstrates the deviation from the actual physical problem. In short, from the analysis of Figs. 7.5–7.7, it can be concluded that the *DuFort–Frankel* scheme is sensitive to grid size, which is

MODELING VISCOUS FINGERING 291

contradictory to what has been claimed earlier in the literature. The *Barakat–Clark FTD* and the *Barakat–Clark CTD* schemes prove to be unconditionally stable for the grid size domain worked in this case study. The stability of the *DuFort–Frankel* scheme decreases with the increase in time step size as shown in these figures. Therefore, it is better to address the stability of these schemes separately with changing time step size.

Time step size was varied from 0.0001 to 0.333 and outlet concentration is observed after time = 1 keeping the spatial grid size constant at 0.1. Some very interesting results are observed. Figure 8.8 shows the comparison of three schemes. The *DuFort–Frankel* scheme is found to be accurate and stable in the range of k = 0.0001 to 0.05 after which it starts to deviate from the actual solution and stable behavior. This scheme becomes fully unstable at k > 0.09. This finding is the same as of Aboudheir et al (1999) in which the *DuFort–Frankel* scheme was found to be unstable at k > 0.05. From Figure 8.8 it can be seen that both of the *Barakat–Clark* schemes are 100% stable for every time step size, although their accuracy can be questioned after a certain range (k > 0.005). The *Barakat–Clark CTD* scheme gives better accuracy with increasing time step size. A better visual comparison of the accuracy of both schemes can be observed from Figs. 7.9 and 7.10, plotted at a smaller scale. Fig. 7.9 shows that both schemes observe the same pattern in accuracy in the range k = 0.0001–0.045, whereas in Fig. 7.10, the difference in accuracy becomes prominent at k > 0.05.

Another important parameter ($\Delta x/\Delta y$) is also studied to see the impact of space discretization in the x and y direction. The outlet

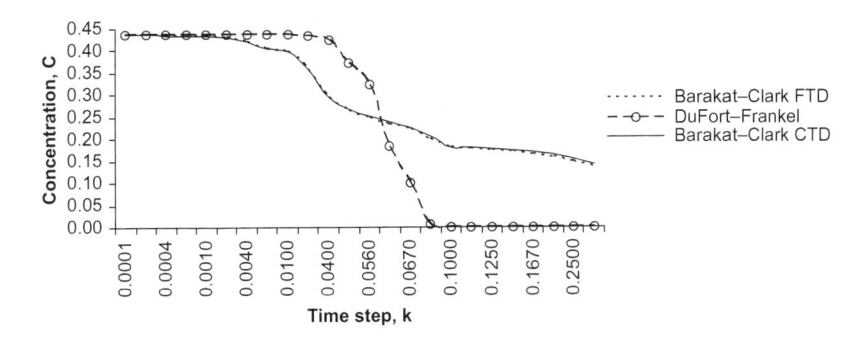

Figure 7.8 Effect of time step size on outlet concentration for different schemes, $t = 1, h = 0.1$.

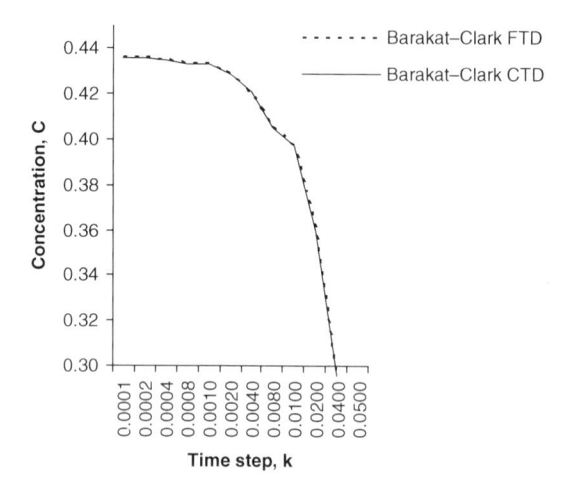

Figure 7.9 Effect of time step, Barakat–Clark, $t = 1$ and $h = 0.1$.

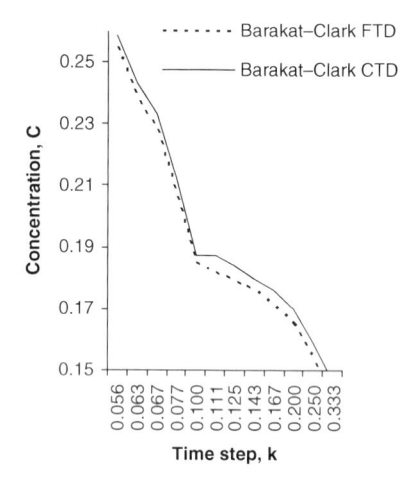

Figure 7.10 Effect of time step, Barakat–Clark, $t = 1$ and $h = 0.1$.

concentration at time =1 is observed for three schemes using a constant time step size of 0.001. Figure 7.11 shows these results. The *DuFort–Frankel* scheme proved to be not only stable for the range of $\Delta x / \Delta y = 0.1$–$12$ but also accurate. Both *Barakat–Clark* schemes also exhibit the stable behavior throughout the range of $\Delta x / \Delta y$ analyzed but with the limitation of accuracy, which shows a decreasing trend with the increase in $\Delta x / \Delta y$ value.

Modeling Viscous Fingering

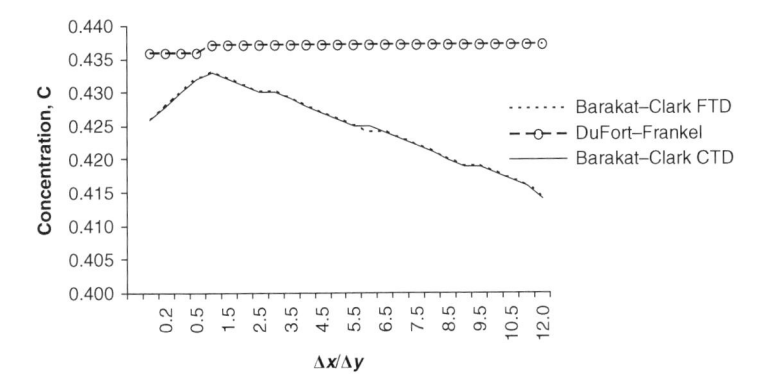

Figure 7.11 Effect of $\Delta x / \Delta y$, $t = 1$ and $k = 0.001$.

Furthermore, both *Barakat–Clark* schemes give almost identical results. This behavior of the *Barakat–Clark* schemes shows their sensitivity towards the spatial grid size, which has already been discussed in previous sections.

7.1.6 Influence of Boundary Conditions

Another very important point that is observed during this numerical study is the extraordinarily sensitive nature of the *Barakat–Clark* schemes towards the selection of the reference time frame for "Neumann Boundary" conditions. It is well known that one has to take specific assumptions while using equations 7.2 and 7.3 in the *Barakat–Clark* scheme due to its limitation to handle the implicit form of these boundary conditions successfully. The forward and the backward marching styles of the *Barakat–Clark* intermediate equations (Equations 7.4 and 7.5) respectively make it impossible to treat both the boundary conditions implicitly. This problem can be addressed either by using explicit forms for both boundary conditions (calculated on the basis of the previous time interval) or treating one boundary condition implicitly (in the present time interval).

Although both of these approaches give stable results for the *Barakat–Clark* schemes they can be a non-favorable choice because the solution may be further away from the accurate one. To illustrate this problem, equation 7.3 is changed to calculate the boundary condition based on the previous time interval in the *Barakat–Clark FTD* scheme. The results of this set of runs are shown in Figs. 7.12 and 7.13. In Fig. 7.12, the sensitivity with space interval (spatial grid size) of this modified boundary condition case is compared with

294　　ADVANCED PETROLEUM RESERVOIR SIMULATION

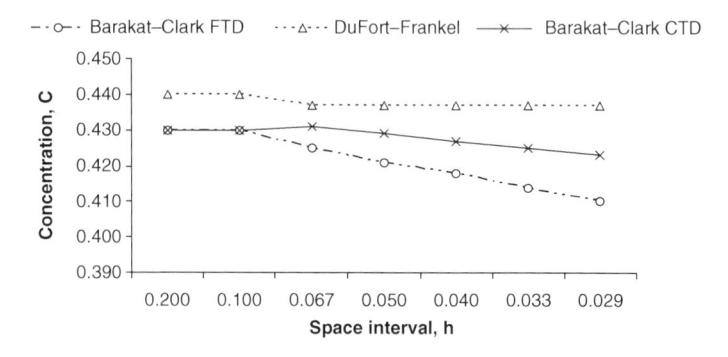

Figure 7.12 Accuracy at different spatial grid sizes of different schemes.

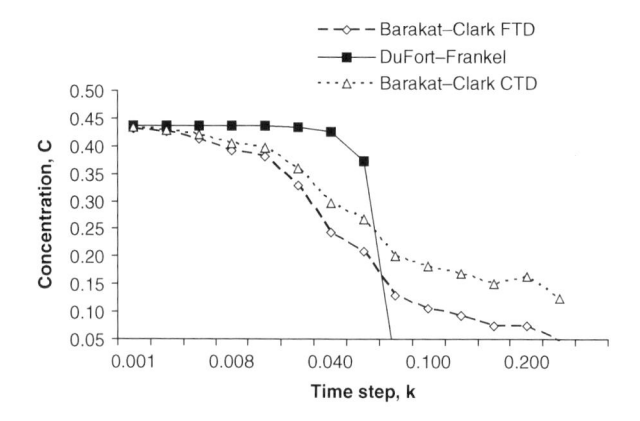

Figure 7.13 Accuracy at different time step sizes of different schemes.

the other two schemes having the same boundary conditions taken in the previous discussion.

It is evident from this figure that the accuracy of the *Barakat–Clark FTD* scheme is the least as compared to the other two schemes. In the same manner, Fig. 7.13 shows the sensitivity of these schemes with the change in time step size. Again it is clear that the *Barakat–Clark FTD* with the changed boundary condition gives the poorest results as compared to the other two cases. The behavior of *Barakat–Clark CTD* and the *DuFort–Frankel* schemes is the reproduction of the discussion given in previous sections. This weak behavior of the *Barakat–Clark – FTD* scheme in accuracy is due to the change in the reference time frame for one boundary condition, therefore, it becomes essential to treat "Neumann Boundary" conditions with special care while modeling the convective dispersion equation by the *Barakat–Clark* scheme.

7.2 Application of the New Numerical Scheme to Viscous Fingering

In this section, the results achieved by running the simulations are presented along with a comprehensive parametric study showing the effects of certain important parameters on the stability of a specific displacement scheme. Stable and unstable displacement simulations were conducted following the stability criterion of Coskuner and Bentsen (1990) and Coskuner (1992). It is important to note that previously no numerical scheme existed that would be able to model both stable and unstable flow.

7.2.1 Stability Criterion and Onset of Fingering

This stability criterion presented by Coskuner and Bentsen (1990) and Coskuner (1992) discussed in detail in Bokhari (2003). This criterion has been followed as a measure of stability of the displacement scheme. Stable and unstable displacements are modeled based on the instability number. Higher values of instability number, I_{sr}, than π^2, were used to model the unstable cases while lower values were used to numerically simulate stable cases. According to this theory if a displacement is unstable ($I_{sr} \geq \pi^2$), the breakthrough recovery will most likely be dominated by viscous fingering.

Modeling the initiation of viscous fingering has always been a challenge for scientists and modelers. It has been found earlier by various scientists that the viscous fingering cannot be initiated in numerical schemes without artificially invoking them. Most of these scientists (Christie et al, 1987, 1989; Tan and Homsy et al, 1986, 1988) used artificial perturbations either in the initial or inlet boundary conditions. The reason for the use of these artificial perturbations is the limitations of the governing equations to address the non linearity and because of this reason all the present methods to solve these equations are limited. Another reason which can be given in this contrast is the lack of heterogeneity in the ideally homogeneous numerical representation of the porous medium used in the numerical schemes. It has been agreed upon by different groups of scientists that the miscible viscous fingering, while possible in ideally homogeneous media, always initiates when the displacing front encounters the local heterogeneity of the porous medium. This concept was utilized in the current work and the local heterogeneity was created by using a random distribution of permeability at the inlet face of the numerical representation of the porous medium. The remaining whole porous medium was considered to be ideally homogeneous and isotropic to model the phenomenon of viscous fingering in a real homogeneous medium.

7.2.2 Base Stable Case

For the purpose of comparison, it is necessary to first select a base case. It has been discussed earlier in the previous section that, in the current numerical model, every flow scheme exhibits stable behavior until the fingering is initiated by using random distribution of permeability in the inlet face of numerical model. This approach is quite realistic and supported by the experimental observation. Therefore, regardless of the instability number and the mobility ratio used, the flow scheme remains always stable. It was tried to use some realistic data to come up to a base stable case. Table 7.1 shows the main parameters used for the base stable case.

Following is the discussion based on the results achieved for the base stable case according to the solution of the governing equations.

Figure 7.14 shows the pressure distribution with time in the porous medium. As the model used is in two dimensions, therefore, for the simplification in plotting the data only the pressure distribution at a horizontally central point is shown here. For the stable case all other points exhibit the same pressure distribution because of the frontal stability of the flow scheme. This plot shows the solution of the Diffusivity Equation as per boundary conditions used. It is clear from the plot that the injection pressure is constant for the whole

Table 7.1 Parameters used in base stable case

Size of Porous Medium ft × ft	P_{in} Injection Pressure psia	Ø Porosity %	k Permeability Darcy	Dx Transverse Dispersion Coefficient ft²/d	Dy Longitudinal Dispersion Coefficient ft²/d
1 × 2	100	50	0.01	0.1	0.5

Modeling Viscous Fingering

time period. Therefore, at each time interval, the pressure starts from a value of 100 psia and it decreases with the increase in the length of the porous medium. It means that the pressure at all points below the inlet face always remain lower than the inlet/injection pressure. With the passage of time, the pressure distribution starts to stabilize. Initial nonlinear behavior, which can also be regarded as "unsteady state behavior" ends at approximately 0.3 days. At this point the pressure distribution starts to exhibit a linear stabilized behavior which can also be regarded as "steady state" behavior.

Figure 7.15 shows the plot of water concentration in the outlet stream of the porous medium with time. This is the typical solution of the mass balance equation in which the outlet concentration increases with an increase in time until breakthrough takes place. After breakthrough, the concentration of water in the outlet stream starts increasing from a value of water concentration in the saturating glycerine solution (zero in this case) till it reaches a value of unity which is the maximum concentration of water in the injection fluid. The reason for this continuous increase in the outlet concentration of water is because of the continuous injection of water to displace the saturating glycerine solution. When the whole saturating fluid is displaced, the concentration of water becomes unity. It means what is being injected at one end is coming out at the other end of porous medium.

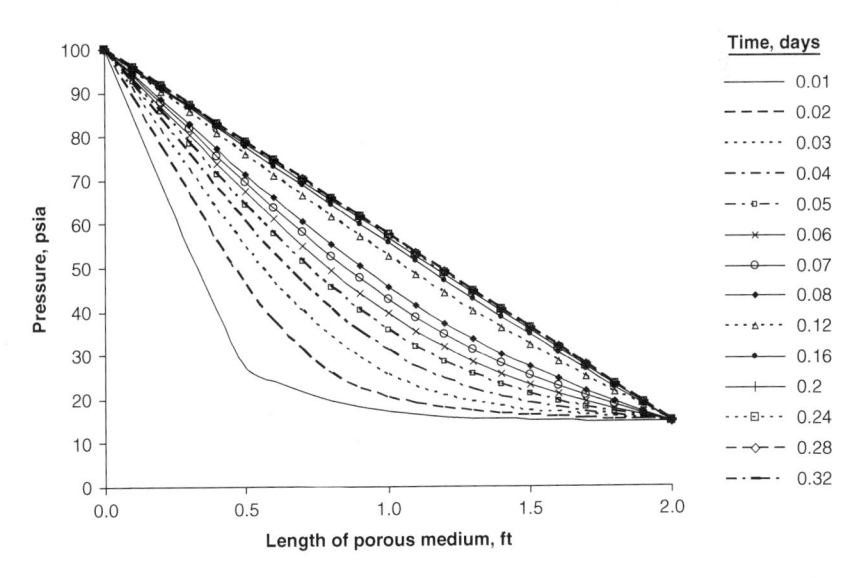

Figure 7.14 Pressure distribution with time at width = 0.5 ft (Base Stable Case).

298   ADVANCED PETROLEUM RESERVOIR SIMULATION

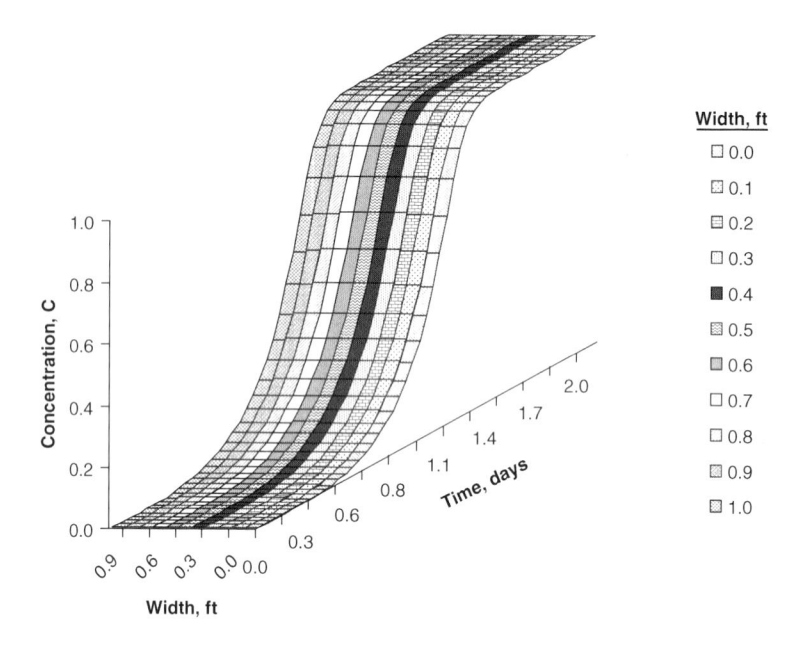

Figure 7.15 Outlet fluid concentration profile with time (Base Stable Case).

Another measure of this recovery phenomenon is the measure of the viscosity of outlet solution coming out of the porous medium. Figure 7.16 shows the viscosity profile of the fluid (glycerine solution) coming out of the porous medium at different time intervals. The figure illustrates that, before breakthrough, the outlet viscosity remains constant as 1410 cp which is the viscosity of pure glycerine. With the passage of time, the two fluids, water and glycerine, start to mix decreasing the effective viscosity of the solution from this fixed value. This viscosity decrease is due to the dilution of glycerine solution coming out of the porous medium. The plot shows that the final viscosity is 1.0 which is the viscosity of pure water. At the time when the outlet fluid viscosity becomes unity, the whole glycerine can be considered already displaced from the porous medium and after that time period further water injection results in the water production only.

Figure 7.17 shows the concentration distribution in the porous medium for this case. As it is quite evident from the plot that each point in the direction of flow behaves similarly with the exception of delay in breakthrough. As the injection is continuous in this case, the displacement from each point follows the same non linear increasing trend in which the concentration of water increases with

Modeling Viscous Fingering

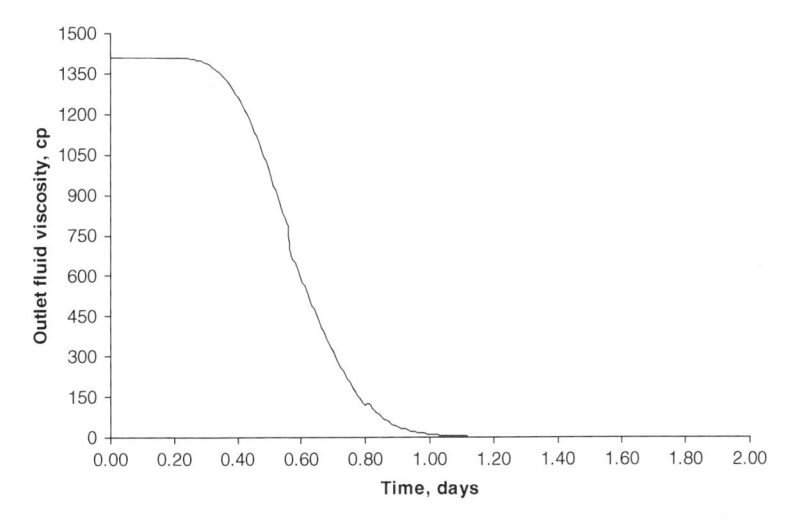

Figure 7.16 Outlet fluid viscosity profile with time (Base Stable Case).

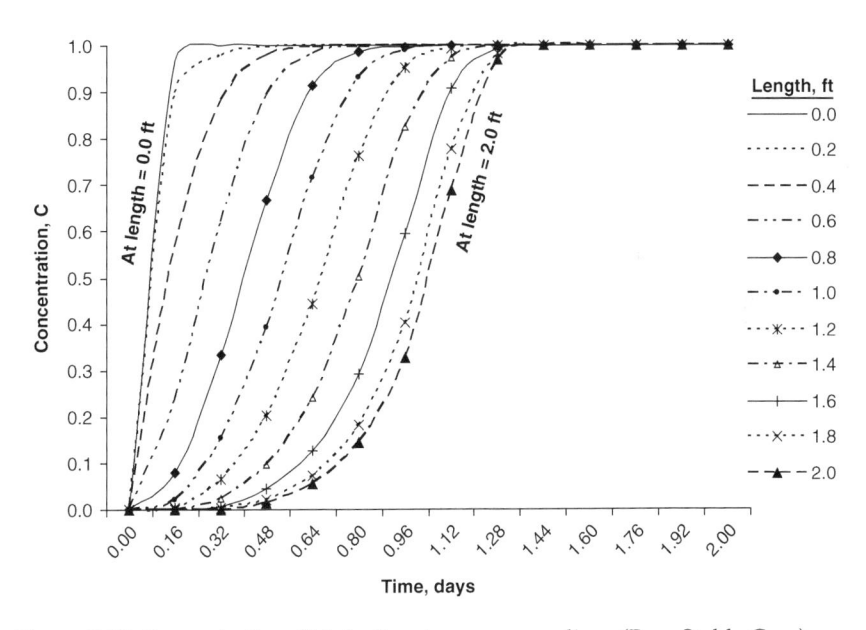

Figure 7.17 Concentration distribution in porous medium (Base Stable Case).

time to reach a maximum value of unity after which further injection results in water production alone. The difference between the shapes of the inlet and outlet concentration curves is due to the fact that the change in concentration at the inlet face is a step change at

300   Advanced Petroleum Reservoir Simulation

which the starting concentration gradient is at its maximum while at the outlet, there is a gradual change in concentration with quite a low concentration gradient as compared to the case of inlet face.

The stability of the displacement front in the base stable case can easily be witnessed in Fig. 7.18. In this figure, the concentration along the width of the porous medium has been plotted at different time intervals at a fixed height in the porous medium. These plots illustrate that the concentration of water increases with time starting from a value of zero to a maximum value of 1.0 after a certain time. The stability of the front can be seen from the straight plain lines of concentration along the whole width of porous medium. This behavior of stability remains intact throughout the whole duration in which concentration becomes unity. These concentration lines dictate the stability of the displacing front as these are the plots giving a measure of the extent of invasion of displacing fluid at a particular time interval in terms of transverse water concentration along the direction perpendicular to the principal direction of flow. The slight bends in these plots are due to the boundary effects (No-Slip) and the velocity distribution in porous medium which remains slightly higher at the centre point in the transverse plane of the porous medium.

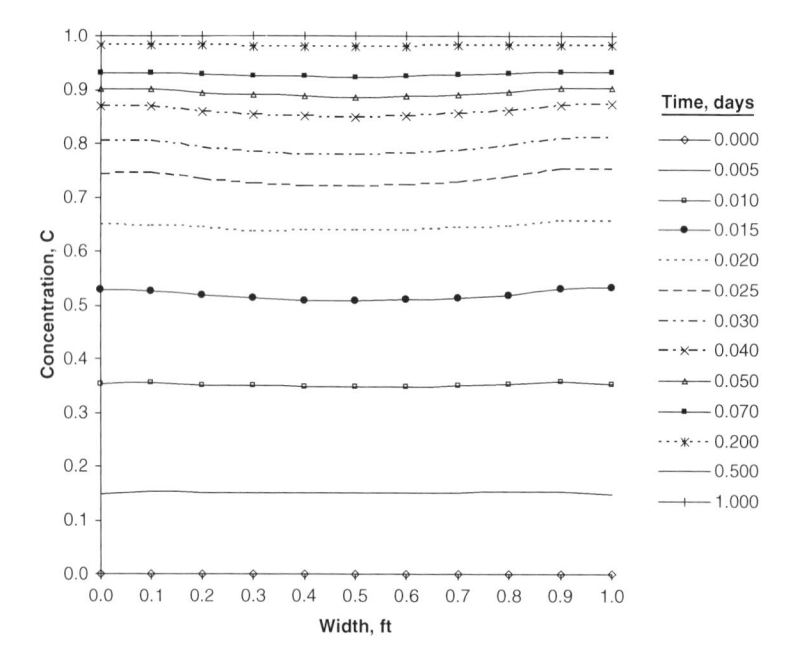

Figure 7.18 Stability of the front with time at height = 0.1 ft (Base Stable Case).

Modeling Viscous Fingering

Similarly, Fig. 7.19 gives the velocity distribution at different heights along the direction of flow. The values plotted are taken at the central point of the whole width of porous medium. The plot shows the solution of the Darcy's equation in the direction of the flow. It is clear that due to the initial unstable period, the velocity shows a quick jump in its value which starts to stabilize with the passage of time. The fluid present at rest at a particular height in the porous medium, starts to move with an increasing velocity. After a certain time this velocity ultimately stabilizes at a fixed value. The same behavior is repeated by the fluid present at different heights in the porous medium with the exception of the time difference in getting its movement started. The reason for this sort of behavior is the difference in time to reach the effect of invading/displacing fluid to that height in the porous medium. The time at which the whole system becomes stabilized or at which the whole fluid starts to move with a same velocity in the porous medium is about 1.4 days in this case. Plots given in Fig. 7.20 show the shape of velocity fronts at different time intervals at a particular height in porous medium.

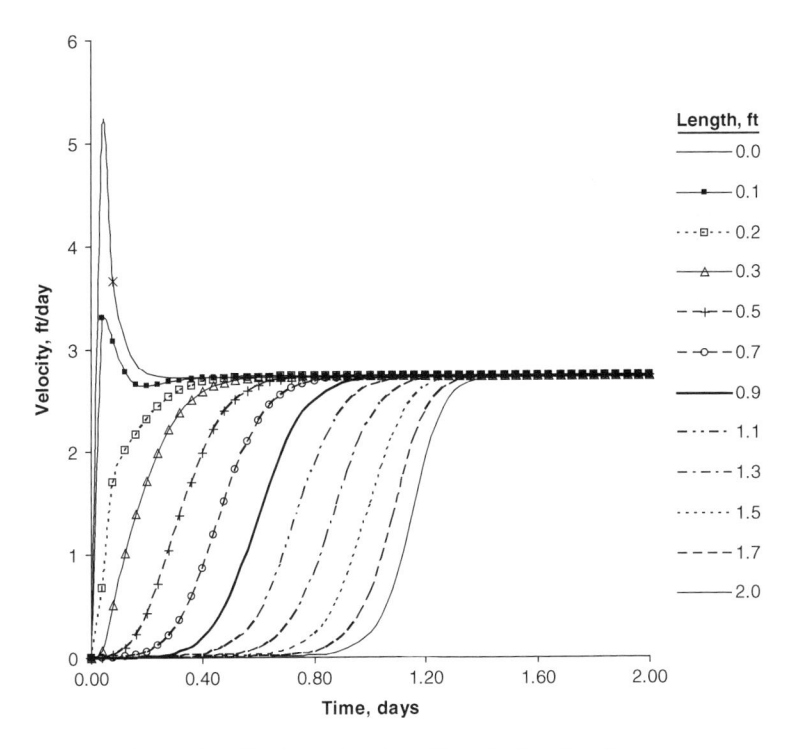

Figure 7.19 Velocity distribution with time (Base Stable Case).

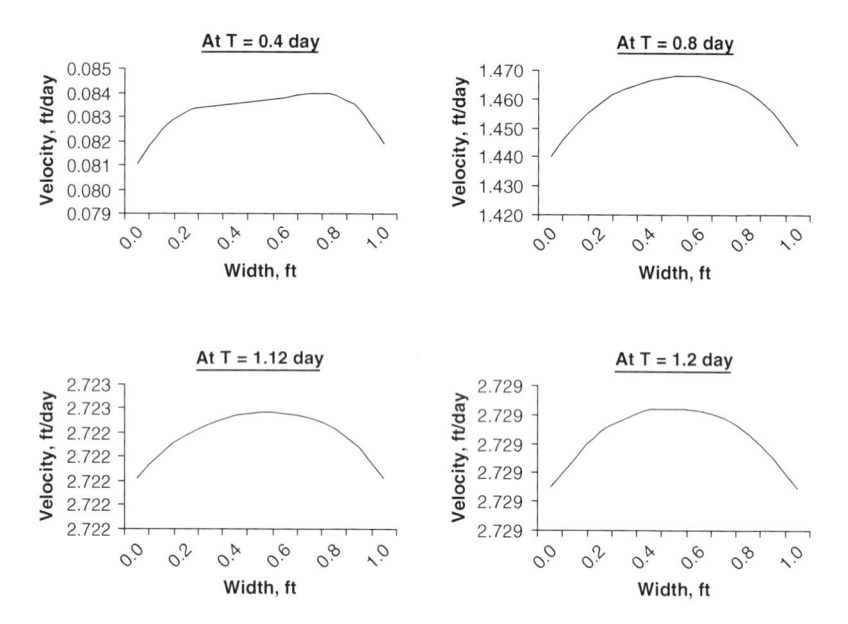

Figure 7.20 Shapes of velocity front at different time intervals.

The concave shape of the fronts is due to the no-slip boundary conditions used in the numerical work. The concaveness of the shapes improves with time due to the increase in stability of displacement.

7.2.3 Base Unstable Case

To study the displacement scheme under the effect of viscous fingering, the stable case presented in the previous section was made unstable as mentioned earlier by the use of random distribution of permeability at the inlet face of the porous medium. Table 7.2 shows the parameter used for the unstable case. It is clear from the values of the parameters that there is no change in the parameters except the use of random permeability distribution at the inlet face only.

In this case also, the porous medium was numerically modeled by a using mesh size of 20 × 40 grids with a value of $\Delta x = \Delta y = 0.05$ ft. The time frame used for the simulation was also kept the same as 2 days with a Δt of 0.001 day.

Figure 7.21 shows the solution of the diffusivity equation for the base unstable case. The plotted data show the pressure distribution in the porous medium at different heights and after different time intervals. Although the solution gives the distribution of pressure in

Modeling Viscous Fingering

Table 7.2 Parameters used in base unstable case

Size of Porous Medium ft × ft	P_{in} Injection Pressure psia	Ø Porosity %	Dx Transverse Dispersion Coefficient ft²/d	Dy Longitudinal Dispersion Coefficient ft²/d
1 × 2	100	50	0.1	0.5

Note: Permeability at the inlet face was varied randomly (10% variation).

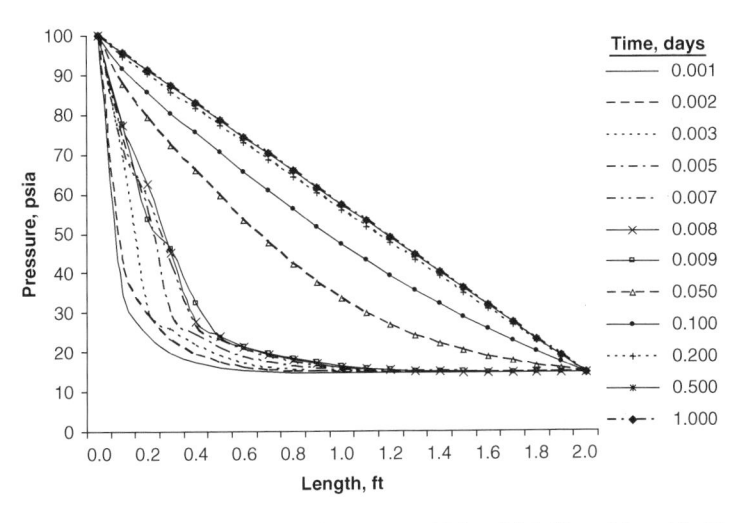

Figure 7.21 Pressure distribution with time at width = 0.5 ft (Base Unstable Case).

two-dimensions but for the sake of simplicity, the longitudinal pressure distribution at a single point in transverse direction, has been plotted. The points used are the width wise central points in horizontal plane of the porous medium. If this figure is compared with the similar Fig. 7.14 of the stable case, two major differences can easily be distinguished. The first difference is the time to reach stability which in this case is about 0.5 day. This relatively long time is justified because of the presence of instability in the porous medium. The second major difference is the disturbed pattern of pressure distribution especially at the initial stages of displacement.

This perturbed pressure region in the unstable zone is the staging area for the fingers to onset because of which some portion of fluid gains a higher velocity to nose ahead from the neighboring fluid. If the overall pattern of this pressure distribution is analyzed, it becomes clear that the same trend is repeated as in the stable case. The high difference of the injection pressure and the initial pressure in porous medium causes the response of initial lengths of the porous medium to become quite precipitous. This steep nature is further enhanced by the presence of a high viscosity contrast which causes this portion of porous medium susceptible for the phenomenon of viscous fingering.

Figure 7.22 is the plot of the concentration distribution in the porous medium for the unstable case. It is obvious from the plot that the concentration distribution follows an extremely disturbed pattern especially at initial lengths. This roughness of the concentration pattern is basically linked with the uneven pressure distribution, which ultimately causes the phenomenon of viscous fingering. The comparison with Fig. 7.17 illustrates that the breakthrough time at similar heights is comparatively low in the unstable case because of the early breakthrough in this case. This early breakthrough is not easily distinguished between the two cases because of the same viscosity contrast used. With the passage of time the concentration increases at all heights until it reaches a value of unity which means the presence of only pure water and this depicts that whole of the saturating fluid (glycerine in this case) has been displaced from that point. Once again, the difference between the shapes of the inlet and

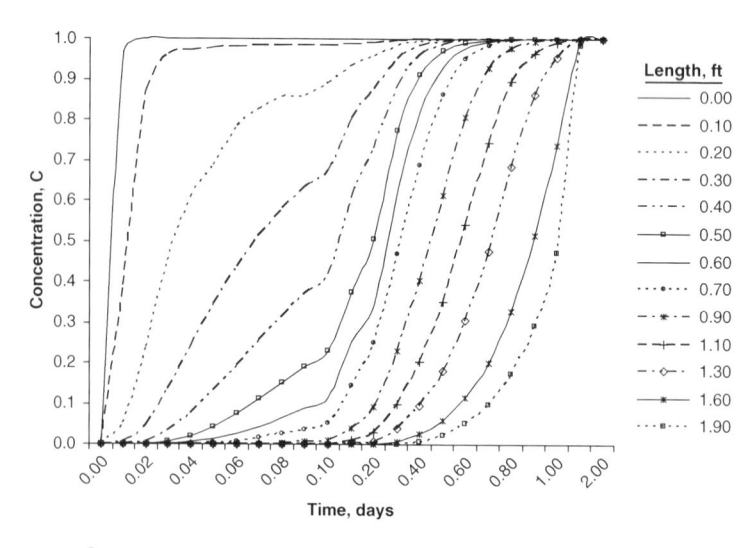

Figure 7.22 Concentration distribution in porous medium (Base Unstable Case).

Modeling Viscous Fingering

outlet concentration curves is due to abrupt change in concentration at the inlet face and gradual change in the case of outlet face.

Figure 7.23 shows the phenomenon of viscous fingering in a very clear manner. The instability of the front, which is another measure of fingering, is plotted in the form of concentration along the width of the porous medium at a distance of 0.1 ft from the inlet face of the porous medium. The change in this concentration distribution with time is also shown in the figure. It is apparent that the concentration, at the initial times along the width of the porous medium, demonstrates a stable behavior that starts to deviate from stability as time progresses. In this figure, the onset of fingering starts after a time interval of 0.04 days. This disturbance in concentration grows with the passage of time and the whole flow scheme becomes dominated by fingers of high concentration wave lengths.

Following Figs. 7.24a through 7.24e show the propagation of fingering with time and also the fingering extent/distribution along the direction of flow. These are the plots showing the fingering in terms of concentration distribution with time along the length of porous medium. Figure 7.24a is the plot of concentration after 0.01 day of injection at different heights in the porous medium.

It is clear from these figures that the extent of fingering at a distance near to the inlet face is more than at a distance which is away from the inlet face. This is quite logical as with the passage of time, the extent of fingering increases at the same height in a porous

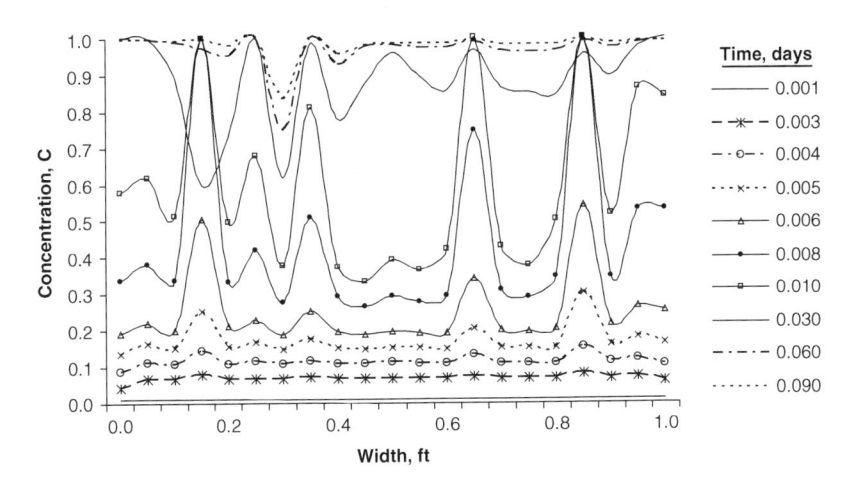

Figure 7.23 Stability of the front with time at height = 0.1 ft.

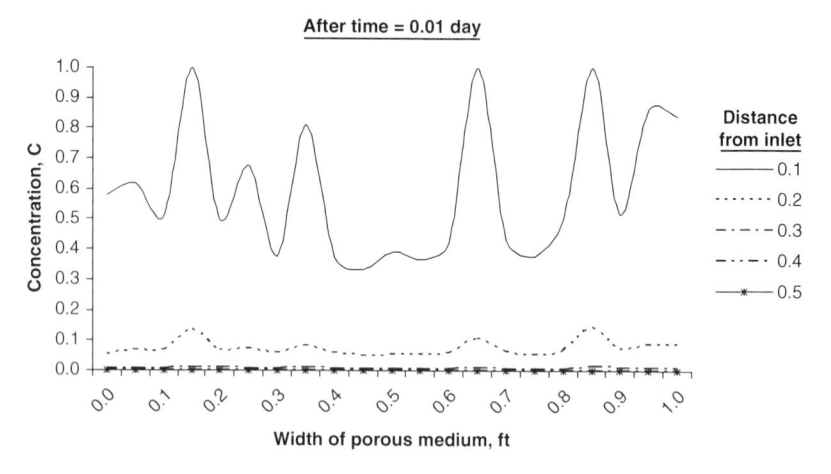

Figure 7.24a Propagation of concentration fingers (t = 0.01 day).

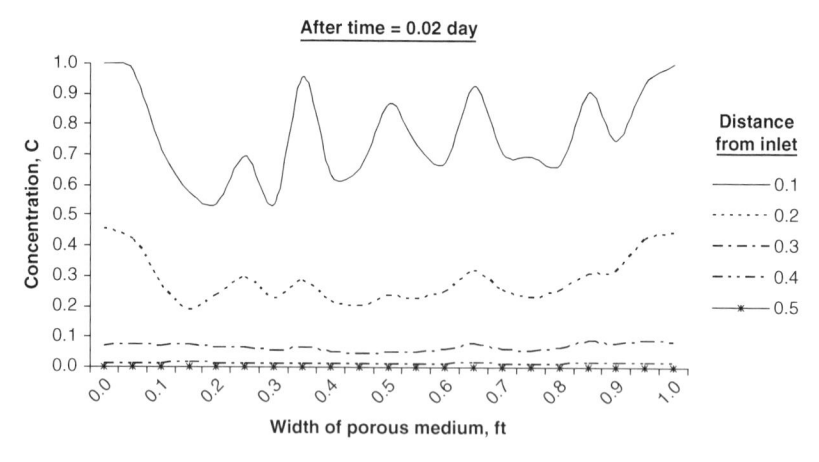

Figure 7.24b Propagation of concentration fingers (t = 0.02 day).

medium because of the continuous injection of the displacing fluid. At a particular time the invasion of displacing fluid is more at the point near to the inlet face as compared to the point which is away from it. Therefore, in Figs. 7.24a through 7.24e, the disturbance in concentration distribution along the traverse direction at a same distance from inlet face increases with time to reach a maximum value of unity. If the concentration plots at the distance of 0.1 ft from the inlet face in all these cases are compared, it becomes obvious that

Modeling Viscous Fingering

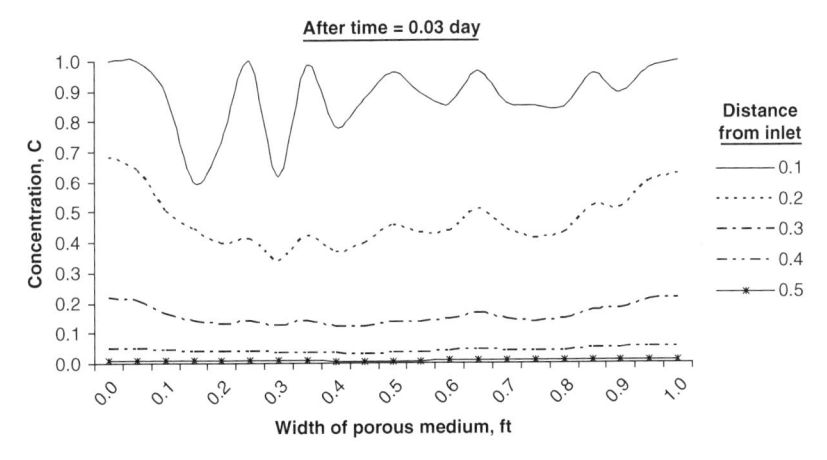

Figure 7.24c Propagation of concentration fingers (t = 0.03 day).

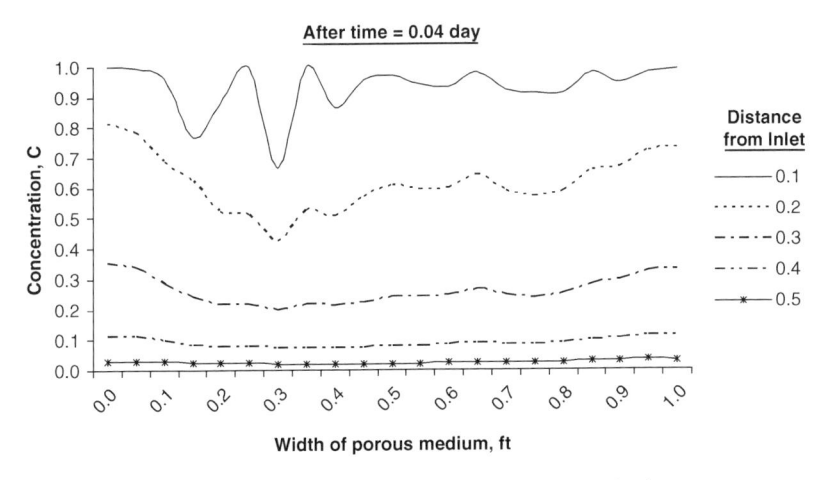

Figure 7.24d Propagation of concentration fingers (t = 0.04 day).

the concentration of water in any specified leg of the front increases with time and at the same time the continuous injection results in the increase of the water concentration in the remaining part of fluid (flat portion of the concentration plots). This increase in water concentration continues till concentration of water in the displaced fluid at the same distance from the inlet face becomes unity and thus that specified point is fully invaded by the displacing fluid. In all these plots, the comparison of concentration stability at any

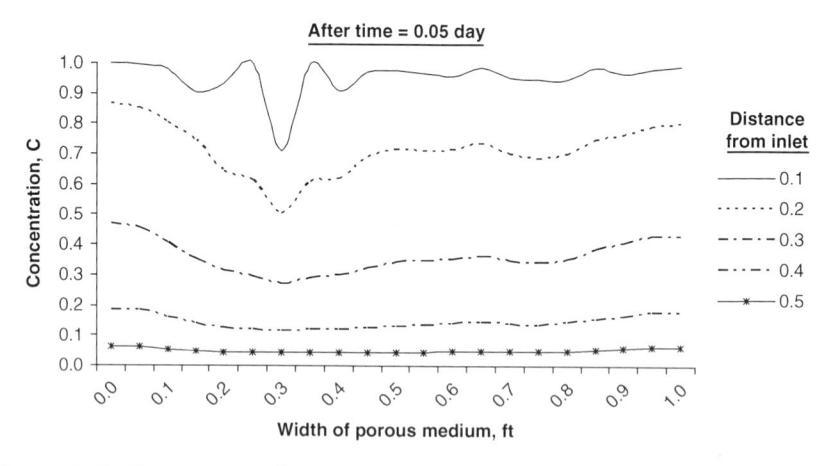

Figure 7.24e Propagation of concentration fingers (At $t = 0.05$ day).

two distances from the inlet face for a same time interval reveals that stability increases with the increase in the distance from the inlet face. This behavior needs some explanation. The reason for this behavior is the diminishing effect of initially encountered local heterogeneity by the fluid as the distance from the inlet face increases. This decline in the instability of the concentration front is due to the ideal homogeneity of the remaining numerical representation of the porous medium, which cannot be found in reality. The phenomenon of tip splitting and merging can be noticed in the form of concentration plots of Figs. 7.24a and 7.24b. In short, the instability of the front, which is known as fingering driven by viscosity contrast, can easily be studied in the form of concentration distribution instability along with its inherent mechanisms of spreading, shielding and merging.

Figure 7.25 gives the velocity distribution with time along the length of the porous medium for the base unstable case. As it has already been discussed that the velocity behavior of a displacement scheme depends upon the pressure distribution in the porous medium, therefore, in this figure, the velocity distribution exhibits a disturbed unstable behavior which gradually progresses towards stability with the passage of time and reaches a steady state of velocity value. At this stage the unstable behavior itself starts to behave stably and thus the front and base of a finger starts to propagate with the same velocity.

Modeling Viscous Fingering

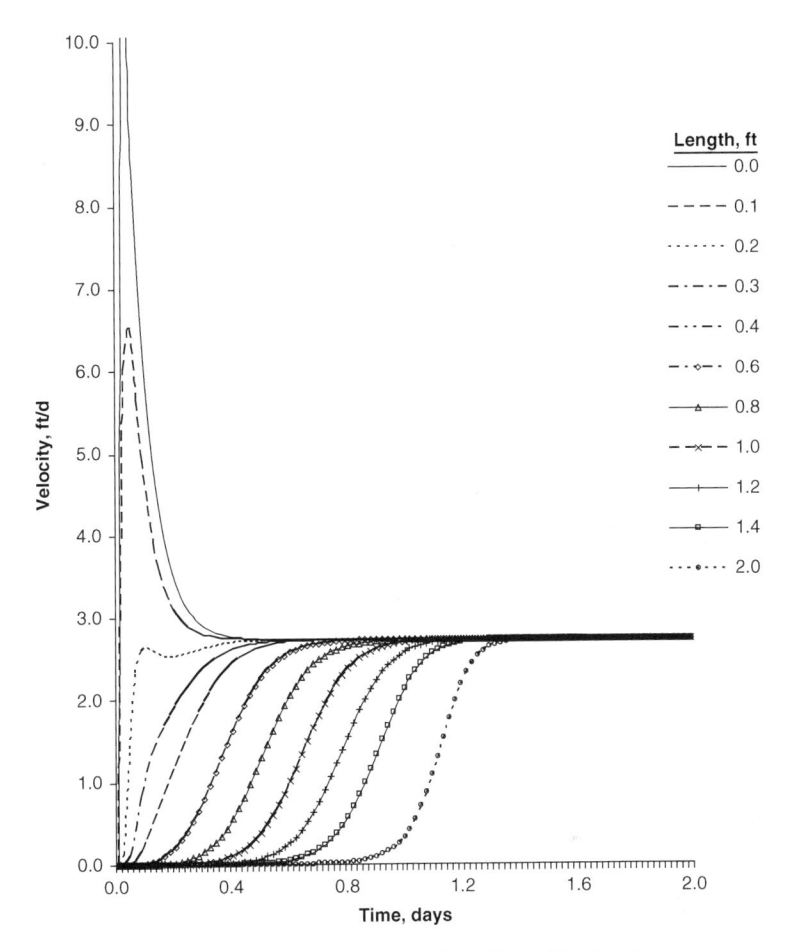

Figure 7.25 Velocity distribution with time (Base Unstable Case).

7.2.4 Parametric Study

The instability number (see Khan and Islam, 2007b for details) consists of various parameters. The magnitude of these parameters governs the magnitude of the instability number, which ultimately renders any flow scheme either stable or unstable. In this parametric study, it tried to study the effect of these parameters individually so that an in-depth understanding of this dependency can be achieved.

7.2.4.1 Effect of Injection Pressure

Injection pressure plays quite an important role in the stability of the displacement front. Increase in the pressure of displacing fluid causes the velocity of the fluid to cross the critical limit and thus the front of the fluid becomes unstable to very small perturbations caused by local permeability distribution.

To study the effect of injection pressure, concentration profile at the same distance from the inlet face of the porous medium has been plotted under the effect of different injection pressures. A velocity dependency is also given in Fig. 7.26 to understand the main driving force behind the difference in the concentration profiles achieved by using different injection pressures.

Figure 7.26 shows the transverse average velocities at the inlet face and the outlet face at different injection pressures. It is quite clear from the figure that at high pressures, the inlet and outlet velocities become the same after some time. At the low pressure ranges, there always remains a difference in these velocities. The logic behind is the presence of viscous resistance to flow which is overcome sooner by high pressure of the injecting fluid as compared in the case of the lower pressures. In all these pressure cases, the initial sharp decrease in the velocity is because of the unsteady

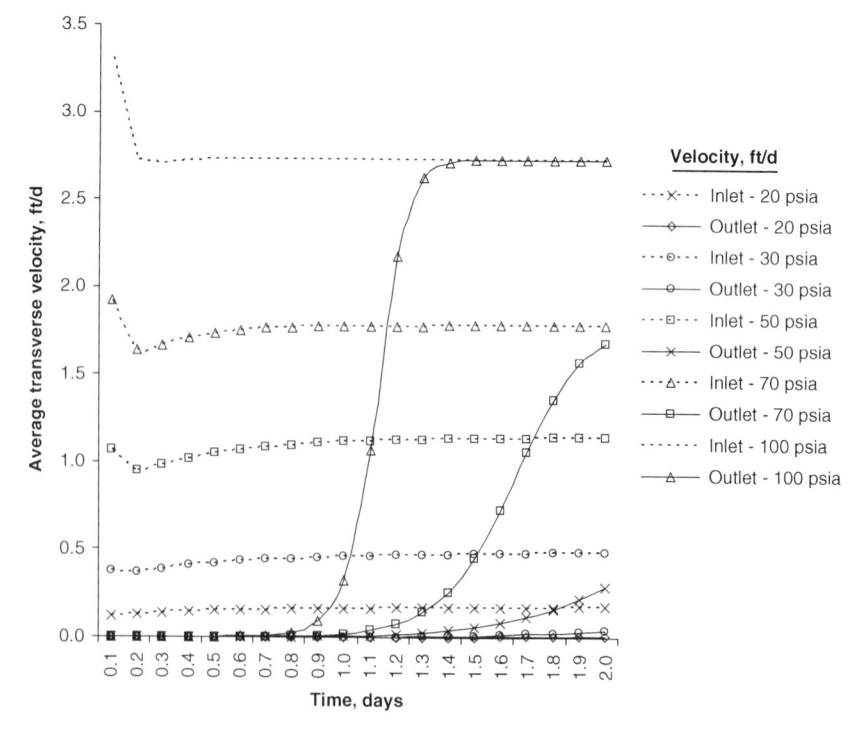

Figure 7.26 Effect of injection pressure on inlet and outlet velocities.

Modeling Viscous Fingering

state pressure behavior of the porous medium which is assumed to be at atmospheric pressure prior to the start of injection. As the difference in the pressure decreases with the passage of time, the velocity at the inlet face becomes constant. The outlet velocities, on the other hand, behaves in a more uniform pattern in which these velocities start to increase gradually with time and ultimately become constant and similar to the inlet velocities. At this point, both the so-called tip and base of concentration fingers move with same velocities and this behavior remains as such till the whole saturating fluid is displaced.

Figures 7.27 through 7.31 show the effect of injection pressure on the frontal instability with the passage of time. The phenomenon of viscous fingering is driven by the injection pressure or the pressure of the displacing fluid in the sense that the increase in injection pressure results in the increase of the intensity (length of finger legs) of the fingers as the fingers grow faster in comparison with the remaining fluid and cause an early breakthrough of the displacing fluid.

Figure 7.27 shows the concentration profile along the whole width of the porous medium at a distance of 0.1 ft from the inlet of face after 0.01 days of injection at different pressures. At a low injection pressure of 20 psia the concentration remains almost stable along the width of the porous medium while in the case of high injection pressure, the breakthrough is visible even after 0.01 days of injection.

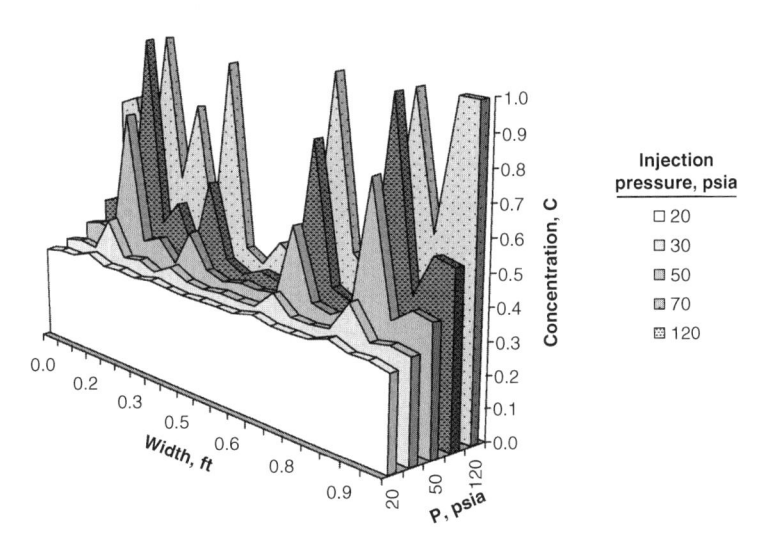

Figure 7.27 Effect of injection pressure on frontal instability after 0.01 day.

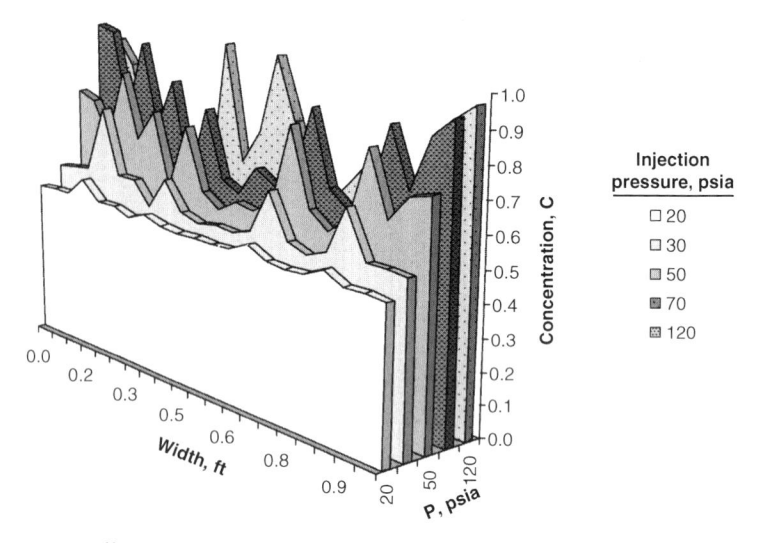

Figure 7.28 Effect of injection pressure on frontal instability after 0.02 day.

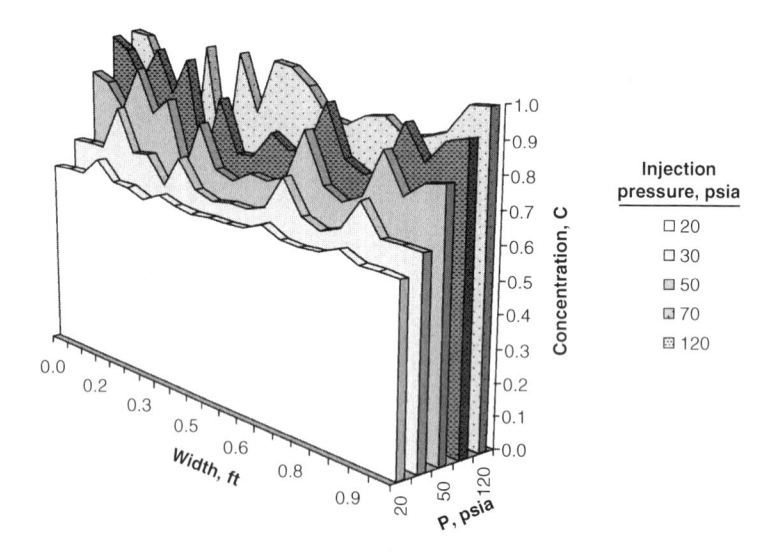

Figure 7.29 Effect of injection pressure on frontal instability after 0.03 day.

If the stability of the front is monitored with time in the case of low injection pressure (Figures 7.27 through 7.31), it becomes evident that the concentration front exhibits an almost stable increasing pattern. This illustrates the stability of the front or non-existence of the phenomenon of viscous fingering at low/favorable injection

Modeling Viscous Fingering

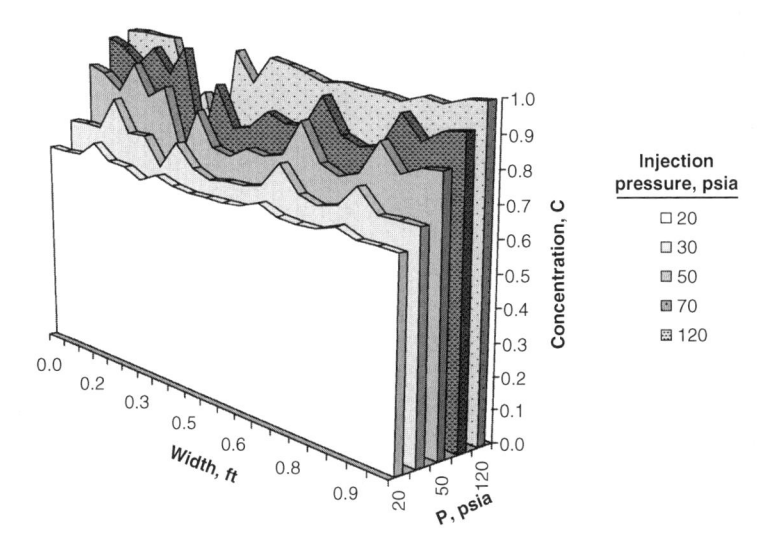

Figure 7.30 Effect of injection pressure on frontal instability after 0.04 day.

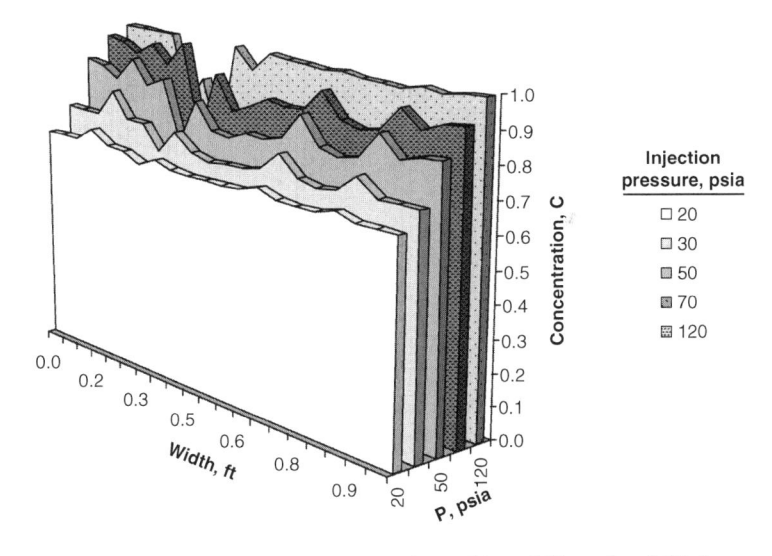

Figure 7.31 Effect of injection pressure on frontal instability after 0.05 day.

pressures. On the other hand, the intensity of the frontal instability increases with the increase in the injection pressure as illustrated clearly in Fig. 7.27. If the propagation of this instability is tracked with time at high pressures at the same distance from the inlet face, it reveals that the concentration contrast between so-called finger

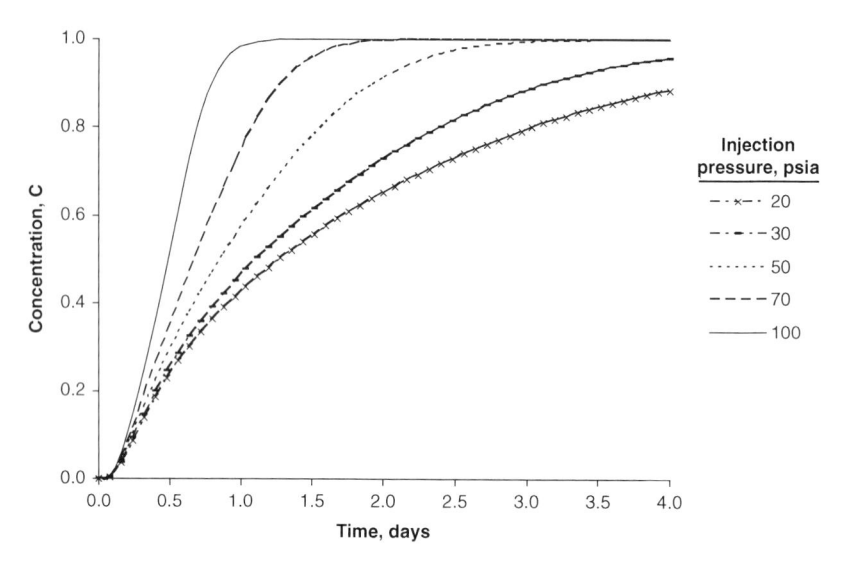

Figure 7.32 Effect of injection pressure on concentration profile with time at 1 ft from the inlet face.

base and tip decreases with time. Figure 7.32 shows the effect of injection pressure on the concentration profile with time. For this purpose, the concentration at a distance of 1 ft from the inlet face has been plotted at the central point along the width of porous medium.

It is clear from the figure that with the increase of injection pressure, time taken to reach the maximum concentration of unity decreases and the steepness of the concentration curve increases.

7.2.4.2 *Effect of Overall Porosity*

The porosity, which is a measure of the hold-up capacity of any porous medium, not only plays a vital role in calculating the quantity of the fluids in the porous medium but is also critical to dictate the stability of a displacement scheme. The role of the porosity parameter in the instability number discussed in an earlier chapter is a limiting one as by the increase in the effective porosity of the instability number decreases. The decrease in the stability number means the decrease in the intensity of viscous fingering. To study this important parameter and its behavior in the concentration distribution, a few simulations are made by varying the porosity in the porous medium and the keeping

Modeling Viscous Fingering

other parameters constant. These parameters are given in following Table 7.3.

It is to be noted that the concentration along the whole width of porous medium at a distance of 0.1 ft from the inlet face has been plotted with time in different runs using different porosities. Figure 7.33, which is a plot showing frontal stability in terms of concentration profile after 0.01 day of injection, illustrates the increase in porosity results in the increase in the frontal stability and vice versa. 8% porosity resulted in the earliest breakthrough while a 70% porosity case exhibited a more stable behavior in terms of the intensity of concentration contrast. This simulated behavior of

Table 7.3 Parameters used to study effect of overall porosity

Size of Porous Medium ft × ft	P_{in} Injection Pressure Psia	Dx Transverse Dispersion Coefficient ft²/d	Dy Longitudinal Dispersion Coefficient ft²/d
1 × 2	30	0.1	0.5

Note: Permeability at the inlet face was varied randomly (10% variation).
Values of Ø used: 8%, 10%, 20%, 30%, 50%, 70%.

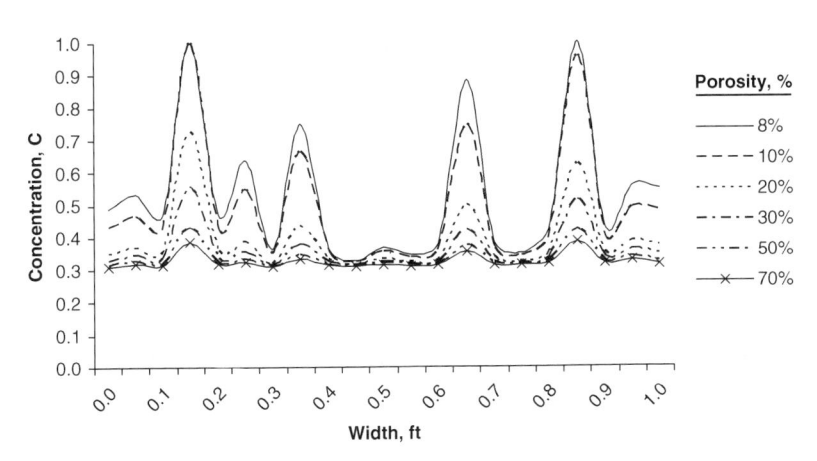

Figure 7.33 Effect of overall porosity on frontal instability after 0.01 day.

porosity is exactly the same as predicted by the role of porosity in the formulae of the instability number.

Figures 7.33 through 7.37 show the propagation of these concentration fronts with time for different cases of porosity. In all these cases, the concentration of each point increases with time causing the concentration contrast of the leading and lagging fluid to decrease. This decrease continues until the concentration at all points along the width of porous medium becomes the same or reaches a value of unity (100% displacing fluid).

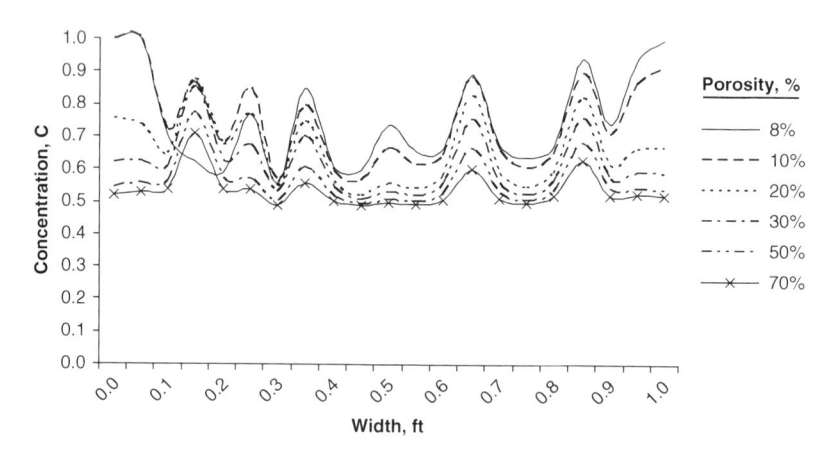

Figure 7.34 Effect of overall porosity on frontal instability after 0.02 day.

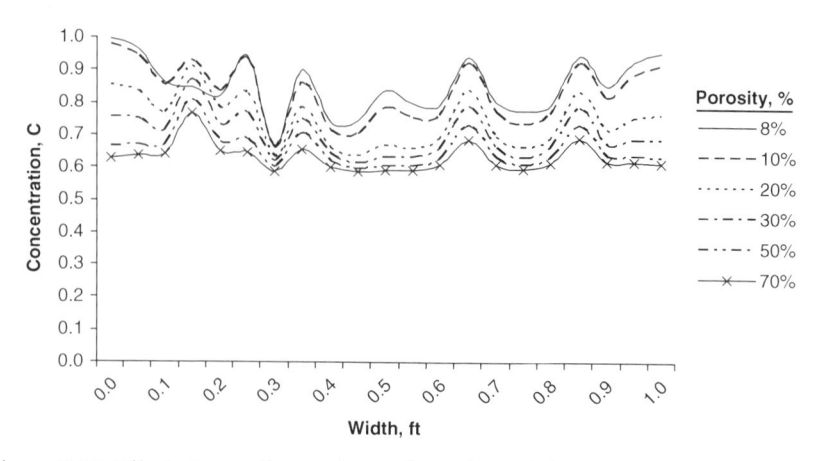

Figure 7.35 Effect of overall porosity on frontal instability after 0.03 day.

Modeling Viscous Fingering

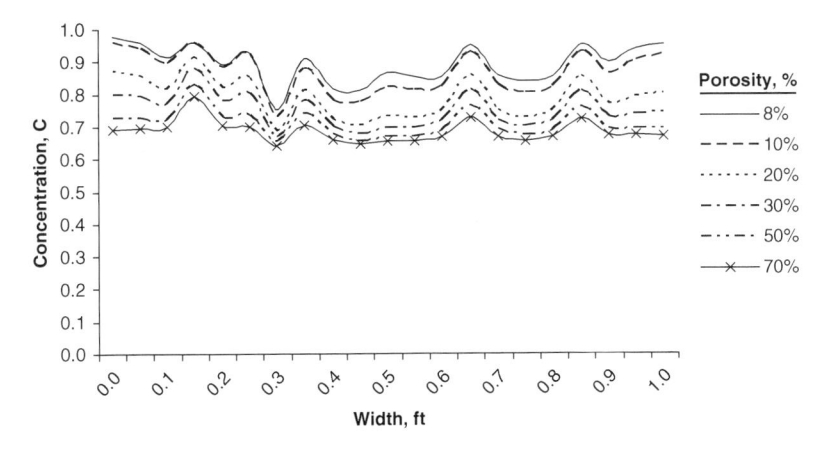

Figure 7.36 Effect of overall porosity on frontal instability after 0.04 day.

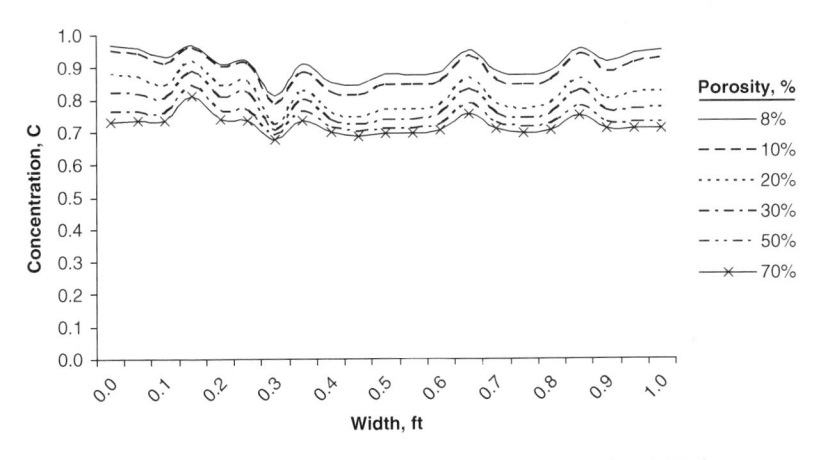

Figure 7.37 Effect of overall porosity on frontal instability after 0.05 day.

7.2.4.3 Effect of Mobility Ratio

For miscible cases, mobility ratio is the ratio of viscosities of the displaced fluid to the displacing fluid. It has already been shown (Blunt and Christie 1992) that the spreading of the solvent front is driven by the mobility contrast between the injected and initial fluid compositions. It was believed previously that the mobility ratio is the only controlling parameter of the phenomenon of viscous fingering but later work showed that there are other parameters, which are critical for the viscous fingering to happen. The later approach was further strengthened by Coskuner and Bentsen (1990). Their instability number covered a number of different parameters but the importance and the governing characteristic of the mobility ratio was still in the instability number.

To study the effect of mobility ratio on the extent of frontal instability of the concentration, different numerical simulations are run. In these simulations, five different mobility ratios are used.

Figures 7.38 through 7.42 show the concentration distribution along the whole width of porous medium at a distance of 0.1 ft from

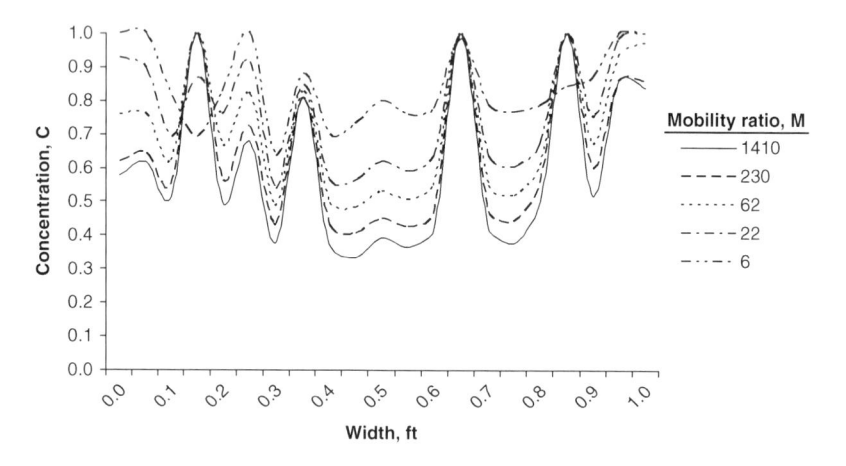

Figure 7.38 Effect of mobility ratio on frontal instability after 0.01 day.

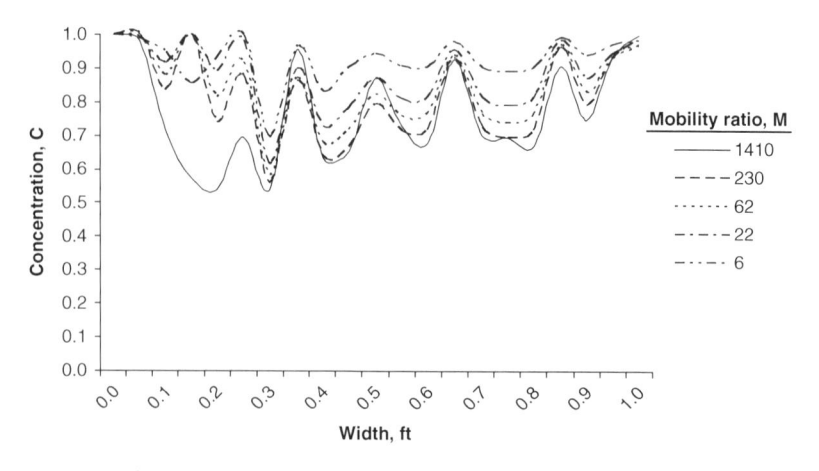

Figure 7.39 Effect of mobility ratio on frontal instability after 0.02 day.

Modeling Viscous Fingering

the inlet face after different time intervals. It is evident from all these figures that the intensity of concentration fingers or the concentration contrasts along the whole width of porous medium increases with the increase in mobility ratio. The maximum mobility ratio used is 1410 while the minimum case is of 6. In the case of maximum mobility ratio, even after 0.01 day of injection, maximum concentration at a few points is unity (breakthrough) while the minimum value of concentration is about 0.35. On the other hand, in the case of lowest mobility ratio the concentration reached unity at some points.

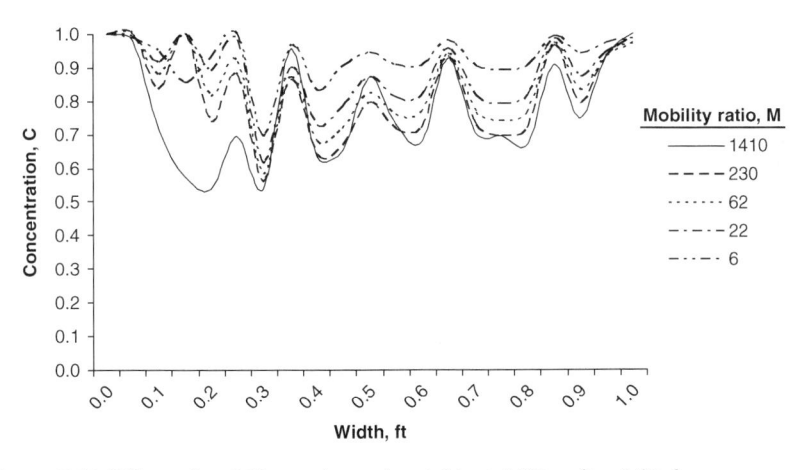

Figure 7.40 Effect of mobility ratio on frontal instability after 0.03 day.

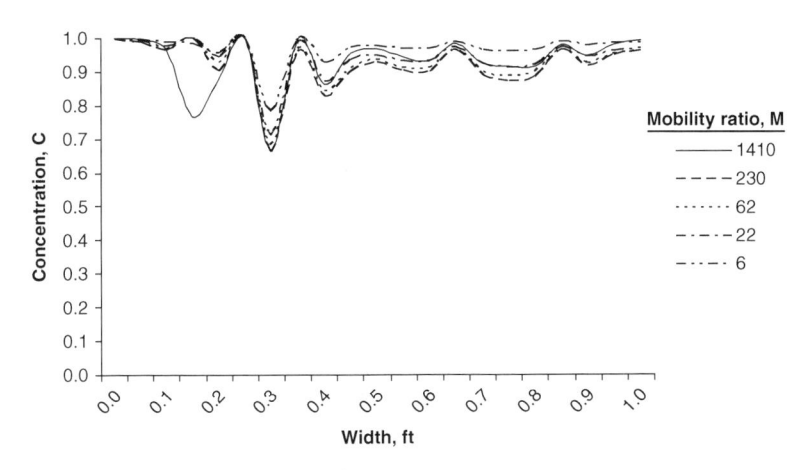

Figure 7.41 Effect of mobility ratio on frontal instability after 0.04 day.

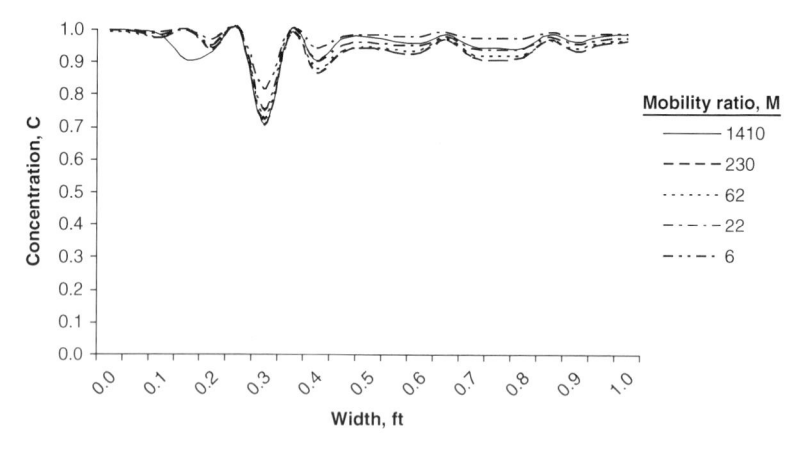

Figure 7.42 Effect of mobility ratio on frontal instability after 0.05 day.

But the minimum concentration along the length of the porous medium was about 0.7 which was quite higher as compared to the maximum mobility ratio case. This low contrast between the higher and lower values of concentration along the width of porous medium indicates that the intensity of fingering or the extent of fingering is low in the case of low mobility ratio or in other words, the difference in the concentration of base and tip of the fingers is quite low in the case of low mobility ratio. The remaining figures show the same phenomenon with the only difference being in the value of concentration, which increases with the passage of time due to the continuous injection of displacing fluid. The intensity of fingers or the instability of concentration along the width of porous medium decreases with the passage of time in the direction of flow. The reason for this numerical stability is the ideal homogeneity of porous medium used in the numerical model. Though the instability invoked by the variation of permeability at the inlet face causes the system to behave unstably but as time passes, the front approaches and encounters homogeneity of the porous medium; therefore, the instability decreases and ultimately the system becomes stable again.

7.2.4.4 Effect of Longitudinal Dispersion

The contribution of longitudinal dispersion to a flow scheme becomes of extreme importance in porous medium where the flow velocities are very low. Even at very low velocities, the whole

MODELING VISCOUS FINGERING

displacement scheme is controlled by the longitudinal dispersion along the direction of flow. Although the role of longitudinal dispersion in any flow scheme is obvious, its contribution towards the phenomenon of viscous fingering needs further discussion. To study the effect of longitudinal dispersion on viscous fingering phenomena, a set of simulations are run varying its numerical value and keeping other variables constant. The parameters, which are kept constant, are given in the following Table 7.4.

The dependency of the displacement efficiency on the magnitude of longitudinal dispersion coefficient is evident in the Fig. 7.43, which is a plot of concentration along the width of the porous

Table 7.4 Parameters used to study effect of longitudinal dispersion

Size of Porous Medium ft x ft	P_{in} Injection Pressure Psia	Ø Porosity %	Dx Transverse Dispersion Coefficient ft²/d
1 x 2	30	50	0.1

Note: Permeability at the inlet face was varied randomly (10% variation).
Values of Dy used: 0.03, 0.07, 01, 0.5, 0.7 ft²/d.

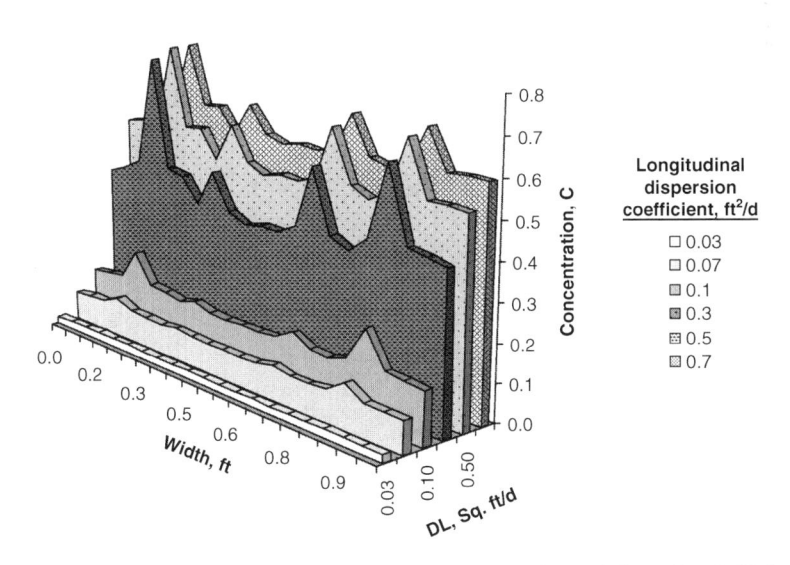

Figure 7.43 Effect of longitudinal dispersion on frontal instability after 0.02 day.

medium at a point 0.1 ft away from the inlet face after 0.01 days of injection. This plot illustrates that the breakthrough time decreases with the increase in the dispersion coefficient. This rapid increase in the concentration with the increase in the longitudinal dispersion is a measure of the effectiveness of the longitudinal mixing of the displacing fluid into the displaced fluid. If the extent of frontal instability is observed in Figs. 7.43 through 7.47 after different

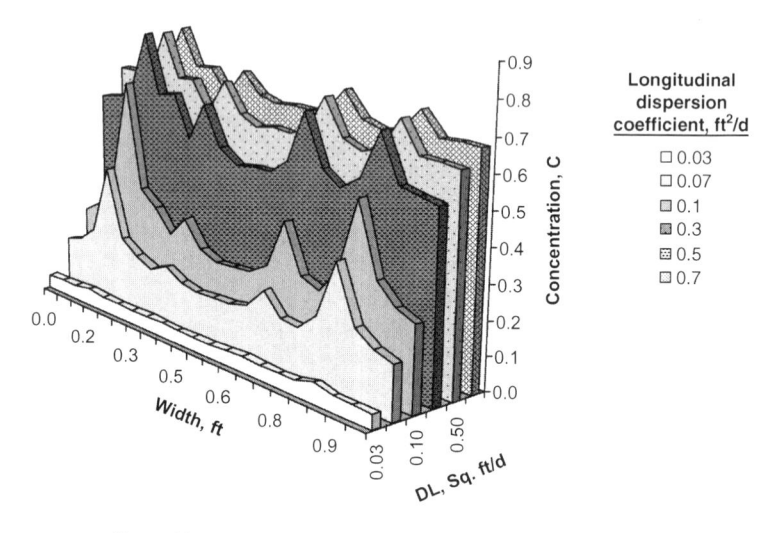

Figure 7.44 Effect of longitudinal dispersion on frontal instability after 0.03 day.

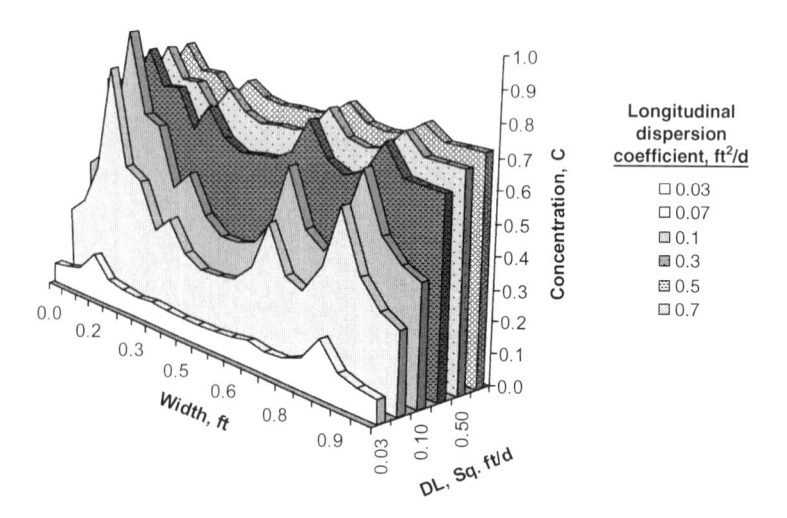

Figure 7.45 Effect of longitudinal dispersion on frontal instability after 0.04 day.

Modeling Viscous Fingering

time intervals, it becomes clear that the intensity of the fingering decreases with the increase in the longitudinal dispersion coefficient along with the breakthrough time.

Figures 7.48 and 7.49 show the effect of dispersion coefficients on the concentration profile under different injection pressures. It is clear in these two plots that the breakthrough time decreases with the increase in longitudinal dispersion coefficient. The increase in pressure also

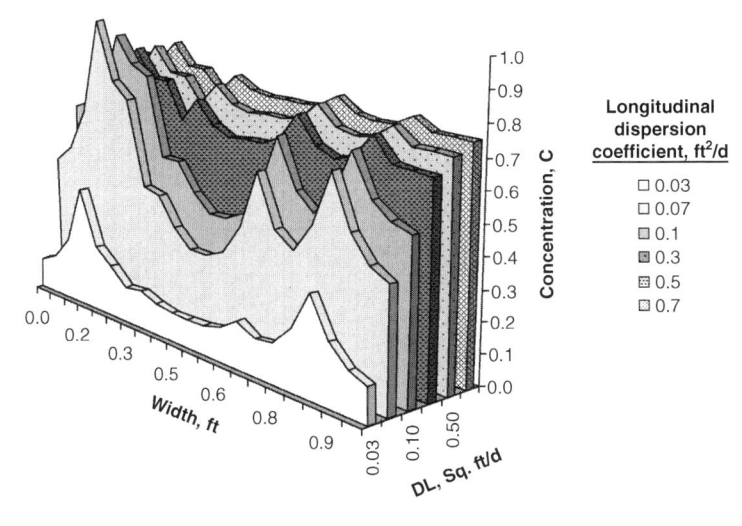

Figure 7.46 Effect of longitudinal dispersion on frontal instability after 0.05 day.

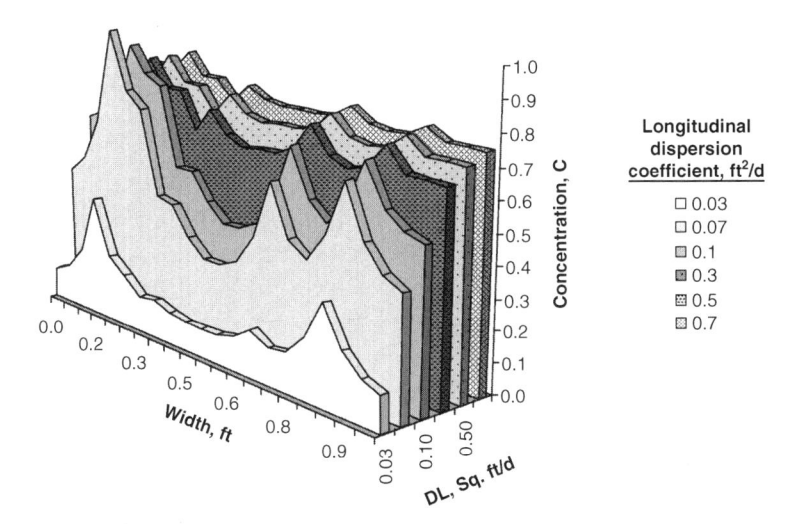

Figure 7.47 Effect of longitudinal dispersion on frontal instability after 0.05 day.

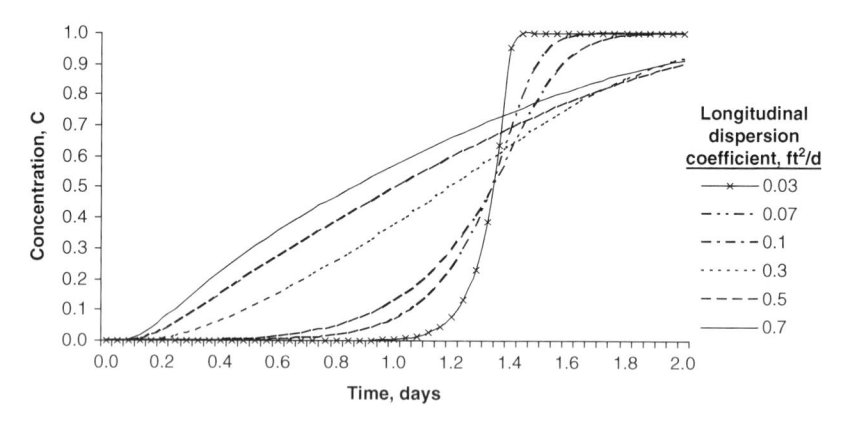

Figure 7.48 Effect of longitudinal dispersion coefficient on concentration profile with time at 1 ft from Inlet face – $P_{Inj} = 50$.

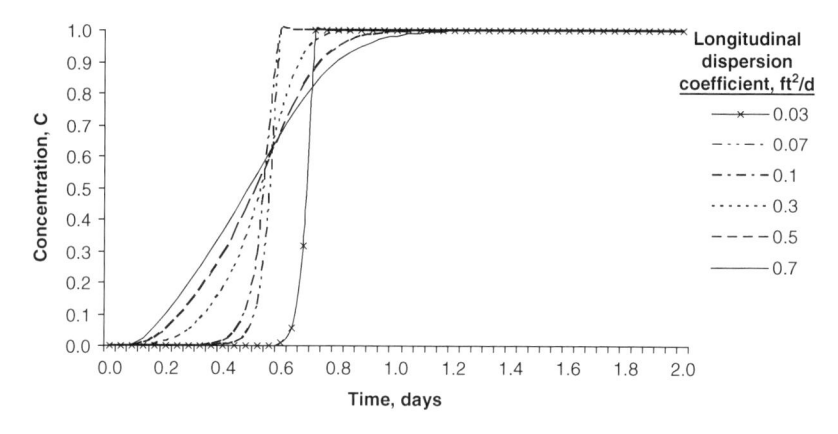

Figure 7.49 Effect of longitudinal dispersion coefficient on concentration profile with time at 1 ft from Inlet face – $P_{Inj} = 100$.

results in decrease in the breakthrough time. At low pressures, the impact of longitudinal dispersion coefficients becomes more dominant because of the decrease in the convective effect. At this point, the displacement scheme is mainly governed by the dispersion forces. The difference in the intensity of curvatures in these plots is due to the difference in the magnitude of dispersion coefficient.

7.2.4.5 Effect of Transverse Dispersion

It has been well established earlier that in most cases, transverse and longitudinal dispersion cannot be scaled because the scaling criterion leads to impractically large models in the lab. Therefore, the importance of a numerical model increases manifolds to accurately predict the behavior of a displacing scheme with such parameters which otherwise cannot be predicted easily. The important role of transverse dispersion in the phenomenon of viscous fingering is its diminishing effect on otherwise sharp fingered patterned flow. Increase in the transverse dispersion causes the spreading and mixing of the two liquids (displacing and displaced) in the transverse direction which results in diffused finger pattern or in other words reduces the extent of fingers of high concentration contrast. Different values of transverse dispersion have been used in this set of simulations to see the impact on the concentration profile achieved. For this purpose, the following parameters are kept as fixed while different values of transverse dispersion coefficients are played with numerically.

Size of Porous Medium ft × ft	P_{in} Injection Pressure psia	Ø Porosity %	Dy Longitudinal Dispersion Coefficient ft²/d
1 × 2	30	50	0.5

Note: Permeability at the inlet face was varied randomly with 10% variation.
Values of Dx used: 0.03, 0.07, 0.1, 0.3, 0.5, 0.7 ft²/d.

Figures 7.50 through 7.53 show the concentration plots along the whole width of porous medium at a distance of 0.1 ft from the inlet face. In Fig. 7.50, the initiation of fingers is visible in form of concentration instability in horizontal direction even after 0.01 day of injection. A very important and interesting observation in these plots is the flattening of the finger tips with the increase in the transverse dispersion coefficient or in other words in all these plots the instability in the horizontal concentration profile of the front decreases with the increase in the dispersion coefficient.

If the change in these concentration profiles is observed over time, it can easily be seen that the effect of the transverse dispersion becomes more dominant. As the numerical representation of the porous medium is heterogeneous at the inlet face only and the whole remaining medium is homogenous, therefore, the instability decreases with the passage of time when pressure distribution in the porous medium starts to become steady.

The transverse dispersion acts against the instability number in the formulation of the instability number given in previous chapters.

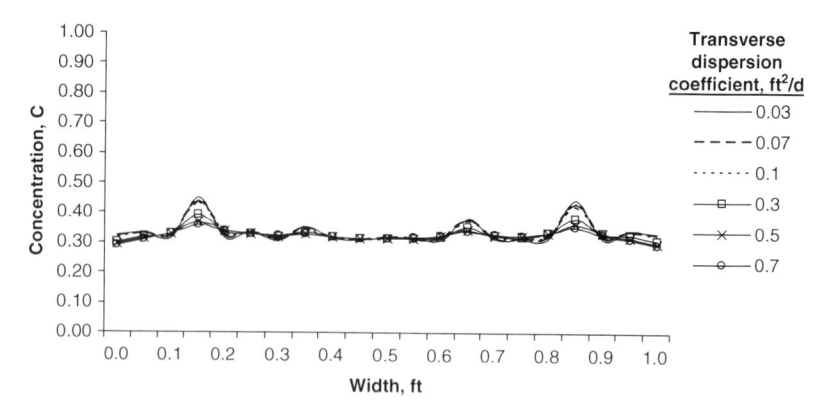

Figure 7.50 Effect of transverse dispersion on frontal instability after 0.01 day.

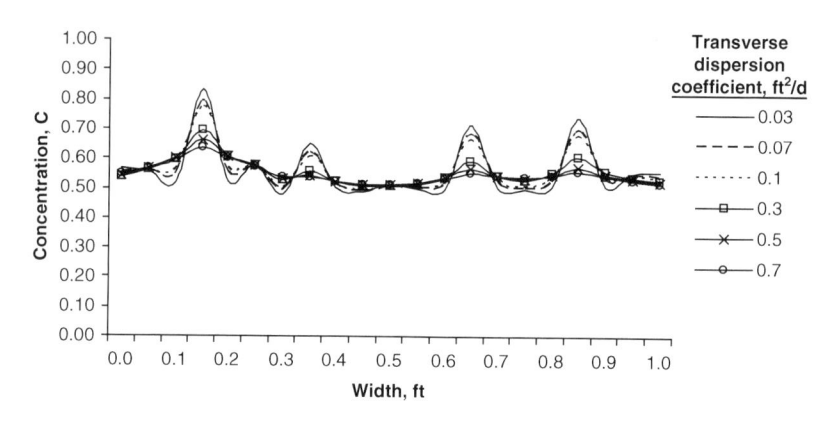

Figure 7.51 Effect of transverse dispersion on frontal instability after 0.02 day.

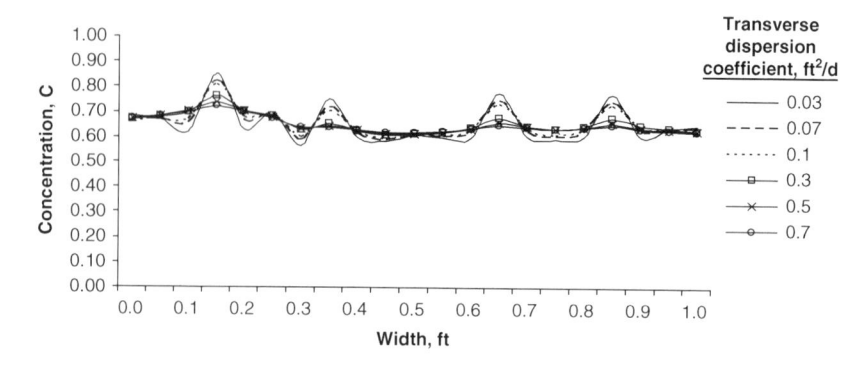

Figure 7.52 Effect of transverse dispersion on frontal instability after 0.03 day.

Modeling Viscous Fingering

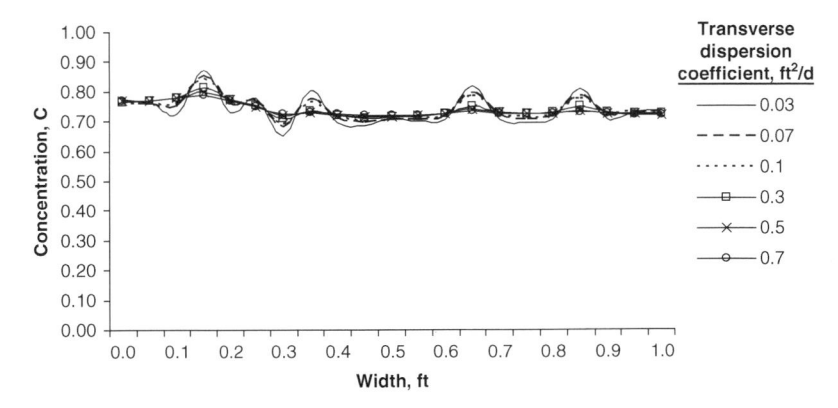

Figure 7.53 Effect of transverse dispersion on frontal instability after 0.04 day.

An increase in this coefficient results in the decrease of the instability number and ultimately the increase in the stability of the displacement front. The same qualitative dependency has been achieved by the numerical model discussed in this section.

7.2.4.6 Effect of Aspect Ratio

The instability number presented by Coskuner and Bentsen (1990) made it possible for the first time to infer explicitly the effect of the length of a porous medium on the instability of a miscible displacement. Because of the importance of dimensions of the porous medium in the instability number, which ultimately renders a flow scheme either stable or unstable, the effect of aspect ratio, the ratio between the height and width of a porous medium, has been studied numerically by running different simulations. Aspect ratio was varied by varying the length of the porous medium while keeping the width constant. Although, because of the memory limitation of the computer used, large aspect ratios could not have been simulated, with the used range of aspect ratios, some interesting results have been observed. In these simulations, 30 psia of injection pressure has been used. It is to be noted that the dependency of the remaining parameters on the dimension of the porous medium has not been included in these simulations. All other parameters are kept fixed while only the dimensions of the porous medium are varied.

Table 7.5 shows the parameters used in this set of simulations. All these parameters are kept constant while only the length of the porous medium is changed to vary the aspect ratio in these simulations.

Table 7.4 Parameters used to study effect of Aspect Ratio

Width of Porous Medium ft	P_{in} Injection Pressure psia	Ø Porosity %	K Permeability Darcy	Dx Transverse Dispersion Coefficient ft²/d	Dy Longitudinal Dispersion Coefficient ft²/d
1	30	50	0.1± 10%	0.1	0.5

Note: Permeability at the inlet face was varied randomly.
Aspect ratios used: 0.5, 1, 1.5, 2, 3.

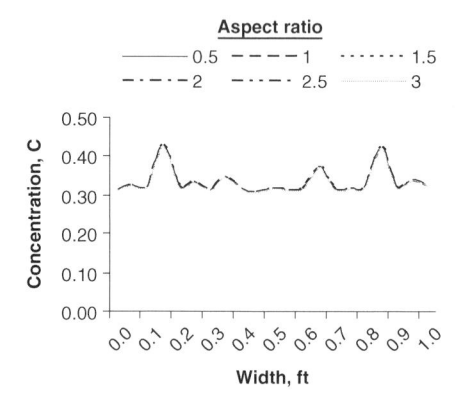

Figure 7.54 Effect of aspect ratio on frontal instability after 0.01 day.

Figure 7.54 is the plot of water concentration at a distance of 0.1 ft from the inlet face of the porous medium after 0.01 day of injection of displacing fluid (water). It is clear that there is no major difference in the concentration curves for different aspect ratios used in this case.

Figures 7.55 through Fig. 7.58 show the concentration of water along the transverse direction (width of the porous medium) at the same distance of 0.1 ft from the inlet face at different time intervals. Although there is no major difference in the propagation of instability, an increasing trend in the difference of the concentrations for two extreme aspect ratios can easily be noticed with the passage of time. This observation implies that fingers propagate faster in the case of higher aspect ratios. This numerical observation is also supported by the role of length of the porous medium in the instability

MODELING VISCOUS FINGERING

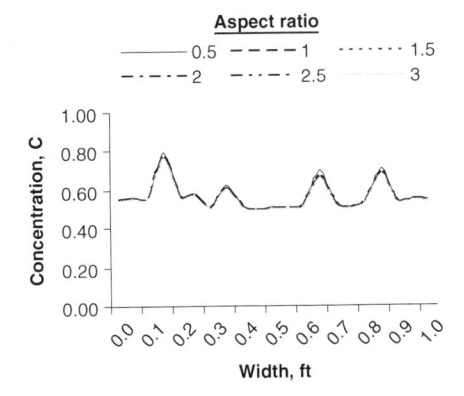

Figure 7.55 Effect of aspect ratio on frontal instability after 0.02 day.

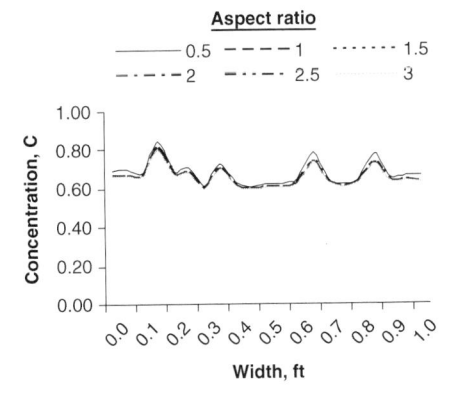

Figure 7.56 Effect of aspect ratio on frontal instability after 0.03 day.

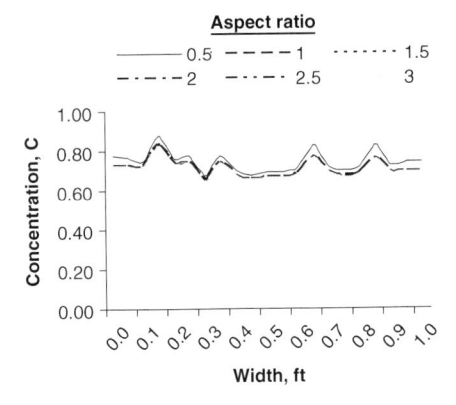

Figure 7.57 Effect of aspect ratio on frontal instability after 0.04 day.

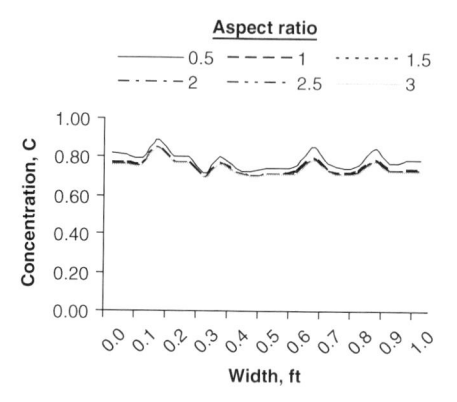

Figure 7.58 Effect of aspect ratio on frontal instability after 0.05 day.

number, in which the instability number increases with the increase in the length of the porous medium. The increase in the instability number means the increase in the frontal instability which is the extent of fingering. Therefore, it can easily be inferred from these numerical results that the length of the porous medium for very large aspect ratios (in field scale cases) plays a vital role in the phenomenon of viscous fingering.

7.2.5 Comparison of Numerical Modeling Results with Experimental Results

7.2.5.1 Selected Experimental Model

For the purpose of comparison between the experimental and numerical results, the first scenario of experimental runs has been modeled numerically. As the main objective of the model is to predict the efficiency of a miscible scheme, therefore, the emphasis has been given to simulate the breakthrough time and the distance travelled by the leading finger to quantify the results. It is important to note that most previous attempts to model viscous fingers focused on the shape of the finger. Our contention is, by comparing the most important outcome of an unstable process, *viz*, breakthrough, we are testing the most important feature of a predictive tool. Here, it suffices to show that the shape of the viscous fingering, as determined numerically, was similar to that observed experimentally.

7.2.5.2 Physical Model Parameters

A 2-D simulation model was applied to the rectangular porous medium. The important physical parameters obtained by experiment are porosity (Ø), permeability (k), injection pressure (P_{inj}) and the dimensions of the porous medium. Some other important parameters were also calculated based upon the data available in literature and the physical parameters measured during experiments. These parameters are shown in Table 7.6. It is to be noted here that the injection pressure for all these runs was kept constant. Porosity was measured for each run and, from data given in the following table, it is clear that the porosity remained almost constant for all experimental runs under this scenario. The instability number has also been calculated for the selected case to see the stability of the displacement scheme according to the basis of instability theory. The data shows that the instability number appropriate parameters attained by either suitable measurements or by authentic data from literature.

According to Whitaker (1967), the ratio between the longitudinal to transverse dispersion coefficient for a homogeneous and uniform medium can be taken as 3. Therefore, for numerical purposes the values of transverse dispersion coefficient have been used as 1/3rd of the longitudinal dispersion coefficient. The diffusion coefficient for the glycerine solution has been taken from the data given by Bouhroum (1985). This data is reproduced in Fig. 7.59. For the simulation purposes, diffusion coefficient for glycerine solution is taken from the data given in Fig. 7.59. Longitudinal dispersion coefficients are calculated for different experimental runs at different mobility ratios

Table 7.6 Simulation parameters – Scenario 1

Glycerine Concentration Fraction	Injection Pressure pisg	Velocity m/sec	Mobility Ratio M	Instability Number I_{sr}	Porosity Ø	Permeability Darcy
0.0	0.5	0.00124	1.0	0	0.522	0.3
0.4	0.5	0.00053	3.7	50	0.514	0.3
0.6	0.5	0.00019	10.8	952	0.515	0.3
0.8	0.5	0.00008	61.6	1073	0.527	0.3
1.0	0.5	0.00004	1410.2	1525	0.508	0.3

Note: Particle size used in Scenario is 100 μm.

[Figure: Concentration and temperature dependence of molecular diffusion coefficients for glycerine solution, with curves for 20 deg. C and 25 deg. C. X-axis: Concentration in mass fraction of glycerine (0.0 to 1.0). Y-axis: Molecular diffusion coefficient D_m (× 10^{-5} cm²/sec) (0.0 to 1.0).]

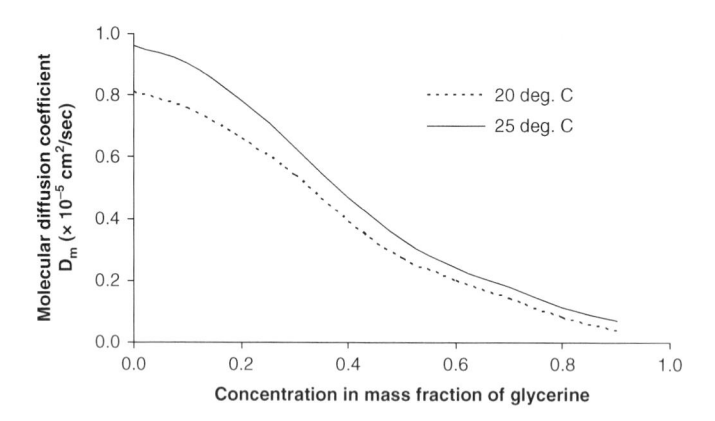

Figure 7.59 Concentration and temperature dependence of molecular diffusion coefficients for glycerine solution *(after bouhroum, 1985)*.

Table 7.7 Input data used for modeling viscous fingering

Glycerine Concentration fraction	Molecular Diffusion m²/sec	Peclet Number, Pe	Longitudinal Dispersion Coefficient, D_L m²/sec	ft²/d
0.0	9.6E-10	129	1.04E-06	0.740
0.4	4.2E-10	127	4.41E-07	0.315
0.6	2.7E-10	71	2.97E-08	0.021
0.8	1.1E-10	69	1.19E-08	0.008
1.0	4E-11	89	4.64E-09	0.003

Note: Particle size used in Scenario is 100 μm.

with the help of calculated Peclet number given in Table 7.7 and using Fig. 7.60. These calculated longitudinal dispersion coefficients are given in Table 7.7.

7.2.5.3 Comparative Study

In this section, a necessary comparison of the experimental work is being presented. The primary purpose of this chapter is to draw

Modeling Viscous Fingering

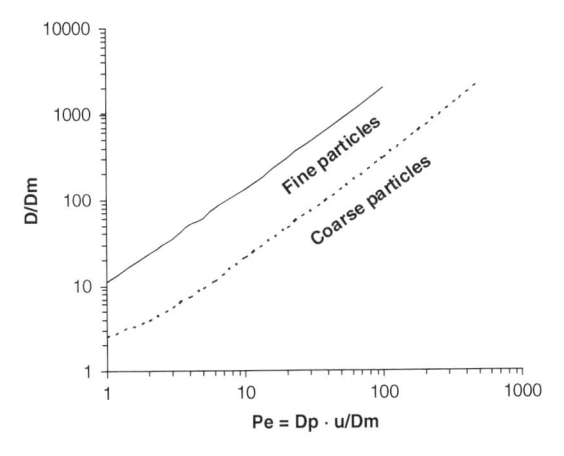

Figure 7.60 Range of dispersion coefficients based on tracer tests for various particle sizes: fine 200–270 mesh, coarse 40–200 mesh (after mannhardt et al, 1994).

some conclusions based on the limited number of runs made during the current study.

The propagation of leading finger tip is simulated for different cases. Figure 7.61 shows the propagation of finger tip with time for the case of mobility ratio of 1410 which is the worst case (I_{sr} = 1525) as far as the stability of the displacement is concerned. A good agreement was found especially in the initial time. At the later times the simulated results showed deviation leading to a late breakthrough as compared to the experimental observation. The apparent reason for this deviation is the homogeneity of the numerical model which causes the effect of fingering to diminish with time. The impact of heterogeneity invoked near the inlet boundary/face by introducing the random distribution of permeability reduces with the passage of time as the disturbance in the system tends to stabilize causing the so-called fingers to suppress.

Numerical results in the 2nd case in which 80% glycerine has been used to reduce the mobility ratio to 62, shows reasonable agreement with the experimental results. The deviation of the numerical results from the observed propagation is less in magnitude as compared to the previously described case. In this case, a better match can be observed in the later times also. The reason of this better comparison is the low mobility ratio between the displacing and displaced fluids which causes the intensity of fingers/concentration contrast to some what low even in the initial length of the model and therefore, with the passage of time the homogeneity of

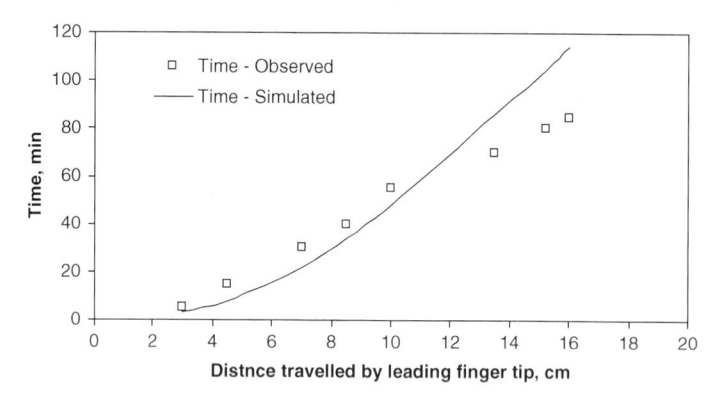

Figure 7.61 Propagation of leading finger tip (100% glycerine solution case).

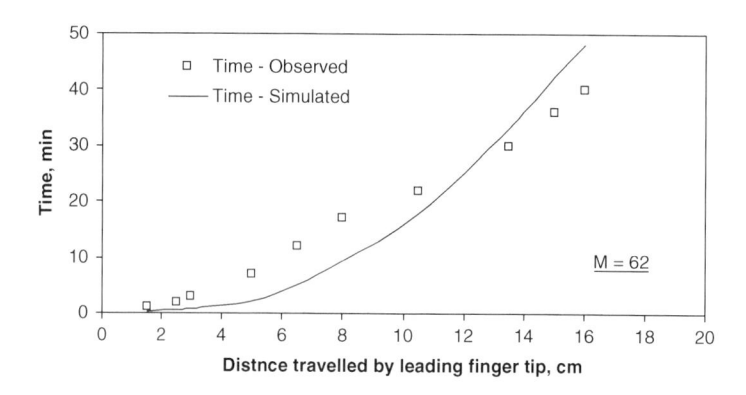

Figure 7.62 Propagation of leading finger tip (80% glycerine solution case).

the remaining porous medium plays a less intense role in stabilizing the flow.

Another important parameter which can cause the deviation of numerical results is the use of fixed longitudinal and transverse dispersion coefficients during one simulation. In fact, dispersion coefficient is a tensor, the value of which changes with the concentration and the viscosity of the fluid. These parameters change throughout the porous medium due to the mixing of displacing and displaced fluids, therefore, the use of constant dispersion coefficients may result in the deviation of numerical results.

The third case for which the propagation of the leading finger tip is modelled is the case of 60% glycerine solution where the mobility

Modeling Viscous Fingering

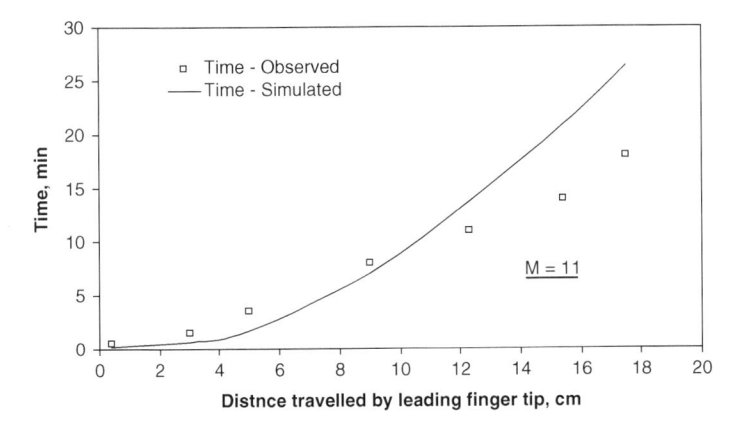

Figure 7.63 Propagation of leading finger tip (60% glycerine solution case).

ratio is 11. The impact of this low mobility ratio is visible in the numerical results shown in Fig. 7.63 where the both experimental and numerical results match almost ideally at the initial time intervals. The deviation of the numerical results from the experimental observations at later times can be a result of the increase in longitudinal dispersion coefficients due to low viscosity contrast because of the continuous injection of displacing fluid. This low longitudinal dispersion coefficient causes the finger to change its velocity and results in early breakthrough. In the numerical simulation, though, this change cannot be captured because of the constant dispersion coefficients used throughout the porous medium.

Another important parameter, which can be used as a measure of the success of any displacement scheme, is the breakthrough time. The breakthrough times for the four cases of first scenario are computed numerically and compared with the breakthrough times observed experimentally. Fig. 7.64 illustrates a simple comparison of the observed and simulated breakthrough times for different displacement runs under different mobility ratios.

It is clear from this figure that a good qualitative agreement is present in these results. The numerical simulations results in slightly higher breakthrough time in all the cases. These deviations increase with the increase in mobility ratio. The reason for these deviations is the same as to what has already been discussed. Microscopic heterogeneity and the change in dispersion coefficient

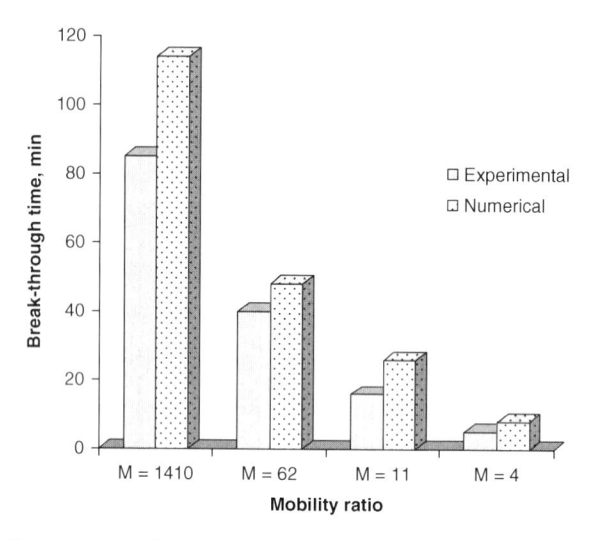

Figure 7.64 Comparison of experimental and numerical breakthrough time - Scenario - 1 (100 μm particle size).

are not accounted for/incorporated in the current model and, therefore, the results exhibit poor agreement at the later times especially in predicting breakthrough.

7.2.5.4 Concluding Remarks

Although, in numerical simulations, effort has been made to use exactly the same parameters which were involved in experimental runs, but there are still limitations because of which the results are not fully comparable with the experimental observations.

One of the reasons is probably the use of improper molecular diffusion coefficients of glycerine solution, which are dependent on both the concentration and temperature of the solution. Although real data has been used, the concept of averaged concentration to have the diffusion coefficient may lead to the discrepancy in the results because of the changing concentration at each point of the porous medium.

Another important limitation of the current numerical model is that it cannot capture or account for the microscopic heterogeneity, which results in the difference between the microscopic fluid velocity and the macroscopic fluid velocity.

The difference between the microscopic and the macroscopic fluid velocities is also a main contributing factor towards the difference

Modeling Viscous Fingering

in the results. This difference is due to the microscopic heterogeneity. This variation in velocity at the microscopic scale develops due to the following three mechanisms:

1. Fluid particles travel at different velocities at different points along a cross section of single pore channel because of the roughness of the pore surfaces and the viscous fluid.
2. Discrepancies in pore dimensions (including dead-end pore) cause some fluid particles to move faster than others.
3. Variations in pore geometry (including tortuosity, branching and fingering) along the flow paths result in fluctuation of the microscopic flow velocity with respect to the macroscopic average flow velocity.

Obviously, the macroscopic average velocity is generally different, both in direction and in magnitude from the real microscopic fluid velocity at each point within the REV. At some points, the difference may be quite large. Therefore this difference between the two fluid velocities causes inevitable spreading of displacing fluid particles beyond what could be associated with respect to the macroscopic average flow.

Because of the branching, spreading occurs both in the direction of bulk flow (longitudinal dispersion) and in directions perpendicular to the flow (transverse dispersion). Not only branching but other molecular variations of pore channels such as tortuosity interfingering and dead end contribute to the velocity difference at the microscopic scale as well.

The second velocity discrepancy exists at the microscopic scale. Because of molecular diffusion (due to concentration gradients), which occurs during the fluid flow and does not depend upon the fluid flow, displacing fluid particles move with different velocities.

8

Towards Modeling Knowledge and Sustainable Petroleum Production

This chapter provides one with a list of challenges in modeling true knowledge. In doing so, the chapter summarizes future research topics that must be included in order to move forward with knowledge-based reservoir simulation. It is shown that with the proper criterion, true knowledge can be identified to help with correct decision making. The nature of knowledge-based decision as opposed to prejudice-based decision making is discussed. It is shown that with the knowledge-based approach, results are significantly different for most of the solution regime. This finding would help determine a more accurate range of risk factors in petroleum reservoir management. With the proposed technique, sustainable petroleum operations can be modeled to yield different results from unsustainable practices. This also allows one to distinguish between materials that cause global warming and the ones that don't. Such distinction was not possible with conventional modeling techniques.

8.1 Essence of Knowledge, Science, and Emulation

In their recent work, Zatzman and Islam (2007) and Zatzman et al (2007a) proposed scientific definition of truth and knowledge. They stated that knowledge can only be achieved with true science. A process or an object is true only if it has three real components, namely 1) origin; 2) process; and 3) end. For instance, for an action to be true, it must have a real origin (true intention); followed by real process (process that emulates nature); and real end (in line with nature that constantly improves with time). How can an intention be real or false? If the intention is to go with nature, it is real. Instead, if the intention is to fight nature, it is false. For a process to be real or sustainable, it must have the real source. This would eliminate all non-natural sources as the feedstock. With this analysis, genetically modified seed is automatically artificial whereas fossil fuel as the source is real. Other than the source, the process itself has to be real, which means it has to be something that exists in nature. For instance, light from direct burning of fossil fuel is real, whereas light from an electric lightbulb is artificial, hence unsustainable. Similarly, nuclear energy is artificial because it uses enrichment of uranium that is inherently anti-nature whereas fossil fuel is real because the process is real (e.g. thermal combustion). This analysis was recently elaborated by Chhetri and Islam (2008).

Zatzman and Islam's work outlines fundamental features of nature and shows there can be only two options: natural (true) or artificial (false). They show that Aristotle's logic of anything being 'either A or not-A' is useful only to discern between true (real) and false (artificial). In order to ensure the end being real, they introduce the recently developed criterion of Khan (2006) and Khan and Islam (2007a). If something is convergent when time is extended to infinity, the end is assured to be real. In fact, if this criterion is used, one can be spared questioning the 'intention' of an action. If in any doubt, one should simply investigate where the activity will end up if time, t goes to infinity. The inclusion of a real (phenomenal) pathway would ensure the process is sustainable or inherently phenomenal.

8.1.1 Simulation vs. Emulation

In mathematics, it matters whether in principle, a solution can be found to a differential equation, but the theorem that demonstrates that a solution or solutions can exist provides neither a method for solving any particular differential equation (PDE) nor whether any solution will be found to some given differential equation by any known or existing method. The more closely a series of PDEs seem to model the results of some natural phenomenon, the more convinced we are of its applicability. Why should this be so? Mathematically (including numerically), for such PDEs, a finite number of solution-classes can be demonstrated, and the possibility of some finite number of singular solutions is not foreclosed. The natural phenomenon being modeled, however, probably has an infinitude of solutions if only because, with time in nature itself being unbounded, both before the period in which study of the phenomenon is undertaken as well as aeons into the future, many other things – anything – could yet happen or has happened with this phenomenon about which our PDEs will suggest nothing about and know nothing of.

Apart from the entire fraught question of the limitations built into the measuring devices used to observe on our behalf the output of some natural phenomenon in the first place, why in principle should it be concluded that a "result" in physical reality that is similar or closely similar to the predictions of a PDE or set of PDEs verifies the mathematical model and – or any of its underlying assumptions?

There is a huge leap of logic taking place here behind the scenes. Consider the syllogism:

"All Americans speak French" – "Nocolas Sarkozy is an American" – "Therefore, Nicolas Sarkozy speaks French"

If one obtains the same result from observing nature as from crunching a PDE to its conclusion, we may decide that result is meaningful and true: "Nicolas Sarkozy speaks French." How this comes to be is another matter entirely. The fact that the PDE gave the same result as nature tells us nothing about the assumptions built into those PDEs. They could have been founded on completely false or incorrect notions (e.g. "All Americans speak French" and "Nocolas Sarkozy is an American").

In nature: EVERYTHING has its consequence, its impact, its solution(s); the output of one stage is the input in whole or in part of some subsequent stage; and no natural phenomenon is actually linear or finite, however much it may appear to be for an external observer applying some arbitrary duration of time to observing the phenomenon.

No formal mathematical functional description can be more, or any better, than a partial analog of any collection of processes in nature that can be meaningfully grouped together as a "function". The analogy of the most fully-applicable or comparable mathematical function used to describe some natural process or processes is similar at most and at best only in FORM.

This similarity in FORM tells us nothing about actual CONTENT, of how the natural process operates. Let us apply this to a very simple but clear example. If, as the result of some natural process, something becomes divided into three roughly equal parts, we can apply to the result the mathematical terminology of "one-third". We say 'roughly' because dividing anything into exact fractions is an absurd concept. There is no physical mechanism known to mankind today that will allow us to divide anything into exactly equal number of parts, making sure that all components, including quarks and all the other particles that are yet to be discovered. Initially, if we have a process that accumulates a result by taking 30% of some initial quantity and then adds increments formed from taking one tenth of the new total at each subsequent step, i.e. an infinite series starting at 0.3 and then adding 0.03, 0.003, 0.0003 etc., I will end up with 0.3333333333333333333333 ..., i.e. one-third. The first process involves division into three. The second process involves repeated division by 10 of an initial 30% division. Mathematically, they give the same result. Physically, they are very different, and totally unrelated, processes. Is a mathematics that applies the same terminology to entirely different processes because their end-result "looks the same" good enough for a science of nature as it actually is? Elsewhere we have asked whether it is true, complete, accurate, or good enough for a proper science of nature to operate according to the assumption that "chemicals are chemicals". Here we similarly have to ask whether it is true that "wholes are wholes" or "fractions are fractions".

8.1.2 Importance of the First Premise and Scientific Pathway

Figure 8.1 shows the importance of the first premise. The solid circles represent phenomenal first premise, whereas the hollow circles represent aphenomenal first premise. The thicker solid lines represent scientific steps that would increase knowledge, bringing one closer to the truth. At every phenomenal node, there will be spurious suggestions that would emerge from aphenomenal root (e.g. bad faith, spurious model approximations), as represented by dashed thick lines. However, if the first premise is phenomenal, no node will appear ahead of the spurious suggestions. Every logical step will increase knowledge, following the true path of science. The thinner solid lines represent choices that emerge from aphenomenal first premise. At every aphenomenal node, there will be aphenomenal solutions, however each will lead to further corruption of the logical process, radically increasing prejudice. This process can be characterized as disinformation – the anti-thesis of science (as a process). At every aphenomenal node, anytime aphenomenal solution is proposed, it is deemed spurious because it opposes the first premise of the disinformation mode. Consequently, these solutions are rejected. This is the inherent conflict between good faith and bad faith starting points. In the disinformation mode, phenomenal plans have no future prospect, as shown by the absence of a node.

Towards Modeling Knowledge

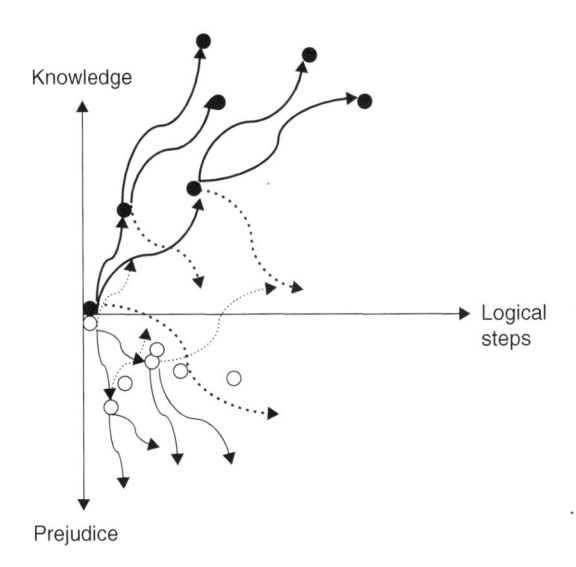

Figure 8.1 The role of first premise in determining the pathways to knowledge and prejudice.

One important feature of the disinformation mode is, as logical steps progress, the true nature of the first premise becomes unraveled, requiring one to invent further disinformation to justify the logical conflicts. In this mode, the more logical steps are taken, the further away one becomes from the truth, consolidating prejudices that only serve to justify aphenomenal conclusions that support the original bad-faith judgment.

As an example, consider the following steps:

Step 1. All Americans speak French {major premise}.
Step 2. Nicolas Sarkozy is an American {minor premise}.
Step 3. Therefore, Nicolas Sarkozy speaks French {aphenomenal conclusion}.
Step 4. Because Nicolas Sarkozy speaks French, Nicolas Sarkozy is an American.
Step 5. Therefore, all Americans speak French.

After Step 3, even the logic becomes aphenomenal. However, if the pathway or the time functions is ignored, this forms a recipe for disinformation. With this mode, all logics end up with confirming the first premise that was aphenomenal. The most important tactic of the disinformation model, therefore, obstructs any attempt to determine the pathway (the time function) of the logical steps.

8.1.3 Mathematical Requirements of Nature Science

Knowledge of etymological origins has to be supplemented in this case with awareness of the shift in conceptions of mathematical knowledge that took place independently and continuously. As far as mathematical conceptions are concerned, the important source languages are Arabic and Sanskrit for operations on numbers. Greek mathematics conceived of mathematical operations as construction and separation of figures in geometric space. Separately and independently of this, there was a usage of notions of addition and multiplication as part of commercial calculations, in trade – largely borrowed or adapted in the Greek world from some Chinese root-words and concepts. Division and subtraction also stem from practices in commercial trade, and with the Romans this is extended to military organization (subtraction is from a Latin word). It is during the European Middle Ages, all the terms for arithmetic operations are replaced with their Latin equivalents, from which we get addition, subtraction, multiplication and division in all the European languages. Until the Renaissance, algebra was considered to be Arab mathematics, geometry was considered to be Greek mathematics. Commercial arithmetic was not considered part of mathematics. These barriers in thinking ended only after the link between algebra and geometry was made explicit by Descartes in the early 1600s (17th century).

At the nexus of the tangible and the intangible stands time-consciousness. This constitutes perhaps the most essential content of human intention. We are impelled to act on an intention according to either time t = 'right now', or some other projection of finite time-duration, or some indefinite time-duration, including an infinite duration of time. This means that time, and consciousness of its passage, are each intimately bound up with every intention. Expanding this discussion that has opened concerning the root meanings of mathematical terms, we may consider computation and calculation, so far as human intentions are concerned, as tools for mapping our position within some process.

In this context, a related question emerges, *viz.*, why does the Eurocentric development of notions of arithmetic (including the terms for the operations of +, −, × and /) always abstract and only in the direction of tangibles, that is, 1 + 1 must give two tangible entities and no other possibility? In fact, however, it is the only interpretation allowed. We are always told: no, it is "two" in the abstract, and the tangible application is just one interpretation. In fact, however, it is the only interpretation allowed. In European languages, this was subtly implanted by treating numerical quantity as the property of the subject or object to which it refers. The Russian language, for example, still retains the feature all the other European languages had at one time, of referring to more than one of anything in the possessive case. Thus, "two books" was actually, grammatically, constructed as the phrase "two of books", etc. The languages of many indigenous native tribes do not treat numbers this way at all, or even distinguish more generally between singular and plural forms. A discussion on what 'addition' means in Sanskrit ("yugas") shows, on the contrary, that it is entirely reasonable to still think abstractly but not have to arrive necessarily back at this tangible "2" at all. This result points to the following conclusion: there is abstraction where one goes outside the Self, and there is abstraction where one isolates oneself from everything else that surrounds the Self.

This latter form of abstraction is a form of Eurocentrism. The notion that solutions, or solution-sets, are complete and/or unique is thus a similarly Eurocentric concept of "solutions". Solutions that are contingent on position or pathway in time and/or in space cannot and will not be unique, let alone complete. All solutions to analytic formulations of phenomena as they occur in Nature must be solutions of this type, and the analytic formulations must be suited to fulfilling this criterion. Non-linear describes one entire broad class of candidate formulations of this kind.

If one is to conduct meaningful and coherent mathematical analysis of phenomena in Nature as they actually occur and exist, it is not only entirely possible, but necessary, to be unable to add or perform any other mathematical operation "properly" in the sense of always arriving at solutions that are complete or unique. The closest the European scientific tradition came so far is Einstein's formulation of time as a fourth dimension. This has been fitted onto all existing mathematics and analysis by saying: we have three spatial dimensions and time as a fourth dimension. Why not reposition the description thusly: we have three dimensions in which Δt can go to zero and a fourth in which Δt can go to infinity?

The philosophic essence of Eurocentric thought is the isolation of Self from the Other. In the European traditions of mathematical and physical sciences, this is expressed in a definite form. That form consists, generally speaking, of conceiving space without any temporal component. What we call tangible is the copy registered at the brain of perceptual data input about the external world, but the dimensional content of this copy is actually "space", taken to represent the phenomenon as it exists in the external world, i.e. lacking any temporal component. This is the essence of Eurocentric mathematics and science. It is based on perception (time = 'right now') instead of knowledge (time = infinity). Even the grade-school definition of mass repeats this idea thus: "Mass is that which has weight and occupies space." Now consider, by way of contrast, the following alternative definition of mass, within Nature (where the temporal dimension cannot be dispensed with or evaded) as that which has weight and occupies time, including all tangible spatial dimensions.

Humans can become conscious of the temporal component. As far as we have been able to determine, however, everything else on earth (excluding humans and their ancestors) is, or was previously, alive without any consciousness of time or its passage. Of course, it is entirely possible for human beings to function in many discrete tasks without any specific or acute awareness of time or its passage. Collective tasks can even be organized and routinized to the point where, at some level or other, many of the participants accomplish their portion of the overall work without any particularly acute time-consciousness. In general, however, the more conscious we are – including time-conscious – the better things go for everyone involved.

Now consider what the operations of conventional arithmetic represent in this context. Addition is a discrete operation, absent any time component. Multiplication is repeated addition, but this repetition is itself stripped of any actual temporal component. Similarly, subtraction is reversed addition (addition of negative numbers) and division is repeated subtraction, both again lacking any temporal component. All that is required is that the operation ends in some result. If the result is not satisfactory, meaning does not fit the intended outcome, the problem is quickly tagged as 'ill-posed'.

Whether the operation can be performed depends not on time, but on the kind of numbers on which the operation is applied: is that number-type closed for that type of operation? That is: when one adds this integer to that integer, is the answer still an integer? Accordingly, collections of numbers (the integers, the rationals, the reals, etc.) are defined, and extended, according to whether they are closed for these operations. For example, subtraction of a greater natural number from a smaller natural number cannot give an answer in the natural numbers, but extending the natural numbers to the integers, all subtractions will produce a result.

Going in the other direction, so to speak, the algebraic structure, or underlying structural properties, of these collections (fields, rings, etc.) can and are defined according to properties of the operator, rather than the properties of the numbers operated upon. Here, still, only the most abstract notion of time, namely, the sequence in which an operation is applied, is even implicitly permitted.

Thus, one typical defining characteristic is whether one or more binary operators give the same result regardless of the order in which they are applied ('communtative law'). For example, the integers are said to form a field in part because there is an operation called addition which is commutative, i.e. $(-13) + 2$ gives the same answer as $2 + (-13)$, and there is a second operation called multiplication which is similarly commutative. The natural numbers are closed for subtractions of the type "$7-3$" or "greater–lesser", but not defined for "$3-7$", i.e. "lesser–greater". When the natural numbers are supplemented to include all the integers, however, "$7-3$" and "$3-7$" now each yield a result, but subtraction is not commutative because the answer to these subtractions is different.

Another consideration at the level of algebraic structure, in which time in the abstract is again implicit and disconnected from any natural reality, is whether the application of the binary operation to a third or any further member(s) of the number-collection yields a different result if the order is changed in which the operation is applied ("associative law"). This sequencing property implies time only in the sense of before or after: thus the integer field has two operators, addition and multiplication which are each associative in the sense that $(3+17) + (-29) = 3 + (17+(-29))$, and the same if we replace the "+" with "×". If this kind of computation is to be applied to quantities or results observed in some natural process, there has to exist other information, independent of this computation, about any further effects of the passage of actual time back in that process on the quantities being used to represent the output of the process or any stage(s) of it.

What if the quantities being used as inputs for these computations are continuing to vary within some natural process? Under such conditions, the fact that there exists a more or less complete and systematic logic for any given computational process, or the algebraic structure(s) in which it is applied, would seem to guarantee nothing about the reliability of results generated by the computation. Even if, for "3" or "17" or "–29", some symbol such as "f(t)" – which clearly points to variation of an output-value with respect to time – is substituted, the results of such computation do not become any more reliable in the absence of further specific information about the natural process that is the originating source of data for the computation.

8.1.4 The Meaningful Addition

"Addition" is one of the five meanings of *Yoga* in Sanskrit. *Yoga* means 1. Addition, 2. Union or integration, 3. Eternity (with reference to time), 4. Action, or Kriya (to do), and 5. The state of oneness between the seeking self and the ultimate Self. For example, in mathematics, what we call "plus," or addition, means *Yoga*. Again, when two or more people meet or unite for any reason, it is *Yoga*. At what time or under which circumstances or state the inner Revelations are revealed, is called *Yoga* or *Yuga* (*Yuga* means two). For example, in the light of *Kriya* (the verb "to do"), the *Yugas* mean the following:

1. *Satyayuga* means when the seeking self is in Oneness with the ultimate Self in between the eyebrows.
2. *Tretayuga* means when the seeker's breath is in between the eyebrows and the throat inside the Spinal Cord.
3. *Dwaparayuga* means when the seeker's breath is in between the navel and the throat inside the Spinal Cord.
4. *Kaliyuga* means when the seeker's breath is in between the coccygeal and lumbar centers inside the Spinal Cord or outside.
5. Over the centuries, astrologers have placed interpretations on the yugas that limited them to a physical and temporal context.
6. It is clear that these Yugas have, at the core of their meaning, a deeply spiritual application. That kind of activity of integration, by which Realization is attained, is called Yoga.
7. Lastly, the state of Oneness between the *jivatma*, individual self, and *Paramatma*, the ultimate Self, is called Yoga. Others, I am still looking for…

With the above meanings in mind, let's consider a simple addition.

Towards Modeling Knowledge

$$1 + 1 = 2 \qquad (8.1)$$

This operation is supposed to be so trivial that 'even a child do it'. However, this simple equation hides a very significant fact and that is: this operation is valid only in a tangible sense. Tangible operations are repeatedly seen as opposite to reality outside a time period in which $\Delta t = 0$, because the most important space, namely, 'time' is absent. This is the essence of Eurocentric mathematics; the most absurd solution being promoted as the 'only solution'. However, if we expand this equation to include physical sense to the equation, a different form emerges. For instance, the moment we include a description (or qualifier, the first intangible) of the numbers, we are faced with a question that cannot be answered in a concrete way. Consider the following:

$$1\,\text{orange} + 1\,\text{apple} = 2\,\text{of what?} \qquad (8.2)$$

Of course, a quick response would be, you cannot add an apple with an orange. Both objects must have the same dimension. If, then, we proceed to using the same type, we obtain:

$$1\,\text{orange} + 1\,\text{orange} = 2\,\text{orange(s)} \qquad (8.3)$$

Now, can we trace back the RHS of equation 8.3 to the LHS? The answer is no. It is because, by letting this equation stand, we have ignored a very important space, that is, time (a similar argument can be made for addition of 1 D of various elements). Another way to look at the equation is to state that the only time one can trace back to LHS is when the two elements are identical. Once this assumption is exposed, we realize that equation 8.3 is aphenomenal. It is so, because of at least two reasons:

1. there are no two objects that are identical
2. even the same object does not remain in its original form, because everything in nature is dynamic, meaning a function of time (time is a space as identified by Einstein).

So, the correct form of equation 8.3 should be:

$$1\,\text{orange}\,(f(t)) + 1\,\text{orange}\,(g(t)) = 1\,\text{orange}\,(h(t)) + 1\,\text{orange}\,(i(t)) \qquad (8.4)$$

In the above equation, all functions, f, g, h, i are different and their exact forms unknown. Of course, this form will change in case the two objects on LHS merge and a phase change occurs.

From the above discussion, it also becomes clear that the equation sign should be replaced by an equivalent sign. Similarly, other signs should be used to express phase change.

8.1.5 "Natural" Numbers and the Mathematical Content of Nature

Could there be other "stuff" going on behind the scenes, so to speak, that could account for the numerous cases of a lack of "fit", of an incoherence or a gap, between "wholes" and "fractions" as abstractly conceived, compared to actual applications or implementations of wholes and fractions? Are these abstracted essences, and their applications, completely divorced and unrelated to one another in fact, or is their very unrelatedness a sign of "something else"?

Thanks to the work of Joseph (2000) and a handful of others, an answer to this question starts to suggest itself. Quite a few non-European ancient civilizations, in the period before they could possibly have been connected with or influenced by European development, developed their own very definite mathematics. Some of these mathematics – from China, Egypt, the Muslim world, India etc. – made their way into European mathematics, starting with the Greeks. These elements, most notably the notation for and uses of zero, influenced the line of development in various branches of European mathematics before the modern era of calculus-based analysis. Some societies – the Inca and the Maya – arrived independently at other quite striking discoveries and techniques without ever impinging on any line or period of development in European mathematics. The notion that mathematics was some kind of "Greek miracle" is thus a Eurocentric and deeply biased view. It is a view that has rendered modern scholarship singularly incapable of properly appreciating the nature and significance of how societies before the Graeco-Roman launch of "European" civilization elaborated mathematical concepts and methods.

Joseph (2000) brings this out especially sharply in his discussions of systems of enumeration and counting. Some, but by no means all, ancient peoples based their mathematics on counting in 10s. This is almost certainly related to 10 fingers or 10 toes. However, many used 20 as their base, rather than 10. This could be related simply to counting fingers and toes together. However, it likely also connected to that natural approach (at the level of the person) to counting in 60s, i.e. three 20s at a time. Counting in 60s seems to have emerged at very different times and places among civilizations that needed to reckon cyclical calendar-type events such as harvests based on some relatively accurate system of astronomical observation. From this, the significances of 30 (which is both half of 60 and approximately the length of a month), 60 and 360 were hit upon early and entirely independently by the Babylonians in the Fertile Crescent and the Maya in Meso-America. Records of mathematics emerge with the establishment of a settled agriculture. There are as well the well-known examples of central river systems with predictable flood cycles, common to the Egyptians and the Babylonians, and hence a serious concern about lunar cycles. Mayan agriculture, however, was bound up with cycles of the Sun, with no special concern about soil renewal by flooding or lunar cycles that might be involved; and the biggest preoccupation of Mayan mathematics was the exact reckoning of the Sun's annual period. (Incidentally, even nomadic tribes without settled agriculture had some mathematical consciousness. From what has been available to piece together from survivors of some preliterate nomadic tribal cultures, there were many independently developed and unwritten systems of binary counting, i.e. counting by twos.).

The roots of mathematics, then, seem to lie not with this or that level of development of material civilization – Babylonian, Indian, Egyptian or Greek – but rather, and more simply, with the emergence of humans living in social groups. Mathematical thinking begins as a way of tackling and comprehending phenomena encountered in the real world that have some impact on existence.

A nature-mathematics suited to a thoroughly renovated nature-science needs to embark on that path as its starting-point. One such starting-point hinted at earlier in this article involves further deconstructing what we mean by "dimension." One of the aspects that give modeling by PDEs so much power and authority is the possibility to incorporate and interrelate physical quantities in "dimensionless" forms, so that, in effect, apples and oranges can indeed be not only added, but integrated and differentiated.

If this is to be accomplished without arriving at distorted or meaningless results, however, it would seem reasonable to propose that the role or significance of the 'dimensionality' of what normally exists in nature itself be examined first. Is it always a good idea, however convenient it may initially appear, to render all such phenomena dimensionless, i.e. reduced to a single common linear dimension? Many absurd linearizations, like those that rank humans by IQ or that correlate height with the Gross Domestic Product, have come via a similar procedure; their absurdity not screened behind the presence of arcane-looking integral-signs and differential operators. Phenomena within nature, being products of time, at least in some part, each possess unique features that may not be reducible to comparison in one and the same dimension. An honest approach would require taking the one attribute of comparison-interest, call it 'a', and not forget to lump all the other attributes not of immediate interest into another variable, say 'b', and assign 'b' its own dimension, say 'i' (or 'j'). Then one compares a1 + ib1 to a2 + ib2 etc., rather than just a1 to a2. The former must be compared in a plane, not on a line, and there is no "greater-than" or "less-than".

As discussed earlier, the whole question of dimensionality has other problems created as a result of previous treatment of the subject. Not only can one "orthogonalize", i.e. add the next dimension phased $\pi/2$ radians to the last, from the first to the second and then from the second to the third dimension, but one can visually represent that orthogonalization. However, excessive fascination with the tangibility of what such an orthogonal relationship allows, e.g., the extrapolation of any pair of lines into a plane, or the extrapolation of any closed two-dimensional plane figure into a three-dimensional "right solid", seems to have diverted attention away from the significance of what absolutely distinguishes one dimension from the next.

The possibility of a plane and of closed figures is possible only starting in two dimensions. It is not possible in one dimension alone. Certain arrangements like the helix (e.g., DNA and possibly other biological structures), which is NOT a right-solid extrapolated from a two-dimensional closed figure, are meaningless in less than three dimensions.

Three dimensions are immediate physical reality, but still lacking any consideration of time t. Nature, on the other hand, exists in and possesses four dimensions. Four-dimensional nature is distinguished thus: everything in it that already exists within at three-dimensional mass possesses this temporal component.

In modeling nature, or natural systems based on two and-or three dimensions, time is misrepresented as uniform and reversible, the so-called "independent variable". Problems arise not from considering phenomena as functions of time t, but rather from treating time t, which is actually a distinct dimension, as an independent variable. Uniformity and reversibility are not characteristic features of the temporal dimension. Any result from such models can only be valid for the shortest possible term, in which Δt goes to zero, i.e. time t = 'right now'.

Nature also includes energy, however. Energy has no fixed spatial component, i.e. the first three dimensions are entirely absent. The dimensional content of energy is thus exclusively temporal. Potential energy is energy-of-position released as some mass(es) transit a passage of time. Kinetic energy is energy released by the motion of mass(es) transiting through time. The Laws of Conservation of Mass, Energy and Momentum acquire their full meaning only in four-dimensional nature, where nothing can be created nor destroyed but only change form because time is present as an actual dimension and is infinite. Kelvin's Second Law of Thermodynamics formulated its prediction of the 'heat death' of the universe in the absence of any conception of time as the fourth dimension. When nature is reconsidered and repositioned as four-dimensional reality, however, negative entropy, i.e. the increasing organization of nature with time, can take its place within nature alongside Kelvin's conception, rendering the one-sidedness of his positive-only entropy absurd.

To intervene in nature based on models formulated as though nature possessed only three dimensions leads to error. Two serious kinds of mistakes can be immediately distinguished. Firstly, in the reality of four-dimensional nature the notion, or premise based on the notion, that mass or energy is finite – something that seemed so apparently and readily demonstrable in three dimensions – becomes palpable nonsense. Secondly, as a consequence of treating time t as a uniform and reversible independent variable (instead of as a dimension in its own right), all such interventions assume implicitly that the long-term is prefigured in the short-term. By that logic, if the model has taken care of the short term, it can and will take care of the long term. In fact, the standpoint within four-dimensional Nature is exactly the reverse: one cannot look after the short-term without first taking care of the long term.

And all this is before we even broach our fifth, "knowledge" dimension. The main content and feature of this dimension is consciousness (in the sense of "awareness") of nature, including conscience about how to proceed with intervening sustainability within nature. Nature seen in the framework of this kind of dimensionality and analysis has a mathematics that comprises Julia sets of phenomena and forms that are self-similar but non-identical, moving in paths that are cyclical but aperiodic. The effort to unravel Nature's secrets according to this outlook makes it possible to depart once and for all from the path of applying purely abstracted and idealized mathematical constructs on the basis that the "results" generated from these abstractions seem to match whatever is actually observed taking place in nature.

8.2 The Knowledge Dimension

This criterion is based on the fundamental assertion that nature is perfect – in the sense of sufficient unto itself. Everything that is natural is time-tested and is good for humankind. Khan and Islam (2007a) have recently defined what a natural process is. This is of particular importance when it comes to what is natural for humankind. It is identified as the use of conscience, the quality that set humans apart. With this definition of the natural, the problem indeed becomes: how to define the natural mathematically and then how to develop technology based on those mathematics. In this, time domain is the most important dimension.

8.2.1 The Importance of Time as the Fourth Dimension

Visualize an electric fan rotating clockwise. This is the truth. It is not a matter of perception or blind faith, truth is whatever is real. If a phenomenon actually occurs, be that existence of God or the non-flat nature of the planet Earth, any model that predicts contrary to the truth is a model based on ignorance. We term the aphenomenal model – aphenomenal as to be non-existent (Khan et al, 2005). As opposed to the aphenomenal model, the knowledge-based model should predict the truth. A knowledge-based model is dynamic as knowledge is infinity and at no time one can assert that all information is available and the description of a process is complete. Every phenomenon will have room for more investigation and as long as the direction of the investigation is correct (can only be ascertained by the intention of the investigator), further facts of about the truth will continue to unveil (Islam et al, 2006). Because everything is in an unsteady state (Islam, 2004), this investigation in time takes a different meaning as the observed phenomenon is constantly changing. Coming back to our electric fan, any model that would proclaim that the fan is rotating counter clockwise would do nothing to the truth, it would simply falsify the perception of the truth. For instance, if we assume that the observation with the naked eye is complete and no other information is necessary to describe the motion of the electric fan, we risk believing in the aphenomenal model. If a strobe light is used, the motion of the electric fan can be shown to have reversed by adjusting the frequency of strobe light. If the information that the observation of the electric fan is carried out under a strobe light is omitted, the knowledge of the actual motion of the fan would be obscured to say the least. Just by changing the frequency, the perception has been made opposite to the reality. How can we make sure that the prediction of counter-clockwise is discarded? If we insisted on mentioning the frequency of light under which the observation was being made, we would have realized that the frequency of the strobe light is such that the fan is being shown to rotate counter clockwise. Even the frequency of light wouldn't be sufficient to re-create the correct image. Only the sunlight guarantees closest-to-the truth image; any other light is a distortion for human brains. It turns out that frequency is inverse of time, making time the single most important parameter in revealing the truth. We argue that all steady state models are aphenomenal. The most important feature of these steady state models is that they are tangible and valid only for a time = 'right now'. All tangible models are inherently aphenomenal and all intangible models are inherently knowledge-based. This is the essence of our intangible model in science, economics, and engineering. In the case of our electric fan, we have two observations conflicting with each other, by knowing the time-dependence (frequency), we could easily discern the truth from falsehood. In the case of economic analysis, the path that a wealth (tangible or intangible) has traveled and the path that it is set out to travel must be known or at least considered before knowledge-based analysis is introduced.

The importance of time as the fourth dimension has been recently highlighted by Zatzman and Islam (2007) in the context of knowledge economics. It was also identified that in all analyses of New Science, notion of time as a continuous function is muted. This absence of discussion of whatever happened to the tangible-intangible nexus involved at each stage of any of these developments is not merely accidental or a random fact in the world. It flows directly from a Eurocentric bias that pervades, well beyond Europe and North America, the gathering and summation of scientific knowledge everywhere. Certainly, it is by no means a property inherent – either in technology as such, or in the norms and demands of the scientific method *per se*, or even within historical development – that time is considered so intangible as to merit being either ignored as a fourth dimension, or conflated with tangible space as something varying independently of any process underway within any or all dimensions of three-dimensional space. Recently, Mustafiz et al (2007) identified the need of including a continuous time function as the starting point of acquiring knowledge. According to them, the knowledge dimension does not get launched unless time as a continuous function is introduced. They further show that the knowledge dimension is not only possible, it is necessary. The knowledge is conditioned not only by the quantity of information gathered in the process of conducting research, but also by the depth of that research, i.e. the intensity of one's participation in finding things out. In and of themselves, the facts of nature's existence and of our existence within it neither guarantees nor demonstrates our consciousness of either, or the extent of that

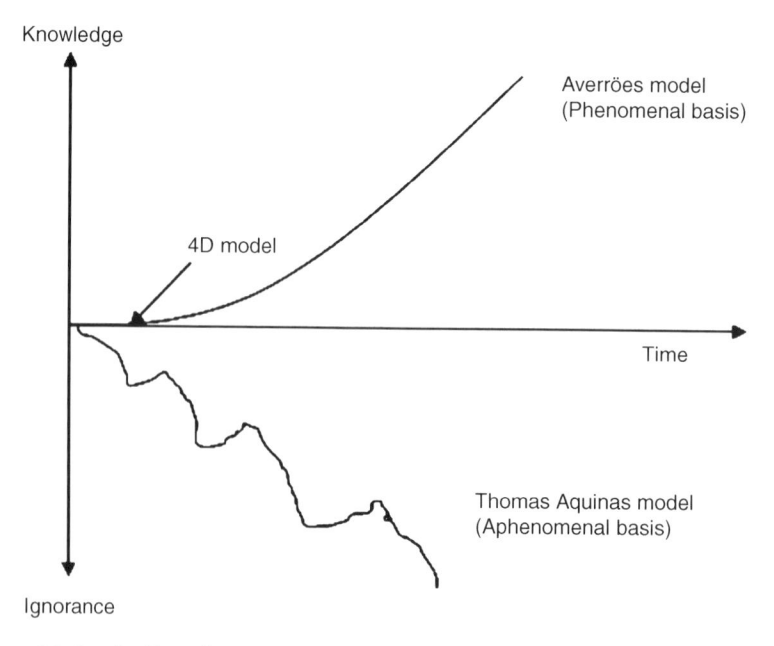

Figure 8.2 Logically, aphenomenal basis is required as the first condition to sustainable technology development. This foundation can be the Truth as the origin of any inspiration or it can be 'true intention', which is the essence of intangibles (modified from Zatzman and Islam, 2007a and Mustafiz et al, 2007).

Towards Modeling Knowledge

consciousness. Our perceptual apparatus enables us to record a large number of discrete items of data about the surrounding environment. Much of this information we organize naturally and indeed unconsciously. The rest we organize according to the level to which we have been trained, and-or come to use, our own brains. Hence, neither can it be affirmed that we arrive at knowledge directly or merely through perception, nor can we affirm being in possession at any point in time of a reliable proof or guarantee that our knowledge of anything in nature is complete.

Table 8.1 shows the characteristic features of Nature. These are true features and are not based on perception. They are true because there are no example of their opposites.

Table 8.1 Typical features of natural processes as compared to the claims of artificial processes (adapted from Khan and Islam, 2007b)

Features of Nature and Natural Materials	
Feature no.	Feature
1	Complex
2	Chaotic
3	Unpredictable
4	Unique (every component is different), i.e. forms may appear similar or even "self-similar", but their contents alter with passage of time
5	Productive
6	Non-symmetric, i.e. forms may appear similar or even "self-similar", but their contents alter with passage of time
7	Non-uniform, i.e. forms may appear similar or even "self-similar", but their contents alter with passage of time
8	Heterogeneous, diverse, i.e. forms may appear similar or even "self-similar", but their contents alter with passage of time
9	Internal
10	Anisotropic
11	Bottom-up
12	Multifunctional
13	Dynamic

Table 8.1 (cont.) Typical features of natural processes as compared to the claims of artificial processes (adapted from Khan and Islam, 2007b)

Features of Nature and Natural Materials	
Feature no.	Feature
14	Irreversible
15	Open system
16	True
17	Self healing
18	Nonlinear
19	Multi-dimensional
20	Infinite degree of freedom
21	Non-trainable
22	Infinite
23	Intangible
24	Open
25	Flexible

Figure 8.3 Knowledge refers to the true state of nature. Shown above natural olive oil and natural rock. They are both extremely heterogeneous and never amenable to solutions with linear solvers.

Without time as the fourth dimension, all models become simulators, focusing on very short-term aspects of natural phenomena (Islam, 2007). In order for these models to be valid in emulating phenomena in the long-term, the fourth dimension must be included. In order for a process to be knowledge-based (precondition for emulation), the first premise of a model must be real, *viz.*, existent in nature. The models of the New Science do not fulfil this condition. It is observed that most of the laws and theories related to mass and energy balance have violated some natural traits by their first premise. These first premises are listed in Table 8.2.

Table 8.2 How the natural features are violated in the first premise of various 'laws' and theories of the science of tangibles (adapted from Zatzman et al, 2008)

Law or Theory	First Premise	Features Violated (See Table 9.1)
Conservation of mass	Nothing can be created or destroyed	None
Lavoisier's deduction	Perfect seal	15
Phlogiston theory	Phlogiston exists	16
Theory of relativity	Everything (including time) is a function of time (concept)	None (concept)
	Maxwell's theory (mathematical derivation)	6, 7, 25 mathematical derivation)
$E = m c^2$	Mass of an object is constant	13
	Speed of light is constant	13
	Nothing else contributes to E	14, 19, 20, 24
Planck's theory	Nature continuously degrading to heat dead	5, 17, 22
Charles	Fixed mass (closed system), ideal gas, Constant pressure	24, 3, 7
Boyles	A fixed mass (closed system) of ideal gas at fixed temperature	24, 3, 7
Kelvin's	Kelvin temperature scale is derived from Carnot cycle and based on the properties of ideal gas	3, 8, 14, 15

Table 8.2 (cont.) How the Natural features are violated in the first premise of various 'laws' and theories of the science of tangibles (Adapted from Zatzman et al, 2008)

Law or Theory	First Premise	Features Violated (See Table 9.1)
Thermodynamics 1st law	Energy conservation (The first law of the thermodynamics is no more valid when a relationship of mass and energy exists)	None
Thermodynamics 2nd law	Based on Carnot cycle which is operable under the assumptions of ideal gas (imaginary volume), reversible process, adiabatic process (closed system)	3, 8, 14, 15
Thermodynamics 0th law	Thermal equilibrium	10, 15
Poiseuille	Incompressible uniform viscous liquid (Newtonian fluid) in a rigid, non-capillary, straight pipe	25, 7
Bernouilli	No energy loss to the sounding, no transition between mass and energy	15
Newton's 1st law	A body can be at rest and can have a constant velocity	Non-steady state, 13
Newton's 2nd law	Mass of an object is constant	13
	Force is proportional to acceleration	18
	External force exists	
Newton's 3rd law	The action and reaction are equal	3
Newton's viscosity law	Uniform flow, constant viscosity	7, 13
Newton's calculus	Limit $\Delta t \to 0$	22

These violations mean such laws and theories weaken considerably or worse, implode, if applied as universal laws and theories. They can be applied only to certain fixed conditions that pertain to 'idealized' conditions yet non-existent in nature. For example, it can be said that the laws of motion developed by Newton cannot explain the chaotic motion of Nature due to its assumptions which contradicts with the reality of Nature. The experimental validity of Newton's laws of motion is limited to describing instantaneous macroscopic and tangible phenomena. However, microscopic and intangible phenomena are ignored. Classical dynamics as represented by Newton's laws of motion, emphasize fixed and unique initial conditions, stability, and equilibrium of a body in motion (Ketata et al, 2007).

Here it is found that time is the biggest issue which, in fact, dictates the correctness of Newton's laws of motion. Considering only instantaneous time ($\Delta t \rightarrow 0$), Newton's laws of motion will be experimentally valid with some error. However, considering the infinite time span ($\Delta t \rightarrow \infty$), the laws can not be applicable.

What Newton did not accomplish was later identified by Einstein, who clearly saw the need of defining everything as a function of time. The omission of this fourth dimension (time) has become the most important flaw in every theory, ranging from Science to Arts, from Engineering to Economics. All of a sudden, Van Gogh's "Starry Night" canvas becomes an imagination, rather than the portrayal, of reality. In 1D, the pyramid looks the same as the butterfly. In 2D, a sleeping person appears to be the same as dancing person. In 3D, a hungry tiger about to devour a prey will appear the same as a content tiger who just finished devouring its prey. Even in 4D, a fan turning clockwise will look the same as the fan turning anticlockwise, depending on the background frequency of light (e.g. solar light replaced by strobe light). Two conclusions arise from this analysis. First, the mistake in omitting a dimension and considering that the solutions are still valid is like comparing a sleeping person with a dancing person and concluding that because they appear the same in two dimensions, they must be performing the same act. Without the full dimension, an otter would look like a rat and the person showing the picture has to be trusted. Secondly, 4D description is a giant leap forward, but is not enough to complete the knowledge dimension. In fact, the knowledge dimension can be complete only after every other object and its natural frequency is considered. It is, therefore, our contention that achieving knowledge remains a continuous process and must constantly accompany serious research work. This aspect has been elaborated further by Zatzman and Islam (2007).

8.2.2 Towards Modeling Truth and Knowledge

In our recent series of papers and books, we have outlined how we can begin to emulate nature. Following is a brief description of our recent efforts.

Novel expert systems embodying pro-nature features are proposed based on natural human intelligences (Ketata et al, 2005a, 2005b). These experts systems use human intelligence which is opposite to artificial intelligence (see Table 8.3). We also attempted to include the knowledge of non-European races who had a very different approach to modeling. Based on Chinese abacus and Quipu (Latin American ancient tribe), we developed an expert system that can be characterized as the first expert system without using the conventional computer counting system. These expert systems provided the basis of an intelligent, robust, and efficient computing tool. Further description can be found in Ketata et al (2006a; 2006b).

Table 8.3 Computer and brain compared

Element	Computer	Brain
Memory	Doubling every 12–18 months (Gordon Moor)	Continuously Renewing
		Crystalized intelligence grows continuously
	Continuously degrading	Even fluid intelligence can grow with more research, i.e., observation of nature and thinking
Size	The bigger the better	No correlation
Thinking	Does not think. All computations are 1-to-1 correlation, between 0 and 1 (linearization of every calculation)	Fuzzy & Variable There is no one-to-one correlation involved
Decision	Based on one number (series)	Decision is through spontaneous or/and thinking

Table 8.3 (cont.) Computer and brain compared

Element	Computer	Brain
Multitasks	No two tasks at the same time	Decision is made after fuzzy logical interpretation of the big picture.
	Quick decision	
Computation/ Math	Quantitative	Qualitative (intangible, $\Delta t = \infty$), nonlinear.
Limit	Limited by programmer	Unlimited
Energy Supply	110 volt converted to 3 volt. When works hard it is heated	µ Volt, use glucose (biochemical energy) When human brain works hard it cools
Hard Ware	Nothing good in it; all kind of heavy metals, non degradable plastics, batteries etc.	100% Degradable
Regeneration Time	Not applicable	Continuous regeneration
Functionality	Crashes, restart takes time, an insane computer is never found	Never crashes, spontaneous, never stops. Insane humans are those who do not make use of it

Because all natural phenomena are non-linear, we argued that any acceptable computational technique must produce multiple solutions. With this objective, we developed a new computational method (Islam, 2006) that finds dynamic derivatives of any function and also solves a set of non-linear equations. More recently, Islam et al (in press) and Mousavizadegan et al (in press), proposed a new technique for finding invariably multiple solutions to every natural equation. These techniques essentially create a cloud of data points and the user can decide which ones are most relevant to a certain application.

8.3 Examples of Linearization and Linear Thinking

The most astounding example can be given from the mathematical modeling and numerical simulation aspect of engineering. Virtually all engineering design comes from mathematical modeling. Practically

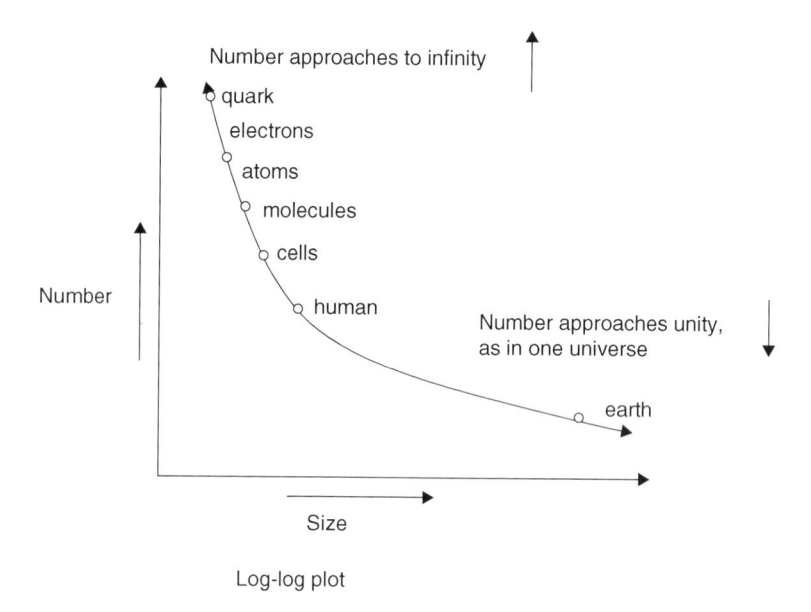

Figure 8.4 Infinity and zero was easily connected to each other by conducting the dimension and frequency of occurrence analysis.

all scientific theories end up with an equation that needs to be solved. Because nature is non-linear, one expects the governing equations describing nature to be non-linear. However, current knowledge of mathematics also makes solutions of non-linear equations impossible beyond trivial solutions that have been found centuries ago. Mathematicians resorted to linearizing the problem by grossly simplifying the boundary conditions and/or initial conditions to a level that the subject of modeling became aphenomenal, rendering solutions irrelevant. The advent of computers glamorized numerical modeling that involved discretization of the solution domain. While speed was high, the solutions remained hopelessly linear. Because computers are also incapable of solving any non-linear equations, scientists resorted to linearizing the governing equations – after initial discretization (Abou-Kassem et al, 2006). The discretization process itself had the assumption that often invoked impossible conditions and the subsequent linearization made the process even more opaque than what mathematicians were doing for the longest time. Efforts have been made to eliminate the discretization of governing equations by introducing the so-called engineering approach (Abou-Kassem et al, 2006).

Towards Modeling Knowledge 365

from discussion in EEC Research Group meeting
Thursday 8 September 2005

All Governing Equations (Non-Linear)

"Engineering Approach" (algebraic equations)
– eliminates the opacity inherent in the Taylor Series expansion approach, but also linearizes to the point of producing a UNIQUE solution

Mathematical Approach (partial differential equations)

ANALYTICAL
Linearization of problem, either by "simplifying" the governing equation, or by simplifying the selection of boundary conditions – to produce a UNIQUE solution

NUMERICAL
Discretization, e.g. Taylor Series expansion, based on assuming $\Delta x < 1$, hence (Δx)-squared $<< 1$, leading again to linearization – only later in the computational process

THE OUTSTANDING QUESTIONS / PROBLEM REMAINS:
is it possible to generate the multiple solutions that one expects from non-linear equation, and recover the original governing equation when any of these solutions is substituted?

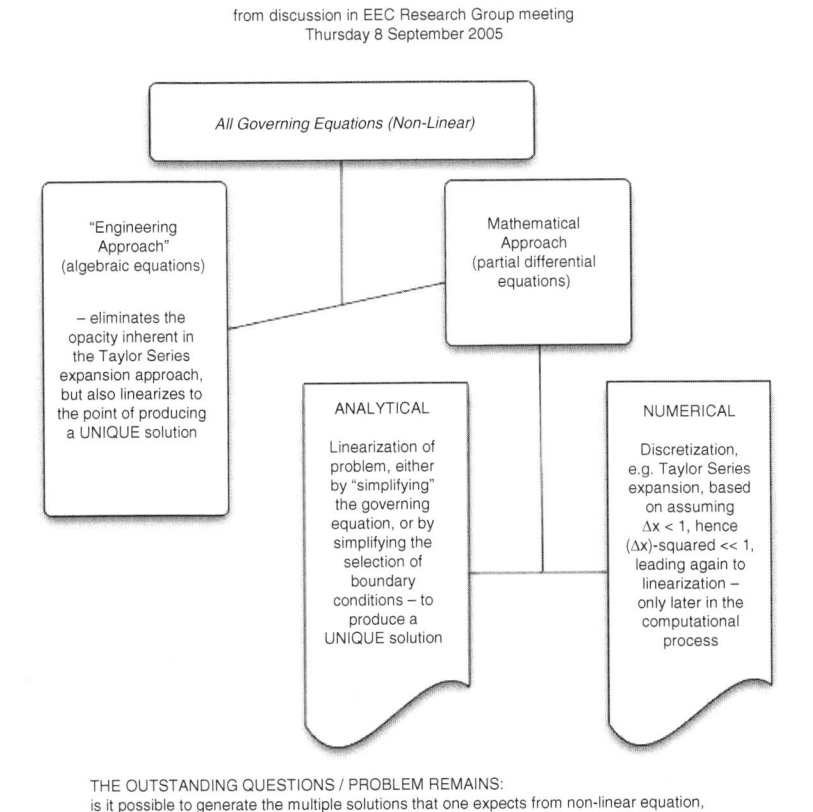

Figure 8.5 Even when non-linear equations are tracked, linearization takes place prior to final solution, making all results questionable.

However, final linearization continues to take place. This is shown in Fig. 8.5. Only recently, our research group (e.g. Mustafiz and Islam, 2005; Mustafiz et al, 2005) have published a series of papers finding solutions to the non-linear governing equations without linearization. However, to-date all solutions found remain unique and no multiple solutions (essence of non-linear equations) have been reported.

8.4 The Single-Parameter Criterion

This criterion is based on the fundamental assertion that nature is perfect – in the sense of being sufficient unto itself. Everything that is natural is time-tested and is good for humankind. Islam (2003) has recently defined the natural. This is of particular importance when it comes to what is natural for humankind. It is identified as the use of conscience, the quality that set humans apart. With this definition of the natural, the problem indeed becomes: how to define the natural mathematically and then how to develop technology based on those mathematics.

8.4.1 Science Behind Sustainable Technology

As stated earlier, taking off the time dimension from every calculation has been the gravest mistake of the previous technology development models. In Nobel Laureate (in Physics) Feynman's words, "Nature cannot be fooled" (Feynman, 1988). There is a need to develop models that are inherently time dependent or truly dynamic. This is possibly the most important trait of nature and that which can be called natural. We must appreciate as humans that we were supposed to safeguard nature rather than molest it. Any anti-nature scheme is inherently anti-human. We defined (Khan et al, 2005) anti-nature as the scheme that works against the characteristic rate of any process. Violating this characteristic rate amounts to perturbing the ecosystem (Islam, 2004). In the past technology development in the modern age, practically all human interventions had an indeterminate influence on nature. However, this influence cannot be determined when there's no clue whether there is a characteristic natural rate in the first place, let alone what that rate is. Nevertheless, utter ignorance on either point - on whether there is a characteristic rate for certain climatic changes, for example, and what that rate or cycle might be - does not stop some people from insisting that climate change of recent years is due somehow to human intervention.

8.4.2 A New Computational Method

8.4.2.1 *The Currently Used Model*

Many consider the use of computers that can only add and use only 0 and 1 as the beginning of linearization in modern computational methods. Our recent investigation shows that the linearization in computational methods precede computers by centuries. The computational technique of the modern age is based on the highly linearized (hence, aphenomenal) model originally proposed by Newton. On the front of scientific work undertaken to investigate and determine laws of motion, Isaac Newton represents the watershed. His elaboration of the general laws of motion of all matter was a huge advance over the incoherent and conflicting notions that prevailed hitherto. Of course, various limitations appeared at certain physically measurable/detectable boundaries – at speeds approaching the speed of light, for example, or within space approaching the measurable minimum limit of (approximately) 10^{-32} m, etc. This led researchers to make important corrections and amendments to Newton's formulae. The fact remains, nevertheless, that Newton's fundamental breakthrough lay in the very idea of summarizing the laws of motion itself, common to all discrete forms of matter understood and observed to that time, i.e. not atomic, molecular or sub-atomic. Equally remarkably, in order to take account of the temporal component attending all matter in motion, Newton invented an entirely new departure in mathematics. A new departure was required because existing mathematics were useless for describing any aspect of change of place while matter was undergoing such change.

Apart from their long standing despite some amendment, this mathematical apparatus used to describe and apply Newton's laws is worth re-examining to get a better understanding of some of the basic tools used throughout scientific work in all fields, including fields far removed from having to deal with laws of motion. Here we have in mind the fundamentals of integral and differential calculus.

Newton's mathematics made it possible to treat time as though it were as infinitely divisible as space – something no one had ever conceived of doing before. This worked extremely well for purposes involving the relative motion of masses acting under the influence of the same external forces, especially force due to gravity and acceleration due to gravity. Extended to the discussion of the planets and other celestial bodies, it appeared that Time throughout nature – Time with a capital "T" – was indeed highly linear. For Newton and for all those applying the tools of his calculus to problems of time and space comprehensible to ordinary human perception, t_{LINEAR} and $t_{NATURAL}$ were one and the same.

To elaborate his method into what he called, in the *Principia Mathematica*, a "theory of fluents and fluxions", Newton built on and refined the implications and tentative conclusions of a number of contemporaries and near-contemporaries who, although lacking an overarching theoretical framework, were already working with processes of infinite summation that converged to some finite value. He proposed differentiation as a method for deriving rates of change at any instant within a process, but his famous definition of the derivative as the limit of a difference quotient involving changes in space or in time as small as anyone might like, but not zero, *viz.*:

$$\frac{d}{dt}f(t) = \lim_{\Delta t \to 0} \frac{f(t+\Delta t) - f(t)}{\Delta t}$$

Figure 8.6 Formulation of Newton's breakthrough idea *(expressed in Leibniz–Cauchy notation)*.

Set the cat among the pigeons. It became apparent that, without further conditions being defined as to when and where differentiation would produce a meaningful result, it was entirely possible to arrive at "derivatives" that would generate values in the range of a function at points of the domain where the function was not defined or did not exist. Indeed: it took another century following Newton's death before mathematicians would work out the conditions – especially the requirements for continuity of the function to be differentiated within the domain of values – in which its derivative (the name given to the ratio-quotient generated by the limit formula) could be applied and yield reliable results.

It was in the period 1740–1820 that the basic theory of differential equations also came to be elaborated. Newton's notation was almost universally replaced by that of calculus' cofounder Leibniz, facilitating the achievement of several further breakthroughs in the theory of analysis for the Swiss mathematician Euler among others. Many notable techniques were developed using the techniques of superposition.

The notion of superposition was an ingenious solution to a very uncomfortable problem implicit in (and left over from) Newton's original schema. Under certain limiting conditions, his derivative would be useful for dealing with what today are called vectors – entities requiring at least two numerical quantities to fully describe them. All the important and fundamental real-world entities of motion – velocity, acceleration, momentum etc. – are vectorial insofar as, if they are to usefully manipulated mathematically, not only their magnitude but also their direction must be specified.

Here there inheres a limiting condition for applying Newton's calculus. So long as magnitude and direction change independently of one another, no problems arise in having separate derivatives for each component of the vector or in superimposing their effects separately and regardless of order. That is what mathematicians mean when they describe or discuss Newton's derivative being used as a "linear operator". At the moment it is not possible to say whether these elements are changing independently, however, a linear operation will no longer hold. Because modeling is always an approximation, this for a long time provided many researchers a license to simplify and relax requirements, to some degree or other, as to just how precisely some part of natural reality had to fit the chosen or suggested model. Naturally, one could generate some sort of model, and results, provided the assumptions – boundary conditions or initial conditions – which were then retrofitted more or less so as to exclude unwanted dependencies. The interior psychology of this act of choice seems to have been that the literalized option would reach a result, therefore it could and should be used. The implication of this choice has been rather more mischievous: everything non-linear has been marginalized either as exceptional, excessively intractable in its "native" non-linear state, or usable only insofar as it may be linearized.

In the actual evolution and development of what became the field of real analysis, of course, every step was taken incrementally. Newton's discoveries were taken up and re-used as tools. Meanwhile, however, the theoretical work needed to explain the conditions under which analytic methods in general, and the derivative in particular, were applicable had not reached the stage of explicit elaboration. Thus, the notion of the derivative as a linear operator, and even aspects of a more generalized theory of linear operators, began to develop and be utilized before the continuity criteria underpinning the entire field of real analysis were made explicit. This led to associating linearity principally with superposition techniques and the possibility of superposition. By the time Cauchy published his work elaborating the importance of continuity, no one would connect continuity with linearization. In real analysis, discontinuity became correlated mainly and even exclusively with undifferentiability.

With the rigorizing of real analysis by Cauchy and Gauss, applied mathematics in the middle third of the nineteenth century developed a powerful impetus and greatly broadened its field of action throughout all the natural sciences, especially deeply in all areas of mechanical engineering. There arose a preponderating interest in steady and-or equilibrium states, as well as in the interrelations between static and dynamic states.

While this was not at all unexpected, it is crucial at this point to make what was actually going on more explicit. Some initial analysis of a deliberately simplified example (see Fig. 8.7) will help illuminate something that often becomes obscured:

Assume some process described by the simple sine function illustrated above. As may be recalled from introductory calculus, using Newton's difference-quotient formula (from Fig. 8.8), the instantaneous rate of change anywhere along the graph-line of this function, which will be continuous anywhere within the interval $(-\infty, +\infty)$, i.e. $-\infty \leq t \leq +\infty$, can be computed stepwise as follows:

This means that, as one moves continuously along the domain t, the *instantaneous rate of change* along the curve represented by the graph for f(t) can be computed by evaluating the cosine of t at that value on the horizontal axis. What is being described is change within the function; the function itself, of course, *has not changed*. As this particular function happens to be periodic, it will cycle through the same values as the operational output described by

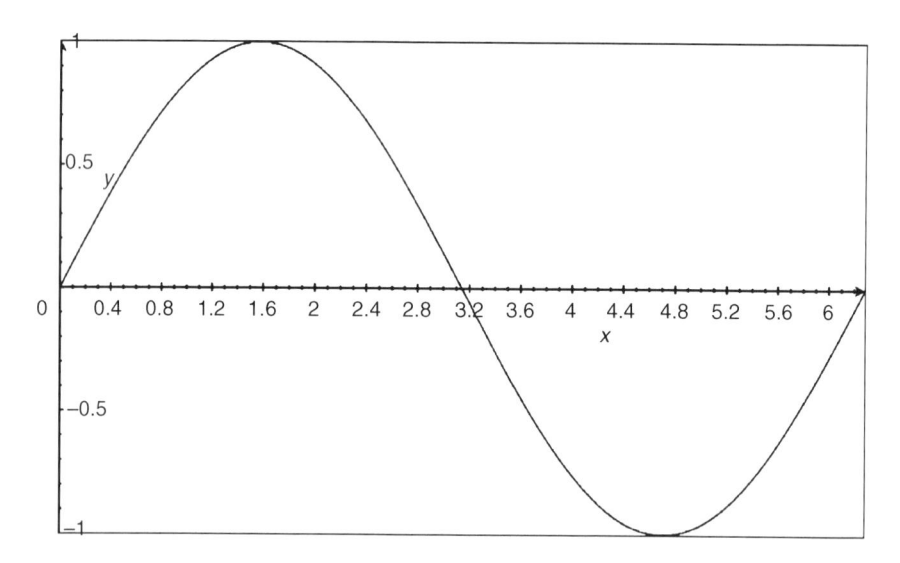

Figure 8.7 Graphic representation, in Cartesian coordinates, of the classic simple function $f(t) = \sin t$.

Towards Modeling Knowledge

$$\frac{d}{dt}f(t) = \lim_{\Delta t \to 0} \frac{\sin(t+\Delta t) - \sin t}{\Delta t} = \lim_{\Delta t \to 0} \frac{\sin t \cos \Delta t + \sin \Delta t \cos t - \sin t}{\Delta t}$$

As Δt approaches 0, $\cos \Delta t$ approaches $\cos 0$, which is 1. Meanwhile, because $\sin x$ approaches x for decreasingly small values of x, the term $\frac{\sin \Delta t}{\Delta t}$, also becomes unity. So:

$$\frac{d}{dt}f(t) = \lim_{\Delta t \to 0}\frac{\sin(t+\Delta t)-\sin t}{\Delta t} = \lim_{\Delta t \to 0}\frac{\sin t \cos\Delta t + \sin\Delta t \cos t - \sin t}{\Delta t} = \lim_{\Delta t \to 0}\frac{\sin t + \sin \Delta t \cos t - \sin t}{\Delta t} =$$
$$\lim_{\Delta t \to 0}\frac{\sin\Delta t \cos t}{\Delta t} = \lim_{\Delta t \to 0}\frac{\sin\Delta t}{\Delta t}\cos t = \lim_{\Delta t \to 0}\cos t = \cos t \lim_{\Delta t \to 0}\frac{\sin \Delta t \cos t}{\Delta t} = \lim_{\Delta t \to 0}\frac{\sin\Delta t}{\Delta t}\cos t = \lim_{\Delta t \to 0}\cos t = \cos t$$

Figure 8.8 Generating the first derivative $f'(t)$ for the function $f(t) = \sin t$ using Newton's difference-quotient formula.

this graph proceeds through subsequent cycles. This makes it quite easy to see that the function itself describes a steady-state condition. In fact, however, even if the function were some polynomial, anything lying on the path of its graph would represent the steady-state operation of that function: steadiness of state is not reducible to some trait peculiar to periodic functions.

Newton's method itself, long described as "Newton's method of tangents" because it could be illustrated geometrically by picturing the derivative as the slope of a straight-line segment tangent to the curve of any function's graph, relies implicitly on the notion of approximating instantaneous moments of curvature, or infinitely small segments, by means of straight lines. This alone should have tipped everyone off that his derivative is a linear operator precisely because, and to the extent that, it examines change over time (or distance) within an already established function, i.e. within a process that has reached its steady state.

The drive to linearize covers a multitude of sins. Thus for example, as bold and utterly unprecedented as Newton's approach, it also contains a trap for the unwary: going backward or forward in space or in time is a matter of indifference. If natural reality is to be modeled as it actually unfolds, however, the requisite mathematics has to close the door on, and not permit the possibility of, treating time as reversible. What use can be made, then, of such mathematics for describing anything happening in nature according to naturally-conditioned temporal factors? To engineer anything in Nature, applying Newton's calculus requires suppressing or otherwise sidelining such considerations, and indeed: it has long been accepted, as a pragmatic matter, that fudge factors and ingenious work-arounds are needed to linearize the non-linear. What has not been clarified, or discussed much if at all up to now, is that this is inherently what they must be about. If this nub of the issue is inherent, then it follows that merely backing up a few steps on the path that brought matters to this stage, back to the point where everything still looked more or less linear and the non-linearities had not yet taken over, is not going to overcome the fundamental difficulty. The starting-point itself contains the core of the problem, which is that Newton's calculus edifice, in its very foundation, is truly anti-Nature. Starting anywhere on this path, one will diverge ever further from Nature.

8.4.2.2   Towards Achieving Multiple Solutions

Our research was based on attempting to solve problems that are truly natural. Because all natural phenomena are also non-linear, we argued that any acceptable computational technique must produce multiple solutions. Using the same example as above, one starting point for a new path might go somewhat as follows: consider as the starting point for modeling such natural processes some series of observations of an ongoing phenomenon for which there is no "analytic" function that fits perfectly or even fits over an extended run of results. Results are needed that are grouped reasonably closely in time (the assumption of continuity must be more-or-less likely or possible to validate). Instead, however, of computing a difference quotient based on evaluating the limit at some arbitrary common value like 0, consider what happens if some positive finite constant real value c were used instead. A new derivative may be defined thus:

Here we are dealing with multiple, in fact: infinite, solutions, as should be expected when modeling problems in Nature. The impossibility until relatively recently, i.e. the last third of the 20th century, of computing such representations efficiently, or at all "within anyone's lifetime" for that matter, and their inherently inelegant appearance as represented by the system of notation available, doubtless drove many away from even considering these phenomena as worthwhile subjects of investigation. Many researchers applying mathematics to modeling real-world phenomena would likely reject, as an extremist position, the militant insistence of the British mathematician G.H. Hardy, briefly the mentor of the Indian mathematical genius Ramanujan, that one should approach and present mathematics as some kind of pure thought-experiment continuous in time, untainted by (and having nothing to do with)

Towards Modeling Knowledge

$$f'(t) = \frac{d_c}{dt}f(t) = \lim_{\Delta t \to c} \frac{\sin(t+\Delta t) - \sin t}{\Delta t} = \lim_{\Delta t \to 0} \frac{\sin t \cos \Delta t + \sin \Delta t \cos t - \sin t}{\Delta t}$$

Now, as Δt approaches c, $\cos \Delta t$ approaches $\cos c$, which is anywhere in the interval $(-1, +1)$. Meanwhile, the term $\frac{\sin \Delta t}{\Delta t}$ may fall anywhere in the interval $\left(-\frac{\sqrt{3}}{2}, +\frac{1}{c}\right)$. Applying these maxima and minima generates the open interval $-(2\sin t + \frac{1}{c}\cos t) \leq \frac{d_c}{dt}f(t) \leq \frac{1}{c}\cos t$, in which:

- at $t = 0$ $(+2k\pi)$, $\frac{d_c}{dt}f(t)$ converges to a single value, viz., $\frac{1}{c}$, which is positive (>0);
- at $t = \frac{\pi}{6}$: $-\left(2 + \frac{\sqrt{3}}{2c}\right) \leq \frac{d_c}{dt}f(t) \leq \frac{\sqrt{3}}{2c}$, which straddles 0;
- at $t = \frac{\pi}{4}$: $-\sqrt{2}\left(1 + \frac{1}{2c}\right) \leq \frac{d_c}{dt}f(t) \leq \frac{\sqrt{2}}{2c}$, which straddles 0;
- at $t = \frac{\pi}{3}$: $-\left(\sqrt{3} + \frac{1}{2c}\right) \leq \frac{d_c}{dt}f(t) \leq \frac{1}{2c}$, which straddles 0;
- at $t = \frac{\pi}{2}$: $-2 \leq \frac{d_c}{dt}f(t) \leq 0$, which is mainly negative (≤ 0);
- at $t = \frac{2\pi}{3}$: $-\left(\sqrt{3} - \frac{1}{c}\right) \leq \frac{d_c}{dt}f(t) \leq -\frac{1}{2c}$, which is entirely negative (<0);
- at $t = \frac{3\pi}{4}$: $-\sqrt{2}\left(1 - \frac{1}{2c}\right) \leq \frac{d_c}{dt}f(t) \leq -\frac{\sqrt{2}}{2c}$, which is entirely negative (<0) and reduces, at $c = 1$, to $-\frac{\sqrt{2}}{2}$.

From here, heading towards $t = \pi$, other features emerge:

- At $t = \frac{5\pi}{6}$, $\frac{d_c}{dt}f(t)$ lies somewhere between $-\left(1 - \frac{\sqrt{3}}{2c}\right)$ and $-\frac{\sqrt{3}}{2c}$, in which:
  - for $c = \frac{\sqrt{3}}{2}$, $-1 \leq \lim_{\Delta t \to c} f'(t) \leq 0$;
  - for $\frac{\sqrt{3}}{2} < c < \sqrt{3}$, while for $c > \sqrt{3}$, $\lim_{\Delta t \to c} f'(t) < 0$; and
  - for $c = \sqrt{3}$, $\lim_{\Delta t \to c} f'(t) = -0.5$;
- At $t = \pi$: $-\frac{1}{c} \leq \frac{d_c}{dt}f(t) \leq \frac{1}{c}$

Figure 8.9 Generating family of first derivatives, $\{f'(t) = \lim_{\Delta t \to c} \frac{d_c}{dt}f(t)\}$, for $f(t) = \sin t$ using modified difference-quotient.

any possible application. Nevertheless, Eurocentric conceptions, stemming from ancient Greek philosophy, of beauty as a function of two-dimensional symmetry, "balance", etc., remain very much part of the expectation of most mathematicians alive and working today on current problems of both pure and applied mathematics. This has served to reinforce a tendency to discard or dismiss as unlikely an "inelegant-looking" result.

This method gives us the first attempt to solve non-linear problems (the only relevant ones) that have inherently multiple solutions. This is possible because we have not invoked any linearization throughout the whole exercise. Our recent work shows that this technique is applicable to practically all real-world scenarios and all domains of science (Zatzman and Islam, 2006).

8.5 The Conservation of Mass and Energy

Lavoisier's first premise was "mass cannot be created or destroyed". This assumption does not violate any of the features of Nature. However, his famous experiment had some assumptions embedded in it. When he conducted his experiments, however, he assumed that the container is sealed perfectly – something that would violate the fundamental tenet of Nature that an isolated chamber can be created (see Item 15, Table 9.1 *supra*). Rather than recognizing the a phenomenality of the assumption that a perfect seal can be created, he 'verified' his first premise (law of conservation of mass) 'within experimental error'.

Einstein's famous theory is more directly involved with mass conservation. He derived $E = mc^2$ using the first premise of Planck (1901). However, in addition to the aphenomenal premises of Planck, this famous equation has its own premises that are aphenomenal (see Table 8.1). However, this equation remains popular and considered to be useful (in the pragmatic sense) for a range of applications, including nuclear energy. For instance, it is quickly deduced from this equation that 100 kJ is equal to approximately 10^{-9} gram. Because no attention is given to the source of the matter nor the pathway, the information regarding these two important intangibles is wiped out from the conventional scientific analysis. The fact that a great amount of energy is released from a nuclear bomb is then taken as an evidence that the theory is correct. By accepting this at face value (heat as the one-dimensional criterion), heat from nuclear energy, electrical energy, electromagnetic irradiation, fossil fuel burning, wood burning or solar energy becomes identical.

In terms of the well-known laws of conservation of mass (m), energy (E) and momentum (p), the overall balance, B, within Nature may be defined as some function of all of them:

$$B = f(m, E, p) \tag{8.5}$$

The perfection without stasis that is Nature means that everything that remains in balance within it is constantly improving with time. That is:

$$\frac{dB}{dt} > 0. \qquad (8.6)$$

If the proposed process has all concerned elements such that each element is following this pathway, none of the remaining elements of the mass balance discussed later will present any difficulty. Because the final product is being considered as time extends to infinity, the positive (">0") direction is assured.

8.5.1 The Avalanche Theory

A problem posed by Newton's Laws of Motion, however, is the challenge they represent to relying upon and using the principle of energy-mass-momentum conservation. This principle is the sole necessary and sufficient condition for analyzing and modeling natural phenomena *in situ*, so to speak – as opposed to analyzing and generalizing from fragments captured or reproduced under controlled laboratory conditions.

The underlying problem is embedded in Newton's very notion of motion as the absence of rest, coupled to his conception of time as the duration of motion between periods of rest. The historical background and other contradictions of the Newtonian system arising from this viewpoint are examined at greater length in Abou-Kassem et al (2008), an article that was generated as part of an extended discussion of, and research into, the requisites of a mathematics that can handle natural phenomena unadorned by linearizing or simplifying assumptions. Here the aim is to bring forward those aspects that are particularly consequential for approaching the problems of modeling phenomena of Nature, where "rest" is impossible and inconceivable.

Broadly speaking, it is widely accepted that Newton's system, based on his three laws of motion accounting for the proximate physical reality in which humans live on this Earth coupled with the elaboration of the principle of universal gravitation to account for motion in the heavens of space beyond this Earth, makes no special axiomatic assumptions about physical reality outside the scale on which any human being can observe and verify for himself/herself (i.e. the terrestrial scale on which we go about living daily life).

For example, Newton posits velocity, v, as a change in the rate at which some mass displaces its position in space, s, relative to the time duration, t, of the motion of the said mass. That is:

$$v = \frac{\partial s}{\partial t} \qquad (8.7)$$

This is no longer a formula for the average velocity, measured by dividing the net displacement in the same direction as the motion impelling the mass by the total amount of time that the mass was in motion on that path. This formula posits something quite new (for its time, *viz.*, Europe in the 1670s), actually enabling us to determine the instantaneous velocity at any point along the mass's path while it is still in motion.

The "v" that can be determined by the formula given in (8.3) above is highly peculiar. It presupposes two things. First, it presupposes that the displacement of an object can be derived relative to the duration of its motion in space. Newton appears to cover that base already by defining this situation as one of what he calls "uniform motion". Secondly, however, what exactly is the time duration of the sort of motion Newton is setting out to explain and account for? It is the period in which the object's state of rest is disturbed, or some portion thereof. This means the uniformity of the motion is not the central or key feature. Rather, the key is the assumption in the first place that motion is the opposite of rest.

In his First Law, Newton posits motion as the disturbance of a state of rest. The definition of velocity as a rate of change in spatial displacement relative to some time duration means that the end of any given motion is either the resumption of a new state of rest, or the starting-point of another motion that continues the disturbance of the initial state of rest. Furthermore, only to an observer external to the mass under observation can motion appear as the disturbance of a state of rest and a state of rest appear as the absence or termination of motion. Within nature, meanwhile, is anything ever at rest? The struggle to answer this question exposes the conundrum implicit in the Newtonian system: everything "works" – all systems of forces are "conservative" – if and only if the observer stands *outside* the reference frame in which a phenomenon is observed.

In Newton's mechanics, motion is associated not with matter-as-such, but only with force externally applied. Inertia on the other hand is definitely ascribed to mass. Friction is considered only as a force equal and opposite to that which has impelled some mass into motion. Friction in fact exists at the molecular level, however, as well as at all other scales – and it is not a force externally applied. It is a property of matter itself. It follows that motion must be associated fundamentally not with force(s) applied to matter, but rather with matter itself. Although Newton nowhere denies this possibility, his First Law clearly suggests that going into motion and ceasing to be in motion are equally functions of some application of force external to the matter in motion; motion is important relative to some rest or equilibrium condition.

Following Newton's presentation of physical reality in his Laws of Motion: if time is considered mainly as the duration of motion arising from force(s) externally applied to matter, then it must cease when an object is "at rest". Newton's claim in his First Law of Motion that an object in motion remains in (uniform) motion until acted on by some external force appears at first to suggest that, theoretically, time is taken as being physically continual. It is mathematically continuous, but only as the independent variable, and indeed, according to (3) above, velocity v becomes undefined if time-duration t becomes 0. On the other hand, if *motion* itself ceases – in the sense of ∂s, the rate of spatial displacement, going to 0 – then velocity must be 0. What has then happened, however, to *time*? Where in nature can time be said either to stop or to come to an end? If Newton's mechanism is accepted as the central story, then many natural phenomena have been operating as special exceptions to Newtonian principles. While this seems highly unlikely, its very unlikelihood does not point to any way out of the conundrum.

This is where momentum p, and – more importantly – its "conservation", comes into play. In classically Newtonian terms:

$$p = mv = m\frac{\partial s}{\partial t} \tag{8.8}$$

Hence

$$\frac{\partial p}{\partial t} = \frac{\partial}{\partial t} m \frac{\partial s}{\partial t} + m \frac{\partial^2 s}{\partial t^2} \tag{8.9}$$

If the time it takes for a mass to move through a certain distance is shortening significantly as it moves, then the mass must be accelerating. An extreme shortening of this time corresponds therefore to a proportionately large increase in acceleration.

However, if the principle of conservation of momentum is not to be violated, either:

(a) *the rate of its increase for this rapidly accelerating mass is comparable to the increase in acceleration* – in which case the mass itself will **appear** relatively constant and unaffected; or
(b) mass itself will be **increasing**, which suggests the increase in momentum will be greater than even that of the mass's acceleration; or
(c) *mass must **diminish** with the passage of time*, which implies that any tendency for the momentum to increase also decays with the passage of time.

The rate of change of momentum ($\partial p/\partial t$) is proportional to acceleration (the rate of change in velocity, as expressed in the $\partial^2 s/\partial t^2$ term) experienced by the matter in motion. It is proportional as well to the rate of change in mass with respect to time (the $\partial m/\partial t$ term). If the rate of change in momentum approaches the acceleration undergone by the mass in question, i.e. if $\partial p/\partial t \to \partial^2 s/\partial t^2$, then the change in mass is small enough to be neglected. On the other hand, a substantial rate of increase in the momentum of some moving mass – on any scale much larger than its acceleration – involves a correspondingly substantial increase in mass.

The analytical standpoint expressed in Equations (8.8) and (8.9) above work satisfactorily for matter-in-general, as well as for Newton's highly specific and indeed peculiar notion of matter in the form of discrete object-masses. Of course, here it is easy to miss the "catch". The "catch" is ... the very assumption in the first place that matter is an aggregation of individual object-masses. While this may well be true at some empirical level at terrestrial scale – 10 balls of lead shot, say, or a cubic liter of wood sub-divided into exactly 1,000 one-cm by one-cm by one-cm cubes of wood – it turns out in fact to be a definition that addresses only some finite number of properties of specific forms of matter that also happen to be tangible and hence accessible to us at terrestrial scale.

Newton's laws of motion work satisfactorily for matter-in-general. Once again, the generalizing of what may only be a special case – before it has been established whether the phenomenon is a unique case, a special but broad case, or a characteristic case – begets all manner of mischief.

To appreciate the implications of this point, consider what happens when an attempt is made to apply these principles to object-masses of different orders and-or vastly different scales, but within the same reference-frame. Consider the snowflake – a highly typical piece of natural mass. Compared to the mass of some avalanche of which it may come to form a part, the mass of any individual component snowflake is negligible. Negligible as it may seem, however, **it is not zero**. Furthermore, the accumulation of snowflakes in the avalanching mass of snow means that the cumulative mass of snowflakes is heading towards something very substantial, infinitely larger than that of any single snowflake. To grasp what happens for momentum to be conserved between two discrete states, consider the starting-point: $p = mv$. Clearly in this case, that would mean in order for momentum to be conserved,

$$p_{avalanche} = p_{snowflakes\text{-}as\text{-}a\text{-}mass} \tag{8.10}$$

which means

$$m_{avalanche} v_{avalanche} = \sum_{snowflake=1}^{\infty} m_{snowflake} v_{snowflake} \tag{8.11}$$

At terrestrial scale, avalanching is a readily-observed physical phenomenon. At its moment of maximum (destructive) impact, an avalanche indeed looks like a train-wreck unfolding in very slow motion. However, what about the energy released in the avalanche? Of this we can only directly see the effect, or footprint – and another aphenomenal absurdity pops out: an infinitude of snowflakes, each of negligible mass, have somehow imparted a massive release of energy. This is a serious accounting problem: not only momentum, but mass and energy as well, are to be conserved throughout the universe.

The same principle of conservation of momentum enables us to "see" what must happen when an electron or electrons bombard a nucleus at very high speed. Now we are no longer observing or operating at terrestrial scale. Once again, however, the explanation conventionally given is that since electrons have no mass, the energy released by the nuclear bombardment must have been latent and entirely potential, stored within the nucleus.

Clearly, then, as an accounting of what happens in nature (as distinct from a highly useful toolset for designing and engineering certain phenomena involving the special subclass of matter represented by Newton's object-masses), Newton's central model of the object-mass is insufficient. Is it even necessary? Tellingly on this score, the instant it is recognized that there is no transmission of energy without matter, all the paradoxes we have just elaborated are removable. Hence we may conclude that, for properly understanding and becoming enabled to emulate nature at all scales, mass-energy balance and the conservation of momentum are necessary and sufficient. On the other hand, neither constancy of mass, nor of the speed of light, nor even uniformity in the passage and measure of time are necessary or sufficient. This realization holds considerable importance for how problems of modeling Nature are addressed. An infinitude of energy and mass transfers take place in Nature, above and to some extent in relation to the surface of the earth, comprising altogether a large part of the earth's "life cycle". In order to achieve any non-trivial model of Nature, time itself becomes a highly active factor of prepossessing – and even overwhelming – importance. Its importance is perhaps comparable only to the overwhelming role that time plays in sorting out the geologic transformations under way inside the earth.

In summary, the above analysis overcomes several intangibles that are not accounted for in conventional analysis. It includes: a) source of particles and energy; b) particles that are not visible or measurable with conventional means; c) tracking of particles, based on their sources (the continuous time function).

8.5.2 Aims of Modeling Natural Phenomena

The inventor of the Hamming code – one of the signal developments in the early days of information theory – liked to point out in his lectures on numerical analysis that "the purpose of computing is insight, not numbers" (Hamming, 1984). Similarly, we can say the aim in modeling natural phenomena, such as Nature, is direction (or – in more strictly mathematical-engineering terms – the gradient). That is, this aim is not and cannot be some or any precise quantity.

There are three comments to add that help elaborate this point. *First*: Nature being the ultimate dynamical system, no quantity however precisely measured at time t_0 will be the same at time $t_0 + \Delta t$, no matter how infinitesimally small we set the value of that Δt. *Secondly*: in Nature, matter in different forms at very different scales interacts continually, and the relative weight or balance of very different forces – intermolecular forces, interatomic forces of attraction and repulsion, and gravitational forces of attraction – cannot be predicted in advance. Since Nature operates to enable and sustain life forms, however, it is inherently reasonable to confine and restrict our consideration to three classes of substances that are relevant to the maintenance or disruption of biological processes. *Thirdly*: at the same time, none of the forces potentially or actually acting on matter in Nature can be dismissed as negligible, no matter how "small" their magnitude. It follows that it is far more consequential for a practically useful Nature model to be able to indicate the gradient/trend of the production, conversion or toxic accumulation of natural biomass, natural non-biomass and synthetic sources of biomass respectively.

As already discussed earlier, the generalizing of the results for physical phenomena observed at one scale to all other scales has created something of an illusion, one reinforced moreover by the calculus developed by Newton. That analytical toolset included an assumption that any mathematical extension, x, might be infinitely subdivided into an infinite quantity of Δx–es which would later be (re-)integrated back into some new whole quantity. However, if the scales of actual phenomena of interest are arbitrarily mixed, leap-frogged or otherwise ignored, then what works in physical reality may cease to agree with what worked for mathematics. Consider in this connection the extremely simple equation:

$$y = 5 \qquad (8.12)$$

Taking the derivative of this expression with respect to some independent variable x yields:

$$\frac{dy}{dx} = 0 \qquad (8.13)$$

To recover the originating function, we perform

$$\int dy = c \qquad (8.14)$$

Physically speaking, (8.10) amounts to asserting that "something" of indefinite magnitude, designated as c – it could be "5", as a special case (e.g. with proper boundary or conditions), but it could well be anything else – has been obtained as the result of integrating (8.11), which itself had output a magnitude of 0, i.e. nothing. This is scientifically absurd. Philosophically, even Shakespeare's aging and crazed King Lear recognized that "nothing will come of nothing: speak again" (Shakespeare, 1608). The next problem, associated to this analysis is that the pathway is obscured, opening the possibility of reversing the original whole. For instance, a black (or any other color) pixel within a white wall will falsely create a black (or any other color, corresponding to the pixel) wall if integrated without restoring the nearby pixels that were part of the original white wall. This would happen, even though mathematically no error has been committed. This example serves to show the need for including all known information in space as well as in time. Mathematically, this can be expressed as:

$$\int_{t=0}^{t=\infty} \int_{s=1}^{s=\infty} mv = cons\tan t \qquad (8.15)$$

The aim of a useful Nature model can be neither to account for some "steady state" – an impossibility anywhere in Nature – nor to validate a mechanical sub-universe operating according to some criteria of an observer external to the process under observation. Dynamic balances of mass, energy and momentum imply conditions that will give rise to multiple solutions, at least with the currently available mathematical tools. When it comes to Nature, a portion of the space-time continuum in which real physical boundary conditions are largely absent, a mathematics that requires $\Delta t \to 0$ is clearly inappropriate. What is needed are non-linear algebraic equations that incorporate all relevant components (unknowns and other variables) involved in any of these critical balances that must be preserved by any natural system.

8.5.2 Challenges of Modeling Sustainable Petroleum Operations

Recently, Khan and Islam (2007a, 2007b) outlined the requirements for rendering fossil fuel production sustainable. This scientific study shows step by step how various operations ranging from exploration to fuel processing can be performed in such a manner that resulting products will not be toxic to the environment. However, modeling such a process is a challenge as the conventional characterization of matter does not make any provision for separating sustainable operations from unsustainable ones. In order to avoid some of the difficulties associated with a conventional approach, Khan et al (2008) recently introduced simultaneous characterization of matter and energy. This renders time a characteristic of matter itself within the overall context of mass-energy-momentum conservation. In other words, time ceases to be mainly or only a derivative of some spatial displacement of matter. In this way, it becomes possible at last to treat time, consistently, as a true fourth dimension – and no longer as merely the independent variable. This description is consistent with Einstein's revolutionary relativity theory, but does not rely on Maxwell's equations as the starting point. The resulting equation is shown to be continuous in time, thereby allowing transition from mass to energy. As a result a single governing equation emerges. This equation is solved for a number of cases and is shown to be successful in discerning between various natural and artificial sources of mass and energy. With this equation, the difference between chemical and organic fertilizers, microwave and wood stove heating, and sunlight and fluorescent light can be made with unprecedented clarity. This analysis would not be possible with conventional techniques. Finally, analysis results are shown for a number of energy- and material-related prospects. The key to the sustainability of a system lies within its energy balance. Khan et al recast the combined energy-mass balance equation in the following form as in (8.15).

Dynamic balances of mass, energy and momentum imply conditions that will give rise to multiple solutions, at least with the currently available mathematical tools. When it comes to Nature, a portion of the space-time continuum in which real physical boundary conditions are largely absent, mathematics that requires $\Delta t \rightarrow 0$ is clearly inappropriate. What is needed are non-linear algebraic equations that incorporate all relevant components (unknowns and other variables) involved in any of these critical balances that must be preserved by any natural system. In this context, Equation (8.15) is of utmost importance. This equation can be used to define any process, for which the following equation applies:

$$Q_{in} = Q_{acc.} + Q_{out} \tag{8.16}$$

In the above equation classical mass balance equation, Q_{in} in expresses Equation (8.15) for inflow matter, Q_{acc} represents the same for accumulating matter, and Q_{out} represents the same for outflowing matter. Q_{acc} will have all terms related to dispersion/diffusion, adsorption/desorption, and chemical reactions. This equation must include all available information regarding inflow matters, e.g. their sources and pathways, the vessel materials, catalysts, and others. In this equation, there must be a distinction made among various matter, based on their source and pathway. Three categories are proposed: 1) biomass (BM); 2) convertible non-biomass (CNB); and 3) non-convertible non-biomass (NCNB). Biomass is any living object. Even though conventionally dead matters are also called biomass, we avoid that denomination as it is difficult to scientifically discern when a matter becomes non-biomass after death. The convertible non-biomass (CNB) is the one that due to natural processes will be converted to biomass. For example, a dead tree is converted into methane after microbial actions, the methane is naturally broken down into carbon dioxide, and plants utilize this carbon dioxide in the presence of sunlight to produce biomass. Finally, non-convertible non-biomass (NCNB) is a matter that emerges from human intervention. These matters do not exist in nature and their existence can be only considered artificial. For instance, synthetic plastic matters (e.g. polyurethane) may have similar composition as natural polymers (e.g. human hair, leather), but they are brought into existence through a very different process than that of natural matters. Similar examples can be cited for all synthetic chemicals, ranging from pharmaceutical products to household cookware. This denomination makes it possible to keep track of the source and pathway of a matter. The principal hypothesis of this denomination is: all matters naturally present on Earth are either BM or CNB, with the following balance:

$$\text{Matter from natural source} + CNB_1 = BM + CNB_2 \quad (8.17)$$

The quality of CNB_2 is different from or superior to that of CNB_1 in the sense that CNB_2 has undergone one extra step of natural processing. If nature is continuously moving to a better environment (as represented by the transition from a barren Earth to a green Earth), CNB_2 quality has to be superior to CNB_1 quality. Similarly, when matter from natural energy sources come in contact with BMs, the following equation can be written.

$$\text{Matter from natural source} + B_1M = B_2M + CNB \quad (8.18)$$

Applications of this equation can be cited from biological sciences. When sunlight comes in contact with retinal cells, vital chemical reactions take place that results in the nourishment of the nervous system, among others (Chhetri and Islam, 2008a). In these mass transfers, chemical reactions take place entirely differently depending on the light source, the evidence of which has been reported in numerous publications (e.g. Lim and Land, 2007). Similarly, sunlight is also essential for the formation of vitamin D, which is in itself essential for numerous physiological activities. In the above equation, vitamin D would fall under B_2M. This vitamin D is not to be confused with the synthetic vitamin D, the latter one being the product of artificial process. It is important to note that all products on the right hand side are of greater value than the ones on the left hand side. This is the inherent nature of natural processing – a scheme that continuously improves the quality of the environment and is the essence of sustainable technology development.

The following equation shows how energy from NCNB will react with various types of matter.

$$\text{Matter from unnatural source} + B_1M = NCNB_2 \quad (8.19)$$

An example of the above equation can be cited from biochemical applications. For instance, if artificially generated UV is in contact with bacteria, the resulting bacteria mass would fall under the category of NCNB, stopping further value addition by nature. Similarly, if bacteria are destroyed with a synthetic antibiotic (pharmaceutical product, pesticide, etc.), the resulting product will not be conducive to value addition through natural processes, instead becoming a trigger for further deterioration and insult to the environment.

$$\text{Matter from unnatural source} + CNB_1 = NCNB_3 \quad (8.20)$$

An example of the above equation can be cited from biochemical applications. The $NCNB_1$ which is created artificially reacts with CNB_1 (such as N_2, O_2) and forms $NCNB_3$. The transformation will be in negative direction, meaning the product is more harmful than it was earlier. Similarly, the following equation can be written:

$$\text{Matter from unnatural source} + NCNB2_1 = NCNB_2 \quad (8.21)$$

An example of this equation is that the sunlight leads to photosynthesis in plants, converting NCBM to MB, whereas fluorescent lighting would freeze that process which can never convert natural non-biomass into biomass.

8.6 The Criterion: The Switch that Determines the Direction at a Bifurcation Point

It is long understood that the decision making process involves asking 'yes' or 'no' questions. Usually, this question is thought to be posed at the end of a thought process or logical train. It is less understood that the 'yes' or 'no' question cannot lead to a correct answer if the original logical train did not start with a correct first premise and if the full time domain (defining logical train) is not fully considered. Consider the following question. Is whole wheat bread better than white bread? One cannot answer this question without the knowledge of the past history. For instance, for organic flour (without chemical fertilizer, genetic alteration, pesticide, metallic grinder, artificial heat sources), whole wheat bread is better. However, if the flour is not organic, then more questions need to be asked in order to first determine the degree of insult caused to the natural process than to determine what would be the composition of the whole wheat if all ingredients (including trace elements from grinder, chemical fertilizers, pesticide, heat source, and others) are considered. In this analysis, one must include all elements in space at a given time. For instance, if trace elements from pesticide or a metallic grinder are neglected, the answer will be falsified. In this particular case, whole wheat non-organic bread is worse than white non-organic bread, but will not be shown as such if one doesn't include all elements in time and space (mass and energy). In summing up these two points, one must consider the full extent of time from the start of a process (including logical train) and one must include all elements in space (for both mass and energy sources), in line with the theory advanced by Khan et al (2008). Each of these considerations will have a question regarding the diversion of the process from a natural process. At the end, anything that is natural is sustainable, and therefore, is good.

Let's rephrase the question:

Q: Is whole wheat bread better than white bread? The conventional answer sought would be either Y or N or True or False. However, without a proper criterion for determining True or

False, this question cannot be answered. In order to search for the knowledge-based answer this question, the following question must be asked:

Q_k: Are both white bread and whole wheat bread organic? If the answer is yes, then the answer to Q is YES. If the answer is no, then the following knowledge-based question has to be asked:

Q_{k1}: Are both non-organic? If the answer is YES, then the answer to Q becomes NO, meaning whole wheat bread is not better than white bread. If the answer to Qk1 is NO, then further knowledge-based question has to be asked:

Q_{k2}: Is the white bread organic? If the answer is YES, then the answer to Q becomes NO, meaning whole wheat non-organic bread is not better than white organic bread. If the answer to Q_{k2} is NO, then the answer to Q is, YES, meaning whole wheat organic bread is better than white non-organic bread.

In the above analysis, the definition of 'organic' has been left to the imagination of the reader. However, it must be stated that a 100% scientific organic cannot be achieved. Scientifically, organic means something that has no anti-conscious intervention of human beings. Obviously, nature itself being continuous in space and time, there no possibility of having a 100% organic product. However, this should not stop one searching for the true answer to a question. At the very least, this line of analysis will raise new questions that should be answered with more research, if deemed necessary. For this particular question, Q, we have only presented the mass balance aspect. For instance, organic bread also means, baked in clay stove with natural fuel. Now, what happens if this energy balance is not respected? This poses another series of questions. Let's call them Energy-related question, QE. This question must be asked in the beginning, meaning before asking Q_k's.

QE: Are both whole wheat and non-whole wheat breads organically baked? If the answer is YES, then the previous analysis stands. If the answer is NO, then the following knowledge-seeking question must be asked:

$QE_{K'}$: Are both of them non-organically (e.g. electricity, microwave, processed fuel, recombined charcoal, steel stove) baked? If the answer is YES, then the previous analysis stands. If the answer is NO, then it is a matter of doing more research. To-date, we do not have enough research to show how whole wheat flour would react with non-organic energy sources as compared to white flour.

It is clear from the above analysis, we come across many knowledge-seeking questions and each question demarks a bifurcation point. At each bifurcation point, the question to ask is: is the process natural? The time frame to investigate is many times the characteristic time of a process. For environmental sustainability, the characteristic time of the process is the duration of human species in existence. This can easily transform into infinity, as originally proposed by Khan and Islam (2007: Gulf Book). The process, then, involves taking the limit of a process as time goes to infinity. If the process is still sustainable, it can be taken as a natural process and is good for the environment. Otherwise, the process is unnatural, therefore, unsustainable. This analysis shows that the most important role of the time dimension is in setting the direction. In Fig. 8.1, we have seen a real starting point that would lead to knowledge, whereas an unreal starting point will lead to prejudice. If the time dimension is not considered on a continuous ('continual' is not enough) basis, even the logical steps cannot be traced back in order for one to verify the first premise. It is not enough to back up a few steps, one must back up to the first premise that lead to the bifurcation between knowledge and prejudice. Zatzman et al (2008) have recently highlighted the need of such considerations of the time domain in order to utilize the time dimension as a switch. It turns out with such considerations, scientists cannot determine the cause of global warming with the science that assumes all molecules are identical, thereby, making it impossible to distinguish between organic CO_2 and industrial CO_2; scientists cannot determine the cause of diabetes unless there is a paradigm shift that distinguishes between sucrose in honey and sucrose in Aspartame (Chhetri and Islam, 2007).

8.6.1 Some Applications of the Criterion

The same logic would indicate, unless the science that includes intangibles, the cause(s) of Colony Collapse Disorder (CCD) cannot be determined, either. What remains uncharted is precisely the role of pathways and the passage of time – something that cannot be followed meaningfully in lab-controlled conditions – in transforming the internal basis of changes in certain natural phenomena of interest. One example has been given by Khan and Islam (2007b) in the context of the use of catalysts. Tangible science says catalysts play no role in the chemical reaction equation because they do not appear in the result/outcome. No mass balance accounts for the mass of catalyst lost during a reaction, and no chemical equation accounts for what happens to the 'lost' catalyst molecules when they combine with the products during extremely unnatural conditions. By using the science of tangibles, one can argue that the following patent is indeed a technological breakthrough (El-Shoubary et al, 2003). This patented technology separates Hg from a contaminated gas stream using $CuCl_2$ as the main catalyst. At a high temperature $CuCl_2$ would react with Hg to form Cu-Hg amalgam. This process is effective when combined with fire-resistant Teflon membranes.

If the science of tangibles that describes the mass balance within the confines of the reactor, this is indeed an effective technology. As a testimony to this statement, there has been a number of patents (all with El-Shoubary as an inventor) on the same topic. They are:

1. Patent# 6,841,513 "Adsorption powder containing cupric chloride" Jan 11, 2005.
2. Patent# 6,589,318 "Adsorption powder for removing mercury from high temperature, high moisture stream" July 8, 2003.
3. Patent# 6,5824,97 "Adsorption powder for removing mercury from high temperature high moisture gas stream" June 24, 2003.
4. Patent# 6,558,642 "Method of adsorbing metals and organic compounds from vaporous streams" May 6, 2003.
5. Patent# 6,533,842 "Adsorption powder for removing mercury from high temperature, high moisture gas stream" March 18, 2003.
6. Patent# 6,524,371 "Process for adsorption of mercury from gaseous streams" Feb 25 2003.

This high level of recognition of the technology is expected. After all, what happens to Teflon at high temperature and what happens to Cu-Hg amalgam is a matter of long term or at least time being beyond the 'time of interest' (Khan, 2006 describes this as 'time = right now').

However, if longer-term time is used for the analysis and a bigger area is considered for the mass balance, it would become clear that the same process has actually added more waste to the environment in the form of dioxins released from Teflon and Cu-Hg both being in a more harmful state than their original states in Teflon, $CuCl_2$, and gas stream, respectively. In the efficiency calculation, nearly 90% efficiency is reported within the reactor. This figure makes the process very attractive. However, if the efficiency calculation is conducted including the entire system in which the heater resides, the efficiency drops drastically. In addition, by merely including more elements, the conversion of Hg in natural gas stream and Cu in $CuCl_2$ solution into Cu-Hg sludge as well as the addition of chlorine in the effluent gas pose the difficult question as to what has been accomplished overall.

Another example can be given from the chemical reaction involving honey and Aspartame®. With the science of tangibles, the following reactions take place.

Honey + O_2 → Energy + CO_2 + Water
Aspartame® + O_2 → Energy + CO_2 + Water

In fact, a calorie-conscious person would consider Aspartame® as a better alternative to honey as the energy produced in Aspartame is much less than that of honey for the same weight burnt. An entirely different picture emerges if all components of honey and Aspartame® are included. In this case the actual composition of water as a product are very different for the two cases. However, this difference cannot be observed if the pathway is cut off from the analysis and if the analysis is performed within an arbitrarily set confine. Similar to confining the time domain to the 'time of interest' or time = 'right now', this confinement in space perverts the process of scientific investigation. Every product emerging after oxidation of an artificial substance will come with long-term consequences for the environment. These consequences cannot be included with the science of tangibles. Zatzman and Islam (2007) detailed the following transitions in commercial product development and argued that this transition amounts to an increased focus on tangibles in order to increase the profit margin in the short-term. The quality degradation is obvious, but the reason behind such technology development is quite murky. At present, the science of tangibles is totally incapable of lifting the fog out of this mode of technology development.

Air → Cigarette smoke, toxic smoke
Water → Coke
Tomato → Ketchup

Egg → Mayonnaise
Milk → Commercial ice cream, cheese cake
Corn, potato → Chips (trans fats!)
Salad + oil → Coleslaw
Human asset → Human liability

On the "social sciences" side, the same drive for tangibles is ubiquitous. In the post-Renaissance world, all sciences have been replaced by the science of tangibles that works uniquely on perception. Consider the following transitions:

History, culture → Entertainment, belly dancing
Smile → Laughter
Love of children → Pedophilia
Passion → Obsession
Contentment → Gloating
Quenching thirst → Bloating
Feeding hunger → Gluttony
Philosophy and True science → Religious Fundamentalism
Science → "Technological development"
Social progress → "Economic development"

By contrast, the science of intangibles includes all phenomena that occur naturally, irrespective of what might be detectable. For the use of catalysis, for instance, it can be said that if the reaction cannot take place without the catalyst, clearly it plays a role. Just because at a given time (e.g. time = 'right now') the amount of catalyst loss cannot be measured doesn't mean it (catalyst loss and/or a role of catalysts) doesn't exist. The loss of catalyst is real, even though one cannot measure it with current measurement techniques. The science of intangibles does not wait for the time when one can 'prove' that catalysts are active. Because nature is continuous (without a boundary in time and in space), considerations are not focused on a confined 'control' volume. For the science of tangibles, on the other hand, all that the absence of molecules of the catalyst in the reaction products means is that one would not find that role there.

Tangible science says: you can't find it in the reaction product, so it doesn't count. The science of intangibles says: obviously it counts, but – just as obviously – not in the same way as what is measurable in the tangible mass-balance. This shows that the existing conventional science of tangibles is incomplete. Precisely to the extent that it remains incomplete, on this basis of disregarding or discounting qualitative contributions that cannot yet be quantified in ways that are currently meaningful, this kind of science is bound to become an accumulating source of errors.

The most serious, most important, most significant, truly acid test of a proposed scientific characterization or analysis of any phenomenon is that it account for everything **necessary** and **sufficient** to explain the phenomenon – its origin, its path and its end-point – thereby rendering it positively useful to human society. The same criterion was used in previous civilizations to distinguish between real and artificial. Khan (2007) introduced a criterion that identifies the end-point, by extending time to infinity. This criterion avoids scrutiny of the intangible source of individual action (namely, intention). However, Zatzman and Islam (2007a) pointed out that the end-point at time t = infinity can be a criterion, but it will not disclose the pathway unless a continuous time function is introduced. Mustafiz (2007) used this concept and introduced the notion of knowledge dimension – a dimension that arises from introducing time as a continuous function. In all these deductions, it is the science of intangibles that offers some hope. It is important to note that the insufficiency just mentioned is not overcome by doing "more" science of tangibles "better". It is already evident that what is not being addressed are intangible components that cannot be winkled, ferreted out or otherwise measured by existing means available within the realm of the science of tangibles.

Any number of examples could be cited from the commercial world of product advertising to further illustrate the nub of the problem; this chapter will introduce some of the more egregious cases to illustrate the trend being noted here. Which discipline(s) from the science of tangibles, for example, could model the following?

"In every sense, a Whitestone Cheese is the embodiment of its environment. Pressed by hand, bathed by hand, turned by hand and packed by hand, it is a product of skill and mystery. Like original works of art, no two are alike. While their styles embrace a faint echo of Europe, Whitestone's cheeses are unto themselves … as unique as the land that created them."

(from http://www.deliciousorganics.com/Products/cheese.htm)

We all know hand-made cheese is better-tasting, mother's milk is the best, but do we have a criterion that should lead us to expect them to be best? How about hand-drawn milk as compared to machine-drawn? How about unpasteurized milk as compared to pasteurized? Do we even have a choice? We truly don't, since commercialization is done *after* engineering calculations are made from the science of tangibles. Then, the economics of tangibles is applied to provide the justification with a guarantee.

Intangibles, which essentially include the **root** and **pathway** of any phenomenon, make the science suitable for increasing knowledge, as opposed to increasing confidence in a conclusion that is inherently false (Zatzman and Islam, 2007a). Zatzman and Islam (2007) introduced the following syllogism to make this point about the science of intangibles:

All Americans speak French (major premise)
Jacques Chirac is an American (minor premise)
Therefore Jacques Chirac speaks French (conclusion-deduction)

If, in either the major or minor premise, the information relayed above is derived from a scenario of what is merely *probable* (as distinct from what is actually known), the conclusion, which happens to be correct in this particular case, would be not only acceptable as something independently knowable, but reinforced as something also statistically likely. This then finesses determining the truth or falsehood of any of the premises and, eventually, someone is bound to "reason backwards" to deduce the statistical likelihood of the premises from the conclusion. Indeed this latter version, in which eventually all the premises are falsified as a result of starting out with a false assumption asserted as a conclusion, is exactly what has been identified and labeled elsewhere as the aphenomenal model (Khan et al, 2005). How can this aphenomenal model be replaced with a knowledge model? Zatzman and Islam (2007a) emphasized the need of recognizing the first premise of every scientific discourse. They used the term "a phenomenality" (as contrast to truth) to describe in general the non-existence of any purported phenomenon or of any collection of properties, characteristics or features ascribed to such a purported but otherwise unverified or unverifiable phenomenon. If the first premise contradicts what is true in nature, the entire scientific investigation will be false. Such investigation cannot lead to reliable or useful conclusions.

Consider the following syllogism (the concept of "virtue" intended here is 'that which holds positive value for an entire collectivity of people', not just for some individual or arbitrary subset of individual members of humanity):

1. All virtues are desirable
2. Speaking the truth is a virtue
3. Therefore, speaking the truth is desirable.

Even before it is uttered, a number of difficulties have already been built into this apparently non-controversial syllogism. When it is said that "all virtues are desirable", there is no mention of time factor (pathway) or intention (source of a virtue). For instance, speaking out against an act of aggression is a virtue, but is it desirable? A simple analysis would indicate that unless the time is increased to infinity (meaning something that is desirable in the long-run) practically all virtues undesirable (even giving out in charity requires austerity in the short-term, defending a nation requires self sacrifice – an extremely undesirable phenomenon in the short-term). In the same way, if giving away in charity is a virtue, would that make giving away stolen goods a charity? Robin Hood may be an acceptable hero in the post-Renaissance culture, but is such a categorization scientifically grounded? Giving away stolen goods can be a virtue only if the history (time function) is obliterated. The third component is in the source of an act. For instance, is giving away with the intention of recovering something in the future a virtue? Is helping an oppressor a virtue? This logic shows the need for highlighting both the source (intention) and the pathway (time function going back to the origin) of an action in order to qualify it as a virtue.

The scientifically correct reworking of this syllogism should be:

1. All virtues (both intention and pathway being real) are desirable for t approaching ∞
2. Speaking the truth is a virtue at all times
3. Therefore, speaking the truth is desirable at all times

The outcome of this analysis is the complete disclosure of source, pathway (time function), and final outcome (t approaching ∞) of an action. This analysis can and does restore to its proper place the rational principle underlying the comparing of organic products to synthetic ones, free-range animals to confined animals, hand-drawn milk to machine-drawn, thermal pasteurization with wood fire compared to microwave and/or chemical pasteurization, solar heating compared to nuclear heating, use of olive oil compared to chemical preservatives, use of natural antibiotics compared to chemical antibiotics, and the rest of a long list of such artificial versus natural outcomes and pathways. When it comes to food or other matter ingested by the human body, natural components are to be preferred because we can expect that the source and pathway of such components, already existing in nature, will be beneficial (assuming non-toxic dosages of medicines and normal amounts of food are being ingested). Can we hold out such confidence when it comes to artificially-simulated substitutes for the natural versions? The pathway of the artificial substitute's creation lies outside any process already given in nature, the most important feature of a food.

With the science of tangibles, the outcome being the same (e.g. an apparently similar chemical composition of both the nature-sourced and artificially-sourced food) provides the basis for asserting that there are no significant differences between the natural and the artificial. If source and pathway are ignored, Hitler was elected and so were George W. Bush, Stephen Harper, George Washington, etc. With the science of intangibles, source and pathway are taken into account, blocking the way to such premature and counterfeit declarations. This is not to say there may not be ways to improve upon the best uses humans can make of whatever was already given in the natural realm. If, however, any such improvements are to be sustainable – the only test of truth in nature being that which eventuates over the passage of time – they cannot simply substitute something "chemically identical" from the laboratory or the synthesizing cycle in a chemical plant, completely discounting or ignoring the source(s) and pathway(s) of the natural original. In this, if the time consideration is not time t approaching ∞, then the validity of any conclusion would depend on the intention. If the intention is not phenomenal (real or consistent with environmental sustainability), every conclusion will be aphenomenal. Here, there is only one phenomenal intention, which coincides with the true first premise. The first premise of a logical train (the process of scientific investigation) is equivalent to the intention of an individual act. An intention may properly be considered "phenomenal" if or when it is driven by conscience (Zatzman and Islam, 2007a). It can also be characterized as 'good', with similar application as 'for good' alluding to infinitely long term (Khan et al, 2007).

8.7 The Need for Multidimensional Study

Mousavizadegan et al (2007) indicated that the ultimate truth can be revealed only with an infinite number of dimensions. Abou-Kassem et al (2008) argued that by invoking Einstein's theory of relativity through expression of any event as a continuous function of time, one forces the solution to include infinite dimension. This argument makes it possible to solve problems without extending them to infinite number of dimensions, which would be impractical at this point of human knowledge. The problem then is reduced to solving with only known factors, irrespective of how little the impact of the variable may be on the outcome of scientific analysis. Kvitko (2007) discredited Einstein's relativity altogether. However, he did not elaborate on the first premise of the theory. Our contention is, Einstein's relativity theory appears to be spurious if processed through the science of tangibles. So far, there is no evidence that the first premise of the theory of relativity, as Einstein envisioned it, is aphenomenal.

It is important to note that the observation of natural phenomena as a continuous function of time, including differential frames-of-reference for component processes, is a matter of documenting and reconstructing the actual pathway and steps of an overall process. Because of its implicit standpoint of the neutral external observer, conventional analysis is not capable of fully sorting out these pathways and their distinctive components. The form in which this standpoint expresses itself is embedded in the conventions that come with the "usual" linearizations, *viz.*, viewing time as the independent variable that varies *in* dependently of the processes being observed. Both Eulerian and Lagrangian approaches have the concept of external observer embedded in them. For the Eulerian approach, the external observer is static – a physically impossible and hence absurd state anywhere within nature. For the Lagrangian approach, the external observer is in motion, but within the same pre-defined pathway (conditions for independent variable). To an external observer, intermediate changes-of-state at the interface of successive sub-processes are "invisible", in much the same way that the third dimension is invisible at the interfaces of processes observed in two dimensions. (This is why analysis based on comparing output to input "works" so well with the most linearized models). Within nature, there is no external observer, a state of affairs that renders the processes of tangible science 'aphenomenal'.

Some researchers have indeed recognized the notion of 'external' being aphenomenal. Rather than discarding this notion, however, they adapted the same principle calling it 'God's eye view' (He, 2005), while using Einstein's relativity (continuous time function) as the 'human eye view'. We consider this process of scientific investigation aphenomenal.

The following corollary is the core of the argument advanced in this section:

Just because an equation, or set of equations, describing the transformation of an overall process from input to output, can or may be decomposed into a set of linear superpositions does not mean that any or each of these superpositions describes or represents any actual pathway, or portion thereof, unfolding within Nature.

Consider the following logical train:

1. Perfect is preferable;
2. Nature is perfect;
3. Therefore, anything natural is preferable.

Seeking perfection being embedded with humanity, the first premise sets the selection criterion for any conscience-driven human action. However, this one does not guarantee a phenomenality of the scientific process as the definition of 'perfect' is linked to the notion of ideal. If the 'ideal' is aphenomenal, the meaning of 'perfect' is reversed. As for the second premise, viz, 'nature is perfect' is intricately linked with what is nature. The case in point is a Stanford professor's argument (Roughgarden, 2005). She argues if more than 400 species are found to be practicing 'part-time homosexuality', it must be natural for humans to engage in similar practices. In fact, this argument can be used to demonstrate 'homosexuality is preferable'. What is the problem with this logic? Only one dimension of the problem is being considered. If another dimension is used, it can also be deduced that an incestuous relationship is natural, hence, preferable. When a generalization is made, one must not violate characteristic features of the individual or group of individuals. Conscience, here, is not to be confused with moral or ethical values that are not inherent to humans or at least that are subject to indoctrination, learning, or training. Humans are distinct from all other creatures that we know, because of the presence of conscience – the ability to see the intangibles (both past and future), analyze the consequences of one's action, and decide on a course of action.

Another example can be given as:

1. Perfect is preferable;
2. Nature is perfect;
3. Earthquakes are natural;
4. Therefore, earthquakes are preferable.

Reverse arguments can be made to curse nature. For example, on CNN, this was precisely the issue in Larry King's interview with two former US presidents following the December 2005 Tsunami in the eastern Indian Ocean and Bay of Bengal. There are two problems with this argument.

First of all, it is not a matter of 'preference'. Anything that takes place without human intervention cannot be preferred or discarded. It is not a matter of intention, it is rather a matter of wish, which doesn't necessitate any follow-up human action. Any natural phenomenon (including disasters and calamities) will take place as a grand scheme of natural order, as a necessary component of total balance. This total balance cannot be observed in finite time or finite space. All that can be observed of such a phenomenon in finite time and space are fragmentary aspects of that balance. The phenomenon may not appear to be balanced at all, or alternatively, there may occur some equilibrium state and because – the observation period is sufficiently finite – the equilibrium state is assumed to be 'normal'.

Secondly: if nature is perfect and dynamic at the same time, nature must be moving at an increasingly better state with time. This logic then contradicts Lord Kelvin's assertion that nature is moving from an active to a passive state, reaching a state of useless 'heat death'. This is in sharp contrast to what has been found by Nobel Prize winning work (2001) of Eric Cornell and others. As Eric Cornell outlined in his most popular invited lecture, titled: "Stone Cold Science: Things Get Weird Around Absolute Zero," Kelvin's concept of nature and how nature functions is starkly opposite to the modern concept. At very cold temperatures, phase changes do occur but it has nothing to do with losing power or strength – as commonly understood by the term 'death'. This is further corroborated by later discoveries (Ginsberg et al, 2007). Once again, unless the long-term is being considered over a large scale in space, this transition in universal order or in a laboratory cannot be observed. This is true for floods, lightning, and every natural phenomenon that we observe.

8.8 Assessing the Overall Performance of a Process

In order to break out of the conventional analysis introduced through the science of tangibles, we proceed to discuss some salient features of the time domain and present how the overall performance of a process can be assessed by using time as the fourth dimension. Time t here is not orthogonal to the other three spatial dimensions. However, it is no less a dimension for not being mutually orthogonal. Socially-available knowledge is also not orthogonal either with respect to time t, or with respect to the other three spatial dimensions. Hence, despite the training of engineers and scientists in higher mathematics that hints, or suggests, or implies that dimensionality must be tied up 'somehow' the presence of orthogonality, orthogonality is not in itself a relationship built into dimensionality. It applies only to the arrangements we have invented to render three spatial dimensions simultaneously visible, i.e. tangible.

Between input and output, component phenomena can be treated as lumped parameters, just as, for example, in electric circuit theory, resistance/reactance is lumped in a single resistor, capacitance in a single capacitor, inductance in a single inductor, electromotive potential/force and current of the entire circuit are lumped at a power supply or at special gated junction-points (such as between the base and emitter of a transistor), etc.

Similarly, in the economic theory of commodity transactions, relations of exchange in the market lump all "supply" with the seller and all "demand" with the buyer – even though in reality, as everyone knows, there is also a serious question of a "demand" (need for money) on the part of the seller and there is a certain "supply" (of cash) in the hands of the buyer.

In Nature, or even within certain highly-engineered phenomena, such as an electric circuit, in which human engineering has supplied all the ambient conditions (source of electrical energy, circuit transmission lines, etc.), even after assuming certain simplifying conditions like a near-zero frequency and virtually direct current flow and very small potential differences, we still have no idea whether the current is continuous or how continuous, nor how stable or uniform the voltage difference is at any point in the circuit. The lumped-parameter approach enables us to characterize the overall result/difference/change at output compared to the input without worrying about the details of what actually happened between the input and the output. Clearly, when natural processes are being considered, such an approach leaves a great deal unexplained and unaccounted for. So long as the computed result matches the difference measured between the input and the output, this approach opens the door to imposing any interpretation to account for what happened.

Closely related to the technique of characterizing the operation of a process by means of lumped parameters is the technique of assessing or describing overall performance of the process under study (or development) according to objective, external, uniform "standards" or norms. In the MKS system of SI units, for example, the meter is standardized as a unit of distance according to the length of some rod of some special element maintained in a vacuum bell at a certain temperature and pressure in some location in Paris, France. Similarly the NIST in Washington DC standardizes the duration of the "second" as the fundamental unit of time according to an atomic clock, etc.

The problem with all such standards is that the question of the standard's applicability for measuring something about the process-of-interest is never asked beforehand. Consider the known, and very considerable, physical difference between the way extremely high-frequency (tiny-wavelength) EM waves on the one hand, and much lower-frequency (much-greater wavelength) audible-sound waves, each propagate. The meter may be quite reasonable for the latter case. Does it follow, however, that the nanometer – recall that it is based on subdividing the meter into one *billion* units – is equally reasonable for the former case? The physical reality is that the standard meter bar in Paris actually varies in length by a certain number of picometers or nanometers just within an Earth year. If the process-of-interest is EM radiation traversing light-years through space, however, variation of the standard metre by 1 nanometre or even 1000 picometres will make nonsense of whatever measure we assign to something happening in the physical universe at this scale.

What the objectivity, externality and uniformity of standards enables is a comparison based on what the human observer can directly see, hear, smell, touch or taste – or, more indirectly, measure, according to standards that can be tangibly grasped within ordinary human understanding. However, is science reducible to that which may be tangibly grasped within ordinary human understanding? If science were so reducible, we could, and should, have spent the last 350+ years since Galileo fine-tuning our measurements of the speed of bodies falling freely towards the Earth – so this feature might then be catalogued for different classes of objects according to Aristotle's principle, seemingly quite reasonable, perfectly tangible yet utterly erroneous, that the speed with which objects fall freely towards the Earth is a function of their mass.

This example hints at the solution to the conundrum. Once the principle of gravity as a force – something that cannot be directly seen, heard, smelt, touched or tasted – acting everywhere on the Earth was grasped, measuring and comparing the free fall of objects according to their mass had to be given up – because it was the attraction due to gravity that was the relevant common and decisive feature characteristic to all these freely-falling objects, *not* their individual masses. So, standards of measurement applied to phenomena and processes in Nature should cognize features that are characteristic to those phenomena and processes, not externally applied regardless of their appropriateness or inappropriateness.

Instead of measuring the overall performance of a process or phenomenon under study or development according to criteria that are characteristic, however, statistical norms are frequently applied. These are compared and benchmark performance relative to some standard that is held to be both absolute and external. Public concern about such standards – such as what constitutes a "safe level of background radiation" – has grown in recent years to the point where the very basis of what constitutes a standard has come into question. Recently, Zatzman (2007) advanced the counter-notion of using units or standards that are "phenomenal" (as opposed to aphenomenal). For those who want a science of nature that can account for phenomena as they actually occur or appear in nature, standards whose constancy can only be assured *outside* the natural environment – under highly controlled laboratory conditions, for example, or "in a vacuum" – are in fact entirely arbitrary. Phenomenally-based standards, on the other hand, are natural in yet a deeper sense: they include the notion of characteristic feature that may be cognized by the human observer. These are standards whose objectivity derives from the degree to which they are in conformity with nature. The objectivity of a natural standard cannot and must not be confounded with the vaunted neutrality of position of some external arbiter.

For all the work on intangibles (the mathematics of, the science of, etc.), one must establish:

1. actual, true source
2. actual, true science, pathway
3. actual, true end-point, or completion

Knowledge can be advanced even if the 'true object' is not the entire truth. In fact, it is important to recognize the whole truth cannot be achieved. However, this should not be used as an excuse to eliminate any variable that might have some role but whose immediate impact is not 'measurable'. All these potential variables that might have a role should be listed right at the beginning of the scientific investigation. During the solution phase, this list should be discussed in order to make room for possibilities that at some point one of the variables will play a greater role. This process is equivalent to developing the model that has no aphenomenal assumption attached to it. There is a significant difference between that which tangibly exists for the five senses in some finite portion of time and space, and that which exists in Nature independently of our perceptual functioning in some finite portion of time and space. Our limitation is that we are not able to observe or measure beyond what is tangible. However, the model that we are comparing should not suffer from these shortcomings.

If we grasp the latter first, then the former can be located as a subset. However, errors will occur if we proceed from the opposite direction, according to the assumption that what is perceivable about a process or phenomenon in a given finite portion of time and space contains everything typical and-or characteristic of the natural environment surrounding and sustaining the process or phenomenon as observed in some given finite portion of time and space.

Proceeding according to this latter pattern, for example, mediaeval medical texts portrayed the human foetus as a "homunculus", a miniaturized version of the adult person.

Proceeding according to the former pattern, on the other hand, if we take phase (or "angle") x as a complex variable, de Moivre's Theorem can be used to readily generate expressions for $cos\ nx$ and $sin\ nx$, whereas (by comparison) if we struggle with constructions of right triangles in the two-dimensional plane, it is a computationally intensive task just to derive $cos\ 2x$ and $sin\ 2x$, and orders of magnitude more difficult to extend the procedure to derive $cos\ nx$ and $sin\ nx$.

In technology development, it is important to take a holistic approach. The only single criterion that one can use is the reality criterion. A reality is something that doesn't change with time going to infinity. This is the criterion, Khan (2007) used to define sustainability. If the ranking of a number of options is performed based on this criterion that would be equivalent to the real (phenomenal) ranking. This ranking is absolute and must be the basis for comparison of various options. This ranking is given in the left most column of Table 8.4. In technology development, this natural (real) ranking is practically never used. Based on other ranking criteria, most of the ranking is reversed, meaning the natural order is turned upside down. However, there are some criteria that would give the same ranking as the natural one, but that does not mean that the criterion is legitimate. For instance, the heating value for honey is the highest. However, this does not mean the process is correct, or – putting it in terms of the syllogism that launched Section 2 – it reaffirms that "all Americans do not speak French", i.e. something we already knew all along. This table is discussed in Section 8 *infra* as a starting-point for establishing a "reality index" that would allow a ranking according to how close the product is to being natural.

In engineering calculations, the most commonly used criterion is the efficiency, which deals with output over input. Ironically, an infinite efficiency would mean someone has produced something out of nothing – an absurd concept as an engineering creation. However, if nature does that, it operates on 100% efficiency. For instance, every photon coming out of the sun gets used. So, for a plant the efficiency is limited (less than 100%) because it is incapable of absorbing every photon it is coming in contact with, but it will become 100% if every photon is accounted for. This is why maximizing efficiency as a man-made engineering practice is not a legitimate objective. However, if the concept of efficiency is used in terms of overall performance, the definition of efficiency has to be changed. With this new definition (called 'global efficiency' by Khan et al, 2007 and Chhetri, 2007), the efficiency calculations will be significantly different from conventional efficiency that only considers a small object of practical interest. As an example, consider an air conditioner running outdoors. The air in front of the air conditioner is indeed chilled, while air behind the device will be heated. For instance, if cooling efficiency calculations are performed on an air conditioner running outdoors, the conventional calculations would show finite efficiency, albeit not 100%, as determined by measuring temperature in front of the air conditioner and dividing the work by the work done to operate the air conditioner. Contrast this to the same efficiency calculation if temperatures all around are considered. The process will be proven to be utterly inefficient and will become obvious the operation is not a cooling process at all. Clearly, cooling efficiency of the process that is actually heating is absurd. Consider now, with

Table 8.4. Synthesized and natural pathways of organic compounds as energy sources, ranked and compared according to selected criteria

Natural (real) ranking ("top" rank means most acceptable)	Aphenomenal ranking by the following criteria			
	Bio-degradability	Efficiency[1], e.g. $\eta = \dfrac{Outp - Inp}{Inp} \times 100$	Profit margin	Heating value (cal/g)
1. Honey	2	4 "sweetness/g"	4	1
2. Sugar	3	3	3	2
3. Saccharine	4	2	2	3
4. Aspartame	1	1	1	4
1. Organic wood	1 Reverses depending on applic'n, e.g. durability	4 Reverses if toxicity is considered	4 Reverses if organic wood treated with organic chemicals	4
2. Chemically-treated wood	2	3	3	3
3. Chemically grown, Chemically treated wood	3	2	2	2
4. Genetically-altered wood	4	1	1	1
1. Solar	Not applicable	5 # Efficiency cannot be calculated for direct solar	5	5 # Heating value cannot be calculated for direct solar
2. Gas		4	4	4
3. Electrical		3	3	3
4. Electromagnetic		2	2	2
5. Nuclear		#	1	#

Towards Modeling Knowledge

Item		Anti-bacterial soap won't use olive oil; volume needed for cleaning unit area		Reverses if global is considered		# 1 cannot be ranked	
1. Clay or wood ash	1		6		6		4
2. Olive oil + wood ash	3		5		5		6
3. Veg oil + NaOH	4		4		4		5
4. Mineral oil + NaOH	5		3		3		3
5. Synthetic oil + NaOH	6		2		2		2
6. 100% synthetic (soap-free soap)	2		1		1		#
1. Ammonia	1	Not applicable		Unknown	3	Not applicable	
2. Freon	2				2		
3. Non-Freon synthetic	3				1		
1. Methanol	1		1	For hydrate control	3	Not applicable	
2. Glycol	2		2		2		
3. Synthetic polymers (low dose)	3		3		1		
1. Sunlight		Not applicable	6		3	Not applicable	6
2. Vegetable oil light			5		2		5
3. Candle light			4		1		4
4. Gas light			3				3
5. Incandescent light			2				2
6. Fluorescent light			1				1

†This efficiency is local efficiency that deals with an arbitrarily set size of sample.
*calorie/gm is a negative indicator 'weight watchers' (that are interested in minimizing calories) and is a positive indicator for energy drink makers (that are interested in maximizing calories)

an air conditioner running with direct solar heating. An absorption cooling system means there is no moving part and the solar heat is being converted into cool air. The solar heat is not the result of an engineered process. What would, then, be the efficiency of this system and how would this cooling efficiency compare with the previous one? Three aspects emerge from this discussion. First, global efficiency is the only one that can measure true merit of a process. Secondly, the only efficiency that one can use to compare various technological options is the global efficiency. Thirdly, if one process involves natural options, it cannot be compared with a process that is totally 'engineered'. For instance, efficiency in the latter example (as output/input) is infinity considering no engineered energy has been imparted on the air conditioner.

No engineering design is complete until economic calculations are performed. Therein lies the need for maximizing profit margin. Indeed, the profit margin is the single-most important criterion used for developing a technology ever since the renaissance that saw the emergence of the short-term approach at an unparalleled pace. As Table 8.4 indicates, natural rankings generally are reversed if the criterion of profit maximization is used. This affirms once again how modern economics has turned pro-nature techniques upside down (Zatzman and Islam, 2007).

8.9 Implications of Knowledge-Based Analysis

The principles of the knowledge-based model proposed here are restricted to those of mass (or material) balance, energy balance and momentum balance. For instance, in a non-isothermal model, the first step is to resolve the energy balance based on temperature as the driver for some given time-period, the duration of which has to do with characteristic time of a process or phenomenon. Following the example of the engineering approach employed by (Abou-Kassem, 2007) and (Abou-Kassem et al, 2006), the available temperature data are distributed block-wise over the designated time-period of interest. Temperature being the driver, as the bulk process of interest, i.e. changes with time, a momentum balance may be derived. Velocity would be supplied by local speeds, for all known particles. This is a system that manifests phenomena of thermal diffusion, thermal convection and thermal conduction, without spatial boundaries but giving rise nonetheless to the "mass" component.

8.9.1 A General Case

Figure 8.10 envisions the environment of a natural process as a bioreactor that does not and will not enable conversion of synthetic non-biomass into biomass. The key problem of mass balance in this process, as in the entire natural environment of the earth as a whole, is set out in Figure 6: the accumulation rate of synthetic non-biomass continually threatens to overwhelm the natural capacities of the environment to use or absorb such material.

In evaluating Equation (8.16), it is desirable to know all the contents of the inflow matter. However, it is highly unlikely to know all the contents, even at macroscopic level. In absence of a technology that would find the detailed content, it is important to know the pathway of the process to have an idea of the source of impurities. For instance, if de-ionized water is used in a system, one would know that its composition would be affected by the process of de-ionization. Similar rules apply to products of

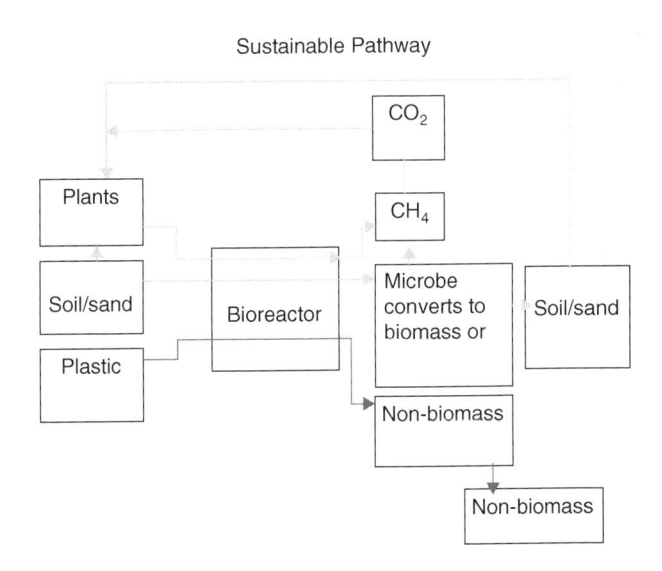

Figure 8.10 Sustainable pathway for material substance in the environment.

organic sources, etc. If we consider combustion reaction (coal, for instance) in a burner, the bulk output will likely to be CO_2. However, this CO_2 will be associated with a number of trace chemicals (impurities) depending upon the process it passes through. Because Equation 16 includes all known chemicals (e.g. from source, adsorption/desorption products, catalytic reaction products), it would be able to track matters in terms of CNB and NCNB products. Automatically, this analysis will lead to differentiation of CO_2 in terms of pathway and the composition of the environment, the basic requirement of Equation (8.17). According to Equation (8.17), charcoal combustion in a burner made up of clay will release CO_2 and natural impurities of charcoal and the materials from the burner itself. Similar phenomenon can be expected from a burner made up of nickel plated with an exhaust pipe made up of copper.

Anytime, CO_2 is accompanied with CNB matter, it will be characterised as beneficial to the environment. This is shown in the positive slope of Fig. 8.11. On the other hand, when CO_2 is accompanied with NCNB matter, it will be considered to be harmful to the environment, as this is not readily acceptable by the eco-system. For instance, the exhaust of the Cu or Ni-plated burner (with

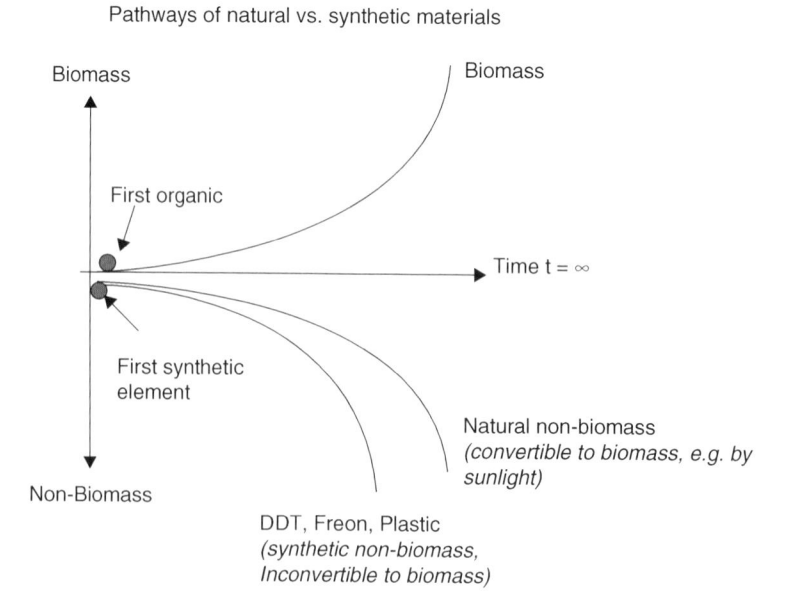

Figure 8.11 Transitions of natural and synthetic materials.

Towards Modeling Knowledge

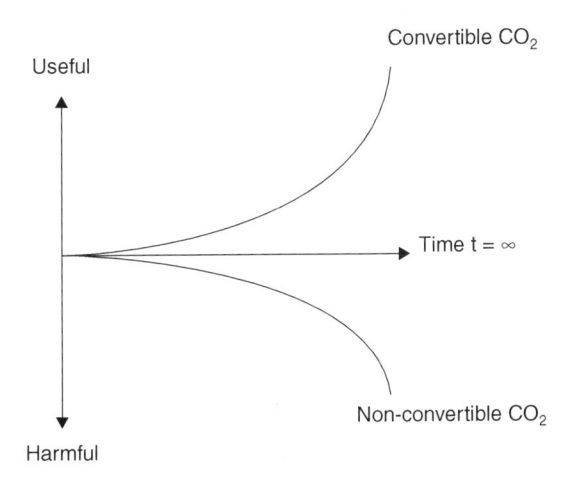

Figure 8.12 Divergent results from natural and artificial.

catalysts) will include chemicals, e.g. nickel, copper from pipe, trace chemicals from catalysts, beside bulk CO_2 because of adsorption/desorption, catalyst chemistry, etc. These trace chemicals fall under the category of NCNB and cannot be utilized by plants (negative slope from Fig. 8.11). This figure clearly shows that on the upward slope case is sustainable as it conforms to the natural eco-system. With conventional mass balance approach, the bifurcation graph of Fig. 8.11 would be incorrectly represented by a single graph that is incapable of discerning between different qualities of CO_2 because the information regarding the quality (trace chemicals) are lost in the balance equation. Only recently, the work of Sorokhtin et al (2007) has demonstrated that without such distinction, there cannot be any scientific link between global warming and fossil fuel production and utilization. In solving Equation (8.16), one will encounter a set of non-linear equations. These equations cannot be linearized. Recently, Moussavizadegan et al (2007) proposed a method of solving non-linear equations. The principle is to cast the governing equation in engineering formulation, as outlined by Abou-Kassem et al (2006), whose principles were further elaborated in Abou-Kassem (2007).

The non-linear algebraic equations then can be solved in multiple solution mode. Mousavizadegan (2007) recently solved such an equation to contemporary professionally acceptable standards of computational efficiency. The result looked like what is pictured in Fig. 8.13.

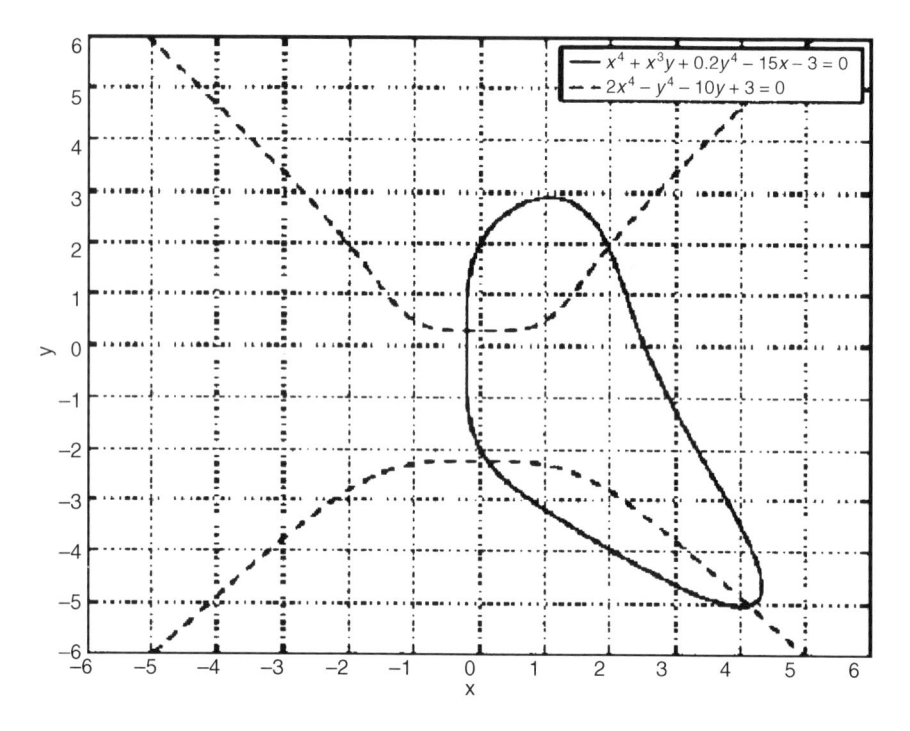

Figure 8.13 The solution behaviour manifested by just two non-linear bivariate equations, $x^4 + x^3y + 0.2y^4 - 15x - 3 = 0$ and $2x^4 - y^4 - 10y + 3 = 0$, suggests that a "cloud" of solutions would emerge.

8.9.2 Impact of Global Warming Analysis

In light of the above analysis shown in Section 8.9.1, consider the problem we encounter in evaluating global warming and its cause, as considered by Chhetri and Islam (2008b).

The total energy consumption in 2004 was equivalent to approximately 200 million barrels of oil per day which is about 14.5 terawatts, over 85% of which comes from fossil fuels (Service, 2005). Globally, about 30 billion tons of CO_2 is produced annually from fossil fuels, which includes oil, coal and natural gas (EIA, 2004). The industrial CO_2 produced from fossil fuel burning is considered solely responsible for the current global warming and climate change problems (Chhetri and Islam, 2007). Hence, burning fossil fuels is not considered to be a sustainable option. However, this 'sole responsibility' is not backed with science (in absence of our analysis above). The confusion emerges from the fact that conventional analysis doesn't distinguish CO_2 from natural processes (e.g. oxidation in national systems, including breathing) from emissions that come from industrial or man-made devices. This confusion leads to making the argument that man-made activities cannot be responsible for global warming. For instance, Chilingar and Khilyuk (2007) argued that the emission of greenhouse gases by burning of fossil fuels is not responsible for global warming and hence is not unsustainable. In their analysis, the amount of greenhouse gases generated through human activities is scientifically insignificant compared to the vast amount of greenhouse gases generated through natural activities. The factor that they do not consider, however, is that greenhouse gases that are tainted through human activities (e.g. synthetic chemicals) are not readily recyclable in the ecosystem. This means when 'refined' oil comes in contact with natural oxygen, it produces chemicals (called non-convertible, non-biomass, NCNB, see Equations 8.19–8.21.

At present, for every barrel of crude oil approximately 15% additives are added (California Energy Commission, 2004). These additives, with current practices, are all synthetic and/or engineered materials that are highly toxic to the environment. With this 'volume gain', the following distribution is achieved (Table 8.5).

Each of these products is subject to oxidation (either through combustion or low-temperature oxidation, which is a continuous

Table 8.5 Petroleum products yielded from one barrel of crude oil in california

Product	Percent of Total
Finished Motor Gasoline	51.4%
Distillate Fuel Oil	15.3%
Jet Fuel	12.3%
Still Gas	5.4%
Marketable Coke	5.0%
Residual Fuel Oil	3.3%
Liquefied Refinery Gas	2.8%
Asphalt and Road Oil	1.7%
Other Refined Products	1.5%
Lubricants	0.9%

From California Energy Commission, 2004

process). As one goes toward the bottom of the table, the oxidation rate is decreased but the heavy metal content is increased, making each product equally vulnerable to oxidation. The immediate consequence of this conversion through refining is that one barrel of naturally occurring crude oil (convertible non-biomass, CBM) is converted into 1.15 barrels of potential non-convertible non-biomass (NCNB) that would continue to produce more volumes of toxic components as it oxidizes either though combustion or through slow oxidation (Islam et al, 1991). Refining is by in large the process that produces NCNB, similar to the processes described in Equations 8.19–8.21. The pathways of oil refining illustrate that the oil refining process utilizes toxic catalysts and chemicals, and the emissions from oil burning also becomes extremely toxic. Figure 8.14 shows the pathway of oil refining. During cracking of the hydrocarbon molecules, different types of acid catalysts are used along with high heat and pressure. The process of employing the breaking of hydrocarbon molecules is the thermal cracking. During alkylation, sulfuric acids, hydrogen fluorides, aluminum chlorides and platinum are used as catalysts. Platinum, nickel, tungsten, palladium and other catalysts are used during hydro processing. In distillation, high heat and pressure are used as catalysts. As an example, just from oxidation of the carbon component,

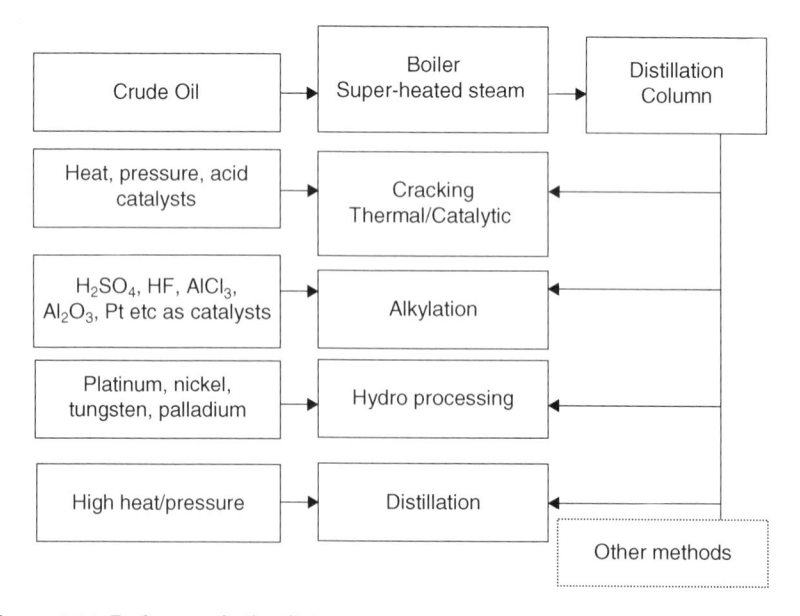

Figure 8.14 Pathway of oil refining process.

1 kg of carbon, which was convertible non-biomass, would turn into 3.667 kg of carbon dioxide (if completely burnt) that is now no longer acceptable by the ecosystem, due to the presence of the non-natural additives. Of course, when crude oil is converted, each of its numerous components would turn into such non-convertible non-biomass. Many of these components are not accounted for or even known, let alone a scientific estimation of their consequences. Hence, the sustainable option is either to use natural catalysts and chemicals during refining or to design a vehicle that directly runs on crude oil based on its natural properties. The same principle applies to natural gas processing (Chhetri and Islam, 2008).

8.10 Examples of Knowledge-Based Simulation

A series of computer simulation runs were performed, using the knowledge-based approach. The first series involved the solution of the 1-D multiphase flow problem. The Buckley-Leverett approach involves neglecting the capillary pressure term, leading to spurious multiple solutions. In order to remedy this problem, the well known 'shock' is introduced. While this approach is a practical solution to the problem, it is scientifically inaccurate. The knowledge-based approach requires that *à priori* simplification and/or linearization be avoided. The governing 1-D multiphase equation, including the capillary pressure term, was solved using Adomian domain decomposition technique. This technique is capable of solving non-linear equations. The details of the approach are available elsewhere (Mustafiz et al, in press). Figure 8.15 shows the results obtained for this case. This figure shows how the spurious multiple solutions disappear in favor of a set of monotonous functions. However, one would easily recall that such solutions also appear with conventional finite difference approach, even though this method does use linearization, albeit at a later stage (during matrix inversion).

In order to determine the role of this linearization, a series of numerical runs was performed using a non-linear solver. No linearization was performed during the solution of these multiphase flow problems.

The flow depends on two types of parameters. The first types are those that are changed with the variation of pressure, such as the fluid formation volume factor, the viscosity of the oil and the porosity of the formation. The second type depends on the variation of water saturation. These are the relative permeability of oil and

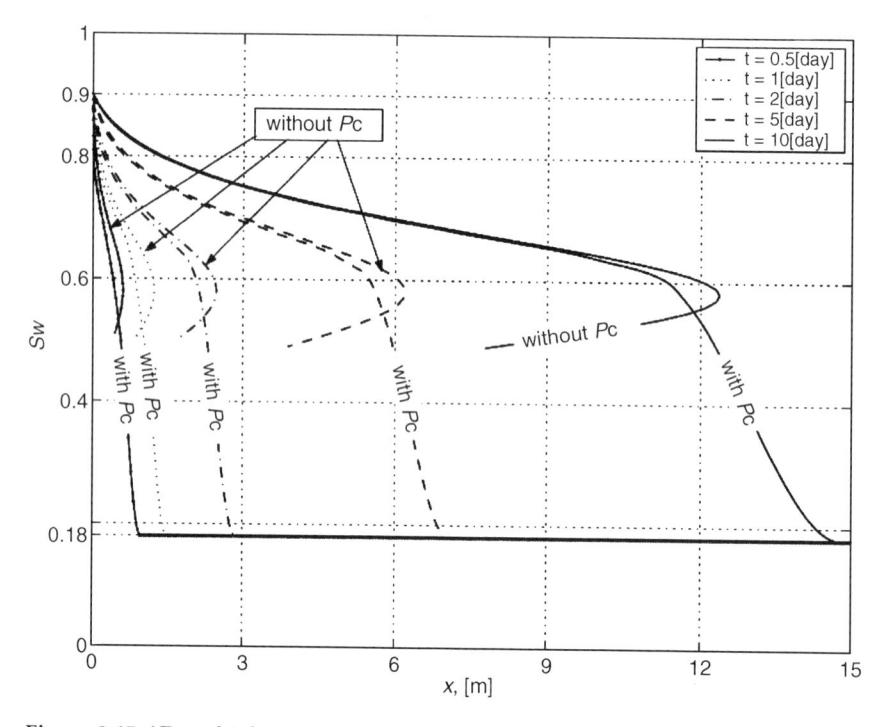

Figure 8.15 1D multiphase problem solved without linearization (classical Buckley Leverett equation).

water and the capillary pressure. The parameters are categorized based on their dependency on the pressure or water saturation and the effect of each group are studied separately. A complete description of these runs is available elsewhere (Islam et al, 2008). In the first series of runs, the value of the fluid formation volume factor in the process of oil production was varied between 1.2063 to 1.22 from the reservoir pressure to the bubble point pressure. We neglect the compressibility of the fluid and take that B_o = 1.22 during the production process. The pressure and water saturation distribution are computed for a year with Δt = 1(day) and n = 256. The result for t = 3, 6, 9 and 12 (months) are given in Fig. 8.16. The obtained results are then compared with the solutions when the fluid formation volume factor is changed with change in pressure. There are no significant differences at early months of production. But, the differences are increased with time and with the increase in production time. However, it shows that the variation of B_o has a minor effect on the distribution of p_o and S_w.

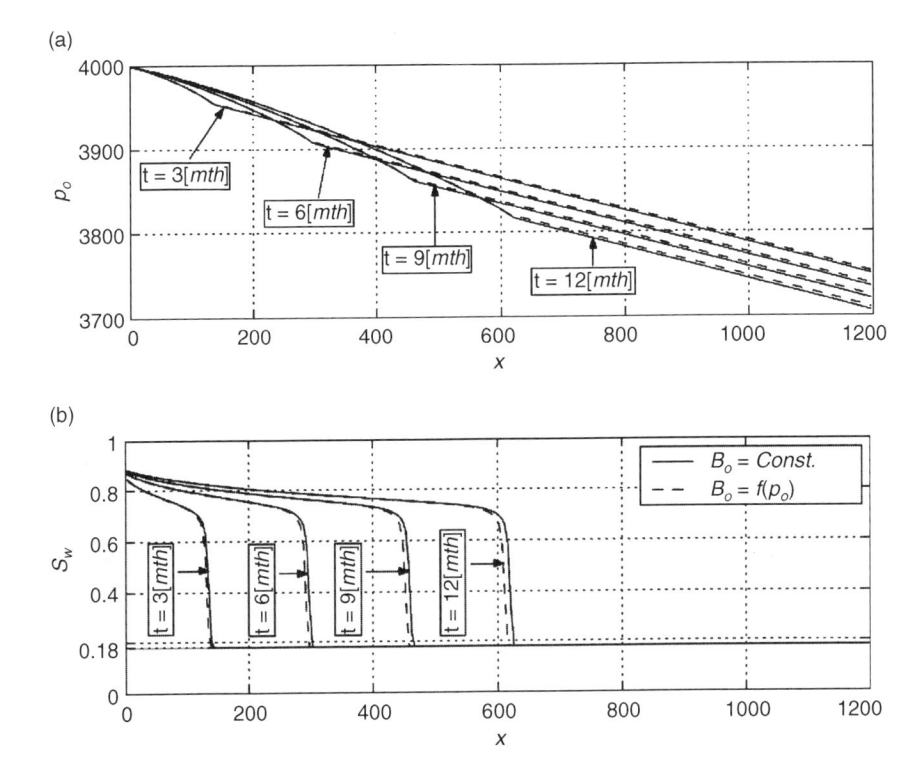

Figure 8.16 Distribution of oil phase pressure and water saturation for constant and variable fluid formation volume factors.

The variation of the viscosity with the pressure is neglected and it is assumed that μ_o = 1.1340 cp during the production process. This corresponds to the viscosity at the initial reservoir pressure, P_R = 4000 psia. The computations are carried out for a year with Δt = 1(day) and m = 256. The pressure and water saturation distributions are shown in Fig. 8.18 for t = 3, 6, 9 and 12 (months). The results are compared with the case when μ_o = f(p_o) as given in (8.4). There are not significant differences between the results when μ_o = Const. and μ_o = f(p_o) for various depicted times. It indicates that the variation of viscosity has not a major effect on the pressure and water saturation distributions.

The porosity variation due to pressure change is neglected in this part. It is assumed that $\varphi = \varphi 0 = 0.27$ during the production process. The computations are carried out for a year with Δt = 1(day) and m = 256. The pressure and water saturation distributions do not demonstrate any difference between two cases of φConst. and φ = f(po). The pressure and water distributions are shown in Fig. 8.18. There is not any difference between the graphs at a certain time with constant and variable φ, respectively, and the diagrams coincide completely.

In order to study the combined effects, a series of runs was conducted. It was assumed that B_o = 1.22, μ_o = 1.1340 cp and φ = φo = 0.27 during the production process. The computations are carried out for a year with Δt = 1(day) and n = 256. The pressure and water saturation distributions are given in Fig. 8.18 for t = 1(year). The results are compared for various cases. Note that all the results were obtained without linearization at any stage. From these results, it appears that the oil formation volume factor, B_o has major effects while the porosity variation has the minor effect on the oil pressure and water saturation distributions. The water permeability of water and oil and the capillary pressure are dependent on the variation of water saturation. The pressure distribution of the single and double

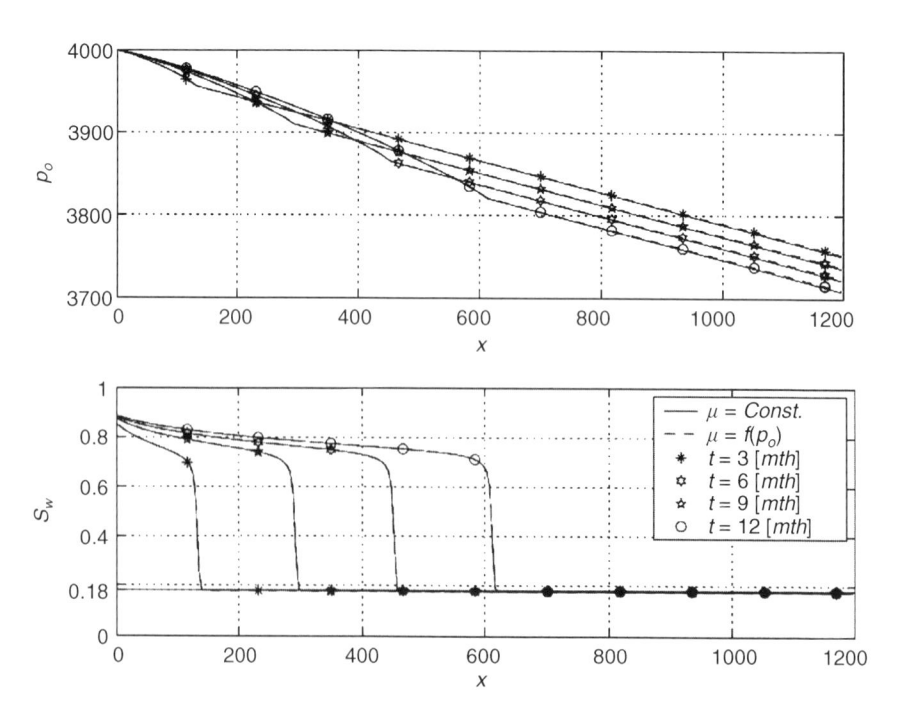

Figure 8.17 The distribution of oil phase pressure and water saturation for constant and variable oil viscosity.

Towards Modeling Knowledge 417

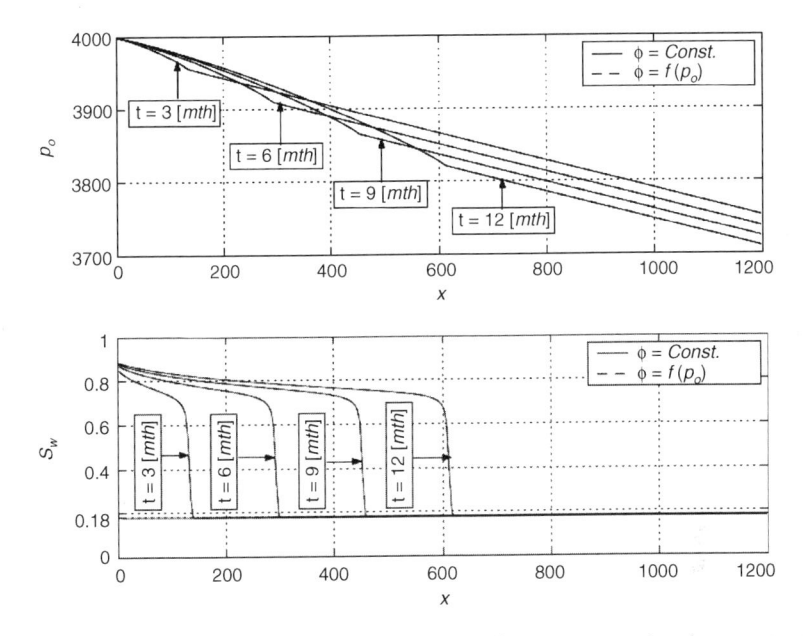

Figure 8.18 Distribution of oil phase pressure and water saturation for constant and variable porosity.

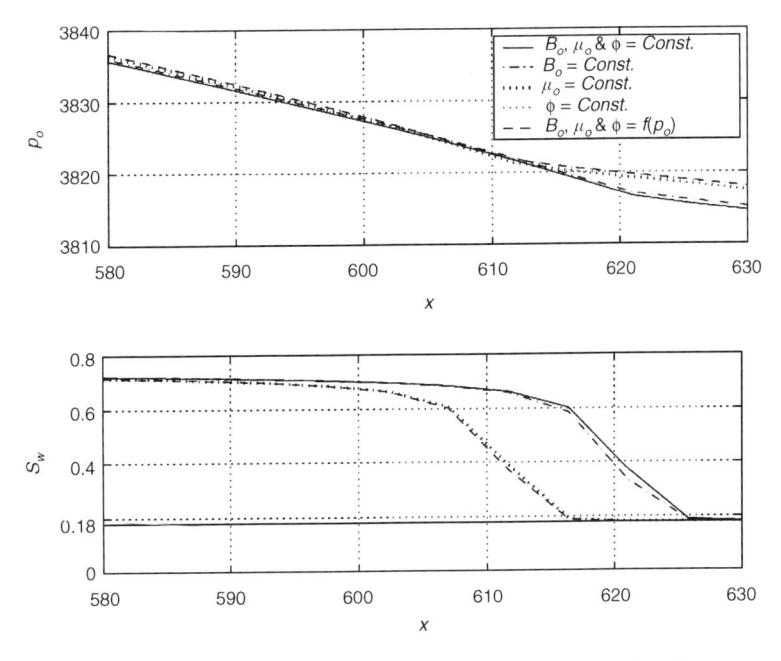

Figure 8.19 Distribution of oil phase pressure and water saturation for constant and variable pressure dependent parameters.

418   Advanced Petroleum Reservoir Simulation

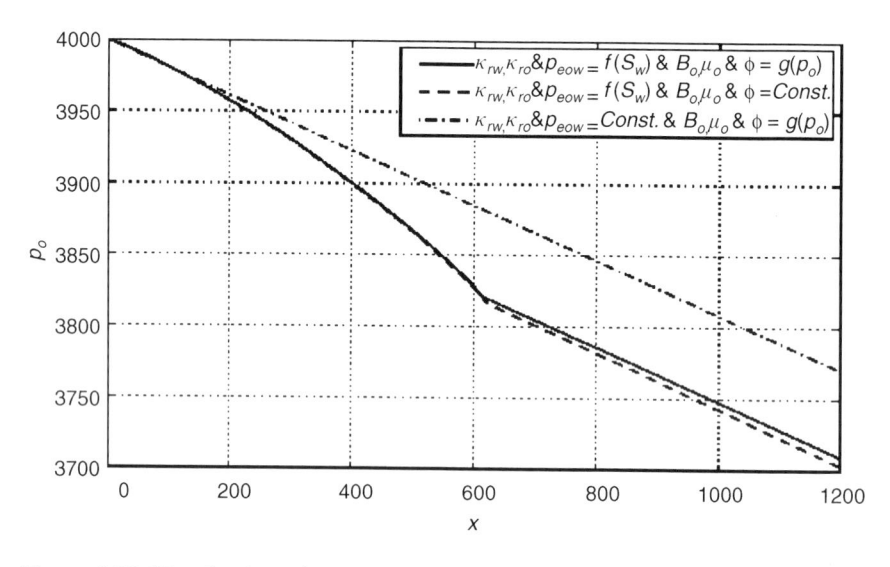

Figure 8.20 Distribution of pressure and water saturation for constant and variable cases.

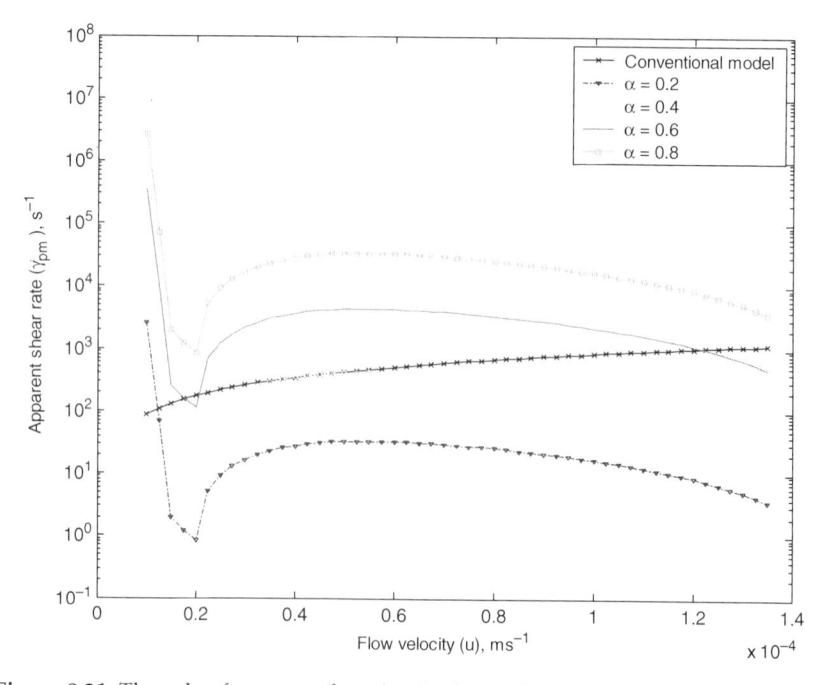

Figure 8.21 The role of memory function in determining apparent viscosity.

Towards Modeling Knowledge

phase flows are given for t = 1(year) in Fig. 8.19. The pressure distribution with neglecting the variation of the pressure dependent parameters is also depicted to provide a comparison and to find out the effect of different parameters on the final results. The effect of the pressure dependent parameters is very small compared with the influence of the water saturation variation.

The effective permeabilities to water and oil and the capillary pressure are dependent on the variation of water saturation. The pressure distribution of the single and double phase flows are given for t = 1(year) in Fig. 8.20. The pressure distribution with neglecting the variation of the pressure dependent parameters is also depicted to provide a comparison and to find out the effect of different parameters on the final results. The effect of the pressure dependent parameters is very small compared with the influence of the water saturation variation.

Finally, results using a recently developed fluid rheology model that includes fluid memory are presented in Fig. 8.21. Once again, fluid memory is considered to be the continuous time function. With the inclusion of the memory, results deviate significantly from the conventional approaches. When this rheology is coupled with the momentum balance equation, classical cases of multiple solutions arise. Because the exact form of the memory is never known, this depiction would give one an opportunity to refine the prediction envelope, rather than putting too much emphasis on a single solution. Further details of this technique are available in recent work of Hossain (2008).

9
Final Conclusions

The most important aspect of petroleum reservoir management is in the prediction of fluid flow behavior during petroleum production. Petroleum reservoir management hinges on accurate prediction of the flow of petroleum fluids. The use of computer simulators is ubiquitous in the petroleum industry. Computer simulation involves the solution of governing equations that have a physical basis for them. These equations are always non-linear. However, it is well known that all currently used commercial computer simulators use linear solvers that produce solutions to the set of linear algebraic equations that are derived by linearizing non-linear governing equations. It is considered that by linearizing governing equations, approximate solutions are close to exact solutions that are not achievable due to the lack of non-linear solvers.

Even though the Information Age has promised realtime data collection, instant data transfer, and ultra-fast computing, these special features alone cannot increase the confidence in reservoir performance prediction. Little has been done in terms of improving the science behind numerical solutions. Yet, flaws in fundamental science and mathematics would account for the most important aspect of accurate prediction of reservoir performance. In this book, fundamental improvements to the solution techniques are proposed with results that show the value of the advanced approach.

In this book, the advanced approach leads to the following major conclusions:

1. The use of linear solvers is not justified for most of the realistic range of petroleum parameters, even for single-phase flow. As much as 30% deviation occurs in terms of pressure values if linearization was avoided altogether. This deviation would translate into large errors in the prediction of petroleum reservoir performance.
2. The impact of linearization is even more intense for multiphase flow. Linearization of governing equations is likely to pervert subsequent results, biasing the decision making process irreversibly. It is recommended

that reservoir model equations be solved without linearization of governing equations.

3. The possibility of multiple solutions is inherent to reservoir simulation problems. For petroleum engineering applications, multiple solutions were observed as early as the 1950's and were correctly dubbed as 'spurious' by the ground-breaking work of Buckley and Leverett. This multiple solution occurred because the capillary pressure term was dropped from the governing equation. In order to avoid this problem, the notion of shock was introduced in place of a realistic transition of the saturation profile. Even though it was recognized that the need for the introduction of the shock is eliminated with the introduction of the capillary term decades ago, only recently it was solved with a non-linear solvers. Non-linear solvers indicate that real multiple solutions are possible.

4. The use of correct criterion is extremely important. The criterion involves the use of full continuous time function. Unless the correct criterion is used, the entire decision making process becomes aphenomenal, leading to prejudice in the cognition process.

5. While modeling enhanced oil recovery (EOR), the correct criterion involves the determination of the flow regime. For an unstable flow regime, solving non-linear equations becomes the most important task. Without the use of rigorous solution techniques, the predicted improvement due to the use of an EOR scheme can be an indication of the numerical error rather than actual improvement in oil recovery.

6. By using comprehensive energy and mass balance equations, engineering schemes suitable for sustainable petroleum operations can be predicted. With the proposed new approach, it is possible to determine the true cause of global warming, which, in turn, would help develop sustainable petroleum technologies.

Appendix A

User's Manual for Multi-Purpose Simulator for Field Applications (MPSFFA, Version 1–16)

Software is available for download from the website www.scrivenerpublishing.com

A.1 Introduction

This manual provides a brief description of a multi-purpose reservoir simulator (MPSFFA) and its capabilities, detailed information on data file preparation, and a detailed description of the variables used in preparing a data file. It also gives instructions for running the reservoir simulator and the graphic post processor on a PC. The primary purpose of presenting this simulator as part of this book is to help MS and PhD graduate students who develop simulation models in their research to check their intermediate and final results and to provide simulation engineers and independent consultants with an advanced tool to run small size field cases and aid in the interpretation of simulation results using the graphic post processor. Educators may use the simulator in their teaching to train undergraduate students.

A.2 The Simulator

The present simulator (MPSFFA) models multi phase fluid, including polymer solution, flow in petroleum reservoirs. Model description (flow equations and boundary conditions), well operating conditions, and methods of solving the algebraic equations of the model are described in detail elsewhere (Luchmansingh, Ertekin, and Abou-Kassem, 1991; Abou-Kassem, Osman, and Zaid, 1996; Abou-Kassem, 1996; Ertekin, Abou-Kassem, and King, 2001; Abou-Kassem, Farouq Ali, and Islam, 2006) The simulator was written in

FORTRAN 95 and the graphic post processor was developed for use with MATLAB version 5.1a and higher.

The simulator is fully implicit, multi-dimensional, and multi-purpose. The model equations are linearized using Newton's iteration and the resulting linear equations are solved simultaneously (for natural ordering and D4 ordering) or iteratively employing advanced Block Iterative method (Behie and Vinsome, 1982) and Nested Factorization method (Appleyard and Cheshire, 1983) accelerated by ORTHOMIN (Vinsome, 1976). The simulator is based on 5-point scheme discertization and uses a block-centered grid. It can be run in one-, two-, or three-dimensional mode in rectangular coordinates (x-y-z space) for modeling field performance and in one- or two-dimensional mode in radial-cylindrical coordinates (r-z space) for modeling single-well performance. The simulator efficiently simulates several processes of practical interest to petroleum engineers including polymer injection, black-oil (oil-water-gas) reservoirs, oil-water reservoirs, oil-gas reservoirs, and gas reservoirs. It can be run as a single-phase, two-phase, or three-phase simulator and as a material balance equation (zero-dimensional simulator). The simulator implements advanced automatic time-step control based on the minimization of truncation error (Sammon and Rubin, 1984). *It should be mentioned that the current version of the simulator handles single-block wells only and it needs further development to handle multi-block wells.*

The simulator has the capacity to model irregular boundary reservoirs with the inactive grid blocks within or outside the reservoir being removed at the matrix level (Abou-Kassem and Ertekin, 1992). The simulator models heterogeneous reservoirs having rock regions with different relative permeability and capillary pressure data and different PVT and viscosity data. It permits performing manual history matching. It has sophisticated and yet simple data inputting procedure (Abou-Kassem et al, 1996). The formulation in the simulator removes, at the matrix level, the polymer equation when simulating a black-oil reservoir, the polymer and gas equations when simulating an oil-water reservoir, the polymer and water equations when simulating an oil-gas-irreducible water reservoir, and the polymer and oil equations when simulating gas reservoirs (Abou-Kassem et al, 1996). This capability considerably reduces the CPU time and memory storage requirements. Although the simulator does not permit multi-block wells, various well operating conditions are simulated.

APPENDIX A

A.3 Data File Preparation

The data required for the present simulator are classified into groups based on how the data within each group are related. A group of data could be as simple as defining a few related variables or as complicated as defining the variables for the well recursive data. These groups of data are classified, according to their format of input procedure, into five categories (A, B, C, D, and E). Categories A, B, and C include 27, 11, and 5 groups of data, respectively, whereas Categories D and E include one group of data each. The data of each category are entered using a specific format procedure; for example, category A uses format procedure A, category B uses format procedure B, etc. Each group of data carries an identification name consisting of the word "DATA" followed by a number and an alphabet character; the number identifies the group and the alphabet character identifies the category and the format procedure. For example, DATA 06B identifies a group of data that belongs to Category B and uses Format Procedure B, and whose variables are defined under DATA 06B in Sec. A.4. *Items of data that are not specific to the problem in question but are requested by the simulator can be set to zero, assigned dummy values, or left unchanged as supplied in the model data file.* Data file preparation, including format procedures and description of variables, follows the work of Abou-Kassem, Osman, and Zaid (1996) for black-oil simulation. The CD that accompanies this book contains examples of data files and output files for four problems reported in the literature: gas injection in a 3D reservoir (Odeh, 1981), water injection in a 2D reservoir (Yanosik and McCracken, 1979), polymer slug injection in a 1D reservoir (Lutchmansingh, 1987), and 2D, 3-phase black oil reservoir (Ertekin, Abou-Kassem, and King, 2001); and the data files and output files for three single-phase problems reported by Abou-Kassem, Faruoq Ali, and Islam (2006). In addition, three more data and output files are included for single-phase water reservoir, 3-phase single-well simulation, and the use of simulator as a material balance equation.

Each format procedure is introduced by a title line (line 1), which includes the identification name and the group of data to be entered, followed by a parameter sequence line (line 2), which lists the order of parameters to be entered by the user. Only format procedure D has an additional parameter sequence line (line 3). The user, in each subsequent data line, enters the values of the parameters ordered and preferably aligned with the parameters shown in the parameter sequence line for easy recognition. Each of format procedures B and E requires a single line data entry, whereas format procedures A, C, and D require multiple-line data entry and terminate with a line of zero entries for all parameters. The various groups of data and any specific instructions for each format procedure are presented in the following sections.

A.3.1 Format Procedure A

This format procedure is suitable for entering data that describe the distribution of a grid block property over the whole reservoir. Such data include block size and permeability in the x, y, and z directions; depth; porosity; fluid saturation; pressure; solution GOR at bubble-point pressure; polymer concentration; rock adsorption; maximum rock adsorption; modifiers for porosity, depth, bulk volume, and transmissibilities in the x, y, and z directions; and block identifiers that label a grid block as being active, belonging to a rock region, belonging to a PVT region, and receiving influx from an aquifer.

Each line of data (for example, line 3) represents a property assignment for an arbitrary reservoir region having the shape of a prism with I1, I2; J1, J2; and K1, K2 being its lower and upper limits in the x, y, and z directions. The data entered by each subsequent line (for example, line 4) are superimposed on top of the data entered by all earlier lines; that is, the final distribution of a property is the result of the superposition of the entire arbitrary reservoir regions specified by all lines of data. This option is activated by setting the option identifier at the beginning of the parameter sequence line (line 2) to 1. This is a powerful method for entering data if a block property is distributed into well-defined (not necessarily regular) reservoir regions. For a homogeneous property distribution, only one line of data is needed (with I1 = J1 = K1 = 1, I2 = n_x, J2 = n_y, and K2 = n_z). If, however, a block property is so heterogeneous that it varies from block to block and regional property distribution is minimal, this method loses its effectiveness. In such cases, the option identifier at the beginning of the parameter sequence line (line 2) is set to 0 and the data for all blocks are entered sequentially in a way similar to natural ordering of blocks along rows (i.e. i is incremented first, j is incremented second, and k is incremented last). In this case, both active and inactive blocks are assigned property values and the terminating line of zero entries is omitted.

APPENDIX A

A.3.2 Format Procedure B

This format procedure is suitable for entering data involving a combination of integer and/or real variables. Groups of data of this type include the options for method of solution, block ordering scheme, restart-up, units of input and output, and interpolation; control integers for printing and debugging options, number of grid blocks in the x, y, and z directions; reservoir reference depth and reservoir orientation in space; oil API and gas and water specific gravities; oil and porosity compressibilities, rock density, and reference pressure for porosity; reservoir initialization data; time step control data, simulation time; number of rock regions; and number of PVT regions. Note that the values of the parameters are entered in line 3. They are ordered and aligned with the parameters shown in the parameter sequence line (line 2) for easy recognition.

A.3.3 Format Procedure C

This format procedure is suitable for entering property tables such as oil-water relative permeability and drainage and imbibition capillary pressure data, gas-oil relative permeability and drainage and imbibition capillary pressure data, PVT data, polymer viscosity data, and resistance factor and residual resistance factor data. The parameter sequence line (line 2) lists first the independent variable followed all other dependent variables. It is important to note that range of change of the independent variable in any table must cover the range that is expected to take place in the reservoir. Each line of data represents one entry in the table of data that corresponds to a specified value of the independent variable. The data of a table are entered in the order of increasing the value of the independent variable. A one-line table is; however, possible if the properties in the table are not functions of the independent variable. Note that the data in each line of data (for example, line 3) are ordered and aligned with the parameters specified on the parameter sequence line (line 2) for easy recognition.

A.3.4 Format Procedure D

This format procedure is suitable for entering well recursive data. As mentioned earlier, this format procedure has two parameter sequence lines. The parameters in the first parameter sequence line (line 2) include a time specification which signals new user's request

(SIMNEW), an over-ride time step to be used (DELT), the number of wells changing operation conditions (NOW), an over-ride printing option (IRITEI), a well economic limit (QOECON), and well production constraints on WOR (WORMAX) and GOR (GORMAX). This line of data can be repeated but each subsequent line must have a time specification larger than the last time specification. The parameters in the second parameter sequence line (line 3) include data for individual wells such as well type (IWT), well identification number (IDW), well coordinates (IW, JW, KW), well operating condition (IWOPC), well geometric factor (GW), the minimum (or maximum) flowing bottom-hole pressure (BHP), the specified maximum production (injection) rate (QSP), the polymer concentration of injected fluid (CONCPI), and the tracer concentration in injected aqua (CONCTI). There must exist NOW lines describing NOW individual wells immediately following the line where NOW specification appears if NOW >0. Using this format procedure, any number of wells can be introduced, shut in, re-opened, re-completed, etc. at any number of key times during both the history matching phase and future performance prediction phase of a study.

A.3.5 Format Procedure E

This format procedure is used to enter one line of information, such as the name of the user and the title of the computer run, consisting of up to eighty alphanumeric characters.

A.4 Description of Variables Used in Preparing a Data File

There are 45 data groups in the data file. The descriptions of variables within each data group are listed under the data group itself. Follows is a list of all 45 data groups starting with DATA 01E and ending with DATA 45D.

Data 01E	Title of Simulation Run
TITLE	Name of user and title of simulation run (one line having up to eighty alphanumeric characters)
DATA 02B	**Restart-up Option and Simulation Time Data**

Appendix A

IPRTDAT	Option for printing and debugging the input data file
	= 0, do neither print nor debug the input data file.
	= 1, print the input data file and activate messages to debug the file.
IRSTRT	Restart-up option
	= 0, simulation starts from initial conditions.
	= 1, simulation continues from a restart-up file that was written previously.
TMTOTAL	Maximum simulation time, D [d]
TMSTOP	Time to stop this simulation run, D [d]
DELT	Time step to be used if IRSTRT = 1, D [d]

Notes:	1) *Detailed output will be printed and restart-up file will be written at TMSTOP.*
	2) *For IRSTRT = 1, the user has the option to modify any or all items of information contained in DATA 01E, DATA 02B, DATA 03B, DATA 04B, DATA 05B, and DATA 45D for the rest of the simulation run.*

DATA 03B	**Units, Interpolation, Time-step Control, and Convergence Options**
MUNITIN	Option for units of input data
	= 1, customary units
	= 2, SPE preferred metric units
	= 3, laboratory units
MUNTOUT	Option for units of output
	= 1, customary units
	= 2, SPE preferred metric units
	= 3, laboratory units
IQUAD	Degree of interpolating polynomial
	= 0, linear interpolation using entries of supplied tables
	= 1, linear interpolation using generated equi-spaced tables
	= 2, quadratic interpolation using generated equi-spaced tables
NINTBLE	Number of entries in generated equi-spaced tables
DTMAX	Maximum allowed time step, D [d]
DTMIN	Minimum allowed time step, D [d]

RERORP	Relative error parameter, $0.0 \leq \text{RERORP} \leq \infty$
DELPO	Desired pressure change over a time step, psi [kPa]
DELRS	Desired solution GOR change over a time step, scf/STB [std m^3/std m^3]
DELSAT	Desired saturation change over a time step, fraction
MATBAK	Convergence check and material balance accuracy
	= 1, normal choice of convergence and material balance accuracy
	= 2, normal choice of convergence and good material balance checks
	= 3, strict convergence and better material balance checks
	= 4, very strict convergence and best material balance checks

> Notes:
> 1) *Users are advised to refer to Ref. 11 (Sammon and Rubin, 1984) should they desire to change the time step control parameters.*
> 2) *Stricter convergence criterion results in better material balance checks. The CPU time, however, increases and smaller time steps are needed. Sometimes stricter convergence may not be achieved leading to difficulties in execution of program.*
> 3) *Use strict convergence for small problems and less strict convergence for large and practical problems.*

DATA 04B	**Control Integers for Printing Frequency and Form of Output**
IPRINT	Form of presenting arrays in output
	= 0, variables are printed in arrays.
	= 1, variables are printed for each grid block.
	= 2, print coefficients of algebraic equations and variables of each grid block for debugging. This option is usually used during simulation development.
IFTSDBG	The number of the first time step at which debugging starts.
ILTSDBG	The number of the last time step to be included in debugging.
INOUTP	Option for printing reservoir initialization
	= 0, do not print initialization.
	= 1, print initialization.

Appendix A

NFRRES	Frequency of printing reservoir summary for whole reservoir
NFRWEL	Frequency of printing detailed well performances
NFRARY	Frequency of printing detailed variable arrays
NWRITE	Frequency of writing restart-up file

> Note: *Frequency means every N time steps.*

DATA 05B **Control Integers for Printing Desired Variable Arrays**

(1, an array is printed; 0, an array is not printed.)

SW	Water phase saturation array
SO	Oil phase saturation array
SG	Gas phase saturation array
SP	Polymer phase saturation array
ROIP	Remaining oil-in-place
RS	Solution GOR array
CONCT	Tracer concentration
P	Pressure array
CONCP	Polymer concentration array
ADS	Polymer adsorption on rock array
MAXADS	Maximum polymer adsorption array
IWC	Well operating condition array

DATA 06B **Components Present in Reservoir**

(1, component is present; 0, component is not present.)

IQW	Water component
IQO	Oil component
IQG	Gas component
IQP	Polymer component

DATA 07B **Reservoir Depth and Orientation in Space**

DEPTH1	Depth of the reservoir outside corner of grid block (1, 1, 1) below a chosen reference level, ft [m].
	If DEPTH1 < -100,000, supply elevation array for top of grid blocks.
SINX	Sine of angle between x-axis and its projection on horizontal plane

SINY	Sine of angle between y-axis and its projection on horizontal plane

> *Note:* *An elevation below the reference level is positive and increases downward.*

DATA 08B	**Aquifer Parameters**
IDAQFR	Type of aquifer
	= 0, no aquifer. Enter zero for all other parameters.
	= 1, pot aquifer
	= 2, infinite steady-state aquifer
	= 3, intermediate size unsteady-state aquifer
AQFH	Aquifer porosity-thickness product, ft [m]
AQKH	Aquifer permeability-thickness product, md-ft [$\mu m^2 \cdot m$]
AQRW	Aquifer internal radius, ft [m]
AQRE	Aquifer external radius, ft [m]
ANGLE	Aquifer Angle coverage of reservoir, degrees
AQCWE	Effective compressibility of aquifer water, psi^{-1} [kPa^{-1}]
DATA 09B	**Reservoir Initialization Data**
INIOPT	Option for reservoir initialization
	= 0, initialization is not performed; therefore the user has to supply initialization. Enter zero for all other parameters.
	= 1, simulator performs initialization.
ICASE	Option for the type of initial fluids distribution in reservoir
	= 1, single-phase oil (specify ZREF, PZREF, and RSO.)
	= 2, single-phase water (specify ZREF and PZREF.)
	= 3, single-phase gas (specify ZREF and PZREF.)
	= 4, two-phase gas-water (specify ZREF for WGC and PZREF.)
	5, two-phase gas-oil (specify ZREF for OGC and PZREF.)
	= 6, two-phase oil-water (specify ZREF for WOC, PZREF, and RSO.)
	= 7, three-phase gas-oil-aqua (specify ZREF for WOC and PZREF, and RSO at OGC.)
ZREF	Depth below sea level of reference elevation (OGC, WOC, or WGC), ft [m]

Appendix A

PZREF	Pressure at ZREF, psia [kPa]
RSO	Gas/oil ratio, scf/STB [std m^3/std m^3]

> Notes:
> 1) *Initialization of single-phase reservoirs involves determination of grid block pressures only, whereas initialization of multi-phase reservoirs involves determination of phase pressures and saturations.*
> 2) *O/W capillary pressure data are needed for options 1, 3, and 5 in order to include irreducible water saturation in initialization.*
> 3) *For multi-phase reservoirs, capillary pressure data for initialization have to be supplied. These data may differ from those used for simulation. For example, water-wet reservoirs require drainage capillary pressure data for initialization, whereas simulation requires imbibition capillary pressure data.*

DATA 10B   **Reservoir Discretisation and Method of Solving Equations**

NX   Number of grid blocks in the x-direction or in the r-direction if NY = 0.

NY   Number of grid blocks in the y-direction. For single-well simulation, set NY = 0.

NZ   Number of grid blocks in the z-direction

METHOD   Method used to solve the linearized finite-difference equations
    = 1, simultaneous solution using natural ordering
    = 2, block iterative using natural ordering
    = 3, nested factorization using natural ordering
    = 4, simultaneous solution using D4 ordering

NXYXORD   Order of axes followed in ordering grid blocks
    = 1, automatic ordering to minimize CPU time and storage requirement for simultaneous solution methods (METHOD = 1, 4)
    = 2, x-y-z (first x-axis followed by y-axis followed by z-axis)
    = 3, x-z-y (first x-axis followed by z-axis followed by y-axis)
    = 4, y-x-z (first y-axis followed by x-axis followed by z-axis)
    = 5, y-z-x (first y-axis followed by z-axis followed by x-axis)
    = 6, z-x-y (first z-axis followed by x-axis followed by y-axis)
    = 7, z-y-x (first z-axis followed by y-axis followed by x-axis)

NUORDR	Option for ordering grid blocks
	= 0, old version
	= 1, new version
NUSLVR	Option for direct solver
	= 0, old version
	= 1, new version
NKRRGN	Number of relative permeability rock regions
NPVTRGN	Number of PVT fluid regions
RADW	Well radius for single-well simulation, ft [m]
RADE	Reservoir external radius for single-well simulation, ft [m]

> *Note:* NXYXORD =2, 3, 4, 5, 6, or 7 *may also aid in obtaining faster convergence, for the nested factorization iterative method, by ordering blocks along the highest transmissibility direction followed by the direction that has lower transmissibility and finally along the direction that has the least transmissibility (Appleyard and Cheshire, 1983).*

DATA 11A **Reservoir Description and Fluid Saturation and Pressure**
DATA 37A **Distributions**

I1, I2	The lower and upper limits in the x-direction of a parallelepiped region.
J1, J2	The lower and upper limits in the y-direction of a parallelepiped region.
K1, K2	The lower and upper limits in the z-direction of a parallelepiped region.
IACTIVE	Block indicator for active and inactive blocks
	= 0, inactive block
	= 1, active block
NKRGN	Block indicator for relative permeability rock region
	= 1, 2, 3, ... if block belongs to rock region 1, 2, 3, ... respectively.
NPVTGN	Block indicator for PVT fluid region
	= 1, 2, 3, ... if block belongs to PVT region 1, 2, 3, ... respectively.
DX	Block size in the x-direction, or in the r-direction if NY = 0, ft [m]
DY	Block size in the y-direction, ft [m]
DZ	Block size in the z-direction (block net thickness), ft [m]

Appendix A

DEPTH	Elevation below sea level of top of grid block, ft [m]
KX	Block permeability in the x-direction, md [µm^2]
KY	Block permeability in the y-direction, md [µm^2]
KZ	Block permeability in the z-direction, md [µm^2]
PHI	Block porosity, fraction
SO	Block oil saturation, fraction
SW	Block water saturation, fraction
SP	Block polymer solution saturation, fraction
RSO	Block solution GOR, scf/STB [std m^3/std m^3]
P	Block pressure, psia [kPa]
CONCP	Block polymer concentration in polymer solution, ppm
CRPMAX	Block maximum allowed polymer adsorption on rock, lbm/lbm [kg/kg]
CRP	Block polymer adsorption on rock, lbm/lbm [kg/kg]
INFBLK	Block indicator for block sides receiving water influx parallel to a given direction.
	= 0, no influx
	= 1, one side, x-direction
	= 2, one side, y-direction
	= 3, two sides, x-and y-directions
	= 4, three sides, two x-and one y-directions
	= 5, three sides, one x-and two y-directions
	= 6, four sides, two x-and two y-directions
	= 7, one side, z-direction
BBLKWT	Weighting factor given to a block receiving water influx, dimensionless
RATIO	Property modifier, dimensionless
	= 0.0, property is not modified.
	> 0.0, property is increased by that ratio.
	< 0.0, property is decreased by that ratio.

Notes:
1) *DX, DY, and DZ have to be supplied for all active and inactive blocks.*
2) *Ratio is the desired fractional change of the property value that was originally entered by the user or calculated by the simulator. Modifiers can be applied to block porosity, block volume, block elevation, and transmissibility in the x, y, and z directions.*

DATA 38C	**O/W Relative Permeability Data Table for Rock Region # 1**
SWT	Water saturation, fraction
KRW	Relative permeability to water, fraction
KROW	Relative permeability to oil in oil-water system, fraction
PCOW	Imbibition oil-water capillary pressure, psi [kPa]
PCOWI	Drainage oil-water capillary pressure used for reservoir initialization, psi [kPa]
DATA 39C	**G/O Relative Permeability Data Table for Rock Region # 1**
SLT	Liquid saturation, fraction
KRG	Relative permeability to gas, fraction
KROG	Relative permeability to oil in gas-oil system, in the presence of irreducible water, fraction
PCOG	Drainage gas-oil capillary pressure, psi [kPa]
PCOG1	Imbibition gas-oil capillary pressure used for reservoir initialization, psi [kPa]

Notes:	1) DATA 38C and DATA 39C are repeated for regions no. 1, 2, 3, ... in that order until relative permeability and capillary pressure data for all NKRRGN regions are entered.
	2) To simulate gas-water reservoirs, data for krw vs. Sw are entered in DATA 38C, whereas data for krg, Pcwg, and Pcwgi are entered in data DATA 39c. In this case SLT, PCOG, and PCOGI stand for Sw, Pcwg, and Pcwgi, respectively.

DATA 40C	**PVT Data Table of Saturated Fluids for PVT Fluid Region # 1**
PRES	Pressure, psia [kPa]
RS	Solution GOR, scf /STB [std m^3/std m^3]
BW	Water formation volume factor, B/STB [m^3/std m^3]
BO	Oil formation volume factor, B/STB [m^3/std m^3]
BG	Gas formation volume factor, B/scf [m^3/std m^3]
MUW	Water phase viscosity, cp [mPa.s]
MUO	Oil phase viscosity, cp [mPa.s]
MUG	Gas phase viscosity, cp [mPa.s]
DATA 41B	**Oil, Water, and Gas Data at SC & Oil properties above PB.**
GAMAW	Formation water specific gravity

Appendix A

GAMAG	Gas gravity (air = 1)
OILAPI	Oil API gravity
CO	Oil compressibility, psi^{-1} [kPa^{-1}]
CMUO	Rate of relative change of oil viscosity w.r.t. pressure above bubble-point pressure, psi^{-1} [kPa^{-1}]

> *Note:* *DATA 40C and DATA 41B are repeated for PVT regions no. 1, 2, 3, ... in that order until PVT data for all NPVTRGN regions are entered.*

DATA 42B	**Rock Properties**
DENR	Rock density, lbm/ft^3 [kg/m^3]
CPHI	Porosity compressibility, psi^{-1} [kPa^{-1}]
PREF	Reference pressure at which porosities are reported, psia [kPa]
DATA 43C	**Polymer Viscosity Data Table**
CONCP	Polymer concentration in polymer solution, ppm
MUP	Polymer solution viscosity, cp [mPa.s]
DATA 44C	**Resistance and Residual Resistance Factors Data Table**
CRP	Polymer adsorption on rock, lbm/lbm [kg/kg]
RF	Resistance factor, dimensionless
RRF	Residual resistance factor, dimensionless
DATA 45D	**Well Recursive Data**
NOW	Number of wells that will change operational conditions
	= 0, no change in well operations
	> 0, number of wells that change operational conditions
IRITE1	Option for writing data to MATLAB files and printing detailed output at SIMNEW
	= 0, output and/or data are not required.
	= 1, output and/or data are required.
DELT1	Time step to be used, D [d]
SIMNEW	Time specification signaling user's new request, D [d]. Well data entered here will be active starting from previous time specification until this time specification and beyond.
QOECON	Economic limit of oil production, STB/D [std m^3/d]
WORMAX	Maximum allowable well producing WOR, B/STB [m^3/std m^3]

GORMAX	Maximum allowable well producing GOR, scf /STB [std m^3/std m^3]
IWT	Index for well type
	= 0, shut-in well
	= –1, production well
	= –2, injection well
IDW	Well identification number. Each well must have a unique IDW.
	= 1, 2, 3, 4, ...
IW, JW, KW	(i, j, k) location of well block
IWOPC	Well operating condition
	IWOPC for Production Well
	= 0, shut-in well
	= 1, oil rate at standard conditions in MSTB/D [10^3 std m^3/d]
	= 2, aqua rate at standard conditions in MB/D [10^3 std m^3/d]
	= 3, gas rate at standard conditions in Mscf/D [10^3 std m^3/d]
	= 4, liquid rate at standard conditions in MSTB/D [10^3 std m^3/d]
	= 5, total rate at standard conditions in MSTB/D [10^3 std m^3/d]
	= 6, oil rate at reservoir conditions in MRB/D [10^3 m^3/d]
	= 7, aqua rate at reservoir conditions in MRB/D [10^3 m^3/d]
	= 8, gas rate at reservoir conditions in Mft^3/D [10^3 m^3/d]
	= 9, liquid rate at reservoir conditions in MRB/D [10^{3}m^3/d]
	= 10, total rate at reservoir conditions in MRB/D [10^{3}m^3/d]
	= 11, specified bottom-hole pressure in psia [kPa]
	IWOPC for Injection Well
	= 0, shut-in well
	= 1, oil rate at standard conditions in MSTB/D [10^3 std m^3/d]

APPENDIX A 439

	= 2, water rate at standard conditions in MB/D [10^3 std m^3/d]
	= 3, gas rate at standard conditions in Mscf/D [10^3 std m^3/d]
	= 4, polymer solution rate at standard conditions in MB/D [10^3 std m^3/d]
GW	Well geometric factor, B-cp/D-psi [m^3.mPa.s/(d.kPa)]
BHP	Minimum BHP of production well or maximum BHP allowed for injection well, psia [kPa]
QSP	Specified well production (injection) rate in thousands of units or BHP in units of psia [kPa]
CONCPI	Polymer concentration of injected polymer solution, ppm
	= 0.0, water injection
	> 0.0, polymer injection
CONCTI	Tracer concentration in injected aqua, ppm

Notes:
1) The NOW line can be repeated for different times but each subsequent line has to have time specification larger than the previous time specification. This line with NOW = 0 can be used to request printing of output and writing MATLAB data files, and to override control time step at specific times.
2) The IWT line enters specifications for one well. This line has to be repeated NOW times provided NOW > 0.
3) To activate maximum injection rate for an injection well, set QSP to a large but realistic value.
4) DELT1 is used only if it is smaller than that determined by the time step selection control. Therefore, to transfer control to automatic time-step selection, set DELT1 to a large value (say, 1000).
5) The specified value of BHP must be within the range of the pressure specified in PVT tables.

A.5 Instructions to Run Simulator and Graphic Post Processor on PC

To use the graphic post processor to examine results generated by the simulator, the user has to run the simulator first. The user of the simulator is provided with copies of ten reference data files (e.g. ref. txt) similar to the one presented in Sec. A.7. The user first copies this file into his own data file (e.g. user-input.txt) and then follows instructions in Sec. A.3 and observe variable definitions given in Sec. A.4 to modify his/her own data file such that it describes the constructed model of the reservoir under study. *It is important to note that a data file must saved as a **Notepad** file.* The simulator can be run any number of times during preparation of the data file to correct errors in data and format. The computer responds with the following statement requesting names (with file type) of input, output, and two restart-up files; and giving the format for reading such names:

ENTER NAMES OF INPUT, OUTPUT, RESTART-UP FILES
input.txt, output.lis, restin.txt, restout.txt

The user responds using the names of four files separated by blanks or commas as follows.

user-input.txt   user-output.lis   r1.txt   r2.txt

The computer program continues execution until completion, which will be signaled by printing on the screen:

END OF MPSFFA RUN

Prior to using the post graphic processor, the user has to run MATLAB and set the path to the appropriate folders (within MALAB, highlight "File", select "Set Path …", then select "Add Folder", "Browse For Folder", and select the folder where the simulation results are stored and the folder for the MATLAB post processor program, push "OK", then push "Save", and finally push "Close"). Once the simulation run is completed successfully, the results (**user-output.lis**) and six other files needed for the post processor (**pvt.lis, krp.lis, well.lis, res.lis, resmat.lis, and rokmat.lis**) are generated by the simulator. The user may then proceed to examine the results using the graphic post processor (**mpsffa_pp.m**) by issuing the following command at the MATLAB command level:

mpsffa_pp

The user then follows instructions to run the post processor. It is important not to delete any window or interrupt the execution of

APPENDIX A

the post processor program. The program gives the user an option to terminate program execution orderly at any time.

A.6 Limitations Imposed on the Compiled Versions

The compiled versions of MPSFFA contained in the accompanied CD is provided here for demonstration and student training purposes. The critical variables were therefore restricted to the dimensions given below. Dimensioning parameters are assigned values such that all methods of solving the linear algebraic equations can be run for the same problem. Advanced users may increase these dimensions by editing the dimensioning FORTRAN program (**sizempsffav1-16.for**) and then re-compiling the main program (**mpsffav1-16.for**).

1. Number of blocks in x-direction ≤ 40
2. Number of blocks in y-direction ≤ 40
3. Number of blocks in z-direction ≤ 40
4. Number of active blocks ≤ 4000
5. Number of Kr rock regions ≤ 30
6. Number of PVT fluid regions ≤ 30
7. Number of entries in any table ≤ 30
8. Number of wells ≤ 200
9. Number of times wells change operational conditions ≤ 100

The following is a list of the values of the parameters, in the dimensioning program (sizempsffav1-16.for), that are used in the compiled versions of MPSFFA.

```
10.       INTEGER, PARAMETER :: I9A=40
11.       INTEGER, PARAMETER :: I9B=40
12.       INTEGER, PARAMETER :: I9C=40
13.       INTEGER, PARAMETER :: I9D=I9A*I9B*I9C
14.       INTEGER, PARAMETER :: I9E=4000
15.       INTEGER, PARAMETER :: I8E=I9E
16.       INTEGER, PARAMETER :: I7E=I9E
17.       INTEGER, PARAMETER :: I9F=30
18.       INTEGER, PARAMETER :: I9G=30
19.       INTEGER, PARAMETER :: I9H=30
20.       INTEGER, PARAMETER :: I9I=30
21.       INTEGER, PARAMETER :: I9J=30
22.       INTEGER, PARAMETER :: I9K=30
23.       INTEGER, PARAMETER :: I9L=30
24.       INTEGER, PARAMETER :: I9M=200
25.       INTEGER, PARAMETER :: I9N=500
26.       INTEGER, PARAMETER :: I1O=MIN(I9A,I9B,I9C)
27.       INTEGER, PARAMETER :: I3O=MAX(I9A,I9B,I9C)
28.       INTEGER, PARAMETER :: I2O=I9A+I9B+I9C-I1O-I3O
29.       INTEGER, PARAMETER :: I9Q=100
30.       INTEGER, PARAMETER :: I9R=4
31.       INTEGER, PARAMETER :: I8R=I9R
```

```
32.         INTEGER, PARAMETER :: I7R=I9R
33.         INTEGER, PARAMETER :: I9S=1+I9R*I9E
34.         INTEGER, PARAMETER :: I9T=I9Q*I9M
35.         INTEGER, PARAMETER :: I9W=I9E
36.         INTEGER, PARAMETER :: I9X=I9W
37.         INTEGER, PARAMETER :: I8X=I9X
38.         INTEGER, PARAMETER :: I9Y=2000000
39.         INTEGER, PARAMETER :: I9A1=1+I9A/10
40.         INTEGER, PARAMETER :: I9A2=1+I9A/20
41.         INTEGER, PARAMETER :: I9B1=1+I9B/10
42.         INTEGER, PARAMETER :: I9C1=1+I9C/10
43.         INTEGER, PARAMETER :: IEE1=1+9/10
44.         INTEGER, PARAMETER :: I3E=15
```

A.7 Example of a Prepared Data File

The *oil-gas-swi-xyz(odeh-1981).txt* data file was prepared for the SPE bench-mark test problem of Odeh (1981).

```
'*DATA 01E* TITLE OF SIMULATION RUN'
'TITLE'              NOTE: SUPPLY A TITLE UP TP 80 CHARACTERS IN LENGTH.
A DATA FILE TO MODEL 3D, 1/8TH OF 5-SPOT PATTERN. AZIZ ODEH'S DATA (JPT  1981).
'*DATA 02B* RESTART-UP OPTION AND SIMULATION TIME DATA'
'IPRTDAT  IRSTRT  TMTOTAL  TMSTOP  DELT'
    1       0     36504.0  3650.4  30.42
'*DATA 03B* UNITS, INTERPOLATION, TIME-STEP CONTROL, AND CONVERGENCE OPTIONS'
'MUNITIN MUNTOUT  IQUAD NINTBLE  DTMAX  DTMIN  RERORP  DELPO  DELRS  DELSAT  MATBAK'
    1       1       0    500   365.04  0.001   0.2   500.0  500.0   0.20      2
'*DATA 04B* CONTROL INTEGERS FOR PRINTING FREQUENCY AND FORM OF OUTPUT'
'IPRINT  IFTSDBG  ILTSDBG  INOUTP  NFRRES  NFRWEL  NFRARY  NWRITE'
    0       0        0        1       0       0       0       0
'*DATA 05B* CONTROL INTEGERS FOR PRINTING DESIRED VARIABLE ARRAYS'
'  SW      SO       SG       SP     STBO     RS    CONCT   P    CONCP   ADS  MAXADS   IWC'
   0       1        1        0       1       1      0      1     0      0     0       1
'*DATA 06B* COMPONENTS PRESENT IN RESERVOIR'
'  IQW     IQO     IQG     IQP'
   1       1       1       0
'*DATA 07B* RESERVOIR DEPTH AND ORIENTATION IN SPACE'
'DEPTH1   SINX    SINY'
 8425.0   0.0     0.0
'*DATA 08B* AQUIFER PARAMETERS'
'IDAQFR   AQFH    AQKH     AQRW     AQRE    ANGLE   AQCWE'
   0      0.0     0.0      0.0      0.0     0.0     0.0
'*DATA 09B* RESERVOIR INITIALIZATION DATA'
'INIOPT   ICASE   ZREF     PREF     RSO'
   0       0      0.0      0.0      0.0
'*DATA 10B* RESERVOIR DISCRETIZATION AND METHOD OF SOLVING EQUATIONS'
'  NX      NY      NZ    METHOD  NXYZORD  NUSLVR  NKRRGN  NPVTRGN  RADW   RADE'
   10      10      3       4        1       1       1       1      0.0    0.0
'*DATA 11A* RESERVOIR REGION WITH ACTIVE OR INACTIVE BLOCK IACTIVE'
1 ' I1     I2      J1      J2      K1      K2     IACTIVE'
    1      10       1      10       1       3       1
    1       1       2      10       1       3       0
    2       2       3      10       1       3       0
    3       3       4      10       1       3       0       NOTE:
    4       4       5      10       1       3       0       FORMAT PROCEDURE A WITH OPTION = 1
    5       5       6      10       1       3       0       MUST HAVE TERMINATING ZERO LINE.
    6       6       7      10       1       3       0
    7       7       8      10       1       3       0
    8       8       9      10       1       3       0
    9       9      10      10       1       3       0
    0       0       0       0       0       0       0
```

Appendix A

```
'*DATA 12A* RESERVOIR REGION THAT BELONGS TO REL. PERM. REGION NKRGN'
0 ' I1   I2    J1   J2   K1  K2   NKRGN'  NOTE: FORMAT PROCEDURE A WITH OPTION = 0
300*1                                           DOES NOT TERMINATE WITH ZEROES.
'*DATA 13A* RESERVOIR REGION THAT BELONGS TO FLUID PVT REGION NPVTGN'
0 ' I1   I2    J1   J2   K1  K2   NPVTGN'
300*1
'*DATA 14A* RESERVOIR REGION HAVING BLOCK SIZE DX IN THE X-DIRECTION'
0 ' I1   I2    J1   J2   K1  K2      DX'
300*1000.0
'*DATA 15A* RESERVOIR REGION HAVING BLOCK SIZE DY IN THE Y-DIRECTION'
0 ' I1   I2    J1   J2   K1  K2      DY'
300*1000.0
'*DATA 16A* RESERVOIR REGION HAVING BLOCK SIZE DZ IN THE Z-DIRECTION'
1 ' I1   I2    J1   J2   K1  K2      DZ'
     1   10     1   10    3   3    20.0
     1   10     1   10    2   2    30.0
     1   10     1   10    1   1    50.0
     0    0     0    0    0   0     0.0
'*DATA 17A* RESERVOIR REGION HAVING TOP OF LAYER ELEVATION BELOW SEA LEVEL'
1 ' I1   I2    J1   J2   K1  K2   DEPTH'
     1   10     1   10    3   3  8315.0
     1   10     1   10    2   2  8330.0
     1   10     1   10    1   1  8325.0
     0    0     0    0    0   0     0.0
'*DATA 18A* RESERVOIR REGION HAVING PERMEABILITY KX IN THE X-DIRECTION'
1 ' I1   I2    J1   J2   K1  K2      KX'
     1   10     1   10    3   3   500.0
     1   10     1   10    2   2    50.0
     1   10     1   10    1   1   200.0
     0    0     0    0    0   0     0.0
'*DATA 19A* RESERVOIR REGION HAVING PERMEABILITY KY IN THE Y-DIRECTION'
1 ' I1   I2    J1   J2   K1  K2      KY'
     1   10     1   10    3   3   500.0
     1   10     1   10    2   2    50.0
     1   10     1   10    1   1   200.0
     0    0     0    0    0   0     0.0
'*DATA 20A* RESERVOIR REGION HAVING PERMEABILITY KZ IN THE Z-DIRECTION'
1 ' I1   I2    J1   J2   K1  K2      KZ'
     1   10     1   10    3   3    50.0
     1   10     1   10    2   2    50.0
     1   10     1   10    1   1    19.23
     0    0     0    0    0   0     0.0
'*DATA 21A* RESERVOIR REGION HAVING POROSITY PHI'
0 ' I1   I2    J1   J2   K1  K2     PHI'
300*0.30
'*DATA 22A* RESERVOIR REGION HAVING OIL SATURATION SO'
0 ' I1   I2    J1   J2   K1  K2      SO'
300*0.88
'*DATA 23A* RESERVOIR REGION HAVING WATER SATURATION SW'
0 ' I1   I2    J1   J2   K1  K2      SW'
300*0.12
'*DATA 24A* RESERVOIR REGION HAVING POLYMER PHASE SATURATION SP'
0 ' I1   I2    J1   J2   K1  K2      SP'
300*0.0
'*DATA 25A* RESERVOIR REGION HAVING SOLUTION GOR AT BUBBLE-POINT PRESSURE RS'
0 ' I1   I2    J1   J2   K1  K2     RSO'
300*1270.0
'*DATA 26A* RESERVOIR REGION HAVING INITIAL PRESSURE P'
0 ' I1   I2    J1   J2   K1  K2       P'
100*4800.0 100*4789.8 100*4783.5
'*DATA 27A* RESERVOIR REGION HAVING POLYMER CONCENTRATION CONCP IN WATER'
0 ' I1   I2    J1   J2   K1  K2   CONCP'
300*0.0
'*DATA 28A* RESERVOIR REGION HAVING HAVING MAX. POL. ADSORPTION CRPMAX'
0 ' I1   I2    J1   J2   K1  K2  CRPMAX'
300*0.0
'*DATA 29A* RESERVOIR REGION HAVING POLYMER ADSORPTION CRP'
0 ' I1   I2    J1   J2   K1  K2     CRP'
300*0.0
```

```
'*DATA 30A* RESERVOIR REGION FOR WATER INFLUX SIDES OF EDGE BLOCKS'
0 ' I1   I2   J1   J2   K1   K2   INFBLK'
300*0
'*DATA 31A* RESERVOIR REGION FOR WEIGHT DISTRIBUTION OF INFLUX TO EDGE BLOCKS'
0 ' I1   I2   J1   J2   K1   K2   BBLKWT'
300*0.0
'*DATA 32A* RESERVOIR REGION WITH BLOCK ELEVATION MODIFICATION RATIO'
0 ' I1   I2   J1   J2   K1   K2   RATIO'
300*0.0
'*DATA 33A* RESERVOIR REGION WITH BLOCK VOLUME MODIFICATION RATIO'
1 ' I1   I2   J1   J2   K1   K2   RATIO'
     1    1    1    1    1    3   -0.5
     2    2    2    2    1    3   -0.5
     3    3    3    3    1    3   -0.5
     4    4    4    4    1    3   -0.5
     5    5    5    5    1    3   -0.5
     6    6    6    6    1    3   -0.5
     7    7    7    7    1    3   -0.5
     8    8    8    8    1    3   -0.5
     9    9    9    9    1    3   -0.5
    10   10   10   10    1    3   -0.5
     0    0    0    0    0    0    0.0
'*DATA 34A* RESERVOIR REGION WITH BLOCK POROSITY MODIFICATION RATIO'
0 ' I1   I2   J1   J2   K1   K2   RATIO'
300*0.0
'*DATA 35A* RESERVOIR REGION WITH X-TRANSMISSIBILITY MODIFICATION RATIO'
0 ' I1   I2   J1   J2   K1   K2   RATIO'
300*0.0
'*DATA 36A* RESERVOIR REGION WITH Y-TRANSMISSIBILITY MODIFICATION RATIO'
0 ' I1   I2   J1   J2   K1   K2   RATIO'
300*0.0
'*DATA 37A* RESERVOIR REGION WITH Z-TRANSMISSIBILITY MODIFICATION RATIO'
1 ' I1   I2   J1   J2   K1   K2   RATIO'
     1    1    1    1    1    2   -0.5
     2    2    2    2    1    2   -0.5
     3    3    3    3    1    2   -0.5
     4    4    4    4    1    2   -0.5
     5    5    5    5    1    2   -0.5
     6    6    6    6    1    2   -0.5
     7    7    7    7    1    2   -0.5
     8    8    8    8    1    2   -0.5
     9    9    9    9    1    2   -0.5
    10   10   10   10    1    2   -0.5
     0    0    0    0    0    0    0.0
'ENTER O/W & G/O RELATIVE PERMEABILITY DATA FOR ALL ROCK REGIONS'
'*DATA 38C* O/W RELATIVE PERMEABILITY DATA TABLE FOR ROCK REGION # 1'
'  SWT      KRW      KROW     PCOW     PCOWI'
 0.0       0.0      1.0      0.0      0.0
 0.120     0.00     1.00     0.00     0.0       NOTE:
 1.000     1.000    0.000    0.00     0.0       FORMAT PROCEDURE C MUST
 0.0       0.0      0.0      0.0      0.0       TERMINATE WITH A LINE OF ZEROES
'*DATA 39C* G/O RELATIVE PERMEABILITY DATA TABLE FOR ROCK REGION # 1'
'  SLT      KRG      KROG     PCOG     PCOGI'
 0.0       1.0      0.0      0.0      0.0
 0.150     0.980    0.0000   0.00     0.0
 0.300     0.940    0.0000   0.00     0.0
 0.400     0.870    0.0001   0.00     0.0
 0.500     0.720    0.0010   0.00     0.0
 0.550     0.600    0.0100   0.00     0.0
 0.600     0.410    0.0210   0.00     0.0
 0.700     0.190    0.0900   0.00     0.0
 0.750     0.125    0.2000   0.00     0.0
 0.800     0.075    0.3500   0.00     0.0
 0.880     0.025    0.7000   0.00     0.0
 0.950     0.005    0.9800   0.00     0.0
 0.980     0.000    0.9970   0.00     0.0
 0.999     0.000    1.0000   0.00     0.0
 1.000     0.000    1.0000   0.00     0.0
 0.0       0.0      0.0      0.0      0.0
'ENTER SATURATED FLUID PVT DATA FOR ALL FLUID PVT REGIONS'
```

Appendix A

```
'*DATA 40C* PVT DATA TABLE OF SATURATED FLUIDS FOR PVT FLUID REGION # 1'
'  PRES      RS       BW       BO      BG         MUW      MUO      MUG'
   14.7      1.00    1.0410   1.0620  0.166666   0.3100   1.0400   0.0080
  264.7     90.50    1.0403   1.1500  0.012093   0.3100   0.9750   0.0096
  514.7    180.00    1.0395   1.2070  0.006274   0.3100   0.9100   0.0112
 1014.7    371.00    1.0380   1.2950  0.003197   0.3100   0.8300   0.0140
 2014.7    636.00    1.0350   1.4350  0.001614   0.3100   0.6950   0.0189
 2514.7    775.00    1.0335   1.5000  0.001294   0.3100   0.6410   0.0208
 3014.7    930.00    1.0320   1.5650  0.001080   0.3100   0.5940   0.0228
 4014.7   1270.00    1.0290   1.6950  0.000811   0.3100   0.5100   0.0268
 5014.7   1618.00    1.0258   1.8270  0.000649   0.3100   0.4490   0.0309
 9014.7   2984.00    1.0130   2.3570  0.000386   0.3100   0.2030   0.0470
    0.0      0.0       0.0      0.0     0.0       0.0      0.0      0.0
'*DATA 41B* OIL  WATER  AND  GAS DATA AT SC  AND OIL PROPERTIES ABOVE PB'
' GAMAW   GAMAG   OILAPI    CO         CMUO'
 1.03784   0.792   48.41   0.147D-04   0.46D-04
'*DATA 42B*  ROCK PROPERTIES'
' DENR    CPHI    PREF'
  0.0    0.3E-05  14.7
'*DATA 43C* POLYMER VISCOSITY DATA TABLE'
' CONCP     MUP'                          NOTE: SUPPLY TERMINATING ZERO LINE
  0.0      0.0                                  IF DATA ARE NOT AVAILABLE.
'*DATA 44C* ADSORPTION VS. RESISTANCE & RESIDUAL RESIST. FACTORS TABLE'
'  CRP     RF     RRF'                    NOTE: SUPPLY ZERO TERMINATING LINE
   0.0    0.0    0.0                            IF DATA ARE NOT AVAILABLE.
'*DATA 45D* WELL RECURSIVE DATA'
' NOW   IRITE1    DELT1      SIMNEW    QOECON   WORMAX    GORMAX'
' IWT    IDW   IW JW KW      IWOPC       GW      BHP       QSP   CONCPI CONCTI'
   2      1      0.001       365.04     0.0      99.0    20000.0
  -2      1    1  1  3         3        5.3038  9000.0   50000.0    0.0    0.0
  -1      2   10 10  1         1        5.3038  1000.0      10.0    0.0    0.0
   0      1      365.04      730.08     0.0      99.0    20000.0
   0      1      365.04     1095.12     0.0      99.0    20000.0
   0      1      365.04     1460.16     0.0      99.0    20000.0
   0      1      365.04     1825.20     0.0      99.0    20000.0
   0      1      365.04     2190.24     0.0      99.0    20000.0
   0      1      365.04     2555.28     0.0      99.0    20000.0
   0      1      365.04     2920.32     0.0      99.0    20000.0
   0      1      365.04     3285.36     0.0      99.0    20000.0
   0      1      365.04     3650.40     0.0      99.0    20000.0
   0      0        0.0         0.0      0.0       0.0        0.0
```

References

Aavatsmark, I.B. (1996). Discretization on non-orthogonal, quadrilateral grids for inhomogeneous, anisotropic media. *Journal of Computational Physics*, vol. 127, pp. 2–14.

Aavatsmark, I. (2002). An introduction to multipoint flux approximations for quadrilateral grids. *Comput. Geosci.* 6, 405–432.

Abdeh-Kolahchi, A., Satish, M.G., Ketata, C., and Islam, M.R. (2009). Sensitivity analysis of genetic algorithm parameters in groundwater monitoring network optimization for petroleum contaminant detection. *Advances in Sustainable Petroleum Engineering and Science*, vol. 1, no. 3, 305–318.

Aboudheir, A., Kocabas, I., and Islam, M.R. (1999). Improvement of numerical methods in petroleum engineering problems. *Proc. of the IASTED International Conference, Applied Modeling and Simulation*. Sept. 1–3, Cairns, Australia.

Abou-Kassem, J.H. (1996). Practical considerations in developing numerical simulators for thermal recovery. *J. Pet. Sci. Eng.*, vol. 15, pp. 281–290.

Abou-Kassem, J.H. and Ertekin, T. (1992). An efficient algorithm for removal of inactive blocks in reservoir simulation. *J. Can. Pet. Tech.*, vol. 31, no. 2, pp. 25–31.

Abou-Kassem, J.H., Osman, M.E., and Zaid, A.M. (1996). Architecture of a multipurpose simulator. *J. Pet. Sci. Eng.*, vol. 16, pp. 221–235.

Abou-Kassem J. and Islam M.R. (2008). Reconstruction of the compositional reservoir simultor using engineering approach. *Journal of Nature Science and Sustainable Technology*, vol. 2, nos. 1 and 2, 68–102.

Abou-Kassem, J. (2007). Engineering approach vs the mathematical approach in developing reservoir simulators. *J. Nature Science and Sustainable Technology*, vol. 1, No. 1, pp. 35–68.

Abou-Kassem, J.H., Farouq Ali, S.M., and Islam, M.R. (2006). *Petroleum Reservoir Simulation: A Basic Approach*. Houston: Gulf Publications Co.

Acs, G.A., Doleschall, S. and Farkas, E. (1985). General purpose compositional model. *Society of Petroleum Engineers Journal*, 543–552.

Adjedj, B. (1999). Application of the decomposition method to the understanding of HIV immune dynamics. *Kybernetes*, vol. 28 (3), pp. 271–283.

Adomian, G. (1984). A new approach to nonlinear partial differential equations. *Journal of Mathematical Analysis and Applications*, vol. 102, pp. 420–434.

Adomian, G. (1986). *Nonlinear Stochastic Operator Equations.* San Diego: Academic Press.

Adomian, G. (1994). *Solving Frontier Problems of Physics.* Boston: Kluwer.

Ahmed, T. (2000). *Reservoir Engineering Handbook.* Boston, U.S.A.: Gulf Professional Publishing, 866 pp.

Ahmed, T. (2002). *Reservoir Engineering Handbook.* 2nd edition, Boston, U.S.A.: Gulf Professional Publishing.

Appleyard, J.R. and Cheshire, I.M. (1983). Nested factorization. *Paper presented at the 1983 SPE Reservoir Simulation Symposium.* 15–18 November, San Francisco.

Araktingi, U.G. and Orr, F.M. (1998). Viscous Fingering in Heterogeneous Porous Media. *the 63rd SPE Annual Technical Conference.* Houston, TX, Oct. 2–5.

Awaland, M.R. and Mohiuddin, M.A. (2006). Autoscan shift: A new core-rock data integration technique to overcome the shortcomings of conventional regression. *J Pet. Sci. Eng.*, vol. 51, issue 3–4, pp. 275–283.

Aziz, K. and Settari, A. (1979). *Petroleum Reservoir Simulation.* Applied Science Publishers Ltd., London, U.K.

Aziz, S. (1989). A Proposed Technique for Simulation of Viscous Fingering in One Unstable Miscible Flows through Porous Media. *Journal of American Institute of Physics*, pp. 127–166.

Aziz, S.O. (1989). A Proposed Technique for Simulation of Viscous Fingering in One Dimensional Immiscible Flow. *SPE Reser. Eng.*, Aug. 2, vol. 4, pp. 304–308.

Baade, H.W., Lopez-Portillo, G.M., and Guillermo, F. (1990). *Texas International Law Journal*, vol. 25(3), pp. 381–387.

Barakat, H.Z. and Clark, J.A. (1966). On the Solution of the Diffusion Equation by Numerical Methods. *J. Heat Transfer, Trans. ASME*, vol. 88, pp. 421–427.

Bear, J. (1972). *Dynamics of fluid in porous media.* New York: American Elsevier Publishing Company, Inc.

Begg, S.H., Carter, R.R., and Dranfield, P. (1989). Assigning effective values to simulator gridblock parameters for heterogeneous reservoir. *SPE Reservoir Engineering, SPE-16754.*

Behie, A. and Vinsome, P.K.W. (1982). Block iterative methods for fully implicit reservoir simulation. *Soc. Pet. Eng. J.*, vol. 22, no. 5, pp. 658–68.

Belhaj, H.A., Mousavizadegan, S.H., Ma, F., and Islam, M.R. (2006). Three-Dimensional Permeability Using Liquid Injection. *SPE-100428, SPE Western Regional Meeting/AAPG - Pacific Section/GSA.* May 8–10, Anchorage, Alaska.

Belhaj, H.A., Mousavizadegan, S.H., Ma, F., and Islam, M.R. (2006). Three-dimensional permeability utilizing a new gas-spot permeameter. *SPE-100427, SPE Gas Technology Symposium.* May 15–17, Calgary.

Benham, A. L. and Olson, R. W. (1963). A Model Study of Viscous Fingering, *Pet. Trans. AIME*, vol. 228, pp. 138.

Bentsen, R. (1978). Conditions under which the capillary term may be neglected. *Journal of Canadian Petroleum technology*, vol. 17(4), pp. 25–30.

Bertsekas, D. and Tsitsiklis, J. (2000). Neuro-Dynamic Programming. *Athena Scientific*. Belmont, MA.

Biazar, J.A. (2005). An approximation to the solution of hyperbolic equations by Adomian decomposition method and comparison with characteristics method. *Applied Mathematics and Computation*, vol. 163, pp. 633–638.

Biazar, J. and Ebrahimi, H. (2008). An approximation to the solution of hyperbolic equations by Adomian decomposition method and comparison with characteristics method. *Applied Mathematics and Computation*, vol. 163. 633–638.

Bjorndalen, N.M. (2005). The effect of irradiation on immiscible fluids for increased oil production with horizontal wells. *ASME International Mechanical Engineering Congress and Exposition (IMECE)*. Orlando, Florida, USA.

Blackwell, R.J. Rayne, J.R., and Terry, W.M. (1959). Factors Influencing the Efficiency of Miscible Displacement. *Pet. Trans. AIME*, vol. 216, pp. 1.

Blunt, M.J. and Christie, M.A. (1992). Theory of Viscous Fingering in Two Phase, Three Component Flow. *SPE-22613, Advanced Technology Series*, vol. 2, no. 2, pp. 52–60.

Bokhari, K. and Islam, M.R. (2005). Improvement in the time accuracy of numerical methods in petroleum engineering problems – A new combination. *Energy Sources*, vol. 27, no. 1, pp. 45–60.

Bokhari, K. (2003). *Experiment and numerical modeling of viscous fingering – liquid-liquid miscible displacement*. MASc Thesis, Dept. of Chemical Engineering, Dalhousie University, Halifax, Canada.

Bokhari, K., Mustafiz, S., and Islam, M.R. (2006). Numerical modeling of viscous fingering under combined effects of thermal, solutal and mixed convection in liquid-liquid miscible displacementsr. *Journal of Petroleum Science and Technology*.

Bouhroum, A. (1985). Beitrag zur Verdrängung mischbarer Flüssigkeiten in porösen Medien unter Berücksichtigung der Dichte - und Viskositäts - Unterschiede. Genehmigte Dissertation von der Fakultät für Bergbau. - *und Viskositäts - Unterschiede. Genehmigte Dissertation von der FakultäHüttenwesen und Maschinenwesen der Technischen Universität Clausthal*.

Brigham, W.E., Reed, P.W., and Dew, J.N. (1961). Experiments on Mixing During Miscible Displacement in Porous Media. *Pet. Trans. AIME*, vol. 222, pp. 1.

Brouwer, D.R., Naevdal, G., Jansen, J.D., Vefring, E.H., and Kruijsdiik, van C.P.J.W. (2004). *Improved reservoir management through optimal control and continuous model updating.* SPE Paper 90149.

Brouwer, D. (2004). *Dynamic Water Flood Optimization with Smart Wells using Optimal Control Theory.* Delft: PhD Thesis, Delft University of Technology.

Bu, T. and Damsleth, E. (1996). *Errors and uncertainties in reservoir performance predictions.* SPE Paper 30604.

Buckley, S.A. (1942). Mechanism of fluid displacement in sands. *AIME*, vol. 146, pp. 187–196.

Cao, H., and Aziz, K. (1999). *Evaluation of Pseudo Functions.* SPE Paper 54589.

Carman, P. C. (1956). *Flow of gases through porous media*, Butterworths Scientific Publications, London.

Chalaturnyk, Rick, and Scott, J.D. (1995). Geomechanics issues of steam assisted gravity drainage. *SPE 30280, presented at the International Heavy Oil Symposium.* Calgary, AB, Canada, June 19–21.

Cheema, T.J. and Islam, M.R. (1995). A new modeling approach for predicting flow in fractured formations. In *Groundwater Models for Resources Analysis and Management* (pp. 327–338). Boca Raton, FL: Lewis Publishers.

Chen, H.Y., Teufel, L.W., and Lee, R.L. (1995). Couple fluid flow and geomechanics in reservoir study-I, theory and governing equations. *SPE annual technical conference and exhibition, SPE 30752.* Dallas, USA., Oct.

Chen, J.C. and Wilkinson, D. (1985). Pore-scale Viscous Fingering in Porous Media. *Phys. Rev. Lett.*, vol. 55, pp. 1892–1895.

Chhetri, A.B. and Islam, M.R. (2007). Reversing Global Warming. *J. Nat. Sci. and Sust.Tech*, vol. 1(1): pp. 79–114.

Chhetri, A.B. and Islam, M.R. (2008). A Comprehensive Pathway of Crude and Refined Petroleum Products. *Energies*, in press.

Chhetri, A.B. and Islam, M.R. (2008). A Global Sustainability of Nuclear Energy. *Energies*, in press.

Chhetri, A.B. and Islam, M.R. (2008). *Inherently Sustainable Technology Developments.* New York, 452 p.: Nova Science Publishers.

Chhetri, A.B. and Islam, M.R. (2008). Problems Associated with Conventional Natural Gas Processing and Some Innovative Solutions. *Petroleum Science and Technology*, vol. 26, Issue 13, pp. 1583–1595.

Chhetri, A.B., Khan, M.M., and Islam, M.R. (2008). Characterization of Some Energy Sources Based on Global Efficiency. *Environmental Geology*, Accepted, pp. 22.

Chilingar, G.V. and Khilyuk, L.F. (2007). Humans are not responsible for global warming. *SPE-109292, Annual Technical Conference and exhibition held in Anaheim.* California, USA,11–14, November.

Choi, I., Cheema, T., and Islam, M.R. (1997). A new dual-porosity/dual permeability model with non-Darcian flow through fractures. *Journal of Petroleum Science and Engineering*, vol. 17, pp. 331–344.

Chouke, R.L., Meurs, P.V., and Poel, C.V.D. (1959). The Instability of Slow, Immiscible, Viscous, Liquid-Liquid Displacement in Permeable Media. *Trans. AIME*, vol. 216, pp. 188–194.

Christie, M.A. and Blunt, M.J. (2001). A Comparison of Upscaling Techniques. *Tenth SPE Comparative Solution Project: SPE Paper 66599*.

Christie, M.A. and Bond, D.J. (1987). Detailed Simulation of Unstable Processes in Miscible Flooding. *SPE Reservoir Engineering*, vol. 2, pp. 514–522.

Christie, M.A. (1989). High Resolution Simulation of Unstable Flows in Porous Media. *SPE Reservoir Engineering*, vol. 4, pp. 297–304.

Christie, M.A. (1996). Upscaling for reservoir simulation. *JPT*, November, pp. 1004–1009.

Christie, M.A., Jones, A.D.W., and Muggeridge, A.H. (1990). Comparison between Laboratory Experiments and Detailed Simulations of Unstable Miscible Displacement Influenced By Gravity. In T. Graham, *North Sea Oil And Gas Reservoirs-Ii* (pp. 244–250). The Norwegian Institute of Technology.

Christie, M.A., Muggeridge, A.H., and Barley, J.J. (1993). 3-D Simulation of Viscous Fingering & WAG Schemes. *SPE Reser. Eng.*, Feb. 8, pp. 19–26.

Coates, D.E., Kirkaldy, J.S. (1971). Morphological stability of α–β phase interfaces in the Cu- Zn- Ni system at 775 C. *Met Trans*, vol. 2, no. 12, December, pp. 3467–77.

Coats, K.H. (1976, October). Simulation of steam flooding with distillation and solution gas. *SPE Journal*, pp. 235–247.

Coats, K.H., Ramesh, A.B., and Winestock, A.G. (1977). Numerical modeling of thermal reservoir behavior. *Paper presented at Canada-Venezuela Oil Sands Symposium*. 27 May–4 June, Edmonton, Alberta.

Collins, R.E. (1961). *Flow of fluid through porous materials*. New York: Reinhold.

Coriell, S.M., McFadden, G.B., Sekerka, R.F., and Boettinger, W.J. (1998). Multiple similarity solutions for solidification and melting. *Journal of Crystal Growth*, vol. 191, pp. 573–585.

Coskuner, G. (1986). *A New Approach to the Onset of Instability for Miscible Displacements*. Ph. D. Dissertation, Univ. of Alberta, Dec.

Coskuner, G. and Bentsen, R.G. (1987). A New Stability Theory For Designing Graded Vicosity Banks. *J. Canad. Petrol. Technol.*, vol. 26, no. 6, pp. 26–30.

Coskuner, G. and Bentsen, R.G. (1987). Prediction of Instability For Miscible Displacements in a Hele-Shaw Cell. *Rev. L' Institut Francais du Petrole*, vol. 42, no. 2, pp. 151–162.

Coskuner, G. and Bentsen, R.G. (1990). A scaling criterion for miscible displacements. *J. Can. Pet. Tech.*, vol. 29, no. 1, pp. 86.

Coskuner, G. and Bentsen, R.G. (1990). An Extended Theory to Predict the Onset of Viscous Instabilities for Miscible Displacements in Porous Media. *Transport in Porous Media*, vol. 5, pp. 473–490.

Coskuner, G. and Bentsen, R.G. (1989). Effect of Length on Unstable Miscible Displacements. *J. Canad. Petrol. Technol.*, vol. 28, no. 4, pp. 34–44.

Coskuner, G. (1993). Onset of Viscous Fingering for Miscible Liquid – Liquid Displacement in Porous Media. *Transport in Porous Media*, vol. 10, pp. 285–291.

Craft, B.C. and Hawkins, M.F., (1959). *Applied Petroleum Reservoir Engineering.* Englewood Cliffs, NJ 07632: Prentice-Hall, Inc.

Dake, L. (1978). *Fundamentals of Reservoir Engineering.* New York, NY 10010, U.S.A.: Elsevier Science Publishing Company Inc.

Dobrynin, V. (1961). Effect of Overburden Pressure on Some Properties of Sandstones. *SPEJ*, December, pp. 360.

DuFort, E.C. and Frankel, S.P. (1953). Stability Conditions in the Numerical Treatment of Parabolic Differential Equations. *Math. Tables Aids Comput.*, vol. 7, pp. 135–152.

Dung, H.T. and Piracha, A.L. (2000). *Journal of Energy and Development*, vol. 25(1), pp. 47–70.

Editorial. (1996). Watching the world: tangible benefits of virtual reality. *Oil & Gas Journal*, vol. 94(13), p. 30.

EIA. *International Energy Outlook. Greenhouse Gases General Information.* Energy Information Administration, Environmental Issues and World Energy Use. EI 30, 1000 Independence Avenue, SW, Washinghton, DC, 20585, 2004.

Eisenack, K., Matthias, K.B., Lüdeke, M.K.B., Petschel-Heldl, G., Scheffran, J., and Kropp, J.P. (2007). Qualitative Modelling Techniques to Assess Patterns of Global Change. In J. A. Kropp, *Advanced Methods for Decision Making and Risk Management in Sustainability Science* (pp. 83–127). Nova Science Publishers.

El-Shoubary, Y., Maes, R., and Seth, S.C. (2003). Carbon-Based Adsorption Powder Containing Cupric Chloride, US patent no 6638347, October 28, 2003.

El-Sayed, S.M. and Abdel-Aziz, M.R. (2003). A comparison of Adomian's decomposition method and wavelet-Galerkin method for solving integro-differential equations. *Applied Mathematics and Computation*, vol. 136, pp. 151–159.

Ertekin, T., Abou-Kassem, J.H., and King, G.R. (2001). *Basic Applied Reservoir Simulation.* Richardson, Texas, 406 pp.: Society of Petroleum Engineers.

Ertekin, T., Abou-Kassem, J.H., and King, G. (2001). *Basic Practical Reservoir Simulation.* SPE Textbook Series Vol. 7, Society of Petroleum Engineers, Richardson, Texas.

Eugene, Y. (1993). Application of the decomposition method to the solution of the reaction-convection-diffusion equation. *Applied Mathematics and Computation*, vol. 56, pp. 1–27.

Farmer, C.L. (2002). Upscaling: a review. *International journal for numerical methods in fluids*, vol. 40, pp. 63–78.
Fayers, F.J. and Sheldon, F.W. (1959). The Effect of Capillary Pressure and Gravity on Two-Phase Fluid Flow in a Porous Medium. *Trans. AIME*, vol. 21, pp. 147–155.
Feature. (1997). Caution expressed over industry optimism on 4D seismic technology. *First Break*, vol. 15(2): pp. 51.
Fetkovich, M.J., Reese, D.E., and Whitson, C.H. (1998). Application of a General Material Balance for High-Pressure Gas Reservoir. *SPE Journal*, March, pp. 3–13.
Fetkovich, M.J., Reese, D.E., and Whitson, C.H. (1991). Application of a General Material Balance for High-Pressure Gas Reservoir. *SPE-22921, SPE Annual Technical Conference and Exhibition*. Dallas, October 6–9.
Feynman, R.P. (1988). *What Do You Care What Other People Think?* WW Norton, NY, 255 pp.
Finley, J.R., Pinter, J.D., and Satish, M.G. (1998). Automatic model calibration applying global optimization techniques. *Journal of Environmental Modeling and Assessment*, vol. 3(1-2): pp. 117-126.
Forchheimer, P. (1931). *Hydraulik, 3rd, ed.* Leipzig, Berlin.
Gerritsen M.G. and Durlofsky, L.J. (2005). Modelling fluid flow in oil reservoirs, *Annual Review of Fluid mechanics*, vol. 37, pp. 211–238.
Ghorayeb, K. and Firoozabadi, A. (2000). Numerical Study of Natural Convection and Diffusion in Fractured Porous Media. *SPE Journal*, March, vol. 5(1), pp. 12–20.
Gill, P.E., Murray, W., and Wright, M. (1982). *Practical Optimization*. New York: Academic Press.
Guellal, S.A. and Cherruault, Y. (1995). Guel Application of the decomposition method to identify the distributed parameters of an elliptical equation. *Mathematical and Computer Modelling*, vol. 21(4), pp. 51–55.
Hadamard, J. (1923). *Lectures on the Cauchy Problem in Linear Partial Differential Equations*. New Haven, London: Yale Univ. Press.
Hall, H. (1953). Compressibility of Reservoir Rocks. *Trans. AIME*, vol. 198, pp. 309–311.
Hamming, R.W. (1984) *Numerical Analysis for Scientists and Engineers*, McGraw-Hill, New York, 2nd Edition, 721 pp.
Harris, D.G. (1975). The role of geology in reservoir simulation studies. *Journal of Petroleum Engineering*, May, SPE Paper 5022, pp. 625–632.
Hassanizadeh and Gray, S. M. (1987). High velocity flow in porous media. *Transport in porous media*, vol. 2(6), pp. 521–531.
Havlena, D. and Odeh, A.S. (1963). The Material Balance as an Equation of a Straight Line. *JPT*, (August) 896–900.
Havlena, D. and Odeh, A.S. (1964). The Material Balance as an Equation of a Straight Line-Part II, Field Cases. *JPT*, (July) 815–822.

Hettema, M., Papamichos, E., and Schutjens, P. (2002). Subsidence delay: field observations and analysis. *Oil and Gas Science and Technology*, Rev. IFP, vol. 57(5): pp. 443–458.

Holmgren, C.A. (1951). Effect of free gas saturation on oil recovery by waterflooding. *Trans. AIME*, vol. 192, pp. 135–140.

Homsy, G.M. (1987). *Annual Rev. Fluid Mech.*, vol. 19, pp. 271–311.

Hossain, M. (2008). *Experimental and Numerical Investigation of Memory-Based Complex Rheology and Rock/Fluid Interactions*. Halifax, Nova Scotia, Canada, 793 p.: PhD Dissertation, Dalhousie University.

Hossain, M.E., Mousavizadegan, S.H., Ketata, C., and Islam, M.R. (2007). A Novel Memory Based Stress-Strain Model for Reservoir Characterization. *Journal of Nature Science and Sustainable Technology*, vol. 1(4).

Hossain, M.E. and Islam, M.R. (2009). *An Advanced Analysis Technique for Sustainable Petroleum Operations*, VDM Publishing Ltd., Germany, 750 pp.

Hossain, M.E., Ketata, C., Khan, M.I., and Islam, M.R. (2009). Flammability and individual risk assessment for natural gas pipelines. *Advances in Sustainable Petroleum Science and Engineering*, vol. 1, no. 1, 33-44.

Hossain, M.E., Mousavizadegan, S.H., and Islam, M.R. (2009). The Effects of Thermal Alterations on Formation Permeability and Porosity. *Petroleum Science and Technology*, vol. 26, no. (10–11), pp. 1282–1302.

Howarth, R.B. and Winslow, M.A. (1994). Energy use and CO_2 emissions reduction: Integrating pricing and regulatory policies. *Energy – The International Journal*, vol. 19: pp. 855–867.

Huang, H.A. and Ayoub, J. (2008). Applicability of the Forchheimer equation for non-Darcy flow in porous media. *SPE Journal*, March, pp. 112–122.

Irmay, S. (1958). On the theoretical derivation of Darcy and Forchheimer formulas. *Trans: American Geophysical Union*, vol. 89 (4), pp. 702–707.

Islam, M.R. (1990). Comprehensive Mathematical Modelling of Horizontal Wells. *Paper no. 69, Proc. of the 2nd European Conf. on the Mathematics of Oil Recovery*. Latitudes Camargue, Arles, France, Sept. 11–14.

Islam, M.R. (2000). Energy State of the Art 2000. In C. Zhou. Charlottesville, VA: Int. Sci. Serv.

Islam, M.R. (2001). Emerging technologies in monitoring of oil and gas reservoirs. *SPE-68804, Proc. SPE Western Regional Conference*, March 26–30. Bakersfield, California.

Islam, M.R. (2002). Emerging technologies in subsurface monitoring of petroleum reservoirs. *SPE 69440, SPE Latin American and Caribbean Petroleum Engineering Conference*. 25–28 March, Buenos Aires, Argentina.

Islam, M.R. (2006). Computing for the information age. *Proc. of the 36th international conference on computers and industrial engineering*. Keynote speech, Taipei, Taiwan, June 20–23.

References

Islam, M.R. and Chakma, A. (1990). Comprehensive Physical and Numerical Modeling of a Horizontal Well. *SPE paper no. 20627, Proc. of the SPE Annual Conf. and Exhib.* New Orleans.

Islam, M.R. and Nandakumar, K. (1990). Transient Convection in Saturated Porous Layers with Internal Heat Sources. *Int. J. Heat Mass Transfer*, vol. 33(1), pp. 151–161.

Islam, M.R., Verma. A., and Farouq Ali, S.M. (1991). In Situ Combustion - The Essential Reaction Kinetics, Heavy *Crude and Tar Sands - Hydrocarbons for the 21st Century*, vol. 4, UNITAR/UNDP.

Joseph, D.D. (1982). Nonlinear equation governing flow in saturated porous media. *Water Resource Res.*, vol. 18, pp. 1049–1052.

Joseph, G.G. (2000). The Crest of the Peacock: Non-European Roots of Mathematics. Princeton University Press, NJ, USA, 416 pp.

Journel, A.G. (1986). Power averaging for block effective permeability. *SPE paper 15128*.

Ketata, C., Satish, M., and Islam, M.R. (2005a). Stochastic Evaluation of Rock Properties by Sonic-While-Drilling Data Processing in the Oil and Gas Industry. *In the proceedings of 2005 International Conference on Computational & Experimental engineering & Science.* 1–6 December, Hyderabad, India.

Ketata, C., Satish, M.G., and Islam, M.R. (2006a). The Meaningful Zero. *Proc. 36th Conference on Computer and Industries.* June, Taiwan.

Ketata, C., Satish, M.G., and Islam, M.R. (2006b). Cognitive Work Analysis of Expert Systems Design and Evaluation in the Oil and Gas Industry. *Proc. 36th Conference on Computer and Industries.* June, Taiwan.

Ketata, C., Satish, M., and Islam, M.R. (2005b). Knowledge-Based Optimization of Rotary Drilling System,. *In the proceedings of 2005 International Conference on Computational & Experimental engineering & Science.* 1–6 December, Hyderabad, India.

Ketata, C., Satish, M.G., and Islam, M.R. (2006c). Multiple Solution Nature of Chaos Number-Oriented Equations. *CIMCA-2006.* Nov., Sydney, Australia.

Ketata, C., Satish, M.G., and Islam, M.R. (2006d). The Meaningful Infinity. *CIMCA-2006.* Nov., Sydney, Australia.

Khan, M.I. and Islam, M.R. (2007). *The Petroleum Engineering Handbook: Sustainable Operations.* Houston, TX, pp. 461: Gulf Publishing Company.

Khan, M.I. and Islam, M.R. (2006). *True sustainability in technological development and natural resources management.* NY, USA: Nova Science Publishers.

Khan, M.I., Lakhal, S.Y., Satish, M., and Islam, M.R. (2009). Towards achieving sustainability: application of green supply chain model in offshore oil and gas operations. *Advances in Sustainable Petroleum Engineering and Science*, vol. 1, no. 1, 46–66.

Kim, K.S. and van Stone, R.H. (1995). Hold time crack growth analysis at elevated temperatures. *Engineering Fracture Mechanics*, vol. 52(3): pp. 433–444.

King, P.R. (1989). The use of renormalization for calculating effective permeability. *Transport in Porous Media*, vol. 4, pp. 37–58.

King, P.R., Muggeridge, A.H., and Price, W.G. (1993). Renormalization Calculations of Immiscible Flow. *Transport in Porous Media*, vol. 12, pp. 237–260.

Kline, M. (1972). *Mathematical Thought from Ancient to Modern Times.* New York, USA, 1238 p.: Oxford University Press.

Kühn, M., Bartels, J., and Iffland, J. (2002). Predicting Reservoir Property Trends under Heat Exploitation: Interaction between Flow, Heat Transfer, Transport, and Chemical Reactions in a Deep Aquifer at Stralsund, Germany. *Geothermics*, vol. 31, pp. 72.

Kvitco, V. (2007). Mathematical Disproof of Lorentz' Mathematics and Einstein's Relativity, Theory *Physics Letters A*, accepted.

Kyte, J.R. and Berry, D.W. (1975). New Pseudo Functions to Control Numerical Dispersion. *SPEJ*, August 1975, pp. 269–276.

Laffez, P.A. (1996). Modelling of the thermic exchanges during a drilling: resolution with Adomian's decomposition method. *Mathematical and Computer Modelling*, vol. 23 (10), pp. 11–14.

Lake, L.W., Srinivasan, S. (2004). Statistical scale-up of reservoir properties: concepts and applications. *Journal of Petroleum Science and Engineering*, vol. 44, pp. 27–39.

Lakhal, S., H'mida, S., and Islam, M.R. (2007). Green Supply Chain Parameters for a Canadian Petroleum Refinery Company. *Int. J. Environmental Technology and Management*, vol. 7, nos. 1/2, pp. 56–67.

Li, R., Reynolds, A.C., and Oliver, D.S. (2001). History matching of three-phase flow production data. *SPE Paper 66351*.

Lohne, A. and Virnovsky, G. (2006). Three-phase upscaling in capillary and viscous limit. *SPE Paper 99567*.

Lutchmansingh, P.M. (1987). *Development and application of a highly implicit, multi-dimensional polymer injection simulator.* PhD Thesis, The Pennsylvania State University.

Luchmansingh, P.M., Ertekin, T., and Abou-Kassem, J.H. (1991). Development and application of a highly implicit multi-dimensional polymer injection simulator. *AIChE Symposium Series*, vol. 87, no. 280, pp. 112–122.

Maugis, P., Hopfe, W., Morral, J., and Kirkaldy, J. (1996). Degeneracy of diffusion paths in ternary, two-phase diffusion couples. *Journal of Applied Physics*, vol. 79, no. 10, pp. 7592–7596.

Merle, H.A., Kentie, C.J.P., van Opstal, G.H.C., and Schneider, G.M.G. (1976). The Bachaquero study - a composite analysis of the behavior of a compaction drive/solution gas drive reservoir. *SPE JPT*, September, pp. 1107–1115.

References

Mishra, S.A. (2005). A computational procedure for finding multiple solutions of convective heat transfer equations. *J. Phys. D: Appl. Phys.*, vol. 38, pp. 2977–2985.

Moses, P. (1986). Engineering Applications of Phase Behaviour of Crude Oil and Condensate Systems. *Journal of Petroleum Technology*, July, pp. 715–723.

Mousavizadegan, S.H., Mustafiz, S.A., and Rahman, M. (2006). The Adomian decomposition method in solution of non-linear partial differential equations. *J. Nature Science and Sustainable Technology*, vol. 1, no. 1, pp. 115–131.

Mousavizadegan, H., Mustafiz, S., and Islam, M.R. (2007). Multiple solutions in natural phenomena. *Journal of Nature Science and Sustainable Technology*, vol. 1, 141–158.

Mufti, A.A., Tadros, G., and Jones, P.R. (1997). Field assessment of fibre-optic Bragg grating strain sensors in the Confederation Bridge. *Canadian Journal of Civil Engineering*, vol. 24 (6): pp. 963–966.

Mustafiz, S. and Islam, M.R. (2008). State of the Art of Reservoir Simulation. *Pet. Sci. Tech.*, vol. 26, no. 11–12.

Mustafiz, S., Belhaj, H., Ma, F., Satish, M., and Islam, M.R. (2005). Modeling Horizontal Well Oil Production Using Modified Brinkman's Model. *Proc. ASME International Mechanical Engineering Congress and Exposition (IMECE)*. Orlando, Florida, USA, November.

Mustafiz, S., Biazar, J., and Islam, M.R. (2005). An Adomian decomposition solution to the modified Brinkman model (MBM) for a 2-dimensional, 1-phase flow of petroleum fluids. *Proc. CSCE, 33rd Annual Conf.* Toronto, ON, Canada, June 2–4.

Mustafiz, S., Mousavizadegan, H., and Islam, M.R. (2008). Adomian Decomposition of Buckley Leverett Equation with Capillary Terms. *Pet. Sci. Tech.*, vol. 26, issue 15.

Mustafiz, S., Moussavizadeghan, H., and Islam, M.R. (2008). The effects of linearization on solutions of reservoir engineering problems. *Journal of Petroleum Science and Technology*, vol. 26, nos. 10–11, pp. 1303–1330.

Naami, A., Catania, P., and Islam, M.R. (1999). Experimental and Numerical Study of Viscous Fingering in a Two-Dimensional Consolidated Porous Medium. *Saskatchewan CIM conference*. Oct., Regina.

Naami, A.M., Catania, P., and Islam, M.R. (1999). Numerical and experimental modelling of viscous fingering in two-dimensional consolidated porous medium. *CIM paper no. 118, CIM conference*. Regina, Oct.

Nguyen, T. (1986). Experimental Study of Non-Darcy Flow through Perforations. *Paper SPE 15473 prepared for presentation at the 61st Annual Technical Conference and Exhibition of the SPE.* New Orleans, Louisiana, 5–8 October.

Nield, D.A. (2000). Resolution of a paradox involving viscous dissipation and nonlinear drag in a porous medium. *Transport in porous medium*, vol. 41 (3), pp. 349–357.

Nixon, D. (1989). Occurence of multiple solutions for the TSD-Euler equation source. *Acta Mechanica*, vol. 80, no. 3–4, pp. 191–199.

Nouri, A., Vaziri, H., and Islam, M.R. (2002). A new theory and methodology for modeling sand production. *Energy Sources*, vol. 24(11): pp. 995–1008.

Nouri, A., Vaziri, H., Kuru, E., and Islam, R. (2006). A Comparison of Two Sanding Criteria in Physical and Numerical Modeling of Sand Production. *J. Pet. Sci. Eng.*, vol. 50, pp. 55–70.

Odeh, A.S. (1981). Comparison of solutions to a three-dimensional black oil reservoir simulation problem. *J. Pet. Technol.*, vol. 33, no. 1, pp. 13–25.

Odeh, A. (1982). An overview of mathematical modeling of the behavior of hydrocarbon reservoirs. *SIAM Review*, vol. 24 (3), pp. 263.

Panawalage, S.P., Biazar, J., Rahman, M., and Islam, M.R. (2004a). An analytical approximation to the solution of reservoir permeability from the porous flow equation. *Int. Math. J.*, vol. 5, no. 2, pp. 113–122.

Panawalage, S.P., Rahman, M., Susilo, A., and Islam, M.R. (2004b). Analytic and numerical approaches of inverse problem in reservoir permeability. *Far East J. Appl. Math.*, vol. 17, no. 2, pp. 207–219.

Pape, H., Clauser, C., Iffland, J. (1999). Permeability prediction based on fractal pore-space geometry. *Geophysics*, vol. 64, no. 5, 1447–1460.

Parvazinia, M.N., Nassehi V., Wakeman R.J. and Ghoreishy M.H.R. (2006). Finite element modelling of flow through a porous medium between two parallel plates using the Brinkman equation. *Transport in Porous Media 63*, pp. 71–90.

Paterson, L., Painter, S., Zhang, X., and Pinczewski, V. (1996). Simulating residual saturation and relative permeability in heterogeneous formations. *SPE paper 36523*.

Peaceman, D.W. and Rachford, Jr. H.H. (1962). Numerical Calculation of Multicomponent Miscible Displacement. *Pet. Trans, AIME*, vol. 219, *SPEJ 327*.

Peaceman, D.W. (1983). Interpretation of well-block pressures in numerical square grid blocks and anisotropic permeability, SPE J, June, 531–543.

Pedrosa, Jr., Oswaldo A., and Aziz, K. (1986). Use of a hybrid grid in reservoir simulation. *SPE 13507, SPE Reservoir Engineering,* Nov. pp. 611–621.

Pickup, G.E., Ringrose, P.S., Jensen, J.L., and Sorbie, K.S. (1994). Permeability Tensors for Sedimentary Structures. *Mathematical Geology*, vol. 26, no. 2, pp. 227–250.

Planck, M. (1901). "Über das Gesetz der Energieverteilung im Normalspektrum" ("On the Law of Distribution of Energy in the Normal Spectrum"), *Annalen Der Physik*, 4, 553.

Portella, F. and Prais, R.C. (1999). Use of Automatic History Matching and Geostatistical Simulation to Improve. *SPE Paper 53976*.

Qi, T. and Hesketh, D. (2005). An analysis of upscaling techniques for reservoir simulation. *Petroleum science and technology*, vol. 23, pp. 827–842.

Rahim, Z. and Holditch, S.A. (1995). Using a three-dimensional concept in a two-dimensional model to predict accurate hydraulic fracture dimensions. *J. Pet. Sci. Eng.*, vol. 13, pp. 15–27.

Rahman, M.A., Mustafiz, S., Biazar, J., and Islam, M.R. (2007). Experimental and numerical modeling of a novel rock perforation technique. *J. Franklin Institute*, vol. 344 (5), pp. 777–789.

Rahman, M.A., Mustafiz, S., Biazar, J., Koksal, M., and Islam, M.R. (n.d.). Investigation of a novel perforation technique in petroleum wells - perforation by drilling. *Journal of the Franklin Institute*, in press.

Rahman, N.M.A., Anderson, D.M., and Mattar, L. (2006). New, Rigorous Material Balance Equation for Gas Flow in a Compressible Formation with Residual Fluid Saturation. *SPE-100563, presented at the SPE Gas Technology Symposium held in Calgary.* Alberta, Canada.

Rahman, N.M.A., Mattar, L., and Zaoral, K. (2006). A New Method for Computing Pseudo-Time for Real Gas Flow Using the Material Balance Equation. *Journal of Canadian Petroleum Technology*, vol. 45 (10), pp. 36–44.

Ramagost, B.P. and Farshad, (1981). P/Z Abnormally Pressured Gas Reservoirs. *SPE-10125, SPE ATCE.* San Antonio, TX, October 5–7.

Rose, W. (2000). Myths about later-day extensions of Darcy's law, *J. Pet. Sci. Eng.*, vol. 26, no. 1–4, May, pp. 187–198.

Roughgarden, J. (2005). *Evolution's Rainbow: Diversity, Gender, and Sexuality in Nature and People*, University of California Press, CA, USA, 474 p.

Sablok, R. and Aziz, K. (2005). Upscaling and Discretization Errors in Reservoir Simulation. *SPE Paper 93372*.

Saghir, M.Z. and Islam, M.R. (1998). Double Diffusive and Marangoni Convection in a Multi-cavity System. *Int. J. Heat and Mass Transfer*, vol. 41, no. 14, pp. 2157–2174.

Saghir, M.Z., Chaalal, O., and Islam, M.R. (2000). Numerical and Experimental Modeling of Viscous Fingering during Liquid – Liquid Miscible Displacement. *J. of Petroleum Science and Engineering*, vol. 26, pp. 253–262.

Saghir, M.Z., Vaziri, H., and Islam, M.R. (2001). Heat and mass transfer modeling of fractured formations. *Journal of Computational Fluid Dynamics*, vol. 15 (4): pp. 279–292.

Saghir, Z., Chaalal, O., and Islam, M.R. (2000). Experimental and numerical modeling of viscous fingering. *Journal of Petroleum Science and Engineering*, vol. 26(1–4), pp. 253–262.

Sammon, P.H. and Rubin, B. (1984). Practical control of timestep selection in thermal simulation. *Paper presented at the 59th Annual Technical Conference and Exhibition.* September, Houston, Texas.

Sarma H.K. and Bentsen R.G. (1987). An Experimental Verification of a Modified Instability Theory for Immiscible Displacements in Porous Media. *J. Can. Pet. Tech.*, July–August, pp. 88–99.

Sarma P., Aziz, K., and Durlofsky, L.J. (2005). Implementation of adjoint solution for optimal control of smart wells. *SPE Paper 92864*.

Sarma, H.K. (1986). Viscous Fingering: One of the Main Factors behind Poor Flood Efficiencies in Petroleum Reservoirs, vol. 48, pp. 39–49.

Schowalter, W.R. (1965). Stability Criteria for Miscible Displacement of Fluids from a Porous Medium. *A.I.Ch.E. J.*, vol. 11, pp. 99.

Service, R.F. (2005). Is it time to shoot for the sun? *Science*, vol. 309, pp. 549–551.

Settari, A., Walters, D.A., Stright, D.H., and Aziz, K. (2008). Numerical techniques used for predicting subsidence due to gas extraction in the North Adriatic Sea. *Petrol. Sci. Technol.*, vol. 26, nos. 10–11, pp. 1205–1223.

Severns, W. (2006). Can Technology Turn the Tide on Decline? *Keynote presentation at the SPE Intelligent Energy Conference and Exhibition*. Amsterdam, Netherlands.

Shapiro, R., Zatzman, G.M., and Mohiuddin, Y. (2007). Towards understanding the science of disinformation. *J. Nature Science and Sustainable Technology*, vol. 1, no. 3, pp. 471–504.

Standenes, S. (1995). Seismic monitoring – a tool for improved reservoir management. *SEISMIC 95*. London.

Stickey, D. (1993). New forces in international energy law: A discussion of political, economic, and environmental forces within the current international energy market. *Tulsa Journal of Comparative and International Law*, vol. 1 (1): pp. 95.

Syahrial, E. A. (1998). A new compositional simulation approach to model recovery from volatile oil resrvoirs. *SPE Asia Pacific Conference* (p. SPE 39757). Kuala Lumpur, Malaysia, 23–24 March: SPE.

Tan, C.T. and Homsy, G.M. (1988). Simulation of Non-linear Viscous Fingering in Miscible Displacement. *Phys. Fluids*, Jun. 33, no. 6, pp. 1330–1338.

Tan, C. T. and Homsy, G. M. (1986). Stability of Miscible Displacement in Porous Media: Rectilinear Flow. *Phys. Fluids*, vol. 29, pp. 3549.

Tharanivasan, A.K., Yang, C., and Gu, Y. (2004). Comparison of Three Different Mass-Transfer Models Used in the Experimental Measurement of Solvent Diffusivity in Heavy Oil. *J. Pet. Sci. Eng.*, vol. 44 (3–4), pp. 269–282.

Tidwell, V.C. and Robert, J.G. (1995). Laboratory investigation of matrix imbibition from a flow fracture. *Geophys. Res. Lett.*, vol. 22, pp. 1405–1408.

Tiscareno-Lechuga, F. (1999). A sufficient condition for multiple solutions in gas membrane separators with perfect mixing. *Computers and Chemical Engineering*, vol. 23, no. 3, pp. 391–394.

REFERENCES

Tortike, W.S. and Farouq Ali, S.M. (1987). A framework for multiphase nonisothermal fluid flow in a deforming heavy oil reservoir. *SPE 16030, The 9th SPE symposium on reservoir simulation.* San Antonio, TX.

Vaziri, H.H., Xiao, Y., Islam, R., and Nouri, A. (2003). Numerical modeling of seepage-induced sand production in oil and gas reservoirs. *Journal of Petroleum Science and Engineering*, vol. 36, pp. 71–86.

Vinsome, P.K.W. (1976). Orthomin, an iterative method for solving sparse banded sets of simultaneous linear equations. *Paper presented at the 1976 SPE Symposium on Numerical Simulation of Reservoir Performance.* 19–20 February, Los Angeles.

Wang, X.D. and Meguid, S.A. (1995). On the dynamic crack propagation in an interface with spatially varying elastic properties. *Int. J. Fracture*, vol. 69, pp. 87–99.

Wazwaz, A. (2001). A new algorithm for calculating Adomian polynomials for nonlinear operators. *Applied Mathematics and Computation*, vol. 111 (1): pp. 33–51.

Wazwaz, A.M. and El-Sayed, S.M. (2001). A new modification of Adomian decomposition method for linear and nonlinear operators. *Applied Mathematics and Computation*, vol. 122 (3): pp. 393–405.

Whitaker, S. (1967). Diffusion and Dispersion in Porous Media. *A.I.Ch.E. J.*, vol. 13, pp. 420.

Wilkinson, D.S., Maire, E., and Embury, J.D. (1997). The role of heterogeneity on the flow and fracture of two-phase materials science & engineering. *A. Structural*, vol. 233, no. 1/2, pp. 145–154.

Yanosik, J.L. and McCracken, T.A. (1979). A nine-point finite-difference reservoir simulator for realistic prediction of adverse mobility ratio displacements. *Soc. Pet. Eng. J.*, vol. 19, no. 4, pp. 253–262.

Zaman, M., Agha, K.R., and Islam, M.R. (2006). Laser based detection of paraffin in crude oil samples: Numerical and experimental study. *Petroleum Science and Technology*, vol. 24 (1): pp. 7–22.

Zaman, M., Bjorndalen, N., and Islam, M.R. (2004). Detection of precipitation in pipelines. *Petroleum Science and Technology*, vol. 22 (9–10): pp. 1119–1141.

Zatzman, G.M. and Islam, M.R. (2007). *Economics of intangibles:.* NY, USA, pp. 407: Nova Science Publisher.

Zatzman, G.M., and Islam, M.R. (2005). Natural gas pricing. In S.S. Mokhatab, *Handbook of natural gas transmission and processing* (p. in press). Elsevier Inc.

Zatzman, G.M., and Islam, M.R. (2007). Truth, Consequences and Intentions: The Study of Natural and Anti-Natural Starting Points and Their Implications. *J. Nature Science and Sustainable Technology*, vol. 1, no. 2, pp. 169–174.

Zatzman, G.M., Khan, M.M., Chhetri, A.B., and Islam. (2009). Delinearized History of Time, Science, and Truth: Towards Modeling Intangibles. *Journal of Information, Intelligence, and Knowldege*, vol. 1, no. 1, pp. 1-46.

Zekri, A.Y., Mustafiz, S., Chaalal, O., and Islam, M.R. (2006). The effects of thermal shock on homogenous and fractured carbonate formation. *Journal of Petroleum Science and Technology*, submitted for publication.

Zhang, P., Pickup, G., and Christie, M. (2006). A new method for accurate and practical upscaling in highly heterogeneous reservoir models. *SPE paper 103760*.

Index

Accumulation terms, 18, 160, 202, 207
Adomian decomposition method, 84, 87, 94–98, 100, 102–104, 108, 114, 413, 445
Anisotropy, anistropic, 15, 135, 155, 156, 130, 357, 445
Aphenomenal, 86, 342, 343, 354–356, 364, 366, 374, 379, 393–396, 401, 402
Aquifer, 17, 31, 35, 65, 66, 70, 190, 192, 230, 233, 247, 248, 265, 268, 426, 432, 441, 455
Avalanche theory, 379

Backward difference, 10, 278, 285, 371
Black oil model or reservoir, 117, 141, 424, 425
Block-centered, 11, 157, 424
Bottom-hole pressure (BHP), 62, 69, 79, 428, 438
Brinkman, 138, 457, 458
Boundary condition, 4, 10, 54, 57, 63, 104, 105, 177, 178, 180, 189–191, 196, 230, 280, 282, 283, 293–295, 364, 369, 423

Capillary pressure, 49, 51, 61, 74, 84, 95, 111–113, 122, 140, 141, 143, 145, 146, 148, 150, 152, 155, 182–184, 191, 230, 233, 237, 242, 413, 414, 416, 419, 422, 424, 427, 435, 451

Effect of, 95, 105, 106, 107, 110, 112, 155, 191, 236
Multiple solutions due to, 104, 111, 112, 155, 192
Cartesian coordinates, 123–125, 156, 370
Cartesian grid block, 16
Central difference, 10, 278, 279, 281, 283, 285, 320
Chaotic, 20, 83, 86, 361
Chaos, 20, 21, 27
Compositional model, 116, 117, 118, 141, 143, 146, 155, 180, 183, 445
Compositional modeling, 117
Compressibility coefficient/factor, 44, 45, 47, 51, 258, 259, 414
Compressibility of rock, 133, 247, 264, 274, 452, 436,
Compressible fluid, 45, 247, 264, 274, 459
Compressible rock, 246, 265, 271, 459
Conservation of energy, 5, 115, 118, 126, 132, 156, 159, 360, 374, 375, 380
Conservation of mass, 85, 86, 116, 118, 123, 139, 141, 145, 156, 159, 208, 209, 359, 374, 375, 380
Constitutive equation, 5, 118, 208,
Control volume, 2, 156, 158–161, 165, 167, 168, 172, 193, 209, 391

463

Crank-Nicolson method, 2, 180, 450
Cylindrical coordinate, 424

Darcy's law, 5, 6, 11, 15, 16, 48, 54, 55, 58, 75, 116, 118, 132, 136–138, 141, 158, 170, 172, 194, 208, 209
  Multiphase flow, 135, 139, 144, 147, 172, 174, 193
Density, 38, 45, 47, 51, 56, 61, 65, 85, 119, 120, 131, 134, 248, 427,
  of rock, 436, 448, 450, 462,
  of a phase, 121
  of a gas, 123
Diffusivity equation, 87, 104, 302
Dimension/dimensional, 35, 36, 56, 57, 63, 75, 115, 116, 135, 137, 156, 170, 172, 174, 180, 183, 208,
  fourth, 24, 86,
  knowledge *see* knowledge dimension
Dimensionality, 88, 178, 179, 351, 352, 354, 399
Dimensionless, 135, 247, 248, 254, 257–259, 271–275, 351, 352, 435, 437
Discretization, 2, 4, 11, 75, 86, 157, 158, 170, 278, 286, 291, 364, 441, 445, 469
DuFort Frankel, 278, 279, 281, 283, 286, 287, 289–292, 294

Elementary volume, *see* control volume
Engineering approach, 4, 155, 156, 364, 406, 445
Equation of state, 118, 121
Equations
  Barakat-Clark, 278, 279, 281, 283–295
  boundary condition, 189, 190–192, 196, 280, 282, 283, 285
  control volume, 2

Crank-Nicolson, 2, 180, 450
Darcy's, *see* Darcy's law
Dufort Frankel, *see* DuFort Frankel
Fictitious well rate, 189
Geometric factor, 176, 209, 231, 428
New fluid flow, 13,
linear algebraic, 4, 421
nonlinear algebraic, 4, 209, 221, 382, 383, 409

Electromagnetic, 374, 404
Explicit formulation, 180
Equilibrium, 29, 46, 61, 89, 115, 117, 120, 121, 140, 141, 148, 150, 153, 173, 360, 370, 377, 398

Fick's law, 118
Fictitious well, 189, 227
Finite difference, 8, 281
Flow equation, 13, 172, 193, 197
Forchheimer model/equation, 16, 136, 137, 138, 451
Formation volume factor, FVF, 210–213
Fracture, 16, 17, 20, 22, 54, 61, 116, 248, 450, 459–462
Fractured reservoir, 12, 15, 116

Gas compressibility factor, 47
Gas-oil ratio, GOR, 43, 44, 62, 70, 143, 224, 232, 426, 428, 430, 431, 435, 436, 438,
Geometric factors, 176, 209, 226, 438
Gridblock, 32, 57, 208, 209, 211, 217, 218, 220, 223, 225, 226, 227, 229–231, 235–237, 244

Heat conduction, 127
Heterogeneous, 14, 47, 52–55, 64, 116, 118, 156, 170, 325, 358, 424, 426, 448

Index

Homogeneous, 14, 39, 40, 47, 54, 54, 55, 73, 120, 134–136, 156, 160, 208, 225, 265, 281, 295, 296, 331, 426, 447

Implicit formulation, 180, 181, 197
Injection
  gas, 18, 25, 117, 273, 425
  steam, 14, 115, 149, 150
  water, 18, 30, 105, 298, 425, 438
Instability number, 295, 296, 309, 314, 316, 318, 325, 327, 330, 331

Knowledge dimension 88, 114

Leibnitz, 8
Linearization, 2, 3, 84, 86, 103, 112, 207, 224, 352, 362–366, 369, 396, 413, 414, 416, 421,
  of boundary conditions, 365

Material balance, 5, 64, 83, 115, 202, 245, 247, 249, 251, 255, 257, 259, 261, 263, 265, 267, 269, 271, 273, 275, 406, 424, 430, 451, 452
Mathematical approach, 4, 447
Matrix, 38, 55, 67, 69, 76, 77, 86, 155, 198, 201, 413, 424
  rock, 40, 49, 127, 129, 131, 132
Memory
  fluid, 201, 214, 378, 419
  rock, 214
  function, 378
Miscible, 20, 21, 117, 277, 295, 317, 327, 330, 449
Mobility ratio, 105, 111, 145, 296, 317–320, 330, 333, 335, 461
Multiphase, 13, 16, 17, 22, 57, 62, 83, 117, 135, 141, 166, 180, 187, 190, 208, 413, 414, 421, 461

Newton's
  approximation, 2, 8
  differential quotient, 10, 114, 367–372
  iteration, 201, 202, 424
  mechanics, 376–378, 380
  law, 133, 360, 361, 375
Non-Darcy, 104, 136, 137, 453

Permeability, 14–16, 18, 19, 24, 35, 38, 40, 49, 67, 73, 74, 76, 80, 136, 137, 157, 217, 219, 225, 249, 250, 296, 302
  effective, 49, 53, 54, 55–57, 105, 146, 455
  relative, 49, 50, 51, 53, 54, 56, 74, 75, 105, 110, 111, 122, 135, 145, 166, 170, 192, 230, 232, 236, 241, 248, 413, 416, 424, 427
  Tensor, 22, 48, 55, 57, 135, 170
Porosity, 14, 19, 35, 38, 41, 47, 48, 52, 53, 67, 76, 80, 106, 110, 133, 136, 145, 157, 225, 227, 232, 238, 241, 242, 246, 249–252, 314–317, 318, 331, 415–417, 426, 427, 432, 435, 450
  Dual, 15, 16, 450
  distribution of, 47
  variation of, 19, 240, 251, 415, 416

Representative Elemental Volume (REV), 12–14, 34, 38, 40–42, 74, 75, 118, 126, 127, 129, 132, 134, 337

Sustainability, 353, 383, 388, 395, 402, 407, 455

Thermal conductivity, 15, 19, 127, 129, 156
Thermal stress, 12, 17–20
Transmissibility, 35, 59, 63, 70, 175, 195, 209, 226, 231, 433, 435

Unsteady, 228, 229, 236, 242, 297, 310, 354, 432

Viscosity, 38, 45, 46, 55, 58, 105, 134, 137, 138, 145, 210, 211, 224, 226, 232, 238, 239, 241, 277, 298, 299, 304, 308, 334, 445, 360, 413, 415, 416, 424, 436, 437

Viscous fingering, 20, 277–337, 445, 445, 450, 451, 452, 457

Volatile oil, 116, 117, 146, 151, 178, 460

Also of Interest

Check out these forthcoming related titles coming soon from Scrivener Publishing

Acid Gas Injection and Carbon Dioxide Sequestration, by John Carroll, May 2010, ISBN 9780470625934. Provides a complete overview and guide on the very important topics of acid gas injection and CO2 sequestration.

The Greening of Petroleum Operations, by Rafiqul Islam, June 2010, ISBN 9780470625903. A breakthrough treatment of one of the most difficult and sought-after subjects of the modern era: sustainable energy.

Energy Storage: A New Approach, by Ralph Zito, July 2010, ISBN 9780470625910. Exploring the potential of reversible concentrations cells, the author of this groundbreaking volume reveals new technologies to solve the global crisis of energy storage.

Formulas and Calculations for Drilling Engineers, by Robello Samuel, September 2010, ISBN 9780470625996. The only book every drilling engineer must have, with all of the formulas and calculations that the engineer uses in the field.

Ethics in Engineering, by James Speight and Russell Foote, December 2010, ISBN 9780470626023. Covers the most thought-provoking ethical questions in engineering.

Zero-Waste Engineering, by Rafiqul Islam, February 2011, ISBN 9780470626047. In this controvercial new volume, the author explores the question of zero-waste engineering and how it can be done, efficiently and profitably.

Fundamentals of LNG Plant Design, by Saeid Mokhatab, David Messersmith, Walter Sonne, and Kamal Shah, August 2011. The only book of its kind, detailing LNG plant design, as the world turns more and more to LNG for its energy needs.

Flow Assurance, by Boyun Guo and Rafiqul Islam, September 2011, ISBN 9780470626085. Comprehensive and state-of-the-art guide to flow assurance in the petroleum industry.